PROCESS THEORIES: CROSSDISCIPLINARY STUDIES IN DYNAMIC CATEGORIES

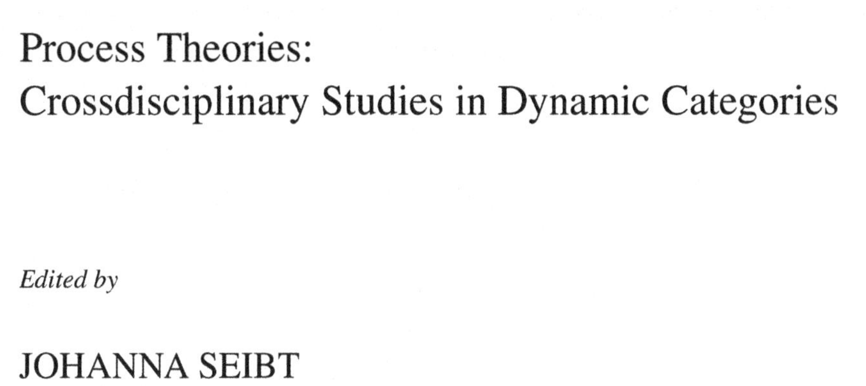

Process Theories:
Crossdisciplinary Studies in Dynamic Categories

Edited by

JOHANNA SEIBT

Philosophy Department
University of Aarhus
Aarhus, Denmark

Partly reprinted from *Axiomathes*, Vol. **14**, nos. 1-3, 2004 (pp. 1-283).

Kluwer Academic Publishers
DORDRECHT / BOSTON / LONDON

A C.I.P. Catalogue record for this book is available from the Libarary of Congress.

ISBN 1-4020-1751-0

Published by Kluwer Academic Publishers,
P.O. Box 17, 3300 AA Dordrecht, The Netherlands.

Sold and distributed in North, Central and South America
by Kluwer Academic Publishers
101 Philip Drive, Norwell, MA 02061, U.S.A.

In all other countries, sold and distributed
by Kluwer Academic Publishers,
P.O. Box 322, 3300 AA Dordrecht, The Netherlands.

Printed in the Netherlands on acid-free paper

TABLE OF CONTENTS

J. SEIBT

PROCESS THEORIES CROSSDISCIPLINARY STUDIES ON DYNAMIC CATEGORIES

1. INTRODUCTION

Processes constitute the world of human experience – from nature to social reality to cognition itself. However, by and large, the centrality of processes does not appear to be reflected in theoretical descriptions of nature and the human domain. Frequently processes are represented in a reductive fashion – in terms of their results, input-output pairs, or sets of state sequences summarized by linear functions. Disciplines without quantitative, algebraic-geometrical tools make do with metaphorical, haphazard, and highly domain-specific classifications of occurrences. Perhaps most remarkable is the general neglect of dynamic entities among ontologists, with a few exceptions, notably Whitehead: dynamic entities are neither properly investigated nor employed as descriptive primitives. Throughout the history of ontological research, the world of human experience has been presented as an assembly of 'static' entities: substances, attributes, relations, facts, ideas – and more recently, tropes, temporally relativized property exemplifications, or four-dimensional expanses.

But the traditional fixation on the permanent is not a requirement of theory construction *per se*. This book contains a collection of thirteen articles showing the fertility of a process-geared perspective. Ten of these articles are based on talks presented at an interdisciplinary research meeting (*Processes: Analysis and Applications of Dynamic Categories*, Sandbjerg Gods, Denmark, June 5–8, 2002) which brought together researchers from widely different fields (formal ontology, cognitive science, linguistics, semiotics, ancient philosophy, ethics, philosophy of music, music theory, theoretical psychology, theoretical biology, philosophy of chemistry, and philosophy of physics). The Sandbjerg 2002 meeting was internationally the first interdisciplinary conference on processes without exclusive commitment to American process metaphysics (Whitehead, Dewey, James). Three further articles were included since they are apt further to contri-

J. Seibt (ed.), Process theories: Crossdisciplinary studies in dynamic categories, vii–xxiii.

bute to the overall theoretical purpose of the collection, namely, to present extant non-reductive theories of processes, and to identify theoretical motivations for the development of process-based or process-geared theories in different areas of application. In contemporary ontology, A.N. Whitehead's 'philosophy of organism' certainly stands out as an example of a non-reductive theory of processes with impressive explanatory scope, and two of the following contributions demonstrate the fertility of Whiteheadian approach. But the collection is primarily aimed to draw attention to non-Whiteheadian research on process within and outside of philosophy.

Like research on 'complexity' and 'emergence', research on 'process' is a topic which calls for a cross-fertilizing interdisciplinary approach. It is certainly an overstatement to say, adapting the Kantian *dictum*, that process ontology must be 'empty' without process-geared theories from other disciplines, or that the latter remain 'blind' without the former. But the most productive strategy for the development of a process-based category theory in ontology is to tailor such a scheme not only to the familiar problems in ontology but also to *descriptive tasks* in the natural and human sciences. And conversely, in order to revise basic assumptions in their field, researches in the natural and human sciences can benefit from new *descriptive tools* furnished by ontologists and philosophers of science. One of the focal points of such cross-fertilizing research is the development of (regional or general) *typologies of processes*, and several of the articles in this volume address this issue explicitly.

Following the two main objectives of the workshop, the contributions to these proceedings have been grouped under two headings: *analysis and application.* Of course, since in most of the following contributions dynamic categories are both analyzed and applied, the division is somewhat arbitrary and reflects merely a topical emphasis. Authors of Part I are mainly concerned with the semantic analysis of our ways of talking about dynamic entities (processes, events, changes, developments, actions, motions etc.); or with the ontological analysis of these entities. Authors of Part II constructively apply extant process-ontological frameworks to specific explanatory tasks (Parkan, Fortescue, Christiaens, Emmeche), or develop process-ontological tools for a specific domain of application (Needham, Manzotti), or show how certain research domains could gain from a process-geared perspective and what kind of descriptive tools are needed for this purpose (Bickhard, W. Christensen). For production-technical purposes the contributions by Manzotti and Emmeche had to be put last, but otherwise the papers in each part are arranged in what appeared the most natural systematic order, and, in particular, with a view to the chronology of research programs reported. The reader should keep in mind that many of the contributions to this volume of *Axiomathes* are primarily designed to serve as research reports; the full picture behind the sketch is to be found in the references.

Part I begins with an exposition of Aristotle's analysis of dynamic categories. This is, in fact, less for historical than for systematic reasons since Aristotle's analysis is not only the first in Western ontology but it is also by no means superseded (Aristotelian criteria for the difference between change and activity can be traced in almost any current formulation of this distinction in philosophy and linguistics). Mary-Louise Gill, one of the leading scholars in her field, presents a reading of the core of Aristotle's metaphysics that centers on Aristotle's analysis of *energeia* (activity). As one might rephrase Gill's main thesis, the primary task of metaphysics in Aristotle's view is to give an account of substance, and precisely this is possible only if the unity of a substance is understood as an activity of sorts. Gill sets out by sketching the problem that Aristotle's analysis of *energeia* is supposed to address. In order to accommodate dynamic phenomena, Aristotle suggests that both the alteration and generation (destruction) of substances can be modelled as the attainment of a form by an underlying subject (matter). As Gill highlights, this simple 'replacement model' of change cannot be combined with Aristotle's main criteria for 'substancehood', in particular the requirements that any substance should be (a) a 'separate this' and (b) unified. Particular objects and living organisms qualify with respect to (a) but as composites of matter and form they fail to be genuine unities. To regain the unity of substance, Gill argues, Aristotle refines his model of dynamic phenomena by introducing the technical terms of potentiality and actuality and distinguishing two senses of each. When someone learns Danish she undergoes a *change*: she actualizes a 'first-level potentiality' of becoming a speaker of Danish and *acquires* a second-level potentiality or capacity to speak Danish; when she *exercizes* this capacity, i.e. when she actualizes the second-order potentiality to speak Danish, she engages in an *activity*. The presence of a form in a substance, Gill argues, is to be understood on the model of an activity, as the expression of a capacity. But this is not all. The capacities expressed in the being of Socrates, this plant, or this table, are highly specific capacities involving both 'formal' and 'material' aspects. The wooden sphere exercises the capacity of being a *wooden* sphere: not just the capacity of being spherical, but also a complex capacity that includes the functional aspects of woodenness which only a certain type of matter can afford. The forms or functional roles a substance can express depends on its matter, and this in turn stands in an "essential relation" to the entire functional role of the substance, as that which affords a certain functional role. On

Gill's reading, then, the concept of an activity (*energeia*), or a capacity's being exercised, allows us to conceive of an essential mutual correlativity of form and matter. The being of substances is the *interaction* between the performance of a certain functional role and that which affords that functional role – an interaction that results in the required type of unity.

The ontological tradition has read Aristotle's investigations into substance in different, simpler ways, largely disconnecting the thesis of the primacy of countable enduring particulars from his analysis of dynamic phenomena. (Mis)guided by the category dualism of predicate logic, early analytic ontologists implemented the most simplistic version of the substance-ontological tradition – an illusory snapshot view of the world sporting particulars connected to universals by mysterious 'ties of exemplification' – which continues to influence theory construction in analytical ontology to the present day. While Whitehead opted for an (almost) wholesale replacement of traditional presuppositions, proponents of so-called 'revisionary' analytical ontologies by and large still pursue the snapshot view, even if this time based on tropes or states of affairs. But recently at least two formal frameworks have been advanced which treat processes as a fundamental category or even as the fundamental category in terms of which other categories are defined. The second and third contribution introduce these two new ontological frameworks.

Seibt's contribution presents a sketch of Free Process Theory (FPT) which she has been promoting since 1990. She begins by briefly setting out the methodological and heuristic background of FPT. First, she suggests that the tasks of ontology are best described if we conceive of ontologies as model theories of certain material inferences ('categorial inferences'). When ontologists offer an 'account' of persons (or relations, properties, states of affairs etc.) they axiomatically define certain entities whose inferential space matches with the inferential space of our commonsense reasoning about persons (or relations, properties, states of affairs etc.). Throughout its history the ontological tradition has preferred construction principles and entities geared to model the inferential space of our concept of things. In order to model the inferential space of our concepts of dynamic entities, she claims, the construction principles of the 'myth of substance' must be dispensed with. To support this claim she presents the inferential space of our talk about activities. Reviewing extant proposals in philosophy and linguistics for a classification of occurrence types denoted by verb phrases or sentences, she suggests that activities (along with the familiar 'Vendler-categories' of accomplishments, achievements, and states) are best conceived of as complex verbal aspects, defined by networks of aspectual implications. Our concept of an activity (accomplishment etc.)

is a way of 'packaging' a predicative content according to its *mode* of occurrence. The inferences that characterize the 'activity mode' imply that our talk about activities should be modeled by ontological entities which are concrete, dynamic, *non- particular* individuals, which Seibt calls 'free processes.' She then sketches the monocategoreal framework called FPT, a mereology with non-transitive part-relation, and outlines a typology of processes. Simple and complex free processes are classified according to (a) their pattern of spatio-temporal automerity (recurrence within a spatial or temporal region); these patterns play a prominent role in the definition of process-ontological correlates for basic common sense categories (activities, things, events, stuffs, heaps). Other classificatory parameters relate to (b) the components of complex free processes ('participant structure'), (c) the type of dynamic composition (including weak and strong emergence), (d) the type of dynamic flow ('dynamic shape'), and (e) the dynamic context of a process.

Heller and Herre outline the relevant part of a comprehensive ontological framework (formulated in the general language GOL) that supports data base maintenance and knowledge representation in a large variety of domains of application, and, in particular those of medical science, where short-term changes and long-term developments are traced over time in variable dynamic contexts. One of the distinctive features of this multicategoreal framework consists in the assumption of 'chronoids' and 'topoids,' i.e., portions of time and space, respectively, which are here treated as entities sui generis. Processes are *projected* onto disconnected or connected portions of time and space (rather than *identified* with filled spatio-temporal regions in the style of recent, four-dimensionalisms). Processes are the basic sort of 'occurrents'; other sorts of occurrents such as histories, states, changes, and locomotions are defined in terms of certain conditions on the temporal and spatial projections of processes. Even though the framework countenances 'substances' and 'moments', these notions designate dependent and derived entities: both substances and moments (called 'endurants') are momentary entities, namely, the 'time boundaries' (three-dimensional cuts of four-dimensional projections) of processes. There is nothing in this framework that is identical through time. Persistence is a matter of momentary entities instantiating a certain universal. Nevertheless, the categorial distinction between enduring entities (endurants) and processes is specified in familiar terms as a difference in the way in which these entities relate to time. Endurants are 'wholly present at a time-boundary', they are 'in time', while processes are 'extended in time' and 'have temporal parts'. Of particular philosophical interest might be the analysis of locomotion where the authors undercut

Zeno's paradox of the non-moving arrow by drawing a distinction between spatial location and extension-space: what exists of a moving body at two coinciding time-boundaries are two temporally coinciding endurants with the same spatial location but different extension spaces. Processes are constituents of larger dynamic contexts called 'situoids'. The authors envisage using these dynamical contexts for a classification of different types of processes that should, among other parameters, relate to the temporal structure of processes and their 'granularity' or degree of structural differentiation.

Each of the three process-geared ontologies introduced in the first three contributions aspires to qualify as an 'integrated ontology', i.e., as a theory that can be used for the ontological interpretation of both commonsensical and scientific descriptions of occurrences. The fourth contribution highlights the particular difficulties that arise for the program of an integrated ontology if the latter is based on dynamic categories. In particular, it casts doubt onto projects where *phenomenological* aspects of perception and action are used to draft descriptive categories which are then used to interpret the ontological structure of observer-independent entities. Svend Østergaard argues that the analysis of our common sense conceptualization of dynamic phenomena might be less of a reliable guide to the structure of the 'real' situation (where 'real' here simply means 'as described by physics'). Following R. Thom and L. Talmy, Østergaard suggests that we conceptualize occurrences in terms of perceived patterns of experientially salient vs. unmarked elements. To show how the difference between salient and unmarked experiental elements might be rendered precise, he briefly sketches how perception is described in catastrophy theory. Here the human brain and its environment are each conceived of as a dynamic system whose behavior can be represented by a phase space trajectory in an n-dimensional space and perception is treated as a relationship between these two ('internal' and 'external') spaces. As Østergaard argues in a second step, this model of perception can be used to describe and even explain our conceptualization of types of occurrences . 'Force-dynamic patterns' consisting of experientially salient, informative discontinuities, and uninformative continuities before, after, or in between such discontinuities, can be used to elucidate the categories of the Vendler classification. For example, an 'accomplishment' in the Vendler classification, such as *topple over* is the pattern of a continuity bounded by two discontinuities. But, Østergaard points out, one might well go beyond the Vendler classification and define the 'schematic meanings' of more fine-grained types of verbal predications. The assumption that many verb phrases express such patterns or schemata of the dynamic relationship would explain the frequent metaphorical usage of verbs ('he fell from power'), which often introduces fictional dynamic elements into the conceptualization of a situation ('the sunlight hit my face'). Such fictional additions are one indication, Østergaard argues in his third step, that the force-dynamic patterns in terms of which we conceptualize dynamic phenomena do not faithfully map real features. Another reason to resist a realist interpretation of such patterns is that these differ in a number of ways from the physical description of real-world dynamics. However, Østergaard observes, these differences militate also against Lakoff's anti-realist claim that all scientific knowledge is owed to the metaphorical projection of those schemes in terms of which we conceptualize our experience of dynamic phenomena. In conclusion Østergaard shows how the distinction of force-dynamic schemes may be used to analyze our conceptualizations of causation and agency.

Part I, devoted to the analysis of dynamic categories, concludes with a contribution from musicology, ostensibly a remote source domain for ontological category theory. Ontologists are likely to understand by the term 'musical process' merely the occurrence of sounds, in contrast with occurrences of light patterns or bodily movements etc. Erik Christensen's contribution conveys, however, that much can be gained, in particular for the ontological analysis of developments, from paying close attention to the implementation of (emergent) dynamicity in music. A 'musical process' as Christensen defines it is a highly complex dynamic phenomenon involving audible structures that evolve in the course of the musical performance. In essence, 'musical processes' are 2nd order developments, i.e., audible developments within audible developments. To come to grips with these subtle phenomena, Christensen approaches them from the composer's point of view, conceptualizing musical processes as products of compositional prescriptions. Working with examples from 20th century composers, Christensen highlights musical processes produced by shifts in repetitive patterns, by expansions of patterns, by compositional algorithms for self-similar scores, or random scores. He suggests that musical processes are best characterized in terms of relations between audible phenomenon and compositional prescription, and offers a tentative classification of musical processes in (a) 'rule-governed', 'goal-directed', or 'non-goal-directed (indeterminate) productions of processes that are (b) either transformations of a given audible structure or generations of new audible structures. (Since musical processes are essentially audible phenomena, the reader is strongly advised to listen to the referenced examples in order to properly understand and appreciate the particular points of Christensen's paper and its dazzling heuristic potential for applications outside music theory.)

Part II presents articles which each promote a process-based or process-geared approach to particular philosophical disciplines or scientific fields: philosophy of mind, cognitive science, linguistics, philosophy of the social sciences, philosophy of chemistry, and philosophy of physics. Four contributions in Part II (the first two and the last two) argue the claim that a process metaphysics offers new and promising routes to a naturalist account of human cognition. The main difficulty for a naturalist account of cognition is how to account for the normativity of 'the mental'. Both Mark Bickhard and Wayne Christensen conquer the problem by meeting it head-on, arguing that normativity already belongs into the naturalist domain: already biological function is normative, but this thesis requires a process-ontological interpretation of biological units.

In "Process and Emergence: Normative Function and Representation" Bickhard argues that the traditional threefold alternative between a (i) dualist endorsement of natural and normative items, (ii) 'anti-naturalism', and (iii) 'anti-normativism' is not exhaustive. The shift to a process ontology opens up the path to a naturalist account of the mind based on emergent normativity. To ease the reader onto this new perspective, Bickhard proceeds in three steps. He first addresses two principled objections against the idea of emergent normativity, namely, Kim's argument that emergent entities are bound to remain causally epiphenomenal, and Hume's claim that normative expressions cannot be defined in terms of descriptive expressions. Bickhard argues that Kim's objection can be undercut by assigning causal efficacy to organization itself and in particular the organization of processes, and that Hume overlooks the possibility of implicit definitions. In a second step Bickhard introduces a naturalist account of biological function based on the theory of dynamic systems. He argues, against Millikan's etiological account, that biological function should be understood as something constituted by the *current* condition of a biological system, rather than the system's history. The key move to a satisfactory account of biological function consists in identifying a causally efficacious process organization that captures the curious 'asymmetry' of the normative: "good and bad are not merely different from each other – good is selected as better or preferred in some sense". This we find in certain dynamic systems, 'self-maintaining systems' (such as a candle flame), that exist far from their thermodynamic equilibrium because they themselves contribute to the special conditions that make their existing far-from-equilibrium possible. In self- maintaining systems functionality – in the fully normative sense of usefulness – emerges. Any causal contribution to the persistence of the system serves a function relative to that system since there is a clear normative standard at play: "contributions to the persistence of the system are functional for that continued existence of the system". Biological functional normativity is, Bickhard suggests, at the base of "a long hierarchy of normative emergences" whose levels include representation, perception, memory, learning, emotions, sociality, language, values, rationality, and ethics. In a third step Bickhard argues this point for the case of representation. In a critical survey of extant accounts of representation (Dretske, Fodor, Cummins) he diagnoses various shortcomings as due to the underlying premise that representation is a form of encoding. Instead, Bickhard argues, representation emerges when a self-maintaining system is sufficiently complex to react to changes in its environment with different strategies or routines of self- maintenance. Any such routine of self-maintenance "functionally presupposes that the environment is (one of) the ways it must be in order for that routine to succeed in its self-maintaining functionality", In that sense, *the occurrence of that routine represents the presupposed conditions for the routine's functional success.* On this 'interactionist' model of representation content remains implicit or intrinsically pragmatically defined – the environmental conditions that "a routine functionally presupposes are implicit in the presupposing. They are not individuated and explicitly symbolized". The interactionist account of representation has three central advantages: first, representational content is internal to the system, not externally attributed; second, since it contributes to the self-maintenance of a self-maintaining system it is inherently normative; third, given that representations are defined via the functionality of triggered behavior, the system itself can detect representational error.

Wayne Christensen's paper sheds additional light onto the systematic significance of normative functions for an account of representation and cognition. While Bickhard argues in essence a thesis about representational content, taking issue with philosophical accounts of representation, the critical focus of Christensen's paper is on treatments of representation and cognition in current research programs of cognitive science. Cognitive science, so one might rephrase Christensen's main thesis, proceeds from some form of 'category mistake' – cognition is just not the kind of structural entity main stream cognitive science takes it to be. Following the computational strategy of the empiricist tradition, Christensen argues, cognition standardly has been understood as a matter of representational structures standing in a correspondence relation (truth) to structures of objects, properties, and facts, with cognitive processes (inference) figuring as rule-governed transformations of representational structures. Well-known problems of cognitive science and AI, such as computational intractablity and context-sensitivity of cognitive relevance (the 'frame problem'), can be traced back, Christensen argues, to the premises of this thoroughly representational and rule-driven, 'cognitivist' approach to cognition. Christensen then discusses two alternative research programs which dispense with cognitivist premises. 'Situated cognition research' or behavior-based robotics replaces internal representations by externally triggered agentive routines, assigning to the environment the job of organizing the system's behavior. 'Pragmatism' operates with representations that *emerge* in the course of interactions with the environment and thus has built-in relevance to the context in which a representation is learned. However, both of these approaches lose cognitive normativity, i.e., the correctness or truth of representations, either – in the case of situated cognition research – since no representations are formed, or – in the case of pragmatism – since the formation of representations is a non-cognitive process. To regain normativity Christensen argues for an even more radical deviation from the cognitivitist approach. Various sources of empirical evidence suggest that cognition does not abide by the principles of traditional computational architecture: (i) there are mutual dependence relations and adjustment by feedback between higher and lower cognitive processes; (ii) representational primitives are created and dissolved; (iii) most importantly, the distinction between content and rule founders since, e.g., the *content* of visual information can affect the *operation* by which visual discriminations are made. In combination these data show "the functional organisation of the system to be tuned by the nature of its activities ... [based on a] regulative loop from activity to architectural change". Christensen's theory of 'self-directed agents' conceives of cognition accordingly as embedded in the activities of an evolutionarily adapted organism. These activities are a functionally constitutive part of an assembly of interacting non-semantic and semantic, non-conscious and conscious processes where "higher order cognitive organizations ... form and dissolve". This approach to cognition "involves representation but is not representation-driven" and derives the normativity of cognition from its biological function.

The contributions by Claus Emmeche and Riccardo Manzotti may be understood as providing (two similar versions of) a radical constructive response to difficulties highlighted in Bickhard's and Christensen's discussions of representation and cognition. Like Bickhard and Christensen, Emmeche and Manzotti pursue a *naturalist* account of representation and conscious experience which is, however, *non-reductionist*, i.e., not committed to the common version of physicalism which views the world as an assembly of particles related by causal impact. Emmeche and Manzotti approach this task at the 'categorial level' by proposing an ontological revision designed to undercut directly the traditional separation of mind and nature. Both Emmeche and Manzotti endorse as basic entity a type of process exhibiting both 'material' and 'mental' aspects. On the basis of a process-based monistic philosophy of mind, both argue, the question about the nature and origin of consciousness can be fruitfully dealt with from a scientific point of view.

Emmeche and Manzotti pursue different strategies, however, to implement the common program of a monistic naturalist approach to consciousness. Manzotti offers a study in what one might call 'experimental metaphysics', presenting an account of conscious cognition together with experimental data from robotics in support of core elements of this account. He begins with the observation that on the basis of an object ontology representation can only be understood as a relation between *separate* entities. Since the physical objects in the brain (neurons or neuron clusters) are not isomorphic to mind-external objects, there must be something about them, namely, associated qualia or abstract functional roles, which establishes that they represent something. In this way, Manzotti argues, object-ontological accounts of representation are, right from the outset, committed to a dualist stance. Instead of curing the symptoms of the problem generated by traditional object-ontological presuppositions Manzotti suggests a replacement of the ontological framework. He introduces a new type of process-particulars, called 'onphenes', which are dynamic, spatiotemporally extended entities that are *representings*, consisting, as it were, of object, content, and act of a representation rolled into one. Within a process-based framework, the mind is no longer a substance 'inside the head' but a species of process, "an activity that reaches beyond the physical area occupied by the brain at a certain time t". All (physical) events which produce phenomenal content are part of this activity. "The traditional picture of a boundary between an internal domain and an external domain is replaced by the image of an immensely complex fabric of processes continuously merging and dividing". Any onphene has an abstract projection as causal chain of events and it is in terms of their causal structure that onphenes can be more concretely characterized. A simple onphene has the structure of reciprocal causation – in the attenuated sense of reciprocity in that the effect E does not literally cause its cause C but brings about the relatedness of entities of the respective kinds of C and E. Motivations, Manzotti claims, can be considered as a special sort of onphenes – they generate not only a present occurrence of their own kind, but also future occurrences of the same kind. Motivation as self-reinforcing activity presents the key to a reconstruction of the self-organization (ontogenesis) of a conscious subject as a causally driven development in which onphenes are bundled into a unified causal structure. Due to the ontological characterization of the onphene, consciousness is not a private affair – "consciousness is how the world is organized due to the presence of a subject". Manzotti then proceeds to sketch a general architecture for the design of physically interactive cognitive systems (robotic systems) in which the occurrence of motivation-based self-organization of onphenes can be elicited. He reports laboratory results showing that on the basis of one single directive, (namely, the principle of motivation or self-reinforcement which demands that any process passing through the system increases the probability of its recurrence) clear examples of data-driven changes in the system's categorizing capabilities can be produced. These results provide the first evidence also for the more comprehensive thesis that an onphene-based (motivation-based) architecture is the right path towards implementing the typical cognitive flexibility we associate with the presence of a conscious mind.

Readers will find it instructive to compare Manzotti's onphene-based approach with the discussion of consciousness offered in the following contribution by Claus Emmeche. Unlike Manzotti Emmeche proceeds from the general semiotic intuition that, due to communalities in dynamic structure, material interactions and interactions between matter and mind can be treated as species of the same processual genus. In combination with a presentation of basic ideas of biosemiotics Emmeche develops an account of causation that allows us to frame an evolutionary explanation of the origin of consciousness. He introduces six principles that a comprehensive account of causation must fulfill. Elaborating on these principles he familiarizes the reader with basic tenets of current biosemiotics. According to semiotics any event – natural, mental, social, or cultural – can be understood as semiosis, the processing of information. In particular, causation in our common understanding as impact causation on material bodies counts as much as a species of semiosis as the partitioning of physical space (inside-outside, center-margin) in the emergent formation of complex structures such as heat convection cells. Biosemiotics, in the version endorsed by Emmeche, applies the general semiotic view of the world as an assembly of signaling processes to the specific task of describing natural evolution in terms of different architectures of semiotic systems. The evolutionary emergence of complex natural systems (from convection cells to biological cells to organisms) can, for instance, be aligned with gradient increases and upward 'jumps' in 'semiosic intensity' (roughly: density of potential information). In a second step Emmeche sketches how the semiotic description of the emergent evolution of complex systems may be used to account for the appearance of consciousness. Emmeche suggests two defining aspects of consciousness. First, consciousness is an instance of self-organization. Any self-organizing system exhibits 'downward causation' in the sense of a system constraint: "the form (pattern, mode) of movement constitutes a higher level which constrains or 'governs' the moments of the entities at the lower level". Consciousness is thus a dynamic "pattern which governs the behavior of neurons or neuron clusters". To this general 'complexity-theoretic' characterization of consciousness Emmeche adds the semiotic thesis that "in macro-evolution experiences are intensified from movement to consciousness". Consciousness presents one of the 'jumps' in the continous development of semiotic intensification that characterizes macro-evolution. Consciousness intensifies experiences in the sense that it changes the ways in which experiences function as signs, increasing their 'degrees of semiotic freedom' (roughly: dimensions of interpretability). In complex living systems of a certain make-up experiences function as 'future-directed' signs, and once such systems in addition correlate experiences of their movements with experiences of changes in the environment a new type of signing arises: the qualitative aspects of experiences are intensified. Consciousness is thus a specific type of semiotic process which involves (i) transformations in neural state space that are typical of self-organisation and (ii) intensified qualitative aspects of experiences – the philosopher's 'qualia.' This form of complexity Emmeche terms 'qualitative complexity' and calls for an *experiential* biology supplementing common research methods in biology in order to accommodate the qualitative, phenomenal aspects of this complexity type.

The following two contributions in part II are not topically but methodologically connected: they each apply the analytic categories of Whitehead's philosophy of organism to a specific reconstructive task, demonstrating once more the extraordinary versatility of a Whiteheadian framework. Both authors constructively modify Whitehead's scheme as set out in *Process and Reality* by treating 'actual occasions' as macrophysical processes of arbitrary spatio-temporal extent. (Readers unfamiliar with Whitehead might want to begin with a look at the 'primer' to basic Whiteheadian notions which Baris Parkan offers in section 2.2.1 of her paper).

Michael Fortescue's contribution presents an illustration of a Whiteheadian approach to the linguistic analysis of real life discourse, which he has laid out in detail in a recent monograph (*Pattern and Process. A Whiteheadian Perspective on Linguistics*, 2001). This approach is here put to test on a particularly difficult piece of discourse, an implicature-laden passage from a detective story where conventional elements of linguistic expression are combined with contextual dependencies and 'on-line reasoning', under constant readjustment and reassessment of the interpretational and communicational strategies of all discourse participants. Fortescue redefines Whitehead's notion of 'subjective aim' so that it can cover more than one speech act, even a speaker's whole discourse participation, and shows how conventions of discourse, Gricean maxims, and constraints known from Leech's 'means-ends diagrams' can be integrated into one framework. One of the remarkable features of this Whiteheadian analysis is that the intuitive reassessments of common knowledge that discourse participants continuously perform can be reconstructed as "directly felt" without the distorting introduction of propositional inferential links at the level of theoretical description. In general, unlike other discourse theories, Fortescue's Whiteheadian framework accommodates a variety of productive factors in the organization of information structure from purely physiological causation (sign recognition) to non-propositional factors (feelings, emotive context) to pre-intellectual compatibility assessments to pragmatic and semantic inferences. It offers an integrated, descriptively economical model of linguistic processing involved in discourse participation, allowing for parallel and hierarchical dynamic organization of very different kinds of constraints on linguistic productions. (Particularly interesting from a practical point of view is section 7 of the paper where Fortescue systematically correlates types of Whiteheadian prehensions with "tasks and phases of language processing", in order to clarify the cognitive underpinnings of language.) The approach to language and meaning that Fortescue presents here is radically different from the traditional model-theoretic or even discourse-representation-theoretic descriptions of meaningful utterances where pragmatic, semantic, and cognitional issues are considered as separable aspects, to be treated within different descriptive frameworks. In contrast, Fortescue's framework affords descriptions of the production of semantically and pragmatically meaningful discourse participation that can be directly linked to recent research on mental modeling and the focus of consciousness.

Baris Parkan applies a Whiteheadian framework to present an ontological analysis for one of the most basic concepts of social science: work. She draws attention to the curious fact that the phenomenon of work, which all working ontologists are intimately familiar with, has not yet become a proper subject of ontological analysis. Among possible reasons that could motivate this neglect, she rejects the argument that the phenomenon is simply too variegated. She discusses various proposed elucidations of the notion and suggests that the traditional accounts as toil (molestia) and accomplishment (opus) naturally combine to yield a simple, unified, yet sufficiently broad characterization of our common sense understanding of 'work' as "the creation of something of value". Proceeding from this characterization she presents a classificatory scheme for types of work based on two dimensions of evaluative parameters, namely, ends of work and modes of work process. In the second part of the paper Parkan argues that work is a phenomenon involving (at least in cases) rather complicated forms of dynamic organization with feedback and goal modification, which traditional ontological tools are quite unsuited to capture. She thus introduces her reader to six basic notions of Whiteheadian process ontology: occasion (actual entity), prehension, subjective aim, eternal object, contrast, and conceptual reversion, and shows how the "dynamic, teleological and self-referential structure of an actual entity" (a process particular) can be used to model the complex structure of a work process. Parkan takes "the worker in interaction with her medium to be *one* process of concrescence", with the worker entering, as it were, an 'entangled state' which gradually dissolves as the product in the making attains determinateness. The suggested Whiteheadian analysis of work not only fits the definition of work suggested in the first part of the paper, but also explains the one-sided historical focus on either toil or accomplishment.

The final two articles in part II are contributions to the ontological interpretation of chemistry and physics, respectively. Paul Needham's paper is first and foremost concerned with an ontological description of macroscopic thermodynamic processes. However, as I shall try to highlight here, quite independently of the specific domain of application, the paper offer a treatment of a number of important systematic problems that are of general interest to any process ontologist. On Needham's approach processes are not changes or causal transitions between different states, which is the common ontological route. Rather, as 'interactions between quantities' (i.e., particular portions of a chemical substance) they are the vehicles of causation that might or might not result in a change of state. As Needham points out, in suitably thermostatically controlled bodies a process of heating may not result in factual changes of temperature; nevertheless the interaction as such is present. In Needham's framework quantities, processes, and continuants stand in relationships of mutual definitional dependence: continuants are essentially involved in chemical processes, quantities essentially depend on processes in their identity conditions, since the amount of a certain substance, as well as the substance kind it belongs to, are determined in terms of their ability to participate in chemical transformations. Needham also addresses the problem of how to determine the region occupied by moving continuants, how to determine continuant and dynamic components of interactions, and he points to the intricacies of determining the spatiotemporal location of a complex process and its components. Also, he observes that the kind of a process is determined by the context of the process: diffusions have different characteristics dependent on where they take place, in which medium they take place and whether that medium is itself a mixture of substances.

The last contribution addresses the question whether process-based ontologies hold out the prospect of providing new avenues to an interpretation of quantum physical theories. Wim Christiaens argues that Free Process Theory (FPT) might be used to make sense of correlations of measurement results in so-called 'Einstein–Podolski-Rosen (EPR)-experiments'. One of the most puzzling phenomena of the quantum domain consists in the fact that the results of simultaneous yet spatially well-separated measurements may exhibit systematic correlations which cannot be explained causally. Such 'EPR-correlations' remain mysterious as long one tries to understand them from within the particularist framework of classical physics where correlations of measurement results are attributed to causal influences between determinately located point masses. The notorious 'logical deviance' of the quantum domain can be substantially reduced, however, if the logic of the macrophysical domain is not identified with the logic of our reasoning about particular entities. Christiaens draws attention to a macrophysical elucidation of EPR-phenomena suggested in 1982 by Diederik Aerts, main proponent of the 'creation-discovery' interpretation of quantum mechanics. Aerts compares the EPR-correlations with correlations of measurement results on parts of a closed fluid system: measuring for certain fluid volumes in one part of the system determines which volumes are measured for in another part of the system. Christiaens argues that FPT may be used to turn this elucidation from "a nice illustrative metaphor ... [into] a literal illustration: a macrophysical case of 'quantum behavior' ". Given that the constructional principles of FPT are inspired by the logic of stuffs and activities, individuals are not fully determinate and determinately located particulars. Rather, they are countable or noncountable in several senses of these terms, and either determinately or indeterminately located. Amounts of free processes are ordinally countable individuals with determinate locations; quantities of processes, on the other hand, are cardinally countable individuals which are indeterminately located. Some amounts of a free process are determinately located compounds of quantities of processes. Based on these conceptual resources of FPT Christiaens shows that we can draft an FPT-interpretation of the quantum-mechanical description of an EPR-experiment that is structurally identical with an FPT-interpretation of measurements on a fluid system.

In perusing the contributions to this volume the reader will find that the analysis and application of dynamic categories amounts to a new research strategy pursued in a many different fields or research. It might even deserve the label of a new 'research paradigm', as some authors suggest. This volume presents not only close-ups of unfamiliar approaches to familiar explanatory tasks, but also vistas onto an unexplored theoretical landscape.[1]

NOTES

[1] I thank Hans Fink, head of the philosophy department of the University of Aarhus, for his active support of research on process ontology, in particular, in his function as co-organizer and official host of the Sandbjerg meeting. I also gratefully acknowledge the financial support of the Danish Research Council in the Humanities.

PART I: ANALYSIS OF DYNAMIC CATEGORIES

MARY LOUISE GILL

ARISTOTLE'S DISTINCTION BETWEEN CHANGE AND ACTIVITY

ABSTRACT. Aristotle's conception of being is dynamic. He believes that a thing is most itself when engaged in its proper activities, governed by its nature. This paper explores this idea by focusing on *Metaphysics* Θ, a text that continues the investigation of substantial being initiated in *Metaphysics* Z. Q.1 claims that there are two potentiality-actuality distinctions, one concerned with potentiality in the strict sense, which is involved in change, the other concerned with potentiality in another sense, which he says is more useful for the present project. His present project is the investigation of substantial being, and the relevant potentiality is the potentiality for activity, the full manifestation of what a thing is. I explore Aristotle's two potentiality-actuality distinctions AND argue that the second distinction is modeled on the first, with one crucial modification. Whereas a change is brought about by something other than the object or by the object itself considered as other (as when a doctor cures himself), an activity is brought about by the object itself considered as itself. This single modification yields an important difference: whereas a change leads to a state other than the one an object was previously in, an activity maintains or develops what an object already is.

1. THE PROBLEM OF SUBSTANCE

Why is primary substance problematic? Against Plato Aristotle insists that being is not a genus.[1] On Aristotle's view there are various sorts of beings, and substance (οὐσία) is the primary sort. Other beings, such as qualities and quantities, are determined as what they are in some relation to it.[2] To understand those other entities, then, one must first understand the being of substance. To judge from Aristotle's *Categories*, this project seems relatively straightforward. Substances are autonomous individuals, such as a particular man or a particular horse, and they are the subjects for various properties, including qualities and quantities, which are located in one or another of the nonsubstance categories. Nonsubstances depend for their existence on the substance to which they belong. The primary substances are themselves individuated by the so-called secondary substances, the species and higher kinds that classify the individuals. Although species and genera determine the primary substances as what they are, the secondary substances, like the nonsubstances, depend for their existence on the primary substances. Remove the primary substances and everything else is

J. Seibt (ed.), Process theories: Crossdisciplinary studies in dynamic categories, 3–22.

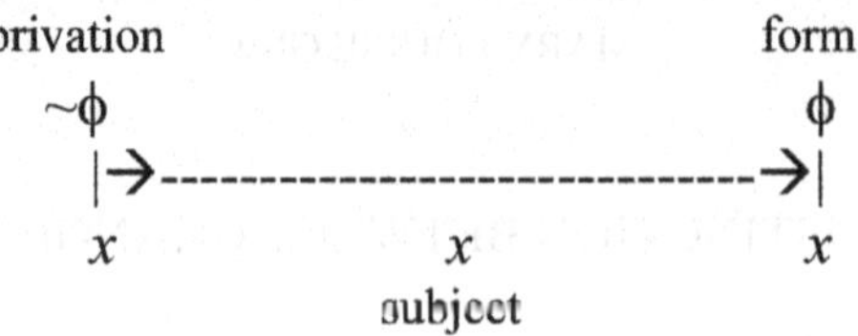

Figure 1. The replacement model of change (*Physics* I.7).

removed as well.[3] Physical objects are primary substances because they are the ultimate subjects. All other things depend on them for their existence.

The problem with this picture is that it fails to accommodate substantial change. The Presocratic philosopher Parmenides had denied the possibility of all change, arguing that change would require the emergence of something from nothing. Aristotle agreed with his predecessor that there is no absolute becoming. But he regarded the existence of change as empirically evident. His task was to account for change without admitting sheer emergence. In *Physics* I.7 he proposed that every change involves three principles: a form (ϕ), an opposed privation ($\sim\phi$), and an underlying subject (x). This account of change is known as the Replacement Model. (See Figure 1).

A change is the emergence of something new, because the form (ϕ) replaces the privation ($\sim\phi$). The change is not a sheer emergence, because some part (x) of the product (ϕx) was there all along, characterized first by the privation and then by the form. For example, when Socrates comes to be musical from being unmusical, Socrates himself survives the change and is characterized first as unmusical and later as musical. The *Categories* accommodates nonsubstantial changes, changes of quality, quantity, or place. In that work Aristotle claims that a distinctive feature of a primary substance is that one and the same individual can survive the replacement of opposed nonsubstantial properties.[4]

But substantial generation and destruction cannot be so easily reconciled with the *Categories*' scheme. In a substantial generation, a substance is the *product* of a change and so cannot be what persists through it. Aristotle's Replacement Model is supposed to provide for this case too, though that provision will entail an ontological scheme more complicated than that envisaged in the *Categories*. A new substance emerges from something else without sheer emergence, because a part of the preexisting item survives in the product. Aristotle calls the continuant *matter*. Matter provides continuity between the preexisting object and the product, and guarantees that the emergence is not from nothing. When a new substance comes into being, the form, which replaces a privation, determines what the emergent substance is. Whereas substances in the *Categories* were treated as simple entities, the analysis of change in the *Physics* reveals them as complex, as *composites* of matter and form.

Now that physical objects are regarded as hylomorphic complexes, what counts as primary substance and on what grounds? These questions are explored in the central books of the *Metaphysics* (ZHΘ). Is the whole complex primary? The complex consists of more basic components, the form and the matter, and so is arguably posterior to them (Z.3, 1029a30–32). Is primary substance the matter (Z.3, 1029a10–27)? Consider a bronze statue, whose matter is bronze. The shape of the statue informs the bronze, and the bronze can survive its removal. The constituent matter seems to satisfy the subject-criterion for substantiality in the *Categories*. If we accept the substantiality of the bronze on grounds of its subjecthood, however, why stop here? The bronze is itself a composite of more basic material ingredients, copper and tin. And these metals are themselves composites of the Aristotelian elements earth and water combined in certain ratios. Why not suppose that the elements are substances, or perhaps some yet more ultimate matter that underlies them? Tradition attributes to Aristotle a belief in prime matter, an ultimate stuff that is nothing in its own right but underlies all material bodies in the sublunary realm. Is some such ultimate matter substance? In *Metaphysics* Z.3 Aristotle excludes such an entity as substance, saying that a substance must be some definite thing in its own right (a τόδε τι) and be separate (χωριστόν) from other things (1029a26–30). An ultimate bare matter satisfies neither of these two conditions. Z.3 shows that if we press the subject criterion for primary substance, we end up with an unknowable object. Such an object is disqualified as substance because it fails to meet the constraints of thisness and separation.[5]

Is primary substance, then, the form of the composite (Z.4–12)? Whereas matter appears to claim existential priority because the form depends for its existence on it, the form appears to claim logical and epistemic priority because it determines the composite as the thing that it is and thus accounts for its knowability. Many scholars think that in the *Metaphysics* Aristotle awards the title 'primary substance' to form, revising the position he advocated in the *Categories*.[6] This solution too is problematic. First, if form is primary substance, what becomes of the subject criterion and the demand that substance be capable of separate existence? Form is predicated of matter and depends on matter for its existence. Second, what becomes of the demand that substance be an individual? Form is something predicable, definable, and knowable, and therefore seems to be a universal.[7] It can also be shared by more than one individual (Z.8, 1034a5–8). In *Metaphysics* Z.13 Aristotle notoriously argues that no universal is a substance. Some advocates of form argue on the basis of Z.13

that form is a particular.[8] Others offer sophisticated readings of the chapter to show that form escapes the objections to the universal, even though it is predicated of matter. They argue that the chapter merely shows that nothing is the substance of that of which it is universally predicated.[9] Form is the substance of one thing (the composite) and predicated of something else (the matter).

In my view both sides of the debate about form are on the wrong track. I have elsewhere argued that the objections to the universal in Z.13 tell against Aristotelian form, whether it is construed as a universal or as a particular.[10] According to Z.13, one criterion of substantiality is being a subject – something of which other things are predicated but not itself predicated of anything else (1038b15–16).[11] If form is predicated of matter, and the nature of the matter is distinct from the form, form is disqualified as substance by that criterion alone.

If I am right about Z.13, form as well as matter faces serious objections in *Metaphysics* Z. These difficulties invite us to reconsider the claims of the composite to be primary substance. My own view is that the *Metaphysics* reconfirms Aristotle's commitment to the primacy of a subset of individual physical objects – living organisms like Socrates and Bucephalus, which he defended as substances in the *Categories*.[12] To preserve their substantiality once they are analyzed into matter and form, however, he must show that living organisms are genuine unities and not accidental compounds. The relationship between form and matter within a substance should not be an accidental relation comparable to that between whiteness and a man who happens to be white. As we shall see, Aristotle believes that the relation between form and matter in those composites that succeed as substances is an *essential* relation.

Aristotle's solution to the problem of the unity of composites depends on his doctrine of potentiality and actuality. The last chapter of *Metaphysics* H takes various people to task for distinguishing between subject and predicate and then introducing a relation, such as participation, to unify them. His own treatment of change in the *Physics*, as well as his treatment of matter and form in *Metaphysics* Z, suggests that he would include himself in this critique.[13] If the relation between matter and form is analogous to that between a physical object and its accidental properties, then composite objects like Socrates and Bucephalus fail to be proper unities and so fail to be primary substances. H.6 proposes a new solution to the problem of unity for forms and composites, and ends with the following celebrated claim:

But, as we have said, the proximate (ἐσχάτη) matter and the form are the same and one, the one in potentiality (δυνάμει), the other in actuality (ἐνεργείᾳ), so that it is like seeking what is the cause of oneness and of being one; for each thing is some one thing, and the thing in potentiality and the thing in actuality are somehow one, so that the cause is nothing else unless there is something that caused the movement from potentiality to actuality. And all those things that have no matter are simply just some one thing. (1045b17–23)

What precisely does Aristotle mean when he says that the proximate matter and the form are the same and one, the one in potentiality, the other in actuality? One main task of *Metaphysics* Θ is to flesh out this statement through its analysis of potentiality and actuality.

2. POTENTIALITY AND ACTUALITY

Aristotle initiates his discussion of potentiality and actuality with a strategic proposal:

Let us distinguish potentiality and actuality, and first let us discuss potentiality which is talked about in the strictest sense, which is not, however, the most useful for our present purpose. For potentiality and actuality extend beyond the cases spoken of only in connection with change. But when we have talked about this, we shall in our distinctions of actuality clarify the other potentialities as well. (1045b34–1046a4)

Aristotle distinguishes two potentiality-actuality models.[14] The first model concerns change, the context in which potentiality applies most strictly. Although he denies that this model is central to the current project, he devotes the first five chapters of *Metaphysics* Θ to it. He states the reason for this attention at the beginning of Θ.6 (1048a28–30): the first model will help to clarify the model that is more useful for the current project.

What is Aristotle's current project? He says quite plainly at the beginning of Θ.1 that his project is the investigation of being, and in the first place being in its primary sense, the being of substance. The beginning of Θ thus reminds us of the investigation in Z-H. The second potentiality-actuality model is designed to solve the problems that emerged in that investigation, and in particular, the unity of composite substances.

Before we turn to the second model, we shall follow Aristotle's lead and consider potentiality in the strict sense, the potentiality concerned with change. The first model provides the tools to understand the second. Let us start by considering the levels of potentiality and actuality distinguished in *De Anima* II.5. The first part of the model (see Figure 2) can be mapped onto the Replacement Model in Figure 1.

In the first place, a child (x) has knowledge of French (ϕ) potentially, because her matter and kind are of an appropriate sort to be in the positive state (417a27). A child is a human being, and some human beings have that knowledge.[15] The child's potentiality is first level, because she actually

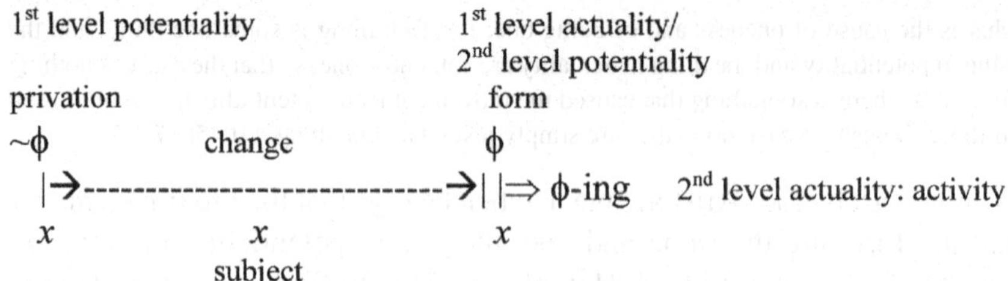

Figure 2. Levels of potentiality and actuality (*De Anima* II.5).

lacks knowledge of French (she is in the privative state ∼ϕ). So an object (x) has a first-level potentiality for ϕ on two conditions: (1) x is the right sort of thing to have ϕ, and (2) x actually lacks ϕ (x is in the privative state ∼ϕ). An object x has a first-level actuality for ϕ, if (1) x is the right sort of thing to have ϕ, and (2) x has ϕ. Thus the person who has learned French and now knows French has a first-level actuality for knowledge of French. The state ϕ is itself a potentiality for something further – speaking and understanding the French language (ϕ-ing). So a first-level actuality is also a second-level potentiality. This second-level potentiality can be realized in an activity when the agent wishes, if nothing external interferes.[16] The activity (ϕ-ing) is a second-level actuality. The transition from a first-level potentiality to a first-level actuality is a change. It takes place across a continuum that can be divided into distinct moments, and it terminates in a state other than the one in which the subject began. The transition from a second-level potentiality to a second-level actuality is not a change. This transition is simply a switch from inactivity to activity, just as the beginning of a change is a switch from rest to motion.

The levels of potentiality and actuality in *De Anima* II.5 apply to what we may call the *patient*, an entity that undergoes a change or exercises its character. Aristotle's account of these motions also includes an *agent*, something that brings about the change or the activity. Let us first consider change.

3. THE FIRST POTENTIALITY-ACTUALITY MODEL: CHANGE

Change involves a mover ($\phi^a y$) and a moved ($\sim\phi x$), each characterized by a special sort of δύναμις, or potency.[17] In *Metaphysics* Θ.1 Aristotle defines the δύναμις that causes change (designated in Figure 3 as ϕ^a) as "the source of change in another thing or [in the thing itself] as other".[18] *Metaphysics* Δ.12 specifies the δύναμις more precisely as a source of active change (ἀρχὴ μεταβλητική).[19] The potency responsible for a change typically belongs to an entity other than the object changed or, in the

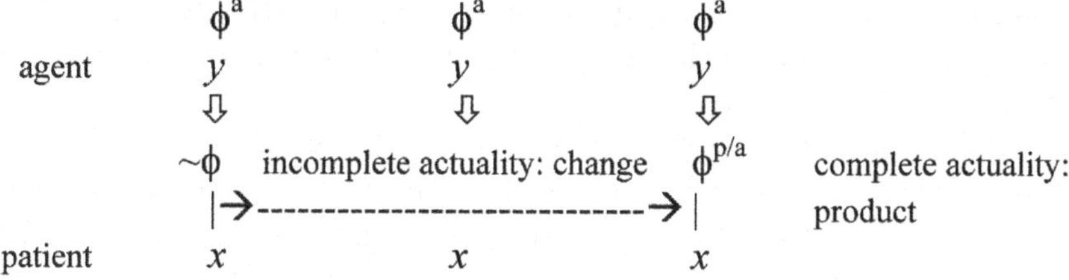

Figure 3. The first potentiality-actuality model: change (*Metaphysics* Θ.1-5).

special case of self-change, as when a doctor cures himself, to the mover itself considered as other.[20] The doctor acts in virtue of his knowledge of health; he undergoes a change in virtue of his privation of health. Aristotle characterizes the potency of the moved (designated in Figure 3 as the privation $\sim\phi$) as a source of passive change (ἀρχὴ μεταβολῆς παθητικῆς) by another thing or by the thing itself as other. How should we understand these two principles?

First, a potency, whether active or passive, is always directed toward a definite end or *actuality*. When the potency is such that its realization requires a change, the actuality is a certain state of the patient. For instance, a doctor has an active, an invalid a passive potency for health. These potencies differ from those of a teacher and student, which are directed toward a certain sort of knowledge, say knowledge of French. Second, as the examples suggest, pairs of active and passive potencies correspond in their goal: both members are directed toward the same state to be realized in the patient. A doctor's active potency for health enables him to bring about health in a suitable patient; a person's passive potency for the same state enables him, when properly affected, to become healthy.

In those changes that are not spontaneous or due to chance, the agent somehow has the character that it induces in the patient. In *Physics* III.2 Aristotle says:

> The mover will always bring a particular form, either substantial, qualitative, or quantitative, which will be the source and cause (ἀρχὴ καὶ αἴτιον) of the change, when it produces change; for instance, a man in actuality produces a man from what is a man potentially. (202a9–12)

At *Physics* VIII.5 he says: "The mover is already in actuality (ἐνεργείᾳ), for example, the hot thing heats and generally *that which has the form* generates" (257b9–10). Whereas in natural contexts the mover's own form is the form it induces in suitable materials, in artificial contexts the form is present in the agent's soul. For instance, a builder who constructs a house has in mind the form of a house: this form is his knowledge of building. The doctor has in mind the state of health: this form is his knowledge of medicine.[21] Aristotle calls the active form 'the maker' (τὸ ποιοῦν) and the source of motion (ὅθεν ἄρχεται ἡ κίνησις) (*Met.* Z.7, 1032b21–23). This active source does not itself bring about the outcome,[22] but its presence in the agent enables the agent to do so.

Aristotle defines *change* in *Physics* III.1 as "the actuality (ἐντελέχεια) of that which is in potentiality, as such (or as potential)" (201a10–11).[23] In *Physics* III.2 he says that the actuality in question is incomplete, since the potentiality of which it is the actuality is incomplete (201b31–33). Take an object (x) with a first-level potentiality for ϕ. Such an object is the right sort of thing to be in the end-state, but it is currently in the privative state ($\sim\phi$). The complete actuality of this object's potentiality is the product ϕx.[24] In his definition of change, Aristotle wants to define not the product of the change but the process that yields the product. The point of the phrase 'as such' or 'as potential' is to specify how the patient is to be regarded. If we aim to specify the change that yields a house, we are not interested in the actuality that the patient is in its own right (say bricks and stones [i.e., x itself]).[25] Nor are we interested in the actuality of the patient's potentially to be a house: the house (ϕx) is that actuality.[26] Rather, we are interested in the actuality of the patient, *as potential*, that is, as potentially a house and not actually a house (i.e., as $\sim\phi x$). The relevant actuality captures the object at all and only moments when it is *potentially* ϕ *but actually* $\sim\phi$. So construed, the definition of change captures the object at all moments up to but excluding the moment at which it is actually ϕ.

Although the definition excludes the final position, it captures more than the change, since the object in its initial state, before the change begins, has not been excluded. Indeed, the definition also fails to capture the change in its dynamic continuity: instead it picks out the object in its initial state and at any stage of its journey. How can Aristotle capture the change as a dynamic whole, excluding the initial as well as the final position? He answers this question in *Physics* III.3. Here Aristotle argues that the change, which the patient undergoes, is the actuality of the agent as well as the patient, though it takes place in the patient. At the end of III.3 he states the definition of change more precisely as "the actuality of that which is potentially productive and of that which can be affected, as such" (202b26–27). By arguing that change is the joint actuality of the agent and the patient, Aristotle can define change as a continuous process toward a goal. The actuality starts to exist at the moment the patient leaves its initial position prompted by the agent, and continues to exist until the last moment before it arrives at the goal.

In contexts of change an agent conveys to a patient a form that the patient is suited to have but actually lacks. Initially the agent and the patient are unlike because the agent's active potency (ϕ^a) is opposed to the patient's lack ($\sim\phi$).[27] By means of the change the agent 'assimilates' the patient to itself; the agent imposes on the patient the form that is the goal of its own active potency.[28] So, once the change has been completed, the agent and the patient are like. Often the resulting likeness is limited to the form that is the goal of an active and passive potency pair, and does not extend to the type of potency the agent has. For instance, if a doctor cures a patient, he conveys to the patient the form for which he has the active potency: health. The patient does not thereby gain an active potency for health (i.e., medical knowledge) but only a second-level passive potency, a potency that enables its possessor to respond to its environment in certain characteristic ways but not to reproduce that form in other things.[29] In some cases, however, an agent conveys to a patient not only the content of its own form, but also a like potency, as when a man generates a man, or when a French teacher conveys to her student the ability not only to use but also to teach French.[30]

4. THE SECOND POTENTIALITY-ACTUALITY MODEL: ACTIVITY

Aristotle has said in *Metaphysics* Θ.1 that potentiality extends beyond situations concerned with change, and that although change is the situation in which potentiality applies most strictly, the potentiality involved in change is not the most useful for the present project, the investigation of being. He discussed potentiality and actuality in connection with change, because that discussion will help to clarify the second model. I believe that the second model resembles the first in all of its basic components.[31] Like the first model, the second employs an active potency and a passive potency; and like the first, the second concerns two main actualities – a motion and a product. But unlike the first, the second model involves an agent and a patient that act and suffer in respect of the same form (ϕ), and unlike the first, the second concerns a motion that is not a change (κίνησις) in the strict sense, but an activity (ἐνέργεια).[32]

Metaphysics Θ.8 specifies a potency, which Aristotle calls a *nature* and contrasts with the active potency familiar from Θ.1. He says:

> I mean by 'potency' not only the one that has been defined, which is called an active source of change in another thing or as other, but generally every source of motion and rest. For *nature* (φύσις) is also in the same genus as potency; for it is a source of active motion, but not in another thing but in the thing itself as itself (ᾗ αὐτό). (1049b5–10)

$$\begin{array}{ccc} & \phi^{a} & \\ \text{complete actuality: product} & \Downarrow|\Rightarrow \phi\text{-ing} & \text{complete actuality: activity} \\ & \phi^{p} & \\ & x & \end{array}$$

Figure 4. The second potentiality-actuality model: activity (*Metaphysics* Θ.6–9).

The single modification, that the source of active motion is in the thing itself *as itself*, yields a scheme quite different from the previous one. In contexts of change the agent acts on a subject deprived of a certain positive character (~ϕ). By means of a change the patient comes to be other than it previously was and is assimilated to the positive state of the agent. On the second potentiality-actuality model the agent and the patient act and suffer in virtue of the same positive character (ϕ), and in natural cases active and passive potencies are located in the same individual.[33] For instance, a living organism has a nutritive soul, which is an active potency (ϕ^{a}), and its body includes a digestive system, which has a corresponding passive potency (ϕ^{p}).[34] When the organism nourishes itself using food, nutrition is a joint operation of its active and passive potencies. Nutrition is not a change, because the motion does not lead the organism into a state other than the one it was previously in. Nutrition and the organism's other natural functions, such as perception, are *activities* (ϕ-ing) which express the organism's nature. By means of activity, the organism preserves and enhances what it already is.[35]

Aristotle does not offer a definition of activity as he does for change. Instead he says that we should be content to grasp the distinction between actuality and potentiality in the second context by observing a series of examples and noting the analogies (*Met.* Θ.6, 1048a35–37). Here are two examples of the relation between actuality and potentiality: that which is awake to that which is asleep, and that which is seeing to that which has its eyes shut but has sight. Being awake and seeing are motions proper to the second model. They are activities. The fundamental difference between change and activity is that, whereas a change is a process from one state to another, an activity is the dynamic expression of a state the subject is already in. *De Anima* III.7 distinguishes the actuality that is an activity from the actuality that is a change in the following way: "Change is the actuality (ἐνέργεια) of something incomplete, but actuality (ἐνέργεια) in the unqualified sense is different, the [actuality] of that which has been completed" (431a6–7). An object that undergoes a change lacks the positive state the change will yield (the object is incomplete in that respect). An object that engages in an activity is already in the positive state (the object has been completed in that respect); its motion is a dynamic expression of that state. Applied to Aristotle's examples of potentiality and actuality on the second model: an animal asleep has various potencies that will be actively expressed when it awakes; the person with sight but eyes shut will see when her eyes open.

Metaphysics Θ.6 lists some differences between the two sorts of motions (*Met.* Θ.6, 1048b18–36). Whereas a change is directed toward an end but is not itself an end, an activity is an end in itself. In cases of change, one cannot say of the same object at the same time that it 'is ϕ-ing' and 'has ϕ-ed', for instance, 'is building a certain house' and 'has built the house'. Since the goal of a change is distinct from the change, the present and perfect verb-forms apply at different times. But an activity, by contrast, is an end in itself, and so one can use the present and perfect verbs-forms of the same object at the same time. For instance, one can say of Socrates at the same time that he 'is seeing' and 'has seen', 'is thinking' and 'has thought', and 'is living' and 'has lived'. An activity is complete as soon as it starts and for as long as it lasts.[36]

5. THE UNITY OF COMPOSITES

As we saw, Aristotle's account of change in the *Physics* threatens the unity of physical objects. In response to Parmenides he argued that all changes involve a continuant, something that remains the same through a replacement of properties. The continuant undermines the unity of the generated object. Since the continuant pre-exists the acquisition of form and can survive its removal, the composite itself is an accidental compound analogous to a white man. Aristotle's doctrine of potentiality and actuality aims to overcome this problem. But the account we have given so far does not overcome it. Take an artificial example (and consider it in terms of Figure 2). A builder builds a house out of bricks, stones, and wood. Before the building takes place these materials have a first-level potentiality to be a house: they are the right sorts of materials (suitable x) to be a house, but while lying in a heap they are deprived of the form of a house ($\sim\phi$). Once the house has been completed, the same materials are organized into a house (ϕx). The materials now have a first-level actuality: they are informed by the form of a house (ϕ). The first-level actuality is also a second-level potentiality for being a house (ϕ-ing). When the house performs its function of protecting goods and bodies, the composite house has a second-level actuality: it is actively being what it is.

Is a house a genuine unity? The bricks, stones, and wood organized in a particular way have a second-level potentiality to be a house. At the same time, those materials are still *actually* bricks, stones, and wood, and can remain what they are when the house is torn down. Let's call these materials the *remnant matter*.[37] If the materials can remain what they are when the organization is removed, the organization that makes the materials be a house is *accidental* to what they are in themselves – namely, bricks, stones, and wood. The house is therefore an accidental unity.

This artificial example is intended to show that the problem for substantial unity is the item designated as '*x*' in the four Figures we have discussed – the item that underlies a change and persists as a component of the generated product. In substantial generations this item is matter. If *x* is something in its own right and can survive a substantial generation and destruction – if, that is, it is remnant matter – then the composite it constitutes is an accidental compound, not a genuine unity. Aristotle's doctrine of potentiality and actuality as so far described does not address this difficulty. The problem is that the remnant matter is *distinct* from the form of the object composed out of the matter, as bricks are distinct from the form of a house. The distinctness of matter and form in artifacts is one reason why artifacts ultimately fail to be Aristotelian substances.

Living organisms are different, but as we shall see, their difference is not enough to overcome the problem. On numerous occasions Aristotle insists that the material parts of living organisms, if separated from the whole, are what they are in name only – homonymously.[38] Call this the *homonymy principle*.[39] For instance, a severed arm is an arm in name only and no better than a sculpted or painted arm. What is true for each bodily part is true for the whole body (*De An.* II.1, 412b17–25). A human corpse is not a human body with the soul removed. It is a human body in name only (*Meteor.* IV.12, 389b31). When an organism dies, what is left is not the organic matter. The matter, as well as the composite, is destroyed when the organism dies. Aristotle's homonymy principle expresses his conviction that the relation between form and matter in living organisms is not an accidental relation. On the contrary, the form determines the matter as the matter that it is. Since the form determines the matter as what it is, the relation between the form and the matter in living organisms is *essential*.[40] Let us call this matter the *functional matter*.

The problem of unity persists, however. Even if the functional matter is determined as what it is by the form of the whole object whose matter it is, the functions belong to some lower level subject, the remnant matter, which survives the destruction of the whole. The functional matter of an animal is destroyed when the animal dies, but some remnant matter is left behind. In *Metaphysics* Z.11, Aristotle says that Callias is destroyed into flesh and bones (1035a18–19, a33). If flesh and bones survive the destruction of a living organism, they undermine the unity of the whole, even if they are functionally organized into an organic body within the animal.

In *Generation of Animals* II.1, Aristotle confronts this difficulty. Flesh, he says, is like a face. After the organism dies, flesh is called flesh only homonymously (734b24–31). Thus he includes the uniform parts, like flesh and bone, with the nonuniform parts, like a face and an arm, as constituents of the functional matter. Still the problem does not go away. What about the matter of flesh – the earth and water of which it is composed? If there is some material continuant that underlies the functional organization and can survive its removal, this is the *x*-factor, the remnant matter, that undermines the unity of the whole. The remnant matter is ontologically prior to both the composite and the form, since the form depends on it for its existence, and the composite consists of the form and the remnant matter as two more basic components.

Aristotle deals with the problem of lower level matter in his account of mixture in *On Generation and Corruption* I.10. He argues that the ingredients of a mixture exist actually before they enter into a combination but are only potentially present in the compound (μίξις).[41] Think of the spongy stuff called 'cake'. The ingredients of cake are eggs, flour, sugar, butter, water, and so on. These exist separately and actually before they are mixed, but once they are combined and the batter is baked, the product is a spongy stuff in which the ingredients are no longer actually present. Aristotle was not an atomist: an analysis of compounds does not yield elemental particles. He proposes that the ingredients are only potentially present in the compound. They are potentially present, because components of that sort are left behind when the compound is destroyed.[42] The original ingredients do make a contribution to the compound: various of its properties are due to the ingredients. For example, the original ingredients of cake account for it flavor, moisture, weight, consistency, and color. The important point is that those properties characterize the higher level compound; there is no remnant matter to which the form of that higher compound belongs.

Aristotle's treatment of *mixis* suggests that he has a new conception of the relation between matter and form which replaces the old predication model. I think that this new model is introduced in *Metaphysics* H.6. Recall that the chapter addresses two outstanding problems, the unity of form and the unity of material composites. Aristotle first asks: Why is form one, even though we specify two items in its definition, a genus and a differentia? For instance, we define man as 'biped animal'. Why is man not two things, animal and biped? The problem goes away, says Aristotle, if we conceive of the one as matter and potential, the other as form and actuality. He says:

But if, as we say, the one is matter and the other form, and the one is in potentiality, the other in actuality, the thing being sought would no longer seem to be a difficulty. (1045a23–25)

The form man is one thing and not two, because animal is something indefinite and potential: it is not some *distinct* thing in addition to biped. Animal is simply an indefinite determinable, which biped determines as man. Since a mention of animal adds no information that is not already contained in biped, and biped exhausts what man is, 'biped animal' specifies just one thing: man. The form man is simply one thing, even though we specify its essence in terms of a genus and a last differentia.[43]

Metaphysics H.6 then extends this solution to the more difficult case of material composites.[44] Take a bronze sphere. Why is a bronze sphere not two things, bronze and sphere? On the old predication model, a bronze sphere is two things specified as 'this in that',[45] form predicated accidentally of remnant matter. But now, he says, the problem of unity disappears. There is matter, and there is form, the one in potentiality, the other in actuality. I have elsewhere argued that in claiming that the problem is the same as the earlier one about form, Aristotle is proposing that we think of the relation between matter and form as comparable to that between genus and differentia.[46] The matter is not an independent subject to which the formal properties belong. Instead, the matter is an indefinite determinable, like a genus, which the form differentiates into a particular object. Instead of remnant matter, Aristotle introduces what I call *generic matter*. There are not two objects in the same place at the same time, the remnant matter and the object it constitutes. There is just one object, which is determined as what it is by its form. To be sure, that object is also characterized by material properties that connect it to its material origins, but those properties do not contribute to the nature of the object. This new relation is what Aristotle describes in his famous statement at the end of H.6:

But, as we have said, the proximate (ἐσχάτη) matter and the form are the same and one, the one in potentiality (δυνάμει), the other in actuality (ἐνεργείᾳ). (1045b17–19)

The new treatment of the bronze sphere reveals that there are not two objects, the bronze and the sphere, each with its own persistence conditions. There is only one object, the sphere, which has been differentiated out of the bronze.

Aristotle explores the new relation between matter and form in greater detail in *Metaphysics* Θ.7, and he contrasts it with the old predication model. The old predicative relation holds as before between nonsubstantial properties (πάθη) and the physical object they characterize. The dependence of the properties on the underlying subject is indicated by the fact that we use adjectives in preference to nouns when we speak of the object's properties (we call the object 'that-en' [ἐκείνινον] with reference to its πάθη). For instance, we call a man 'musical', not 'musicality'. But in the case of matter and form, we no longer have the old predicative relation, but something new. We indicate the dependence of the matter on the object by using adjectives in preference to nouns when we speak of the object's matter (we call the object 'that-en' with reference to its matter). For instance, we call a sphere 'bronz-en' not 'bronze'. This usage indicates that the form is not a property of the matter (if it were, it should be perfectly correct to call the sphere 'bronze', just as we call a musical man 'man'). Instead, the matter is something indefinite and determinable, which the form differentiates into the object. The matter is a property (or collection of properties) of that higher object. Aristotle says:

> In the case of things that are not so [i.e., not related as a physical object to its nonsubstantial properties], but the thing predicated is some form and 'this', the final thing is matter and material substance. And calling [a thing] 'that-en' with reference to its matter and its affections turns out to be quite correct. For both are indefinite (ἀόριστα). (1049a34–b2)

By conceiving of the *x*-factor in this new way, as an indefinite determinable rather than as a definite subject to which properties belong, Aristotle solves the problem of the unity of material compounds like flesh and bronze(n) spheres. More importantly, living organisms like Socrates and Bucephalus turn out to be unified objects after all. For now the nature of the functional matter is exhausted by the form of the organism. Predicating the form of the functional matter is not a case of predicating one thing of another, since the nature of the matter is exhausted by the form. Indeed, the form just is the organism's active potency (ϕ^a) and its matter is its functional body, which is essentially (and exhaustively) determined by the corresponding passive potency (ϕ^p). An organism acts on itself as itself and the motion is its characteristic activity, its mode of living.

The residual material properties (with reference to which parts of the organic body are called *x*-en) remain independent of the form. These properties still have a crucial role to play. Recall that in the key passage in H.6 Aristotle said: "For each thing is some one thing, and the thing in potentiality and the thing in actuality are *somehow* one" (ἕν πώς ἐστιν) (1045b20–21). Why merely 'somehow' one? I have argued that the form and the matter of living organisms are characterized by active and passive versions of the same functional properties. But because organisms are generated out of simpler matter and will be destroyed into simpler matter, they also possess various dispositional properties inherited from the lower level ingredients. For this reason complex organisms easily degenerate into simpler stuff. The residual material properties tend to undermine the unity of the whole, with the result that the unity is unstable and must be constantly maintained.[47] The residual material properties are responsible for the fact that material substances grow tired, weaken, and finally collapse.[48] Because an organism tends to degenerate, staying the same is considerable work. An organism's characteristic activity is consequently more than an expression of what it is. Such activity is also its means of self-preservation and renewal. This dynamic preservation is the joint manifestation of its active and passive potencies, and that activity preserves the organism as the thing that it is.

6. THE PRIORITY OF ACTIVITY

Activity claims a special priority in Aristotle's ontological scheme. A thing's function ((ἔργον, what it does) is its final cause. The function determines what the object is and many other things about it. Consider once more an artificial example, an ax. The function of an ax is to chop things. This function determines which potentialities of an ax are essential to it. An ax has many potentialities that are not essential to it – the potentiality to fall downward when dropped, the potentiality of some of its parts to burn and of others to melt when subjected to intense heat, the potentiality to be painted yellow, and so on. It displays these behaviors and undergoes these changes under certain conditions. But these potentialities do not make the object an ax. The object is an ax because it has the potentiality to chop, and the content of that potentiality is determined by the activity (function) for which it is the potentiality. The function not only determines what the thing is. It also determines those features of the object needed to carry out the function, including its material constitution and the way the material components are organized. For instance, the blade of an ax must be sharp and rigid if it is to chop things, the handle must be solid, convenient to grasp, and sufficiently light for a person to use.

What distinguishes living organisms from artifacts is that artifacts depend for the performance of their function on an external user – the ax depends on the person who wields the ax, the violin on the violinist. An organism is itself both the user and the instrument, and so has an autonomy that artifacts lack.[49] An organism is most itself and most unified when it is engaged in those activities in which its matched active and passive potencies cooperate. These activities constitute its particular life and preserve it as the substance that it is. In this way an organism's proper activity accounts both for its peculiar composition and for its enduring unity.[50]

[1] *Met.* B.3, 998b22–27.

[2] This is Aristotle's doctrine known as focal meaning (πρὸς ἓν λέγεται), which is spelled out in *Met.* Γ.2.

[3] *Cat.* 5, 2a34–b6.

[4] *Cat.* 4a10–21.

[5] I have oversimplified. I think that thisness and separation are introduced as constraints on the subject-criterion. So a bare matter cannot succeed even as a proper subject, let alone as a substance. See Gill (1989), ch. 1. I also argue that Aristotle was not committed to the entity the tradition calls 'prime matter'. See Gill (1989), ch. 2 and Appendix.

[6] But see Wedin (2000), who argues that the *Categories* and *Metaphysics* are compatible. The two works, he thinks, are engaged in different but complementary projects. The project in the *Categories* is ontological, whereas the project in *Metaphysics* Z is explanatory. The *Categories* asks what is ontologically primary and opts for individual physical objects; *Metaphysics* Z asks what is explanatorily primary and opts for substantial form. Wedin could answer the questions I pose in this paragraph, because on his interpretation form need not be ontologically basic.

[7] For form as definable, see esp. *Met.* Z.11, 1036a28–29; cf. Z.15, where Aristotle argues that individuals cannot be defined or known as individuals: definitions are general. For form as predicable, see esp. *Met.* H.2, 1043a5–7. Form is also said to be predicated of matter at Θ.7, 1049a34–36, but I argue below (sec. 5) that Aristotle is talking about a different relation between form and matter.

[8] E.g., Frede and Patzig (1988), I.36–57, II.241–63; Irwin (1988), sec. 140; and Witt (1989), 155–62.

[9] There are many varieties of this basic idea. See, e.g., Woods (1967), Driscoll (1981); Loux (1991) ch. 6; Lewis (1991), ch. 11; Wedin (2000), ch. 9.

[10] Gill (2001).

[11] Cf. the more precise formulation of the subject criterion in Z.3: "that of which other things are predicated but which is not itself predicated of anything other (μηκέτι κατ' ἄλλου)" (1028b36–37). On the more precise formulation, cf. Irwin (1988), sec. 115.

[12] See Gill (1989).

[13] But see Loux (1995) and Lewis (1995), who argue that Aristotle's predication model (form accidentally predicated of matter) is not in fact subject to the critique in H.6. This is not the occasion to discuss their defense.

[14] See the classic paper by Kosman (1984); I discuss the topic in Gill (1989), chs. 6 and 7.

[15] Cf. *Met.* H.4, 1044a27–30: The matter for a particular product must have suitable dispositional properties – one cannot make a saw out of wood or wool. Cf. *Met.* Θ.7, 1048b37–1049a12.

[16] *De An.* II.5, 417a27–28. In this passage Aristotle focuses on what he elsewhere calls a rational potency (such as knowledge of French). Desire is a necessary condition for actualizing a second-level rational potency. If an object has a second-level nonrational potency (say it is inflammable), and it comes into contact with an appropriate causal agent (say fire), the potency will be actualized barring external interference. Obviously desire is not an issue in the nonrational cases. See *Met.* Θ.2 and Θ.5.

[17] I translate the same Greek word δύναμις sometimes as 'potentiality' and sometimes as 'potency.' I use 'potentiality' in discussing Aristotle's distinctions between δύναμις and

ἐνέργεια or ἐντελέχεια ('actuality'). I prefer 'potency' in discussing an object's source of active or passive change, because it lends itself more naturally than 'potentiality' to active and passive construal.

18 1046a10–11: ἡ ἀρχὴ μεταβολῆς ἐν ἄλλῳ ἢ ᾗ ἄλλο.

19 *Met.* Δ.12, 1019b35–1020a6.

20 *Met.* Θ.1, 1046a11–26; cf. Δ.12, 1019a20–23.

21 *Met.* Z.7, 1032a32–b6, 1032b11–14.

22 The active potency is an unmoved mover. See *GC* I.7.

23 Cf. 201b4–5. On Aristotle's definition of change, see Kosman (1969); and Gill (1989), 183–94.

24 Note that Aristotle sometimes speaks as though the form ϕ is the actuality and sometimes as though the ϕ thing (ϕx) is the actuality. For the form, see, e.g., *De An.* II.1, 412a9–11, 412a21–28; *Met.* H.2, 1043a5–6, 1043a12–28; H.3, 1043a29–36. For the ϕ thing, see the discussion of levels of potentiality and actuality in *De An.* II.5, 417a21–b16. Cf. *Met.* Z.9, 1034a16–18; Θ.6, 1048b3–9. In *Met.* Θ.8, the examples of things in actuality that are prior to things in potentiality in account, time, and substance are material composites or immaterial beings, like the prime mover.

25 See *Phys.* III.1, 201a29–34.

26 See *Phys.* III.1, 201b9–13.

27 See *GC* I.7, 323b29–324a5.

28 *GC* I.7, 324a9–11.

29 See *GC* I.7, 324b14–18.

30 Because the resulting form can be active or passive, Figure 3 designates it with a superscript '$^{p/a}$'.

31 See Gill (1989), 214–227.

32 Aristotle's text is confusing, because he uses κίνησις (and its cognates) sometimes as a generic term that applies to both changes and activities, and sometimes as a specific term that applies to change as defined in *Phys.* III.1-3, a motion that takes an object from a privative to a positive state, or from one location to another. When he speaks of κίνησις in this specific sense, he distinguishes it from ἐνέργεια, a motion that is complete as soon as it starts. I translate κίνησις as 'motion' when I take the term to be generic. He also uses μεταβολή and its cognates for change in the strict sense, so I translate this term too as 'change'. To make matters worse, Aristotle defines κίνησις as a sort of ἐνέργεια or ἐντελέχεια. We have looked at his definition of change as an ἐντελέχεια (an incomplete ἐντελέχεια) in *Physics* III.1; cf. *Phys.* VIII.5, 257b8–9. In the doublet of *Phys.* III.1 in *Met.* K.9, he gives the same definition but replaces ἐντελέχεια with ἐνέργεια (1065b16). Both ἐντελέχεια and ἐνέργεια can mean 'actuality' (first or second level), or 'activity' (ἐνέργεια is his more usual choice for 'activity'). Aristotle's use of these key terms is fluid, but his distinctions are quite precise.

33 In Figure 4 the entity that is both agent and patient is designated as the whole complex: $\phi^a\phi^p x$. The second potentiality-actuality model can also apply to artificial cases in which the agent and the patient are distinct individuals but act and suffer in respect of the same positive character ϕ. For instance, a violinist and her violin act and respond in respect of the same positive character, and violin-performance is the joint second actuality of both.

34 For Aristotle's account of roles of nutritive soul and nutritive body (and food) in the activity of nutrition, see *De An.* II.4, esp. 416b20–23.

35 See *De An.* II.4, 416b13–17.

[36] Cf. the discussion of pleasure in *EN* X.4. Pleasure is an activity, which Aristotle contrasts with changes.

[37] I owe this vivid label to Wedin (2000).

[38] See, e.g., *Met.* Z, 10, 1035b24–25; *GA* I.19, 726b22–24, II.1, 734b24–27; *Meteor.* IV.12, 389b32–390a2 and 390a10–13.

[39] For a recent discussion of the homonymy principle, see Shields (1999), ch. 5.

[40] The classic discussion of this topic is Ackrill (1972–1973).

[41] See esp. *GC* I.10, 327b22–31.

[42] In fact, the components extracted are typically not of the sort used in its production but elements – earth, water, air, and fire – that composed the original ingredients. See Aristotle's cyclical model of generation and destruction in *Met.* H.5, 1044b29–1045a6.

[43] Cf. *Met.* Z.12., 1038a25–34.

[44] I discuss this topic in Gill (1989), ch. 5. I develop my interpretation further and respond to some objections raised by Loux (1995), Lewis (1995), and Harte (1996), in Gill (forthcoming).

[45] Aristotle's favorite example of this sort of case is the snub. See Z.5. Cf. Z.11, 1036b21–32, 1037a29–b7.

[46] See above n. 44.

[47] I discuss this topic in Gill (1989), ch. 7.

[48] See Aristotle's discussion of the heavenly bodies in *Met.* Θ.8. Because they do not have the same sort of matter as sublunary objects, they never tire of their proper activity, as do perishable things. Matter of perishable things is the cause of tiring and perishability (1050b20–28).

[49] Note that in *De An.* II.1 Aristotle defines the soul as "the first actuality of a natural instrumental (ὀργανικοῦ) body" (412b5–6). Cf. II.4, 415b17–20.

[50] I read a version of this paper at the conference, "Process: Analysis and Application of Dynamic Categories", in Sandbjerg, Denmark. I am very grateful to Johanna Seibt for organizing that rich and memorable event and to all the participants for the stimulating discussion of process.

REFERENCES

Ackrill, J. L.: 1972–1973, 'Aristotle's Definitions of *psuchē*', *Proceedings of the Aristotelian Society* **73**, 119–133. Reprinted in J. L. Ackrill, *Essays on Plato and Aristotle*, Oxford: Oxford University Press, 1997, pp. 163–178.

Driscoll, J.: 1981, 'EI ΔH in Aristotle's Earlier and Later Theories of Substance', in D. J. O'Meara (ed.), *Studies in Aristotle*, Washington, D.C.: The Catholic University of America Press, pp. 129–159.

Frede, M. and G. Patzig: 1988, *Aristoteles Metaphysik Z. Text. Übersetzung und Kommentar*, 2 vols. München: Verlag C. H. Beck.

Gill, M. L.: 1989, *Aristotle on Substance: The Paradox of Unity*, Princeton: Princeton University Press.

Gill, M. L.: 2001, 'Aristotle's Attack on Universals', *Oxford Studies in Ancient Philosophy* **20**, 235–260.

Gill, M. L.: forthcoming, 'The Unity of Substances in *Metaphysics* H.6', in the Conference Proceedings of the 2002 conference on Aristotle on Substance and Predication in São Paulo, Brazil.

Harte, V.: 1996, 'Aristotle's *Metaphysics* H6: A Dialectic with Platonism', *Phronesis* **41**, 276–304.

Irwin, T.: 1988, *Aristotle's First Principles*, Oxford: Oxford University Press.

Kosman, L. A.: 1969, 'Aristotle's Definition of Motion', *Phronesis* **14**, 40–62.

Kosman, L. A.: 1984, 'Substance, Being, and *Energeia*', *Oxford Studies in Ancient Philosophy* **2**, 121–149.

Lewis, F. A.: 1991, *Substance and Predication in Aristotle*, Cambridge: Cambridge University Press.

Lewis, F. A.: 1995, 'Aristotle on the Unity of Substance', *Pacific Philosophical Quarterly* **76**, 222–265. Reprinted in F. A. Lewis and R. Bolton (eds.), *Form, Matter, and Mixture in Aristotle*, Oxford: Blackwell, 1996, pp. 39–81.

Loux, M. J.: 1991, *Primary Ousia*, Ithaca, N. Y.: Cornell University Press.

Loux, M. J.: 1995, 'Composition and Unity: An Examination of *Metaphysics* H.6', in May Sim (ed.), *The Crossroads of Norm and Nature: Essays on Aristotle's Ethics and Metaphysics*, Lanham, Md.: Rowman and Littlefield.

Shields, C.: 1999, *Order in Multiplicity: Homonymy in the Philosophy of Aristotle*, Oxford: Oxford University Press.

Wedin, M.: 2000, *Aristotle's Theory of Substance: The Categories and Metaphysics Z*, Oxford: Oxford University Press.

Witt, C.: 1989, *Substance and Essence in Aristotle*, Ithaca, N. Y.: Cornell University Press.

Woods, M. J.: 1967, 'Problems in *Metaphysics* Z, Chapter 13', in J. M. E. Moravcsik (ed.), *Aristotle: A Collection of Critical Essays*, Notre Dame, Ind. Notre Dame University Press, pp. 215–238.

JOHANNA SEIBT

FREE PROCESS THEORY: TOWARDS A TYPOLOGY OF OCCURRINGS

ABSTRACT. The paper presents some essential heuristic and constructional elements of Free Process Theory (FPT), a non-Whiteheadian, monocategoreal framework. I begin with an analysis of our common sense concept of activities, which plays a crucial heuristic role in the development of the notion of a free process. I argue that an activity is not a type but a *mode of occurrence*, defined in terms of a network of inferences. The inferential space characterizing our concept of an activity entails that anything which is conceived of as occurring in the activity mode is a concrete, dynamic, *non-particular* individual. Such individuals, which I call 'free processes', may be used for the interpretation of much more than just common sense activities. I introduce the formal theory FPT, a mereology with a non-transitive part-relation, which contains a typology of processes based on the following five parameters relating to: (a) patterns of possible spatial and temporal recurrence (automerity); (b) kinds of components (participant structure); (c) kinds of dynamic composition; (d) kinds of dynamic flow (dynamic shape); and (e) dynamic context. I show how these five evaluative dimensions for free processes can be used to define ontological correlates for various common sense categories, and to draw distinctions between various forms of agency (distributed, collective, reciprocal, entangled) and emergence (weak, strong, as 'autonomous system' (Bickhard/Christensen)).

The study of processes and process-based theory formation is a surprisingly short chapter in the history of ontology. Despite Aristotle's sophisticated investigations into change and interactive development, despite Whitehead's bold attempt at a comprehensive process metaphysics, the traditional research focus in ontology has been on 'static' entities such as objects (substances), properties, relations, and facts. Those contemporary ontological schemes which are deemed 'revisionary' demote the primacy of 'substances' merely to turn still to other types of 'static' particulars such as tropes or states of affairs.

As I have shown elsewhere, current analytical ontology still abides by the (about twenty) presuppositions that characterize the "myth of substance", research paradigm that has been so dominant in ontology, and I have argued that the 'substance paradigm' hampers ontological explanations of identity, qualitative sameness, and persistence.[1] In the following I sketch some elements of 'Free Process Theory', a new (non-Whiteheadian) process-ontology which, to my knowledge, abandons more substance-ontological principles than any other revisionary scheme hitherto pro-

J. Seibt (ed.), Process theories: Crossdisciplinary studies in dynamic categories, 23–55.

posed, including Whitehead's. Even though Free Process Theory operates with a minimum of constructional basics, the scheme is sufficiently versatile to support a rich typology of processes, including process-ontological truth-makers for our talk about allegedly 'static' entities such as things and stuffs. Free processes are a *novelty* in ontological category theory, with Broad's/Sellars' 'subjectless' or 'pure' processes and Zemach's 'types' as their closest systematic cognates.[2] Free processes are modeled on (subjectless) activities, i.e. on the denotations of sentences such as 'it is snowing', 'it is getting warmer', 'the light is shining' etc. In section 1 I discuss Vendler's and Kenny's notion of an activity and argue, based on a look at recent aspectological research in linguistics, that the well-known distinction between activities, accomplishments (performances), achievements, and states, is not a distinction in occurrences *per se* but a distinction in *modes of occurrence*, i.e. in the way in which we take occurrences to occur. The observation that different lexical contents can be conceptually 'packaged' in different ways is crucially important for ontological category theory, since it can be taken to show that *individuality is not tied to spatio-temporal boundedness*. In section 2 I introduce the category of a free process, an entity that is dynamic, concrete, individual, and automerous or repeatable (i.e. non-particular!). In section 3 I give an overview over the main constructional elements of the mereological system used to formulate Free Process Theory. Free processes stand in part-whole relationships, but the part-relation on free processes is not transitive and thus the axioms of classical extensional mereology do not apply. Due to its non-transitivity the part-relation on processes can be diversified with respect to the degree of mereological embedding (parts, parts of parts etc.). This has a variety of definitional advantages, some of which become apparent in section 4 where I present a tentative list of classificatory parameters for a general typology of processes.

1. WHAT IS AN ACTIVITY?

The task of ontology is to explore various ways of structuring the referential domain of a language.[3] More precisely, an ontological theory has the form of the quadruple $\langle M, T_M, f, L \rangle$: it specifies an assignment f which correlates the elements of a class L of L-sentences with "truth-makers": structures of the domain of interpretation M as described by a domain theory T_M. The assignment function f (which remains mostly implicit) should be chosen in such a way that it can be used to *explain* why L-speakers are justified in drawing *categorical inferences*, a certain type of material inferences, which define the meaning of the *ultimate genera terms* of L (e.g., 'thing', 'property', 'person' etc.). This explanation takes the following form: via subsumption under a certain category with certain *category features*, the material inference is turned into a formal inference.[4] Furthermore, given that the structural descriptions of ontology are to serve explanatory purposes, ontological categories – just like theoretical concepts in science – need a *model* or canonical illustration to serve their explanatory function. In physics, a water current or an ideal spring serve as cognitive models for electrical current or an harmonic oscillator, respectively. Similarly, in ontology substances are modeled on things, and monads and Whiteheadian occasions on minds. An ontological model must be familiar to L-speakers or, as I say, '*founded*' in their agentive experience.[5]

According to the postulate of foundedness any new candidate for an ontological category must be shown to have a model, i.e. to be directly associated with a concept which L-speakers are agentively familiar with. Instead of first introducing the category of a free process and then its model I will here turn matters around and begin with reflections on our concept of an activity. As we shall see, if we pay proper attention to our common sense understanding of activities – i.e. to the categorial inferences licensed by those sentences that we commonly accept as being about actitivies – we find that we are already well familiar with the concept of recurrent, non-determinate, dynamic individuals to be introduced in the next section.

1.1. *'Action types': an unsuccessful start*

Aristotle may have been the first to distinguish between occurrence types in terms of inferential and linguistic criteria, sorting occurrences into *energeiai* and *kineseis*.[6] In 1949 G. Ryle revisited the Aristotelian distinction and distinguished 'try it' verbs or 'verbs of activity or process' such as *search*, *kick*, or *treat* from 'got it' verbs or 'achievement words' such as *find*, *score*, or *heal*.[7] Inspired by Aristotle and Ryle, yet independently of each other, Z. Vendler and A. Kenny developed more differentiated classificatory schemes of 'action verbs' or 'action types'.[8] Vendler 's classification is essentially based on four criteria, which I dub the dynamicity condition (C1), the unboundedness condition (C2), the distributivity condition (C3), and the homeomerity condition (C4). For example, verb phrase 'V' is an 'activity verb' iff all of the following four conditions are met (where 'N' stands for a noun-phrase in L):

(C1) *Dynamicity*: 'N is V-ing' is a well-formed L-sentence.

(C2) *Unboundedness*: 'x finished V-ing' is not a truly applicable L-predicate.

(C3) *Distributivity*: For every temporal interval $[t]$, if 'A V-ed during $[t]$' is true then 'A V-ed during $[t']$' is true for every period $[t']$ that is part of $[t]$.

(C4) *Homeomerity*: Any temporal part of the denotation α of 'V' is of the 'same nature' as the whole of α.

Vendler uses these four conditions to set up a fourfold division of 'action verbs' and 'associated time schemata':

TABLE I

Activity verbs: *run, walk, swim, push etc.* • dynamic • unbounded • distributive • homeomerous *Time schema*: 'N was V-ing at t' is true means that t is on *a* time stretch throughout which N was V-ing	**State verbs**: *have, possess, like, hate, desire, want, dominate, rule etc.* • not dynamic • unbounded • (strictly) distributive • homeomerous *Time schema*: 'N V-ed between t_1 and t_2' is true means that at any instant between t_1 and t_2 N V-ed
Accomplishment verbs: *paint a picture, build a house, grow up, recover from illness, run-a-mile etc.* • dynamic • bounded • not distributive • anhomeomerous *Time-schema*: 'N was V-ing at t' is true means that t is on *the* time stretch in which N V-ed	**Achievement verbs**: *start, reach the summit, win the race, be born/die, find, recognize etc.* • not dynamic, instantaneous • unbounded • (trivially) distributive • (trivially) homeomerous *Time schema*: 'N V-ed between t_1 and t_2' is true means that the instant at which N V-ed is between t_1 and t_2

Kenny champions a threefold classification into 'states', 'activities' and 'performances' (i.e., roughly speaking, a class comprising Vendler's accomplishments and achievements). Like Vendler he operates with a mixture of inferential and linguistic criteria, but pays greater attention to verb forms than Vendler. For activities, for example, he adds an additional inferential criterion familiar from Aristotle's characterization of *energeia*, which I call the completeness condition[9]:

(C5) *Completeness condition*: 'N is V-ing' implies 'N has V-ed'.[10]

Verbs which license the entailment specified in the completeness condition Kenny calls 'activity verbs', those which fail to license the entailment but fulfill the dynamicity condition (C1) are 'performance verbs'. "For instance, if a man is building a house, he has not yet built it; if John is deciding whether to join the army, he has not yet decided to; if Mary is cutting the cake, she has not yet cut it."[11] The idea behind Ryle's, Vendler's, and Kenny's approach is to distinguish types of verb denotations by distinguishing types of verbs. However, as has been pointed out in the literature, there are a number of difficulties with the suggested classification.[12] The following two shortcomings are decisive. (a) If at all, the classifications work for whole sentences rather than for 'verbs'; not verbs (verb phrases) but whole sentences (whole predications) carry the different inferential roles that dovetail with 'action types'.[13] (b) While Vendler and Kenny thought that lexical meanings of verb phrases could be sorted into four (or three) types of occurrences, the non-lexical, aspectual meaning of verb phrases plays an essential role for any such classification. Aspectual meanings or 'verbal aspects' are expressed by special morphological elements ('verbal aspect markers') or periphrastically. The difference in aspectual meaning accounts, for instance, for the difference in meaning between the sentences *Tom crossed the street* and *Tom was crossing the street.*[14] Vendler exploits aspectual meaning in formulating the 'dynamicity condition' (C1), but fails to observe that changes in aspectual meaning systematically affect his classification; for example, sentences with 'accomplishments verbs' in the continuous form, such as *Tom is crossing the street*, support an 'activity reading' rather than an 'accomplishment reading.' Kenny makes extensive use of aspectual information (see below) but takes it to be *strictly linked* to lexical information. Both classifications thus neglect the 'phenomenon of type shift', i.e. the fact that a shift in the aspectual meaning of a sentence can effect a shift in the occurrence type denoted by the sentence.[15]

1.2. *Aktionsarten and verbal aspect: aspectological lessons*

Vendler's and Kenny's project received little attention in contemporary ontology but was followed up in linguistic research on verb semantics, partly within the field of formal semantics,[16] partly constituting a new research area that nowadays is labeled 'aspectology.' Aspectology is concerned with information about the dynamic properties of the denotations of verb phrases or sentences (predications) in general i.e. both with (i) *aspectual* meaning or verbal aspects as well as with (ii) *lexically* coded information. Occasionally aspectologists use the term 'aspectual meaning' to cover both (i) and (ii), but I will use that term in the narrow sense of (i) and otherwise speak of 'processual information.'

There are two general aspectological research strategies to be discerned.[17] The older, 'lexical strategy', as one might call it, largely pursues the Vendler/Kenny project of trying to sort verb phrases or predications into various occurrence types or 'Aktionsarten'.[18] Unlike Vendler and Kenny, however, proponents of the lexical strategy are careful to define

'Aktionsarten' in terms of purely lexical criteria, i.e. in terms of possible adverbial modifications or inferences engendered, and without reference to aspectual information expressed by the English Continuous form or Perfect tense. There are various systems of Aktionsarten, mostly set up by means of feature combinations of a three binary parameters such as '+-dynamic', '+-control', '+-telicity.' The resulting classifications of occurrence types differ somewhat and are differently diversified, but in each case they include Vendler's criteria (C2) through (C4) and, fairly consistently, reproduce his four action types.[19] Most significant for our purposes is the fact that all systems of 'Aktionsarten' postulate a type of dynamic and non-development or 'atelic' situations or 'eventualities', but have conspicuous difficulty in defining (in purely lexical terms!) the 'dynamicity' of such situations. While there are ample criteria for the 'atelicity' and 'telicity' of a predication, i.e. for the difference between predications denoting non-developmental or developmental dynamic situations, the distinction between 'static' and 'dynamic' situations poses a genuine problem.[20]

The 'lexical strategy' not only falls short of making proper sense of 'dynamicity', it also has little to say about the phenomenon of 'type shift' mentioned above. Both of these difficulties can be successfully addressed once 'processual information' about occurrence types is taken to be conveyed by the 'interaction' or composition of a predication's lexical meaning and aspectual meaning. This is the second and currently favoured strategy of aspectological research, where aspectual meaning plays an integral part in determining the occurrence type denoted by a predication.[21]

Proponents of the 'compositional strategy' claim that the lexical meaning of predications provides the schema of a dynamics, a temporal extension structure ("phase structures", "temporal schema"), which is further 'operated' on by the processual information of verbal aspects. Of primary importance in this regard is the contrast between the perfective aspect and the imperfective or progressive aspect, respectively. The precise contribution in processual meaning provided by the perfective and imperfective is apparently not easily stated. Linguistic characterizations of this difference exhibit considerable conceptual strain and frequently resort to spatial metaphors: the perfective presents an occurrence 'from the outside', as a 'bounded unit', while the imperfective presents an occurrence 'from the inside'; the imperfective is said to present an occurrence in the way in which a parade is experienced from the point of view of a "person marching along in a parade" who is "in the middle of the parade and unable to oversee ... the full length of the parade", while the perfective presents it like the parade is experienced "from the point of view of a spectator who, from an elevated vantage point, can oversee the entire extension of the parade".[22] Elsewhere I have argued that the best sense we can make of these descriptions is in terms of dynamicity: the perfective presents the factual content of an occurrence; the imperfective highlights the dynamicity of the occurrence or presents the situation as *occurrence*; and the progressive adds to the information that the situation occurs at the reference point of the utterance.[23]

In order to formalize the difference between perfective and imperfective 'focus' of an occurrence, proponents of "selection theories" of processual meaning represent temporal extension structures (phase structures) by means of two elements: the situation boundary and the transition ('phase') between the boundaries. For example, a verb phrase may be lexicalized with the temporal extension structure '$[(\tau)\phi(\tau)]$' (a phase and unspecified boundaries) or with '$[\tau]$' (just the boundary) or with '$[\phi]$' (just the phase) or '$[\phi\tau]$' (phase and specified boundary) etc. As recognizable from these examples, four of these phase structures dovetail with Vendlerian action types. The function of the imperfective and perfective aspect is to select for denotation different elements of a phase structure: the imperfective selects for the phase and the perfective selects for the boundaries.

However successful such compositional theories of processual meaning might be as linguistic semantics, they are mainly theories of denotation and tell us little about the conceptual role of processual information beyond topological implications.[24] But they furnish an important lead for the clarification of our concept of an activity by treating aspectual information as the decisive factor in the classification of the occurrence types denoted by a predication. As shown in the following subsection, the differences in processual information that led Vendler and Kenny to stipulate different occurrence types can be articulated as *differences in complex aspectual information*. The 'viewpoint' of the imperfective (progressive) aspect is inferentially particularly fertile: all that matters about activities (and the other three Vendlerian action types) can be expressed in terms of different implicational roles of sentences with imperfective (progressive) aspect.

1.3. *Activity as mode of occurrence*

Unlike Vendler, Kenny included among the criteria for his 'action types' also inferential schemata involving English verb forms with aspectual differences (cf. for example condition C5 above). To my knowledge, neither the ontological nor the aspectological discussion has paid any attention to Kenny's inferential characterization of action types, mainly because these inferences can be shown to fail for particular instances of the verb types they are supposedly defining. But once we dispense with the idea that occurrence types are fixedly denoted by verbs, Kenny's observations about aspectual implications can be put to good use. Instead of claiming with Kenny that activity verbs are those that fulfill the schema '*N* is *V*-ing implies *N* has *V*-ed' (only to discover that for almost any verb we can think up circumstances where such an implication does not hold), we will take the implication as an indication that a certain occurrence denoted by a predication is *conceived of* as an activity. Instead of claiming that *run* or *read* are activity verbs in any context, and *run a mile* or *read through* are accomplishment verbs in any context, etc., we need to concentrate on the inferential role of a predication in a certain context and say that a predication denotes a certain type of occurrence *just in case it fulfills, within that context, a certain inferential role.*

More concretely, I suggest the following redefinition of Vendler's four action types in terms of sets of implications involving changes in aspect or 'aspectual implications' for short; in formulating these implications I take my bearings from Kenny, replacing English verb forms with general, language-independent references to predications With progressive aspect ('prog(P)'), habitual aspect ('hab(P)'), neutral aspect ('neut(P)'), and in the form of the experiential or resultative perfect ('e-perf(P)' and 'r-pref(P)').[25]

[M1] A sentence *S* with verbal aspect *VA* and predication *P* denotes a *state* iff

for *VA* = r-perf:	$r\text{-perf}(P) \Rightarrow \text{neut}(P)\ \&\ \neg(r\text{-perf}(P) \Rightarrow \text{prog}(P))$
and for *VA* = e-perf:	$\neg(e\text{-perf}(P) \Rightarrow \text{neut}(P))$
and for *VA* = neut:	$\neg(\text{neut}(P) \Rightarrow e\text{-perf}(P))$

[M2] A sentence *S* with verbal aspect *VA* and predication *P* denotes an *accomplishment/development* iff

for *VA* = r-perf:	$r\text{-perf}(P) \Rightarrow \neg\text{prog}(P))$
and for *VA* = e-perf:	$e\text{-perf}(P) \Rightarrow \neg\text{prog}(P))$
and for *VA* = neut:	$\text{neut}(P) \Rightarrow \text{hab}(P)$
and for *VA* = prog:	$\text{prog}(P) \Rightarrow \neg r\text{-perf}(P)$ and $\neg e\text{-perf}(P)$
	$\text{prog}(P) \Rightarrow r\text{-perf} + \text{prog}(P)\ \&\ e\text{-perf} + \text{prog}(P)$

[M3] A sentence *S* with verbal aspect *VA* and predication *P* denotes an *activity/process* iff

(i) for *VA* = r-perf:	$r\text{-perf}(P) \Rightarrow (\text{prog}(P)) \vee \neg\text{prog}(P))$
(ii) and for *VA* = e-perf:	$e\text{-perf}(P) \Rightarrow (\text{prog}(P)) \vee \neg\text{prog}(P))$
(iii) and for *VA* = neut:	$\text{neut}\ (P) \Rightarrow \text{hab}(P)$
(iv) and for *VA* = prog:	$\text{prog}(P) \Rightarrow r\text{-perf}(P)\ \&\ e\text{-perf}(P)$
	$\text{prog}(P) \Rightarrow r\text{-perf} + \text{prog}(P)\ \&\ e\text{-perf} + \text{prog}(P)$

[M4] A sentence *S* with verbal aspect *VA* and predication *P* denotes an *achievement/result* iff

for *VA* = r-perf:	$\Rightarrow (r\text{-perf}(P) \Rightarrow \text{neut}(P))$
and for *VA* = e-perf:	$\neg(e\text{-perf}(P) \Rightarrow \text{neut}(P))$
and for *VA* = neut:	$\text{neut}(P) \Rightarrow r\text{-perf}(P)\ \&\ e\text{-perf}(P)$

Since the clauses of these four definitions are conjuncted, an occurrence type is defined via an entire network of aspectual implications. A sentence like *Tom is mending his shirt* is thus ambiguous, for example. It denotes an accomplishment just in case we take it to imply that *Tom has mended his shirt* is false, that *Tom has mended his shirt*, if true, would imply that *Tom is mending his shirt* is false, and so forth for all implications in [M2]. The very same sentence denotes an activity, however, just in case we understand it as implying that *Tom has mended his shirt* is true (e.g. in the sense of: Tom has done some 'shirt-mending'), that *Tom has been mending his shirt* is true, and so forth for all implications in [M3]. Occurrence types, this is the core intuition underlying the suggested definition, are ways of 'conceptually packaging' an occurrence. Just as we 'package' the lexical content of a predication differently by combining it either with an imperfective aspect or a perfective aspect, so we can 'package' a content also with a more complex processual information consisting of a network of aspectual implications. On this approach, *occurrence types are denoted by complex verbal aspects*. Since verbal aspects in general are processual information about a mode of occurrence (e.g. the progressive presents the occurrence as unbounded and ongoing, the perfective as a bounded unit), also complex verbal aspects such as the 'activity network' or 'accomplishment network' denote *modes of occurrence*. In sum, on the *modal interpretation of occurrence types* that I am suggesting here, the answer to the question 'what is an activity?' runs as follows. Activities are not a certain class of occurrences but they are modes of occurrences. Occurrences are not fixedly linked to any of the 'dynamic shapes' that the Vendler classification identifies. Occasionally we treat typical Vendlerian 'accomplishments', 'achievements', or 'states' as activities, and vice versa; treating an occurrence as activity, accomplishment etc. means to conceptualize the occurrence as something that happens *in a certain way*. Our concept of an activity is the concept of *how* something is happening – it consists in the set of inferences specified in [M3i–iv] which we can draw when we know that something has happened in this way. To reformulate the conceptual knowledge encoded in [M3i–iv], if an occurrence is an activity, then it occurs in a mode of occurrence that is characterized by the following four features:

(C5) *Completeness condition:* According to (M3-iv), from '*N* is *V*-ing' we can infer '*N* has *V*-ed', i.e. *activities are always already completed once going on.*

(C6) *Resumability condition:* According to (M3-i and ii), from '*N* has *V*-ed' we cannot conclude '*N* is still *V*-ing' nor '*N* is no longer *V*-ing', i.e. *activities can be suspended and resumed.*[26]

(C7) *Recurrence condition:* According to (M3-iii), from '*N Vs*' we can infer that there are multiple recurrences of the same activity, i.e. *activities can recur.*

(C8) *Dynamicity condition:* According to (M3-iv), from '*N* is *V*-ing' we can infer '*N* has been *V*-ing', i.e. *any present going on of an activity is the outcome of its past going on.*

2. FREE PROCESSES: A NEW CATEGORY

If an occurrence occurs as activity, then it is conceived of as an entity that fulfills conditions (C5) through (C8) which correspond to the categorial inferences licensed by activity sentences. If free processes are to be the truth-makers of (among others) sentences about activities, then the category features of free processes must entail [M3i–iv] or (C5) through (C8). But which category features will accomplish this task? Consider the following two mereological features:

[1] *Likepartedness:* An entity of kind K is likeparted iff *some* of its spatial or temporal parts are of kind K.

[2] *Strict likepartedness:* An entity of kind K is likeparted iff *all* of its spatial or temporal parts are of kind K.

If we take the completeness condition (C5) to state that for *most* times *t* during *N*'s *V*-ing it holds that *N* has *V*-ed at *t*, then the truth-maker (C5) is compatible with both of these conditions. On the other hand, if we read (C5) as saying that for *all* times during *N*'s *V*-ing it holds that *N* has *V*-ed, then only strictly likeparted or strictly self-contained entities can qualify as truth-makers for sentences about activities. Vendler took 'activities' to be strictly likeparted (cf. C3 above); others criticized this decision because, strictly speaking, in every occurrence – even a running, reading, sliding, or seeing of something – we *can* identify phases: every occurrence *can* be shown to contain change.[27] Indeed, we *can* easily argue that not every part of running is a running, just as not every part of water is again water – the homeomerity of 'activities' has been compared to the homeomerity of stuffs and taken to hold only for a certain 'grain-size.' In my view this discussion is partly misguided; when we discern phases in an occurrence we are no longer conceiving of it as an activity but view one part of it as an accomplishment – all that the discussion displays is possible 'type shift.' Nevertheless, the debate about the 'granularity' of 'activities' can be used to motivate the following generalization of the predicate of homeomerity:

(HOM) *Homeomerity:* An *n*-dimensional ($1 \leq n \leq 4$) entity *E* of kind *K* is homeomerous in dimension *n* iff (a) *all* or (b) *some* or (c) *none* of *E*'s parts in dimension n are of kind K. In case (a) E is *maximally likeparted*, in case (b) *likeparted*, and in case (c) *minimally likeparted.*[28]

An entity E that is homeomerous in this general sense can be the truthmaker of sentences implying (C5) (in case E is maximally likeparted), or of sentences about accomplishments (in case E is minimally likeparted), and of sentences about sequences of accomplishments (in case E is likeparted).

An entity that is homeomerous in the sense of (HOM) can be the truthmaker of sentences about any extended occurrence. But this is not all. As already intimated above and frequently observed in linguistic semantics, there are striking inferential symmetries between sentences about activities and sentences about stuffs or 'masses', and correspondingly between sentences about accomplishments and sentences about things. Homogenous masses (such as gold or water) are maximally likeparted in space, heterogenous masses or mixtures (such as fruit salad, furniture, or H_2O) are likeparted in space, and things or 'countables' are minimally likeparted in space. Where ontologists have taken note of these inferential symmetries, they have failed to observe, however, that the denotations of nouns display a similar flexibility in category implications as the denotations of verbs. Just as the distinction between activity sentences and e.g. accomplishment sentences is best considered a matter of complex verbal aspects (represented by networks of aspectual implications), so the distinction between 'mass readings' and 'count readings' of noun phrases is determined by 'nominal aspects'.[29] Most nouns in English are lexicalized either with the category implications for masses ('water') or the category implications for countable items ('book'). But again, as in the case of verb phrases, these lexicalized meanings specify merely defaults readings; supported by morphosyntactic nominal aspect markers and, in particular, the context of interpretation, speakers may shift the categorial implications from mass to count or *vice versa.* In classifier languages where nouns are lexically underspecified with respect to their categorial implications such nominal type shifts are the rule – different classifiers 'package' nominal content in different ways, e.g. as denoting a mass or as denoting a countable item. But the phenomenon is also frequent in English; for instance, relative to sentence context and utterance context, we can read *chicken*, *car*, or *grapes* with the category implications for countable individuals or with the category implications for a certain (functional) stuff (cf. *I'd like some of the chicken, get more car for your money, a pound of grapes contains about 50 grapes*). Even proper nouns or pronouns may be given a 'mass' interpretation – you may wish that someone were less George and more president, or join the American army to "be all you can be". In short, the inferential symmetries observed between activities and masses, or accomplishments and things, respectively, suggest that there are two types of conceptual 'packaging': contents of both verbs and nouns may be packaged with the category implications of non countable items or with the category implications of countable items. Non-countable items are maximally likeparted or likeparted, countable items are minimally likeparted.

But free processes are not merely homeomerous entities, they fulfill an even stronger mereological predicate. Let us contrast likepartedness as above in [1] with *self-containment*, and homeomerity as in (HOM) with *automerity*:

[3] *Self-containment:* An entity E is self-contained iff the spatiotemporal region in which all of E occurs has some spatial or temporal parts in which E (i.e., all of E) occurs.

(AUT) *Automerity:* An n-dimensional ($1 \leq n \leq 4$) entity E of kind K is automerous in dimension n iff (a) *all* or (b) *some* or (c) *none* of E's parts in dimension n are an occurrence of E. In case (a) E is *maximally self-contained* in case (b) *self-contained*, and in case (c) *minimally self-contained.*[30]

As witnessed by (C7) above, if we conceptualize an entity as an activity, we conceptualize it as a *recurrent entity*, and the same holds for masses: if swimming laps is taken to occur as an activity, and if 'grapes' is taken to specify a stuff, then swimming laps – the very same activity – and grapes – the very same stuff – can both occur now and then, or here and there at the same time. The feature of self-containment is one way to articulate in mereological terms an ancient ontological intuition, namely, that there are entities which occur multiply as the *numerically* (not just qualitatively!) *same* entity. Self-contained entities are obviously not particular entities, i.e. entities that occur in a unique spatiotemporal location at any one time, and as long as one adheres to the presuppositions of the substance-paradigm one is forced to class an entity that recurs as the *numerically same* entity as a universal.[31] At first glance this seems smooth enough: universals are said to occur multiply in space, their identity conditions do not involve its spatiotemporal location(s), and they are not countable by means of spatiotemporal extent. But in order to occur spatiotemporally, traditional universals, whether abstract or concrete, always depend on a particular that is the logical subject of their qualitative determination. In contrast, we may well think self-contained entities as multiply occurrent without being 'attributed' to any particular – you and I, we may have the same wine in our glasses and when we drink from them we engage in the same activity, but there is no 'logical subject', not even the spacetime region, to which the wine or the drinking need to be *attributed* to.

Moreover, a self-contained entity can well be considered an individual, i.e. something we are able to refer to. What we refer to must be 're-referrable-to', and this requires, *vide* Strawson, transtemporal sameness. But, *pace* Strawson, referential reidentification does not imply particularity as required by spatial co-ordinatization. Co-ordinatization in arbitrary parameter spaces or functional individuation is all that is needed. Anaphoric references as in *yesterday I saw in Tom's sailboat the wood we used to have in the kitchen* or *That sport has been practiced since the middle ages* indicate that self-contained entities easily qualify as individuals. Functionally individuated entities are *determinable* entities, at least with respect to their spatio-temporal location; once we admit that self-contained entities are individuals, we thus break away from what might be considered the core presupposition of the myth of substance: that all individuals are particular and fully determinate entities.

Altogether, then, free processes are concrete, individual, automerous entities; since they are automerous, they are by implication functionally individuated and determinable. While traditional ontology has prioritized countable particulars (things), in free process theory countability and particularity are merely the limiting cases of non-countability (self-containment) and determinability. Some free processes are minimally self-contained – things and developments (taken as functional wholes) do not occupy any proper part of the spatial region they occupy, and that is all there is to countability. Some free processes have a functionally individuating description which happens to include the specification of a unique spatiotemporal location (e.g., snowing-at-t-in-location-s), and that is all there is to particularity. But note that particularity as such does not imply full determinateness in the sense of a Leibnizian *infima species*. Free processes are determinable 'functional stuffs' even when occurring as particulars: as detailed below the *amount* of a free process α is itself a free process β which has determinate spatiotemporal location but might be indeterminate in other respects.

To my knowledge all extant ontological schemes of process ontologies conceive of processes as *concrete particulars*, that is, as filled spacetime regions or the discrete particular fillings of spacetime regions.[32] The paradigm example of a process is taken to be either the particular performance of a human activity, such as a particular running or reading, or, the particular occurrence of a 'subjectless', 'absolute', or 'pure process' (C.D. Broad; W. Sellars): a snowing or thundering in a particular spatiotemporal region. In contrast, free processes are not only 'free' in the sense that they are not alterations in a subject, they are also free in the sense that they are not 'bound to' a specific spatiotemporal region. Intuitively speaking, free processes are *goings-on* as expressed by 'feature-placing statements' of a more or less 'placing' sort: *it is raining now, it is itching here, there's good sailing all along the coast, a photon is traveling from the sun to the earth.*

So far I have mentioned category features of free processes that are implied by or compatible with conditions (C5) and (C7) above, but I have not commented on the category feature 'dynamicity' which dovetails with conditions (C6) and, in particular, (C8) above. Elsewhere I argue, based on an analysis of Aristotle's notion of *energeia*, the process-ontological core thesis, namely, that being and going on or dynamicity as 'self-production' are co-intensional concepts.[33] The notion of dynamicity supplied there distinguishes free processes from the static expanses of 'four-dimensionalism', and allows us to stipulate free processes even as truth-makers for many Vendler 'states', reading predications such as *the ball's being red and circular* as though they were to involve the progressive.

The task of Free Process Theory is to show that the notion of a free process is wide enough to accommodate the inferential roles of a large number of classificatory terms (thing, event, action etc.); that is, to show that the truth-makers of English and its translation equivalents consist of *nothing else but free processes*: goings-on, automerous in this or that dimension, simple and complex, slow and fast, evenly and with culmination, creating (inferential stand-in's for) particular τοδε τι's 'on the go'.

3. FPT: A MEREOLOGY ON FREE PROCESSES[34]

Classical Extensional Mereology (CEM) is geared to the reading of the part-whole relationship that applies if the parts and wholes in question are geometrical regions. But, as has been noted in the literature, in application to other 'arguments' the relation axiomatized in CEM does not capture equally well our intuitive usage of 'part-whole' concepts. Mostly this deficiency of CEM is attributed to a failure of the extensional identity principle of CEM (Proper Parts Principle).[35] In contrast I suggest dropping the transitivity axiom.[36] The relation modeled by classical mereology (is a piece of, is an extension part of) is *not* the basic sense of the 'is part of' relation which has functional overtones that disturb the transitivity of the relation.[37] (The fact that the transitive 'is a piece of' relation has taken centerstage in the theoretical modeling of parthood reflects nicely the particularist bias of the substance paradigm.) Rather, the role of articulating the fundamental sense of parthood should better be assigned to the non-transitive part-relation holding among stuffs and activities – 'is part of' (belongs to, comes with, is involved in) is more basic than 'is *a* (spatiotemporal) part of.' In order to contrast 'is part of' in this widely functional, non-spatiotemporal sense from the classical geometric reading, I refer to it as the 'common sense part relation.' But note that the non-transitive common sense part relation can be used to define the familiar transitive relation of parthood as holding between geometric regions or 'extensions' ('extent part', abbreviated as '$<_{ext}$' and diversified into '$<_{temp}$ and '$<_{sp}$').[38]

The common sense part relation (read: 'is part of'; abbreviated '$\subsetneq$') of FPT is defined on the field of free processes as an asymmetric and irreflexive relation:

(Ax1) $x \subsetneq y \rightarrow \neg y \subsetneq x$

(Ax2) $\neg x \subsetneq x$

Parthood-or-identity is then defined as

(D1) $x \underline{\subsetneq} y \leftrightarrow x \subsetneq y \vee x = y.$

In FPT parthood as expressed by '$\subsetneq$' reaches only into the immediate or 'first-level' parts of a process. To express part relations with ancestors a notion of n-part is defined (which corresponds, for variable n, to the transitive closure of '$\subsetneq$'; abbreviated by '${}^{n}\subsetneq$'):

(D2) $\forall n \in N$; if $n = 1$ then $x\ {}^{n}\underline{\subsetneq}\ y \leftrightarrow x \underline{\subsetneq} y$; if $n \geq 2$ then $x\ {}^{n}\underline{\subsetneq}\ y \leftrightarrow$ $\exists z\ (x \underline{\subsetneq} z\ \&\ z\ {}^{n-1}\underline{\subsetneq}\ y)$

The expression '$+m$-part' refers to all n-parts with $m \geq n$. FPT is 'extensional' in the sense that the Proper Parts Principle (PPP) supplies the identity principle for free processes, but note that it is defined in terms of an n-part, i.e. only as axiom schema:[39]

(Ax3) $\forall z (z \subsetneq_n z \leftrightarrow z \subsetneq_n y) \leftrightarrow x = y.$

Since '$\subsetneq$' is not the relation of extent part, the well-known counter-examples to the (PPP) (e.g. undesired identifications of functional units with material constituents) are avoided. The relation of 'overlap' can be defined as usual,

(D3) $x \circ y \leftrightarrow \exists z\ (z \underline{\subsetneq} x\ \&\ z \underline{\subsetneq} y),$

but due to the non-transitivity of ‘$\mathrel{\subset\!\!\!\to}$’ the overlap of two entities we receive two types of non-overlap, discreteness (D4) and disjointness (D5):

$$\text{(D4)} \quad x \int y \leftrightarrow \neg\exists z\, (z \mathrel{\underline{\subset\!\!\!\to}} x \;\&\; z \mathrel{\underline{\subset\!\!\!\to}} y)$$

$$\text{(D5)} \quad x \mid y \leftrightarrow \neg\exists z\, (z \;{}^{n}\!\mathrel{\underline{\subset\!\!\!\to}} x \;\&\; z \;{}^{n}\!\mathrel{\underline{\subset\!\!\!\to}} y)$$

The supplementation principle in FPT is defined in terms of discreteness rather than disjointness:

$$\text{(A4)} \quad x \mathrel{\subset\!\!\!\to} y \rightarrow \exists z (z \int x \;\&\; z \mathrel{\subset\!\!\!\to} y)$$

This allows for wholes whose parts cannot be completely ‘disentangled’, such as, for example, activities performed simultaneously by a living organism (walking and jumping puddles) which each involve the same component activity (lifting the left leg). Similarly, due to the nontransitivity of ‘$\mathrel{\subset\!\!\!\to}$’ product and sum, defined in the usual way:

$$\text{(D6)} \quad \text{prod}(x, y) =_{\text{df}} \iota z (\forall w (w \mathrel{\underline{\subset\!\!\!\to}} x \;\&\; w \mathrel{\underline{\subset\!\!\!\to}} y) \leftrightarrow w \mathrel{\underline{\subset\!\!\!\to}} z)$$

$$\text{(D7)} \quad \text{sum}(x, y) =_{\text{df}} \iota z (\forall w (w \circ z \rightarrow w \circ x \vee w \circ y))$$

deviate to some extent from their classical meanings.[40] But product and sum are associative and commutative.[41] The inverse of the sum of free processes, their difference, is defined as:

$$\text{(D8)} \quad \text{dif}(x, y) =_{\text{df}} \iota z (z \mathrel{\underline{\subset\!\!\!\to}} x \;\&\; z \int y).$$

The sum of two free processes is itself a free process with arbitrarily scattered spatiotemporal parts; no causal interaction is implied. For instance, walking and chewing gum is a sum of free processes, but so is a phone conversation, the French revolution, or inflation. Since the sum operation does not ‘reach’ below the level of 1-parts, we can distinguish a large number of different process combinations in terms of additional conditions on the structure of +2-parts (see section 4.2 below). Some process combinations ‘create’ +2-parts, others ‘cancel’ existing ones (phenomena of ‘emergence’ and ‘suppression’). The general term used in FPT for all process combinations which ‘effect’ a change in +2-parts is ‘interference’ (abbreviated by ‘$I(x, y)$’). To highlight that the interference of two free processes α and β is itself a free process γ, read ‘$\gamma = I(\alpha, \beta)$’ as ‘$\gamma$ is the interfering of α and β’.

The field of ‘$\mathrel{\subset\!\!\!\to}$’ consists of more and less specific free processes (e.g., *lifting the right leg* $\mathrel{\subset\!\!\!\to}$ *running*, *nodding one’s head* $\mathrel{\subset\!\!\!\to}$ *greeting*, *photosynthesis* $\subsetneq$ *plant growth*). Among these is spacetime (ρ_0) and its specifications representing different, more or less determinate spatiotemporal regions or 'spacetimings' (ρ_i).[42] If process α is located in a determinate region, then there is a minimally self-contained process β which is the interfering of α and ρ; β is called an *amount* of α, abbreviated '[α]':

(D9a) $\forall x \neq \rho_i$: $[x] =_{\text{df}} \iota z\ (\exists \rho_i\ (z = I(x, \rho_i))$

Amounts of processes also 'carve out' amounts of spacetime:

(D9b) $\forall \rho_i$: $[\rho_i] =_{\text{df}} \iota z\ (\exists x\ (z = I(x, \rho_i)\ \&\ \neg(\rho_0 \subsetneq x))$[43]

Amounts of processes have a determinate location – they are located in (i.e. they are inferences with) an ultimately specific space-time region.[44] In contrast, quantities of processes are located indeterminately – they are interferences with a less specific spatio-temporal region (*in town*, *in the accelerator*).[45] For all processes it holds that that they interfere with ρ_0, i.e. that they occur *somewhere* in spacetime. This is the *occurrence axiom*:[46]

(Ax5) $\forall x[\exists y(x \subsetneq y \vee y \subsetneq x) \rightarrow \exists z, \rho_i\ (z = ((x, \rho_i))]$

That a free process occurs in a certain spatiotemporal region (for instance, that partying is going on at Roger's house tonight) means in FPT that two processes α and β (*partying*, *being-the-extension-of-Roger's-house-tonight*) are superposed to form an amount of each; their sum is an *interfering*, i.e. a third complex process γ, which overlaps with both *partying* and *being-the-extension-of-Roger's house-tonight*) in its 1-parts but differs in its +2-parts from the +2-parts of either of them. For instance, assuming that the 1-parts of *partying* are *communicating* and *human group presence*, we might find that *joking* and *discussing literature* are part of the communicating that goes on at Roger's house tonight, but these are not +2-parts of *communicating* or *human group presence*. The idea behind definitions (D9a) and (D9b) is thus simply this: the occurring of a free process α in a region ρ_i is a process β which is a more specific version of α as well as of ρ_i. It does not hold, however, that every specification of a free process α results in an amount of α. To express the connection between amount and specification of a free process more clearly, we first need to look at the FPT-definition of *extent-part*, i.e. the basic relation of Classical Extensional Mereology:

(D10) $[x] \leq [y] \leftrightarrow \forall \rho_i(I(x, \rho_i) \rightarrow I(y, \rho_i))$

The transitivity of '$\leq$' is warranted by

(Ax6) $[x] \leq [y]\ \&\ [y] \leq [z] \rightarrow [x] \leq [z]$.

On the basis of '$\leq$' the FPT-predicate 'x is a *phase* in y' can be introduced; a phase of α is a part of α an amount of which occurs in every amount of α:

$$\text{(D11)}\quad \text{phase}(x, y) \leftrightarrow x \subsetneqq y \;\&\; \forall z(z = [y] \rightarrow \exists w\,(w = [x] \;\&\; w < z)).$$

A free process α with phases can then be said to be a *specification* of a process β iff every phase of α is an amount of β:

$$\text{(D12*)}\quad \text{specif}(x, y) \leftrightarrow \forall z(\text{phase}(z, x) \rightarrow z = [y])$$

For example, since all the phases performed in cooking lasagne are amounts of cooking, cooking lasagne is a specification of cooking. For processes without phases, such as playing guitar or being a poodle, the quantification is over the spatiotemporal parts of any amount of the specific process. So the full version of the FPT-definition of specification runs as follows:

$$\text{(D12)}\quad \text{specif}(x, y) \leftrightarrow \{\forall z(\text{phase}(z, x) \rightarrow z = [y]) \vee \forall z, x(z \leq [x] \rightarrow z = [y])\}$$

If Max is a poodle, wherever there is an amount of *being-Max*, there is an amount of *being-a-poodle*. An *ultimate specification* of a process is a process for which there is no further specification within the domain of FPT.

$$\text{(D13)}\quad \text{ult-specif}(x, y) \leftrightarrow \text{specif}(x, y) \;\&\; \neg\exists z(\text{specif}(z, x))$$

It is crucial to realize that in FPT an amount of a process α is a specification of α, but *not necessarily* an ultimate specification of α. On the other hand, a process α that is an ultimate specification of a process β must be specified with respect to spatiotemporal occurrence, i.e. it must be an amount of $[\beta]$. This follows from definitions (D9b), (D12), (D13), and the *occurrence axiom* (A5). Ultimate specificity implies not only spatiotemporal occurrence but also ultimate specificity of spatio-temporal occurrence. Thus we can state the following *Particularity Theorem*, which says that ultimate specifications can fulfill the inferential role of traditional particulars, since they have an ultimately specific and thus unique spatiotemporal location:

$$\text{(T1)}\quad \forall x\,[\exists y\ \text{ult-specif}(x, y) \rightarrow \exists z, \rho_i\ (x = I(z, \rho_i) \;\& \neg\exists \rho_j, v\ (\text{specif}(\rho_j, \rho_i) \;\&\; v = I(z, \rho_j)))]$$ [47]

A third basic relationship between processes, besides phase and specification, is that of a *stage* of a process. When we say that applying the primer is part of varnishing the cabinet, that tadpoles become frogs, or that being an adolescent is part of being human, we imply that some but not all of the classificatory predicates that characterize the whole development, apply also to a spatiotemporal part of it. In FPT this is expressed as the requirement that for a process amount $[\alpha]$ and a spatiotemporal part $[\beta]$ of $[\alpha]$, $[\beta]$ and $[\alpha]$ have some but not all specification relationships in common:

(D14) $\text{stage}(x, y) \leftrightarrow \forall z(z = [x] \rightarrow \exists w(w = [y] \,\&\, z < w) \,\&\, \exists z\, (\text{specif}(x, z) \,\&\, \text{specif}(y, z)) \,\&\, \exists z, v\, (\text{specif}(x, z) \leftrightarrow \neg\text{specif}(y, z) \,\&\, \neg\text{specif}(x, v) \leftrightarrow \text{specif}(y, v))$

Axiom (A5) ensures that (D14) is not vacuously fulfilled. Since according to (D14) anything that has stages is 'traceable under some sortal' (i.e., in the FPT idiom: is a specification of some process), a dynamic entity is something that can be viewed as both changing and transtemporally identical.[48]

All entities in the field of '$\subsetneq$' are dynamic, as postulated in the *Process Axiom*:

(Ax7) $\forall x \exists y(x \subsetneq y \vee y \subsetneq x) \rightarrow \text{dyn}(x)]$.[49]

Axioms (Ax5) and (Ax7) in combination imply that all free processes have spatiotemporal parts, i.e., the extent-part relation '$<$' has no atoms in FPT.

4. TOWARDS A TYPOLOGY OF PROCESSES

4.1. *Classification by automerity pattern*

Free processes can be coarsely classified into five types, according to certain characteristic mereological conditions. Recall the distinction between three types of automerity, maximal self-containment, self-containment, and minimal self-containment (cf. (AUT) above). In terms of these three predicates various automerity patterns can be differentiated and associated with basic process types.[50]

(1) *Type 1 processes are temporally maximally self-contained and spatially unmarked*; their prime examples are activities, e.g., a running, raining, or reading taken to occur in the activity mode. The completeness condition (C5) above requires for activities maximal temporal self-containment; but spatially there may be maximal self-containment (falling), self-containment (raining), or minimal self-containment (Tom and

Kim's pair-dancing). In the formal idiom of FPT the characteristic automerity condition for type 1 processes can be restated as follows. (Recall that '$I(\alpha, \rho_i)$' stands for 'an amount of α occurs in spatiotemporal region ρ_i').

(D16) α is type 1 iff $\forall x(x <_{\text{temp}} I(\alpha, \rho_i) \rightarrow x = I(\alpha, \rho_j)$.

(2) *Type 2 processes are temporally minimally self-contained while their spatial self-containment is unmarked*; prime examples of such processes are occurrences in the accomplishment mode (developments, events). The performance of Mozart's Clarinet Concerto, for instance, is temporally minimally self-contained (no part of the performance of the Concerto is itself a performance of the Concerto) and spatially minimally self-contained; other events, however, like the communal singing of the national anthem, are equally temporally minimally self-contained but do have spatial parts that fulfill the description of the whole. To restate:

(D17) α is type 2 iff $\forall x(x <_{\text{temp}} I(\alpha, \rho_i) \rightarrow \sim (x = I(\alpha, \rho_j)))$.

(3) *Type 3 processes are spatially minimally self-contained and temporally maximally self-contained*; their prime examples are entities conceived of as things and discrete expanses of matter. According to our common ways of talking, a table or piece of rock do not take time but 'endure', i.e. every temporal part of them fulfills the description we apply to the temporal whole. To restate:

(D18) α is type 3 iff $\forall x(x <_{\text{temp}} I(\alpha, \rho_i) \rightarrow x = I(\alpha, \rho_j))$ & $\forall x(x <_{\text{sp}} I(\alpha, \rho_i) \rightarrow \sim (x = I(\alpha, \rho_j))$.

(4) *Type 4 processes are spatially and temporally self-contained.* Their prime examples are oscillatory sequences of developments as well as heaps, e.g. running conceived of as a sequence of repetitive movements or water taken as assembly of molecules. When authors argue for the limited homogeneity of activities or masses, then they actually speak about oscillatory sequences and heaps, assigning to an occurrence like running a phase that is not a running (lifting the left foot), or to a heap (coarse-grained mixture) like H_2O a spatial part (oxygen molecule) that is not H_2O. Note, however, an important asymmetry between oscillatory sequences and mixtures: the latter are always also temporally maximally self-contained, since they are taken to 'endure' like things. Formally restated:

(D19) α is type 4 iff $\exists x\ (x <_{\text{temp}} I(\alpha, \rho_i)$ & $x = I(\alpha, \rho_j))$ & $\exists x(x <_{\text{sp}} I(\alpha, \rho_i)$ & $x = I(\alpha, \rho_j))$.

(5) *Type 5 processes are spatially and temporally maximally self-contained.* Prime examples of such processes are entities conceived of as masses or stuffs proper. Once water is not viewed as a mixture but as stuff, it is spatially maximally self-contained, and, since it endures, it is temporally maximally self-contained. Taking a 'substantivist' view on space and time (as this is done in the current version of FPT), space, time, and spacetime are also type 5 processes; further candidates are perhaps secondary qualities such as colors or sounds.

(20) α is type 5 iff $\forall x\ (x <_{\text{temp}} I(\alpha, \rho_i) \rightarrow x = I(\alpha, \rho_j))\ \&\ \forall x\ (x <_{\text{sp}} I(\alpha, \rho_i) \rightarrow x = I(\alpha, \rho_j))$.

It is important to note, however, that the conditions stated in (D16) through (D20) figure only as *necessary* conditions in the FPT-interpretation of common sense classificatory predicates such as 'activity', 'event', 'thing', 'stuff', and 'quality'. The FPT-definition of thinghood, for example, requires type 3 processes to be spatially unified, transportable, and to possess functional shape, which distinguishes them from discrete expanses of matter.[51]

4.2. *Classification by dynamic parameter analysis*

A more fine-grained classification of free processes can be achieved within a four-dimensional parameter space based on the following evaluative dimensions: (a) participant structure (types and roles of participants); (b) dynamic composition; (c) dynamic shape; (d) dynamic context.

4.2.1. *Participant structure*

From an Aristotelian point of view, *talking* involves one substance, *arguing* two, and the 'elemental transformation' *snowing* none; whether *chopping* involves more than one substance remains unclear, due to the inclusive status of composite artifacts. Such differences in the participant structure, i.e. the types and roles of participants in occurrences, are often encoded linguistically: partly in the explicit argument structure of verbs (transitive vs. intransitive), partly they are expressed by prepositions (compare *hit the door* vs. *hit at the door*), partly they are signaled by special verb forms (e.g., passive, accusative, ergative forms).[52] In general, however, it is not possible to use syntax as a reliable indicator of the participant structure. Many sentences with intransitive verbs do not involve exactly one agent. Some involve implicit references to locations, times, or observers – cf. 'he disappeared', 'the wedding took place', 'the game was disappointing' – others express qualifications of dynamic shape or dynamic context of a process, such as 'the rain increased', 'the rate of change remained constant', 'this development could not be stopped', etc. Sometimes only context can determine the participant structure of a sentence with an intransitive verb: 'Fred stumbled.' The framework of FPT allows us to express rather subtle differences in participant structure. Compare for instance the following six distinctions:

(1) A process α is a *one-agent process* iff there is an amount of α that is part of a temporal part of exactly one process γ of type 3 or 4 (thing, piece of matter, or stuff), and no part of α is part of a temporal part of another process of this type at that time.[53] For example, walking or rolling are one-agent processes insofar as there are amounts of walking or rolling that are a part of a temporal part of Fred or this ball, respectively, and of no other thing (piece of matter or stuff) at that time.

(D21) *one-agent process* (α) iff $\exists[\alpha], [\beta], x, \exists^1\gamma_{3,4}([\alpha] \subsetneq [\beta] \;\&\; x = [\gamma] \;\&\; [\beta] \leq_{\text{temp}} [\gamma] \;\&\; \forall\delta_{3,4}([\beta] <_{\text{temp}} [\delta] \rightarrow \delta = \gamma))$[54]

(2) A process α is a *one-agent-one-patient* process iff there is an amount of α that is part of a temporal part of process γ, and there is a stage δ of α which interferes* a temporal part of a process ε (where γ and ε are not specifications of α). For example, *touching, picking up, looking at, receiving* are processes with stages that are part of the temporal parts of the 'patient' involved. The set of agents and patients is here unrestricted as far as type is concerned – activities, events, things, stuffs, and phenomenal properties can shatter, decorate, disturb, overshadow, or accompany activities, events, things, stuffs, and phenomenal properties.[55]

(D22) *one-agent-one-patient process*(α) iff $\exists[\alpha], [\beta], [\delta], [\phi], x, y, \exists^1\gamma, \varepsilon([\alpha]) \subsetneq [\beta] \;\&\; x = [\gamma] \;\&\; [\beta] \leq_{\text{temp}} [\gamma] \;\&\; \text{stage}(\delta, \alpha) \;\&\; [\delta] \subsetneq [\phi] \;\&\; y = [\varepsilon] \;\&\; [\phi] \leq_{\text{temp}} [\varepsilon \;\&\; \neg(\text{specify}(\gamma, \alpha) \vee \text{specif}(\varepsilon, \alpha)))$.

(3) Since more than one agent may collectively or separately affect one or more patients collectively or separately, we can diversify (D22) further into many types of (collective)-n-agent-(collective)-m-patient processes: (a) *collective n-agent m-patient* processes (as expressed in the sentence 'Between the two of them, Max and Tom ate three pizzas') (b) *n-agent-collective-m-patient* processes (as in: 'Max and Kim each carried three pizzas'), (c) *collective-n-agent-collective-m-patient* processes (as in: 'Max and Kim are assembling the parts') and (d) *n-agent-m-patient* processes (as in: 'While Max played guitar, Kim read a book'). Collective

agency/patiency is in FPT expressed in terms of process interference and separate agency/patiency in terms of process sums.[56]

(4) Collective agency and patiency may further be differentiated according to the *the degree of entanglement of agents and patients involved.* Collective agency (patiency) as defined in terms of interference expresses merely the type of togetherness resulting from spatiotemporal co-occurrence. But in many cases co-occurrence is accompanied by dependence relationships holding among the involvements of the agents or patients. In (a) *concerted* (or *weakly entangled*) *n-agency* (as in: 'Max and Tom lifted the piano' or 'sticklebacks perform a complicated courting ritual') or *concerted* (or *weakly entangled*) *n-patiency* (as in: 'Max drank the gin-and-tonic', 'the measurement showed an anti-correlation of the electron's spin and the positron's spin') the agent/patient process is an *interference* (weakly emergent product, see below) of its 1-parts (component actions/passions such as 'Max's pulling the piano up', 'the male stickleback's zig-zag dance', or 'the electron's spin being measured', 'the gin's being drunk', resp.) which can be performed independently. In contrast, in (b) *reciprocal* (or *deeply entangled) n-agency/patiency* (as in: 'Max and Kim discussed the issue', or 'the roof is supported by intricate timbering', or, for patiency: 'measurement M effects a superposition of states with spin 1/2'), the interference concerns not only the 1-parts of the composite action. Here the 1-parts (e.g., 'Max's discussing the issue', 'the ridge beam's lying horizontally above the ground', 'spin 1/2') are themselves interferences of a certain kind and cannot occur independently of what they are interferings of. They are either weakly emergent products as just introduced, or strongly emergent products, i.e., processes that are interferings of a collection of processes which each contain at least one other member of the collection as part (more on weak and strong emergence below).

(5) Finally, participant structures are classed according to *the types and identities of agents and patients involved*, which allows, for example, for useful distinctions between: (a) bodily movements, (b) reflexes and afflictions, (c) undertakings, i.e. deliberate bodily movements such as push-ups, where a human agent is acting on herself '*qua* other' (i.e. human body or mind), and (d) basic actions, where the agent is acting on herself '*qua* self.'

4.2.2. *Dynamic composition*

The evaluative dimension called *dynamic composition* addresses dynamic phenomena discussed under headings such as 'generation', 'destruction', 'emergence', 'downward causation', 'self-organisation', and 'complexity.'

Just as in classical mereology, in FPT the sum γ of two items α and β is the item γ that is overlapped by anything which overlaps α or β. However, due to the non-transitivity of the part-relation in FPT this definition leaves room for further differentiations in 'additivity types.'

> (Add-1) A process γ is *simply additive* iff $\gamma = \text{sum}(\alpha_i)$, and there are no interferences* on the n-parts of the α_i, i.e. the +2-parts of γ are the +2-parts of all α_i.

Intuitively, a simple addition of processes consists of two non-interacting processes, such as my walking in Texas and radioactivity on Jupiter, where (presumably) none of the constituent processes of my walking in Texas or of radioactivity on Jupiter cancel each other out by interference, nor any new processes are generated.

> (Add-2) A process γ is *linear additive* iff $\gamma = \text{sum}(\alpha_i)$ and for any $+n$-part β of γ, β there is a despecification of β which is also a despecification of some $+n$-part of every α_n, $1 \leq n \leq i$.

Linear additivity generates complex processes which, intuitively speaking, have the same kind of features as their components but different magnitudes, the prime example being the vector addition of forces.

> (Add-3) A process γ is *weakly emergent* on processes α_i iff γ is a +2-part of $\beta = \text{sum}(\alpha_i)$ and γ is not a +2-part of any α_i and no n-part of γ is also an n-part of some α_i.

A weakly emergent process has functional features that none of its component processes has. A wheel has the weakly emergent feature of being circular which none of its components has, a chord has a harmonical role which none of its component sounds has, a simple feedback cycle has a control function which none of the components have.

> (Add-4) A process γ is *strongly emergent* on processes α_i iff γ is weakly emergent on processes β_i which are each weakly emergent α_i.

The wheel's rolling for example is strongly emergent on the structural components of the wheel and of the supporting surface, since it is weakly emergent on the wheel's circularity and on the inclination of the surface; self-organizing systems such as biological organisms and economies are strongly emergent on the specific physiological and economical processes occurring here and now since they are weakly emergent on hierarchies of functional units whose levels are related by weak emergence.[57]

Non-linear addivity can take many forms. On the one hand, there are many ways to combine and reiterate weak and strong emergence. For example, 'musical processes' in E. Christensen's sense (cf. his contribution to this volume) are, I believe, an illustration of processes that are strongly emergent on musical units (phrases, harmonies) which themselves are strongly emergent on tone sequences. On the other hand, we can distinguish different types of non-linear dynamic composition according to the dependence structures among the n-parts of the emergent process γ. For example, consider Bickhard/W. Christensen's notion of an 'autonomous system'.[58] An autonomous system – a 'generalization of the concept of autocatalysis' – is characterized by a very specific form of organizational interdependence of its structural components: it "interactively generates the conditions required for its existence".[59] This involves interdependence of the system's components at different degrees of directness:

(D23) Process γ is an autonomous system iff γ is strongly emergent on a set of processes α_i and for any 1-part β of γ there is an n-part α of γ such that $\beta \hookrightarrow \alpha$ and $\alpha \hookrightarrow \beta$.

In FPT with its non-transitive part-relation we can – unlike in systems of classical mereology – painlessly express the fact that an organism's $(= \gamma)$ blood flow $(= \alpha)$, cellular metabolism $(= \beta)$, and motor activity $(= \delta)$ mutually condition each other: $(\alpha, \beta \hookrightarrow \delta)$ & $(\beta, \delta \hookrightarrow \alpha)$ & $(\alpha, \delta \hookrightarrow \beta)$, turning γ into an autonomous system.

4.2.3. *Dynamic shape and dynamic context*

The details of the third and the fourth dimensions of evaluative parameters in FPT are still very much under construction. 'Dynamic shape' relates to differences with respect to the 'course' or flow of a process, as these are expressed, for instance, by verbal aspects. Most verbal aspects (e.g., perfective, progressive, repetitive, ingressive, egressive) correlate naturally with simple shapes of trajectories in phase space, or parts of such trajectories and thus admit of mereo-topological definitions on phase space. But processes may also be differentiated with respect to how they affect their *dynamic context*. The dynamic context of a complex process may differ with respect to (a) *linear causal consequences*, (b) *non-linear causal consequences* in their immediate dynamic environment, such as disturbances in the air flow around a kite, and those with (c) *non-linear consequences in the generative environment* of the process, such as changes in ecosystems that alter selection pressures.[60]

5. Conclusion

I have sketched here some elements of the heuristics and construction of Free Process Theory, a monocategoreal process-ontological scheme which operates with dynamic, multiply occurrent individuals. Abandoning the traditional ontological bias in favor of countable particulars, FPT situates all entities on a gradient scale of determinability and automerity where particularity and countability merely present the limiting cases (minimal determinability and minimal automerity, respectively). FPT was originally developed in 1990 in the course of an attempt to expose and undercut substance-ontological presuppositions that hamper satisfactory solutions to the classical problems of individuation, universals, and persistence. Later studies work out FPT's built-in (dis)solutions of these three classical problems,[61] develop an account of thinghood, and explore the scheme's possible feasibility for the interpretation of quantum-physical entities.[62] My primary goal here was to draw attention to two facts. First, as our reasoning about activities displays, we are agentively well-familiar with instances of 'free processes', a new ontological category. Second, a non-transitive part-relation between free processes can be used to define various specific part-relations (extent part, functional part, material part, construction part etc.), the spatiotemporal location of processes, distinctions between common sense 'categories' such as thing, event, stuff etc., and many different types of fine-grained constituent relationships, including those involving internal interdependencies, dynamic modifications, and emergent complexity.

NOTES

1 Cf. Seibt 1990b, 1995, 1996a, 1996b, 1996c, 1999, 2000d, 2002, 2003.

2 Cf. Sellars 1981, Zemach 1970.

3 This at least is the view of the – currently predominant – Carnapian tradition in analytical ontology. On historical and methodological issues cf. Seibt 1996d, 1997, 2000a, 2000c, 2002a, where I also explain in which sense ontologies are language-transcendent (i.e., applicable to the class of functional equivalents of L) – a point which is here omitted for reasons of simplicity.

4 That is, for its basic types of entities (categories) the domain theory specifies certain features (category features) with suitable explicit definitions. The inferences licensed by a sentence S are justified if they can be shown to follow from (the definitions of) features of categories that are part of the truth-maker of S. For example, consider the English sentences:

(1) 'My Ford is a rather old car'.
(2) 'A car is a means of transportation'.

Sentence (1) licenses the inference:

(3) 'What is left of my Ford is not my Ford',

but (2) does not license the sentence analogous to (3):

(4) 'What is left of a car is not a car'.

That (1) licenses (3) (while (2) fails to license (4)) has traditionally been explained by the fact that the ontological correlate of 'my Ford' (but not of 'a car') is a primary substance, which has the category feature of particularity or unique occurrence in space defined as entailing (3).

[5] This is one way to read the Carnapian postulate of *foundedness* in the *Aufbau*, cf. Seibt 2000a. For a more detailed description of the relationship between category and model cf. 2000c. Note that the 'model' of an ontological category is denoted by an *ultimate genera term* of L. One of the primary difficulties for theories of tropes (or 'moments') consists, incidentally, in the fact that the category 'trope' lacks a model in this sense – in English there is no ultimate genera term expressing the genus of 'this red' versus 'that red.' This lacuna is covered up by the tropist's quick move to technical jargon like 'property instances' or 'exemplifications of attributes', which does not *found* the category in the required sense.

[6] Compare the contribution of M.-L. Gill in this volume.

[7] Ryle 1949: 149.

[8] Cf. Vendler 1957, Kenny 1963.

[9] Kenny 1963: 172.

[10] Note that (C1) through (C5) are already used by Aristotle to distinguish *energeia* from *kinesis*.

[11] *Ibid.*

[12] The critical interaction with Vendler's and Kenny's classification begins with Mourelatos' seminal discussion from 1978. Cf. also Taylor 1985.

[13] Different sentential contexts can change the 'action type' of the verb; compare: *Tom smoked a cigarette – Tom smoked cigarettes – Mary pushed the cart to the shop – Mary pushed the cart for hours.*

[14] Spell-bound by the 'logical grammar' developed by native speakers of Germanic languages, analytical ontologists largely continue to overlook the inferential significance of verbal aspects. This is not surprising since in Germanic languages there are few, if any, verbal aspect markers; differences in 'aspectual meaning' are mostly expressed by periphrasis. In Romance and Slavic languages aspectual meaning features more prominently; compare the difference between the French *Il regnait trente ans* (he reigned for thirty years, Imperfect, expressing the so-called imperfective aspect) and *Il regna trente ans* (he had a reign for thirty years, Past Definite, expressing the so-called perfective aspect). Aspectual meaning consists in information about the dynamic organization of an occurrence or its relationship to other occurrences as backgrounding or incident, i.e. verbal aspects characterize an occurrence as ongoing (progressive), attempted (conative), about to begin (prestadial), just begun (ingressive), in the middle (continuous), about to end (egressive), frequently recurring (iterative), habitual, or as a factual unit (perfective) etc. Cf. Comrie 1976 and Dik 1997.

[15] In the linguistic literature occasionally both the shift of verbal aspects and its effect are referred to as 'verbal aspect shift.'

[16] Cf. for example Dowty 1979.

[17] Cf. Sasse 2002.

[18] Cf. for example Dowty 1979, Bache 1995, Dik 1997, Smith 1997, Kearns 2000, Rijksbaron 1989.

[19] The exception being achievements; compare for instance Dik's classification (1997), to my knowledge the most diversified, which countenances 10 categories: situations (which are either states or positions), events (which are, if processes, either dynamisms or changes), and actions (which are either activities or accomplishments).

[20] For example, a predication *P* is said to be telic iff it can be combined with 'almost' or 'within *n* minutes (hours/days/years)' or used in constructions with 'finish' or 'stop.' Altogether there are 9 linguistic criteria of telicity, which determine three slightly different notions of telicity: mereological, topological, and modal telicity. For this and a detailed discussion of extant linguistic definitions of dynamicity cf. Seibt, forthcoming, section 2.4. In contrast, none of the definitions of dynamicity offered are successful: they are either circular, or blur the distinctions between telic and atelic dynamicity, or between 'static' and 'dynamic' atelicity. In short, there is a strong intuitive difference between two classes of situations – Vendler called them 'states' and 'activities' – which both fulfill the condition of distributivity and homeomerity (cf. C3 and C4 above), but this difference, it appears, cannot be defined at the lexical level alone without involving aspectual information.

[21] Cf. for example Smith's 'two component theory' (1997) and, in particular, so-called 'selection theories of aspect', as collected in Breu 2000.

[22] Cf. Comrie 1976: 3 and 18; Dik 1997: 222, reformulating a comparison by A. Isačenko.

[23] Cf. Seibt forthcoming.

[24] In particular, selection theories tell us little about whether verbal aspect shift indeed amounts to a *conceptual* type shift. All we know about the processual meaning of a sentence such as *Tom was crossing the street* is that it denotes the phase of an occurrence with phase and specified boundary; but it remains unclear whether our conception of such a phase differs from the phase denoted by the sentence *Tom was running*, i.e., whether the going on of an 'accomplishment' and the going on of an 'activity' are the same type of going on, i.e. roughly speaking, whether accomplishments are just 'activities with boundaries.'

[25] The neutral aspect consists in the lack of aspectual information, cf. Smith 1997. The perfect (not to be confused with the perfective aspect!) is a mixture of tense and aspect; the experiential perfect indicates that an occurrence has happened in the past and is over, the resultative perfect conveys that an occurrence has happened in the past but has still current relevance (compare the difference between *Bill has been to America* and *Bill has gone to America*, cf. Comrie 1976).

[26] Note that this feature provides some difficulty for those interpretations of Aristotle's energeia that take the latter to be 'physis', that which constitutes the living unity of an organism, and present physis as a special kind of activity. On this issue cf. Gill in this volume.

[27] Cf. Kathleen Gill 1993.

[28] Less than four-dimensional homeomerous entities are, for example, sensory impressions and unbounded or bounded surfaces or boundaries. Among concrete entities, there is no 'contrast class' to homeomerous entities; the purpose of such a general definition of homeomerity is obviously to allow for a unified perspective of comparison.

[29] Cf. Rijkhoff 2002. To highlight the connection to masses, I used to refer to free processes as 'dynamic masses.'

[30] In previous expositions of FPT I failed to distinguish clearly between homeomerity and automerity, and characterized free processes nominally merely as 'homeomerous' entities.

[31] Cf. Seibt 2001b.

[32] Cf. Heller 1990, Needham 1999, Rescher 2000, Stout 1997, Whitehead 1928, Zemach 1970. That processes are non-particular has been (implicitly) suggested first in Sellars 1981 (or so I argue in Seibt 2000b) and I myself have promoted this idea explicitly in Seibt 1990b, 1995, 1997a, 1999, 2000d, 2001a,b. In his contribution to this volume Paul Needham stipulates that processes have unique temporal occurrence in time but are not located (i.e., also not multiply located) in space. This is a third position between the classical particularist approach to process ontology and the non-particularist approach of Free Process Theory. Also, note that in the approach of Heller/Herre in this volume *projections* of processes (onto portions of space and time) are particulars; I am unclear on whether processes themselves are particulars in this framework.

[33] Seibt forthcoming, ch. 3.

[34] The following two sections contain an (I hope) improved version of an earlier exposition of FPT (then called 'APT') in Seibt 2001.

[35] Cf. Simons 1987.

[36] For a first sketch of the formal framework FPT cf. Seibt 1990b; 1995, 1996c, 1999, 2000d, and 2001 contain versions of FPT at different stages of development.

[37] For example, in [1] and [2] sentences (a) through (b) or (a) through (b), respectively, do not entail sentence (C). [1] (a) Changing diapers is part of being a parent. (b) Opening the box with wipes is part of changing diapers. (c) Pressing your thumb upward is part of opening the box with wipes. (C) Pressing your thumb upward is part of being a parent. [2] (a) The door is part of a house. (b) The hinge is part of the door. (c) The hinge is part of a house.

[38] Moreover, it also offers economical ways to model our reasoning about part-whole relationships in artifacts pertaining to function, construction, material constitution, and maintenance. For the latter compare Simons/Dements (1996).

[39] To appreciate the usefulness of such a schematic approach to identity consider the following example, involving partitions on actions: If Kim plays violin at 9am at home – she performs her entry exam to the conservatory at home at 9am – she is awakening her neighbor Tom (a nocturnally working philosopher). Assume that Kim intended to perform the entry exam and by doing so, to disturb her neighbor Tom. Performing an entry exam is to play the violin in a place where an examiner is listening. If we define the identity of actions in terms of 1-parts, direct intended consequences then the making music today is identical to the performance of the entry exam but not to the disturbance of Tom. In legal contexts, however, we might choose to define the identity of actions in a more fine-grained fashion, e.g., in terms of an action's 1-part and 2-parts, its instrumentally intended consequences. In this case Kim's playing on this occasion will be identical with disturbing Tom on this occasion. In other words, FPT's identity condition is as flexible as our actual reasoning requires.

[40] For example, the following two theorems do not hold (with '$\cap$' and '$\cup$' as abbreviations for 'prod' and 'sum'): (1) $(w \subsetneq (x \cap y)\ \&\ v \subsetneq (x \cap y)\ \&\ w \circ v) \rightarrow (w \cap v) \subsetneq (x \cap y)$. (2) $(w \subsetneq (x \cup y)\ \&\ v \subsetneq (x \cup y)) \rightarrow (w \cup v) \subsetneq (x \cup y)$.

[41] It holds that $w \subsetneq (x \cap y) \leftrightarrow w \subsetneq x\ \&\ w \subsetneq y$. Thus $x \cap (y \cap v) = \iota z\ (\forall w\ (w \subsetneq x\ \&\ w \subsetneq (y \cap v) \rightarrow w \subsetneq z) = \iota z\ (\forall w\ (w \subsetneq x\ \&\ w \subsetneq y\ \&\ w \subsetneq v \rightarrow w \subsetneq z) = (x \cap y) \cap v$. Similarly for the sum operation because of $x \circ (y \cup z) = x \circ y \vee x \circ z$.

[42] I am assuming here that spacetime (either of physics or of common sense) is nonhomogeneous in the sense that different regions of spacetime are qualitatively different and thus, according to the (PPP) (i.e., Ax3), different spacetimings. A notational convention:

Variables in lower case Greek range over items in the field of '⊊' but (unlike 'x, y, z' etc.) they are never co-denoting; variables 'ρ' with subscripts are reserved for 'spacetimings.'

43 The last clause of (D9b) postulates that the instantiations of variable 'x' are not in the 'genus of spacetime', i.e., that they are free processes which are not themselves spacetimings. In addition, variables z in D9a and x in D9b, respectively, are restricted to minimally self-contained processes. Note that minimal self-containment can be defined without making explicit reference to part of spatiotemporal regions (as this was done above for expository purposes): a minimally self-contained entity has no parts identical to itself.

44 An ultimate specific space-time region is a space-timing which is not a (functional, not spatiotemporal!) part of any other space-timing.

45 Note that this is not the notion of 'quantity' used in Needham's contribution to this volume – quantities in Needham's (Cartwright's) sense are uniquely and determinately located, i.e. they are particulars.

46 In the concluding paragraph of his contribution to this volume Wim Christaens abandons (Ax5) and suggests that FPT could be used to talk about entities that do not exist in space (as opposed to: do not have a determinate spatial location).

47 In the compressed presentation I can offer here it is easy to overlook the larger significance of the particularity theorem. In effect, FPT makes room for indeterminately located indeterminate individuals and yet accommodates, as their limit case, determinately located determinate individuals (particulars). This makes FPT a *prima facie* promising candidate for the interpretation of quantum physical entities, cf. Seibt 2002 and Christiaens in this volume.

48 Cf. Seibt 1996c.

49 With a unitary element characterized by dynamicity only ('empty process') and P being the field of '⊊', the Structure $W = \langle P, \text{sum} \rangle$ forms an Abelian group. The definition of 'dynamicity' is presently under construction – in an earlier version of FPT (2001a) it is defined as '$\text{dyn}(x) \leftrightarrow \exists z, y$ (stage (z, x) & stage (y, x) & $z \neg y$' but this links dynamicity to change. In Seibt (forthcoming) I define dynamicity as self-production, inspired by Aristotle's characterization of *energeia*.

50 That spatiotemporal characteristics may be used to distinguish between basic ontological categories (objects, events, properties, processes) is not new (cf. Mayo 1961 and, closest in spirit to FPT, Zemach 1970). New is the idea that the latter can be considered as *species* of one basic category.

51 For details cf. Seibt 2000d.

52 On transitivity alternations and 'oblique subject' alternations cf. Levin 1993: 25–45 and 79–82.

53 Events and properties may not be agents of *one-agent* processes, since sentences with events and properties in subject position make implicit reference to other agents or patients. This is partly obscured by lexical ambiguities, cf. (1) 'the avalanche is racing down' and (2) 'the avalanche came suddenly.' In sentence (1) the subject term 'avalanche' denotes a discrete expanse of matter, while in sentence (2) the subject term refers to an event, relating its occurrence to an implicit patient for whom the occurrence was sudden.

54 Indexed quantifiers '$\exists^1$', '$\exists^2$', ... etc. abbreviate the common definitions of numerals by means of existential quantification. Subscripts of variables, e.g., '$\alpha_{1,2}$', indicate type restrictions on the quantification, relative to process types 1 through 5 as defined in 4.1.

55 For the latter compare for instance: 'Lifting the lid only intensified the smell', 'the tornado changed the color of the sky'.

56 For details compare Seibt 2001a, forthcoming.

[57] Cf. Wilson/Lumsden 1991.

[58] Cf. Christensen/Bickhard 2002.

[59] *Ibid.*

[60] In formulating these distinctions in dynamic context I take my bearings here from Bickhard/Campbell 2000: 343.

[61] (a) On individuation cf. Seibt 1995 and 1996a. (b) Since free processes can recur, the qualitative identity of two regions can be accounted for in terms of the numerical identity of a free process recurrent in both. Roughly speaking, 'a is F and b is F' is made true by three processes, α, β, and γ, where γ is a $+n$-part of α and β. Cf. Seibt 1990a, 1990b, 2000b. (c) The temporal recurrence of free processes opens up a new avenue to a solution to the problem of persistence between endurance and perdurance, the so-called 'recurrence view' of persistence (cf. Seibt 1996c, 2003a). The recurrence theory of persistence endorses the main tenet of the endurance theory that persistence is sameness over time. But such sameness is not the numerical sameness of a particular, nor, as recently championed in Simons 2000a, the sameness of a universal. This (Whiteheadian!) idea is already discussed in Carter/Hestevold 1994 who – rightly I think – dismiss it as implying that our assertions about persistence are about abstracta. FPT offers the option to think of such sameness as the numerical identity of a concrete, recurrent individual.

[62] Cf. Seibt 2000d and 2002; cf. also Christiaens' contribution to this volume.

REFERENCES

Andersen, P. et al (eds.): 2000, *Downward Causation: Mind, Bodies, and Matter*, Aarhus: University of Aarhus Press.

Bache, C.: 1995, *The Study of Aspect, Tense and Action – Towards a Theory of the Semantics of Grammatical Categories*, Frankfurt a. M.: Peter Lang.

Bickhard, M. and D. Campbell: 2000, 'Emergence', in Anderson et al. 2000, 322–349.

Breu, W.: 2000, *Probleme der Interaktion von Lexik und Aspekt (ILA)*, Tübingen: Niemeyer.

Campbell, K.: 1990, *Abstract Particulars*, Oxford: Blackwell.

Christensen, W and D., M. Bickhard: 2002, 'The Process Dynamics of Normative Function', *The Monist* **85**, 3–28.

Carter, W. and H. Hestevold: 1994, 'On Passage and Persistence', *American Philosophical Quarterly* **31**, 269–283.

Comrie, B.: 1976, *Aspect*, Cambridge, UK: Cambridge University Press.

Dik, S.: 1997, *The Theory of Functional Grammar*, 2nd revised edition by Kees Hengeveld, Berlin: Mouton de Gruyter.

Dowty, D.: 1979, *Word Meaning and Montague Grammar*, Dordrecht: Reidel.

Gill, K.: 1993, 'On the Metaphysical Distinction Between Processes and Events', *Canadian Journal of Philosophy* **23**, 365–384.

Gill, M.-L.: 1989, *Aristotle on Substance. The Paradox of Unity*, Princeton: Princeton University Press.

Gill, M.-L.: 1994, 'Individuals and Individuation', in T. Scaltsas et al. (eds.), *Unity, Identity, and Explanation in Aristotle's Metaphysics*, Oxford University Press.

Heller, M.: 1990, *The Ontology of Physical Objects*, Cambridge, UK: Cambridge University Press.

Kearns, K.: 2000, *Semantics*, Houndmills: Macmillan.

Levin, B.: 1993, *English Verb Classes and Alternations*, Chicago: University of Chicago Press.
Mayo, B.: 1970, 'Objects, Events, and Complementarity', *Philosophical Review* **70**, 340–361.
Mourelatos, A.: 1978, 'Events, Processes and States', *Linguistics and Philosophy* **2**, 415–434.
Needham, P.: 1999, 'Macroscopic Processes', *Philosophy of Science* **66**, 310–331.
Pelletier, F.J.: 1974, 'On Some Proposals for the Semantics of Mass Terms', *Journal for Philosophical Logic* **3**, 87–108.
Pelletier, F.J.: 1978, *Mass Terms. Some Philosophical Problems*, Dordrecht: Reidel.
Rescher, N.: 2000, *Process Philosophy*, Pittsburgh: University of Pittsburgh Press.
Rijkhoff, J.: 1991, 'Nominal Aspect', *Journal of Semantics* **8**, 291–309.
Rijkhoff, J.: 2002, *The Noun Phrase*, Oxford: Oxford University Press.
Rijksbaron, A.: 1989, *Aristotle, Verb Meaning and Functional Grammar. Towards a New Typology of States of Affairs*, Amsterdam: J.C. Gieben.
Sasse, H.: 2002, 'Recent Activity in the Theory of Aspect: Accomplishments, Achievements, or just Non-Progressive State?', *Linguistic Typology* **6**, 199–273.
Seibt, J.: 1990a, *Properties as Processes. A Synoptic Study in W. Sellars' Nominalism*, Reseda, CA: Ridgeview.
Seibt, J.: 1990b, *Towards Process Ontology: A Critical Study in Substance-Ontological Premises*, Pittsburgh: University of Pittsburgh doctoral dissertation. Published as microfiche with UMI-Dissertation Publication, Michigan.
Seibt, J.: 1995, 'Individuen als Prozesse: Zur prozess-ontologischen Revision des Substanzparadigmas', *Logos* **5**, 303–343.
Seibt, J.: 1996a: 'Non-countable Individuals: Why One and the Same Is Not One and the Same', *Southwest Philosophy Review* **12**, 225–237.
Seibt, J.: 1996b, 'The Myth of Substance and the Fallacy of Misplaced Concreteness', *Acta Analytica* **15**, 61–76.
Seibt, J.: 1996c, 'Existence in Time: From Substance to Process', in J. Faye, U. Scheffler, and M. Urs (eds.), *Perspectives on Time, Boston Studies in Philosophy of Science*, Dordrecht: Kluwer, pp. 143–182.
Seibt, J.: 1996d, 'Der Umbau des "Aufbau": Carnap und die analytische Ontologie', *Deutsche Zeitschrift für Philosophie* **44**, 807–835.
Seibt, J.: 1997, 'The "Umbau": From Constitution Theory to Constructionism', *History of Philosophy Quarterly* **14**, 305–351.
Seibt, J.: 1999, 'Dinge als Prozesse', in R. Hüntelmann and E. Tegtmeier (eds.), *Neue Ontologie und Neue Metaphysik*, Köln: Academia Verlag, pp. 11–41.
Seibt, J.: 2000a, 'Constitution Theory and Metaphysical Neutrality: A Lesson for Ontology?', *The Monist* **83**, 161–183.
Seibt, J.: 2000b, 'Pure Processes and Projective Metaphysics', *Philosophical Studies* **101**, 253–289.
Seibt, J.: 2000c, 'Ontology as Theory of Categorial Inference', in D. Greimann and C. Peres (eds.), *Wahrheit–Sein–Struktur. Auseinandersetzungen mit Metaphysik*, Hildesheim: Olms Verlag, pp. 272–297.
Seibt, J.: 2000d, 'The Dynamic Constitution of Things', in J. Faye, U. Scheffler, and M. Urchs (eds.), *Facts and Events, Poznañ Studies in Philosophy of Science* **72**, 241–278.
Seibt, J.: 2001a, 'Formal Process Ontology', in C. Welty and B. Smith (eds.), *Formal Ontology in Information Systems: Collected Papers from the Second International Conference*, ACM Press: Ogunquit, 333–345.

Seibt, J.: 2001b, 'Processes in the Manifest and Scientific Image', in U. Meixner (ed.), *Metaphysik im postmetaphysischen Zeitalter/Metaphysics in the Post-Metaphysical Age*, Wien: öbv&hpt, pp. 218–230.

Seibt, J.: 2002, 'Quanta, Tropes, or Processes: On Ontologies for QFT beyond the Myth of Substance', in M. Kuhlmann, H. Lyre and A.Wayne (eds.), *Ontological Aspects of Quantum Field Theory*, Singapore: World Scientific, pp. 53–93.

Seibt J.: 2003, 'Process and Particulars', in M. Weber (ed.), *Process Metaphysics*, Munich: Philosophia, forthcoming.

Seibt, J.: forthcoming, *Free Process Theory: A Study in Revisionary Ontology*, Habilitationsschrift at the University of Konstanz, Germany.

Simons, P.: 1987, *Parts: A Study in Ontology*, Oxford: Oxford University Press.

Simons, P.: 2000, 'Continuants and Occurrents', *The Aristotelian Society*, Supplementary Volume **LXXIV**, 78–101.

Simons, P. and C. Dements: 1996, 'Aspects of the Mereology of Artifacts', in R. Poli and P. Simons, (eds.), *Formal Ontology*, Dordrecht: Kluwer, pp. 255–276.

Smith. C.: 1997, *The Parameters of Aspect*, Dordrecht: Kluwer Academic Publishers.

Stout, R.: 1997, 'Processes', *Philosophy* **72**, 19–27.

Taylor, B.: 1985, *Modes of Occurrence: Verbs, Adverbs, and Events*, Oxford: Blackwell.

Whitehead, A.: 1928, *Process and Reality*, New York: Macmillan.

Wilson, E. and Ch. Lumsden: 1991, *Biology and Philosophy* **6**, 401–412.

Zemach, E.: 1970, 'Four Ontologies', *Journal of Philosophy* **23**, 231–247.

BARBARA HELLER and HEINRICH HERRE

ONTOLOGICAL CATEGORIES IN GOL

ABSTRACT. *General Ontological Language* (GOL) is a formal framework for representing and building ontologies. The purpose of GOL is to provide a system of top-level ontologies which can be used as a basis for building domain-specific ontologies. The present paper gives an overview about the basic categories of the GOL-ontology. GOL is part of the work of the research group *Ontologies in Medicine* (Onto-Med) at the University of Leipzig which is based on collaborative work of the *Institute of Medical Informatics (IMISE)* and the *Institute for Computer Science (IfI)*. It represents work in progress toward a proposal for an integrated family of top-level ontologies and will be applied to several fields of medicine, in particular to the field of *Clinical Trials*.

1. INTRODUCTION

In recent years research in ontology has become increasingly widespread in the field of information systems science. Ontologies provide formal specifications and computationally tractable standardized definitions of the terms used to represent knowledge of specific domains in ways designed to enhance communicability with other domains (Gruber 1995). The importance of ontologies has been recognized in fields as diverse as e-commerce, enterprise and information integration, qualitative modelling of physical systems, natural language processing, knowledge engineering, database design, medical information science, geographic information science, and intelligent information access. In all of these fields a common ontology is needed in order to provide a unifying framework of communication. The GOL-project started in 1999 as a collaborative research project of the *Institute for Medical Informatics* (IMISE) and the *Institute for Computer Science* (IfI). The project is aimed, on the one hand, at the construction of an ontological language powerful enough to serve as a formal framework for building and representing complex ontological structures, and, on the other hand, at the development and implementation of domain-specific ontologies in several fields, especially medical science (Heller *et al.* 2003b).

The term *Formal Ontology* has its origin in philosophy (Husserl) but here we use it in a special sense to designate a research area in theoret-

J. Seibt (ed.), Process theories: Crossdisciplinary studies in dynamic categories, 57–76.

ical computer science which is aimed at the systematic development of formalized axiomatic theories of all forms and modes of being, and at the elaboration and design of formal specification tools to support the modeling of complex structures of the real world. Ontologies have different levels of generality, and thus the question arises whether top-level ontologies, i.e. ontologies of the most general level, are needed in applications. Some people believe that top-level ontologies are important, others prefer to focus on domain-specific ontologies which are intuitively adequate for the needs of a special group or community. We assume as a basic principle of our approach that every domain-specific ontology must use as a framework some upper-level ontology which describes the most general, domain-independent categories of reality.

General Ontological Language (GOL) is a formal framework for building and representing ontologies. The purpose of GOL is to provide a system of formalized and axiomatized top-level ontologies which can be used as a framework for building more specific ontologies. GOL consists of a syntax, and of an axiomatic core which captures the meaning of the introduced ontological categories. The system of top-level ontologies of GOL is called GFO (General Formal Ontology). There is a debate whether the top-level ontology should be a single, consistent structure or whether the top-level ontology should be considered as a lattice of theories each of which may be inconsistent with theories that are not situated on the same path. There are arguments for and against the lattice approach. The arguments for a lattice of theories are, first, that there are multiple, incompatible, and – under certain assumptions – equally acceptable views on how to describe the world. Second, it seems to be possible that the adequateness of a top-level ontology depends on the domain of application. Against a multiple ontology one might argue that such lattices are more difficult to maintain and to use.

On the lattice approach ontologies are distinguished in two ways. On the one hand, ontologies may differ with respect to the basic categories of entities postulated. On the other hand, even if two ontologies use the same basic categories they may differ with respect to the axioms pertaining to these categories. Our general approach is to admit a restricted version of the lattice approach. We restrict the selection of top-level ontologies with different systems of basic categories but we are more liberal with respect to the admitted systems of axioms within a fixed system of ontological categories. In our opinion the investigation of a system of axioms with respect to its possible consistent extensions is an important research topic for its own.

In what follows we will discuss the ontologically basic entities and certain basic relations between them. The main distinction we draw is between *urelements*, *sets* and *classes*. Sets, classes and urelements constitute a metamathematical superstructure above the other entities of our ontology, but we also consider them to be entities in the world rather than mere formal tools. At the bottom of the class hierarchy we have the class U of urelements conceived as the realm of existing things in the world which are not sets.

Sets and Classes. The entities of the world are classified according to type. Sets and urelements are entities of type 0, and $C[0]$ is the class of all entities of type 0. Let $\tau_1, \ldots, \tau_n$ be types, and $C[\tau_i]$ the class of all classes of type τ_i, respectively. Then $C[\tau_1, \ldots, \tau_n]$ is the class of all classes of relations whose arguments are classes of types $\tau_1, \ldots, \tau_n$, respectively. A class is of finite type if it can be generated by a finite number of iterative steps. Let C_{FT} be the class of all classes of finite type. In our class hierarchy, C_{FT} is the top-most node.

Urelements. Urelements are entities which are not sets. Urelements form an ultimate layer of entities lacking set-theoretical structure in their composition. Neither the membership relation nor the subclass relation can reveal the internal structure of urelements. (Degen et al. 2001)

Lists. Let U be a class of entities. Then `List`(U) is the smallest class containing the empty list [] and closed with respect to the following condition: if $l_1, \ldots, l_k \in$`List`(U) ∪ U then $[l_1, \ldots, l_k] \in$ `List`.

We shall assume the existence of three main categories of urelements, namely *individuals, universals*, and *spatio-time entities*. Besides urelements there is the class of *formal relations*. We assume that formal relations are not universals, but classes of certain types. Besides these entities there are language-depended entities such as *definable relations* and *definable predicates*.

An *individual* is a single thing which is in space and time. A universal is an entity that can be instantiated by a number of different individuals. The individuals which instantiate a universal are similar in some respect. We assume that the universals exist in the individuals (*in re*) but not independently from them; thus our view is Aristotelian in spirit. A universal can also be understood as a *content of thought*.

For every universal U there is a set $Ext(U)$ containing all instances of U as elements. We assume the following axioms: that the class of urelements is the disjoint union of the class of individuals, the class of universals, and the class of space-time entities.

There are some refinements of the ontology of universals. We may assume that there are universals which can be instantiated by universals. Such *meta-universals* are of practical importance; an example is the concept of a *power class* in UML (Booch *et al.* 1999). In GOL meta-universals (and universals of higher order) are presented by classes of higher order, and the class hierarchy of GOL allows for arbitrary finite towers of meta-universals; these are needed in Software Engineering (Welty 1999).

3. SPACE-TIME

There are several basic ontologies about space and time. In the first top-level ontology of GOL which is reviewed in this paper chronoids and topoids represent kinds of urelements. Chronoids can be understood as connected temporal intervals, and topoids as spatial regions with a certain mereotopological structure. Chrono-topoids are four-dimensional space-time manifolds. On one version of our theory chronoids and topoids have no independent existence; they depend for their existence in every case on the situoids which they frame.

We assume that time is continuous and endorse a modified and refined version of an approach which is sometimes called the *glass continuum*. Chronoids are not defined as sets of points, but as entities *sui generis*. Every chronoid has boundaries, which are called time-boundaries and which depend on chronoids, i.e. time-boundaries have no independent existence and every chronoid has exactly two time-boundaries. The class TE of temporal entities consists of two disjoint sub-classes: the class Chr of chronoids and the class TB of time-boundaries; thus $TE = Chr \cup TB$. Every chronoid has inner time-boundaries which arise from proper sub-chronoids of a chronoid; $TB(c)$ denotes the class of all time-boundaries of the chronoid c. By a *temporal structure* we understand a sub-class of TE, i.e. the class TS of all temporal structures is defined by $TS = \{K : K \subseteq TE\}$. We assume that temporal entities are related by certain formal relations, in particular the *part-of relation between chronoids*, the relation of *being a time-boundary of a chronoid*, and the relation *of coincidence between two time-boundaries* which is denoted by $coinc(x, y)$. In this approach to an ontology of time we are adapting ideas of Brentano (1976) and Chisholm (1983) and advance and refine the theory of Allen *et. al.* (1989).

A class K of chronoids is bounded if there is a chronoid c which contains every member of K as a temporal part. We stipulate a *continuity axiom* stating that for every bounded class K of chronoids there exist a least unique chronoid c containing every member of K as a temporal part. A *generalized chronoid* is the mereological sum of a class of chronoids. The *part-of relation* between chronoids is naturally extended to a part-of relation between generalized chronoids. There are two kinds of time boundaries: time-bondaries *looking in the future* and time boundaries *looking in the past.* We use the term *time-point* to denote entities consisting of two coinciding time boundaries: a *future boundary* and a *past boundary.* A *now* can be considered as a time point of this kind because from a *now* we may look in the future and in the past. There is the following branching point for axioms. One axiomatic system claims that there are no atomic chronoids, another system assumes that every non-atomic chronoid has an atomic part.

Our theory of topoids uses ideas from Brentano (1976), Chisholm (1983), Smith *et al.* (2000). Similar as in Borgo *et al.* (1996) we distinguish three levels for the description of spatial entities: the *mereological level* (mereology), the *topological level* (topology), and the *morphological level* (morphology). Topology is concerned with such space-relevant properties and relations as connection, coincidence, touching, and continuity. Morphology (also called qualitative geometry) analyses the shape, and the relative size of spatial entities. To describe the form of an object we adopt a relation of *congruence* between topoids holding between topoids with the same shape and size. For every topoid t we introduce a universal $U(t)$ whose instances are topoids that are congruent with t. This leads to a theory of shapes of pure topoids, separated from the theory of substances.

4. BASIC CATEGORIES OF INDIVIDUALS

Individuals are entities which are in space and time. That means that there are certain dependency relations relating individuals to spatio-temporal entities. Individuals can be classified with respect to their relation to space and time. The main distinction in the (first) GOL-ontology is between endurants and processes.

4.1. *Endurants and Processes*

There is a debate among philosophers concerning the distinction between processes and objects. According to the *endurantist* view there is a categorical distinction between objects and processes, while, according to the *perdurantist* view there are only processes in the most general sense of four-dimensionally extended entities. Endurantism and perdurantism have their respective advantages and disadvantages. One of the advantages of perdurantism is its simplicity; on the other hand, the advantage of endurantism is that it captures the intuitive distinction between objects and processes. In the top-level ontology of GOL reviewed in the current paper we assume the endurantist point of view. However, given our pluralist research commitments, we are also exploring perdurantist versions of top-level ontologies, as well as the 'recurrence view of persistence', a third option between endurance and perdurance (Seibt 1997, 2003).

The difference between endurants (elsewhere called 'continuants') and processes is their relation to time. An endurant is an individual which is in time, but of which it makes no sense to say that it has temporal parts or phases. Thus, endurants can be considered as being wholly present at a time-boundary. For endurants time is in a sense a *container*, thus endurants are *in time*, and endurants may be indexed by time boundaries. We use a relation $at(x, y)$ with the meaning *the endurant x exists at time-boundary y*. Let *Endur* be the class of all endurants and TB the class of all time-boundaries. Then we stipulate that *at* is a functional relation from *Endur* into TB, i.e. we assume the following axioms:

$$\forall x(Endur(x) \rightarrow \exists y(at(x, y))$$

$$\forall xy(at(x, y) \rightarrow Endur(x) \wedge TB(y))$$

$$\forall xyz(at(x, y) \wedge at(x, z) \rightarrow y = z)$$

These axioms raise the question of what it means that an endurant persists through time. We pursue an approach which accounts for the persistence of endurants by means of a suitable universal whose instances are endurants. Such universals might be called *abstract endurants*. A similar idea is pursued by Simon (2000) where he considers a continuant as an abstractum over occurrents under a certain equivalence relation.

Processes, on the other hand, have temporal parts and thus cannot be present at a time-boundary. For processes time *belongs to them* because they *happen in time* and the time of a process is built into it. The relation between processes and temporal structures is determined by a projection function $prt(x, y)$ saying that *the process x is projected onto the temporal entity y* or that y is the *temporal projection* of x. We assume that the temporal entity which a process is projected onto is a mereological sum of chronoids, i.e. is a generalized chronoid. Again, $prt(x, y)$ is a functional relation from the class *Proc* of all processes into the class GC of generalized chronoids, and we say also that *y frames x*. Thus,

$$\forall xyz(prt(x, y) \wedge prt(x, z) \rightarrow y = z).$$

There are yet two other projection relations, one of them projects a process p to a temporal part of the framing generalized chronoid of p. The relation $pr(p, c, q)$ has the meaning: p is a process, c is a temporal part of the chronoid which frames p, and q is the projection from p onto c. q can also be understood as the restriction of the process p to the generalized sub-chronoid c. The temporal parts of a process p are exactly the projections of p onto temporal parts of the framing generalized chronoid of p. The other relation projects processes onto time-boundaries; we denote this relation by $prb(p, t, e)$ and call the entity e onto which p is projected the *boundary of p on t*. Let *be* $B(p, t) = e$ if and only if $prb(p, t, e)$. As a bold basic tenet of the present version of GOL we postulate that the projection of a process to a time-boundary is an endurant.

Processes belong to a category which we call *occurrents*. The above projection relation $prt(x, y)$ will be generalized to arbitrary occurrents x; then y is – in the most general case – a temporal structure. Other types of occurents are: *histories, states, change, locomotion*, and *boundaries of processes*. Boundaries of processes are projections of processes to time-boundaries. Histories are families of endurants which are indexed by time-boundaries. Drawing on abstract endurants, histories and projections of processes to time-boundaries we explicate systematically the most important relationships between endurants and processes. All these entities will be considered in section 4.4.

4.2. *Substances*

In our ontology the notion of substance plays – in relation to time – three different roles: when we speak of *substances* simpliciter, we refer to endurants; *abstract substances* are universals which have substances as instantiations; finally, by *substance-processes* we refer to processes of a certain type.

Substances are individuals which satisfy following conditions: they are endurants, they are bearers of properties, they cannot be ‘carried by’ other individuals, and they have a spatial extension. The expressions ‘x carries y’ and ‘x is carried by y’ are technical terms which we define by means of an ontologically basic relation, the *inherence relation* which connects properties to substances. Inherence is a relation between individuals which implies that inhering properties are themselves individuals. We call such individual properties *moments* and assume that they are endurants. Moments include qualities, forms, roles and the like. Examples of substances are: an individual person, a house, the moon, a tennis ball (all of them considered at a time-boundary).

Every substance S has a spatial extension, which is called the *extension-space of* S, and occupies a certain spatial entity which is called the *spatial location of* S. Here we use a formal relation $occ(x, y)$ which means *the substance x occupies the spatial location y*. We consider the extension-space of S and the spatial location of S as different entities. The extension-space e of a substance S can be – in a sense – understood as an individual moment of S, similar as for example the individual weight or individual form of S. The formal relation $exsp(S, e)$ has the meaning: *e is the extension-space of* S, and we assume the following condition about this relation: $\forall Se(exsp(S, e) \wedge exsp(S', e) \rightarrow S = S')$.

We assume that the spatial location occupied by a substance is a topoid which is a 3-dimensional space region. A *physical object* is a substance with unity, and a closed substance is substance whose unity is defined by the strong connectedness of its parts. Substances may have (substantial) boundaries; these are dependent entities which are divided into surfaces, lines and space-points. Every (substantial) surface is the boundary of a substance, every line is the boundary of a surface, and every spatial point is the boundary of a line (Brentano 1973). We emphasize that the boundary of a substance is not the same entity as the boundary of its spatial location. Boundaries of two different substances are touching if parts of the boundaries of their occupied spatial locations coincide. In our *theory of substantial boundaries* two 'bona fide boundaries' may touch which is impossible in the approach of Smith *et al.* (2000). A topoid T *frames* a substance S if the location which is occupied by S is a part of T. We introduce the *convex frame f* of a substance S, denoted by the relation *convf(S,f)*, as the convex closure of the spatial location which is occupied by S.

Substances are related to time by the relation *at* (S, t) having the meaning that S *exists at time-boundary* t. But there is yet another relation between substance and time. What does it mean that a *substance persists through time* or that a *substance has a life-time*? To clarify the problem let us consider a term *John* which denotes a certain individual person *den(John)*. What kind of entity is *den(John)*? If we consider it as a substance then there is a time-boundary t such that *at(den(John),t)*. Because $at(x, y)$ is assumed to be a functional relation the entity *den(John)* depends on the time-boundary t, i.e. we have to add the parameter t to *den(John)*, which we denote by *den(John)(t)*. Obviously – with respect to this interpretation – the term *John* denotes an endurant for certain time-boundaries. Let *TB(John)* be the class of all time-boundaries t at which the term *John* denotes the endurant *den(John)(t)* and let $E(John) = \{den(John)(t) : t \in TB(John) \text{ and } at\ (den(John)(t),t)\}$.

To ensure that all these different endurants *den(John)(t)* present the *same* John we introduce a ontologically basic relation with the meaning that the substances x and y are *ontically connected*, and a universal *endur(John)* whose instances are just all elements of the class *E(John)*. Then we stipulate that two endurants $e(1)$ and $e(2)$ are equivalent with respect to *John*, i.e. represent the *same John*, if $ontic(e(1), e(2))$ and both $e(1), e(2)$ are instances of *endur(John)*. The relation $ontic(x, y)$ should satisfy – at least – the conditions of *spatio-temporal continuity* which are discussed by Le Poidevan (2002).

The *history of John*, denoted by *history(John)*, is defined by the class $\{(den(John)(k): k \in TB(John)\}$ and a time-ordering $<$ on the set *TB(John)*. We say that a universal *U persists through the history of John* if every term of *history(John)* instantiates U. In this framework the *substance John* can be understood as an abstractum *endur(John)* which – by definition – persists trough time. We call the universal *endur(John)* an *abstract substance*. We hold a similar position as Simons (2000) that abstract substances are invariants amid diversity, and that what is true of them is true of their associated class of concrete instances.

Many opponents of endurantism claim that those terms that traditionally have been taken to denote substances denote processes. Thus Lewis (1983, pp. 76 ff.) claims that human beings have temporal parts. In our opinion there is no real incompatibility between substances as we define them and processes. Thus, the term *John* designates also a process having temporal parts, denoted by *process(John)*. We call processes of this kind *substance-processes*. As noted above, the projection of *process(John)* to a time-boundary t is a substance *John(t)*; projections of a process p to time-boundaries are called hereafter 'boundaries of p.' By definition the class of all boundaries of *process(John)* coincides with the full history of *John*. The connection between the time-boundaries is given, then, by the process itself. In general, we assume that time-indexed histories of endurants are entities which depend on processes. There is a close relation between the three kinds of entities representing the notion of substance: *substance-processes, substances (as endurants), and abstract substances (as universals)*. This relation is considered in more detail in section 4.4.

Every substance-process x has a temporal projection which is a chronoid y. The temporal projection y of a substance-process x can be understood – in a sense – as the lifetime of x. The formal relation *lifetime(x,y)* has the meaning *x is a substance-process and y is the temporal projection of x*. We use the functional abbreviation $lf(x)$ which is defined by the condition $lf(x) = y \Leftrightarrow lifetime(x, y)$. If x is an abstract substance then the lifetime of x is defined as the minimal chronoid containing all time-boundaries which are associated to the instances of x.

4.3. *Moments*

Moments are endurants; in contrast to substances, moments are entities which can exist only in another entity (in the same way in which, for example an electrical charge can exist only in some conductor). Moments are property particulars: this color, this weight, this temperature, this thought. According to one version of our ontology moments have in common that they are all dependent on substances, where the dependency relation is realized by inherence. Some moments are one-place *qualities*, for example of color or temperature, but there are also relational moments – for example relators founded on kisses or on conversations – which are dependent on a plurality of substances. Moments can be classified in qualities, forms, roles, relators, functions, dispositions and others.

Every endurant is either a substance, or a moment, or a more complex entity as for example a situation. We call substances or moments *primitive endurants* and suppose that the inherence relation connects primitive endurants only. Obviously, substances are those individuals which do not inhere in any endurant.

As for substances there is a relation $at(m, i)$ stating that the moment m exists at time-boundary t. Also there are classes of moments indexed by time-boundaries which we call histories, and analogously to abstract substances there are *abstract moments*, i.e., moment-universals.

The relation of moments to space seems to be more involved. In general, we may say that a moment is *located at a region* which is itself related to the spatial location occupied by the substance bearing this moment. For example, the spherical form inheres in a ball and is located at the surface-boundary of the ball. But where are color and weight located?

Similar as for substances the notion of moment plays also the role of process of certain type, for example an individual red inhering in an apple during one hour. Such a *moment-process* is connected to a substance-process by the inherence relation.

4.4. *Occurrents*

To restate, we use the notion of *occurents* to cover several categories of individual entities related to processes. *Occurrents* comprise *processes, histories, locomotions, changes, boundaries of processes*, and *states*. A *connected process* is an individual which has temporal parts and whose projection onto time is a chronoid (which is a connected time-interval). An important subcategory of connected processes is formed by the class of *co-*

herent processes. A process p is coherent if, intuitively, its boundaries (and temporal parts) are ontically connected by the basic relation $ontic(x, y)$ and if there are causal relationships between the temporal parts of p. The category of coherent processes needs an elaboration in the spirit of the approach of Le Poidevan (2002). A characterization of coherent processes is beyond the scope of the present paper.

Boundaries of processes are dependent entities, they depend on the processes they bound. Let $B(p)$ the class of boundaries of the process p. A process p is not the aggregate of its boundaries; hence, boundaries of a process are different from entities which are called sometimes *stages* of a process. A process cannot be understood on the basis of its boundaries. Hence, our theory of process-boundaries differs from the stage-theory in the versions of Lewis (1983) and of Sider (2001). In addition, there are some differences between boundaries and stages. A boundary is not a temporal part of a process p, because every temporal part of p has a temporal projection which contains a chronoid. According to Lewis a stage – in contradistinction to a boundary – has a temporal duration but 'only a brief one, for it does not last long'. A stage begins to exist abruptly, and it abruptly ceases to exist soon after. Finally, a stage d' in the sense of Lewis or Sider – seems to have a relatively independent existence, which is impossible for boundaries.

A boundary of a process is – in general – the beginning or the ending of a process. An entity e is an inner boundary of a process p if e is the beginning or the ending of a temporal part of p whose framing chronoid is properly included in the chronoid which frames p. Two boundaries of a process *meet* if their associated time-boundaries coincide. The boundaries of a process are – in general – parts of situations (to be considered in section 5).

A *change* is a pair $(e1, e2)$ of meeting boundaries where one of them is the ending of a past process and one the beginning of a future process, and $e1, e2$ instantiate different universals. To be more precise, changes are relative to a basic universal u such that the change is exhibited by certain proper sub-universals of u. Take for example the universal *color* as the basic universal and *red* and *blue* as discriminating sub-universals. Then, obviously, the change in color from blue to red can be understood in this framework. Changes proper we call also extrinsic changes. In order to understand the essence of a change the above mentioned relation of ontological connectedness $ontic(x, y)$ must be extended to moments and defined in such a way as to exclude the possibility that an individual color may change to an individual temperature. We hold that *changes* are entities which depend on processes. Note that in our approach universals do persist, and change means instantiation of different proper sub-universals by ontically connected endurants. The movement of a body is an example of an *intrinsic* change. Intrinsic changes cannot be captured by universals. We introduce the process category of *locomotions* covering the most important type of processes based on intrinsic changes.

A *state* is a connected process without any extrinsic or intrinsic changes. Obviously, this is a relative notion, because changes are related to universals. It might be that processes are states with respect to certain universals, but with respect to others they contain changes.

Histories are (arbitrary) classes of endurants which are indexed by their time-boundaries at which they exist. Thus, histories h may be presented as partial functions from *Endur* into the class TB of time-boundaries realized by the relation $at(x, y)$. Let $TB(h)$ be the class of time-boundaries which are associated to the history h. Then we need – in addition – a time-ordering $<$ between the elements of $TB(h)$. Thus, a history is more precisely specified by a pair $(h, (TB(h), <))$. We assume that every endurant is contained in a boundary of a process or *is* the boundary of a process. Not every history in this very general sense is a reasonable entity, because no connection between the constituents of such histories is postulated.

We now clarify – on the basis of our framework – some relations between the endurants, processes and space. By assumption every boundary of a process is an endurant, i.e. for every process $p \in Proc$ the condition $B(p) \subseteq Endur$ is satisfied. Let *Mom* be the class of all moments (as endurants), and let *Subst* be the class of all substances (as endurants). Then $Mom \cup Subst$ is a proper sub-class of *Endur*. A process p is said to be *substance-process* if every boundary of p contains a substance, it is said to be *moment-process* if $B(p) \subseteq Mom$. How are processes related to space? Let e be an endurant and let S be the collection of all substances carrying the moments which occur in e. S is said to be the *substantial closure* of e, and the relation which associates S to e is denoted by $substcl(e, S)$. This relation may be extended to processes. If p is a process and $substcl(p, q)$ then q is a process with the same temporal projection as p and such that for every time-boundary t of p the process-boundary $q(t)$ is the substantial closure of the process-boundary $p(t)$; the process q is called the substantial closure of p. Let f be the convex frame of the localization of S; the association between S and f is denoted by the above introduced relation $convf(S, f)$. Then we may introduce a topoid T which is defined as the convex closure of the class $\{f : convf(S, f) \textit{ and } substcl(S, e) \textit{ and } e \in B(p)\}$. We say that the process p is projected onto T and denote this accociation by the relation $prs(p, T)$.

What does it mean to say that a *substance participates in a process*? If (a) by substance we mean an endurant then a substance s participates in a process p if there is boundary of p whose substantial closure contains s, or (b) if 'substance' means 'abstract substance' then the abstract substance S participates in a process p if the substantial closure of every boundary of p contains an instance of S.

We conclude this section with an analysis of the movement of a solid body b; processes of this type are called *locomotions*, i.e. movements in space. The movement is a process p such that the associated history of p is a sequence of time-indexed endurants $\{b(t) : t \textit{ time-boundary of } p\}$. What is a boundary of this process, i.e. the projection of p onto a time-boundary? It is a substance $b(t)$ – a body at this time-boundary – which occupies a certain topoid denoted by $tp(t)$. If we consider two coinciding time-boundaries t_1 and t_2, then there is no universal to discriminate the endurants $b(t_1)$ and $b(t_2)$ (we assume that there are no extrinsic changes concerning qualities of $b(t_1), b(t_2)$). How is it possible that the body moves without extrinsic changes? There are three possible ways to conceive of movement within our setting:

(1) The spatial locations $tp(t_1)$ and $tp(t_2)$ of $b(t_1)$ and $b(t_2)$ are different; then there is a jump from $tp(t_1)$ to $tp(t_2)$.
(2) The spatial locations $tp(t_1)$ and $tp(t_2)$ are different but they are considered as coinciding boundaries (in the spirit of Brentano (1976)).
(3) The spatial locations $tp(t_1)$ and $tp(t_2)$ are the same, but the extension-spaces $exsp(b(t_1)$ and $exsp(b(t_2))$ are different.

Case (1) is not a viable option if we are to assume the continuity of space. Case (2) would be possible if the spatial locations could be taken to be boundaries. Since spatial locations are 3-dimensional they would have to be boundaries of 4D-manifolds. This is problematic, however, because there is an asymmetry between time and space. Topoids are not endurants and they cannot move like a body. This leaves us with case (3). In this case $exsp(b(t_1))$ is contained in the end of a process and $exsp(b(t_2))$ is contained in the beginning of a process and both endurants meet (i.e. the associated time-boundaries coincide). From our point of view case (3) is reasonable; there is – at least – a change of the extension-space at a pair (t_1, t_2) of two coinciding time-boundaries. This change is not a proper change because their meeting boundaries can not be discriminated by universals. But this kind of change is also true for a motionless, resting body b, because the extension-space of b considered as a process – i.e. extended in time – is a moment-process and the substantial closure of this moment-process exhibits the same kind of changes. Thus, we are facing the problem how to distinguish a motionless body from a moving body.

This Zeno-like problem may be analysed as follows. The movement of a body in space cannot be understood locally, i.e. with respect to time-boundaries, but only with respect to chronoids, i.e. extended temporal entities. The following condition holds: for every chronoid c which is a part of the temporal projection of p and the associated boundaries t_1 and t_2 it is $tp(b(t_1)) \neq tp(b(t_2))$. Processes of this kind could be called *continuous* movements in space, and it does not make sense to say that a body moves at a time-boundary or between coinciding time-boundaries. Our analysis of Zeno's problem is not surprising in our framework: a process is not the aggregate of its boundaries. On the other hand, Zeno's basic assumption is that time is continuous and that time is the sum of its time-points.

The most important ontological category treated in this section is the category of connected processes. In applications many other categories of occurrents are relevant. In formalizing the notion of blood-pressure Heller *et al.* (2003c) used histories. There are processes, which are disconnected, i.e. processes whose temporal projection is not a (connected) chronoid. Disconnected processes may be used to describe, for example, diseases as Malaria. All these kinds of occurrents may be analysed in the framework of a particular category of entities, the category of situoids.

5. SITUOIDS, SITUATIONS, AND CONFIGURATIONS

The entities discussed in the preceding sections have no independent existence. Substances and moments presuppose another, and both constitute complex units or wholes of which they are aspects. Such integrated wholes of substances and moments are themselves endurants, and we call them *configurations*. Configurations are classified in *simple* and *non-simple*. A simple configuration is a unit which is made up from one substance and only monadic moments inhering in that substance. A configuration is said to be non-simple if it is made up from more than one substance and relational moments connecting them. A *situation* is a special configuration which can be comprehended as a whole and satisfies certain conditions of unity imposed by certain universals associated with the situation. Situations present the most complex endurants of the world. In the world of endurants they have the highest degree of independence. The convex frame of a situation s is defined by the convex closure of the localizations occupied by the substances which occur in s. There are differences between our *Ontological Theory of Situations* and the *Situation Theory* of Barwise *et al.* (1983). Situations in our sense are build up from substances, universals and from material and formal relations; these notions are missing in situation theory.

On the other hand, according to the basic assumptions of GOL , endurants have no independent existence, they depend on processes. Since configurations are endurants they, too, depend on processes. We call such processes *configuroids*. They are – in a sense – integrated wholes made up from substantial processes and moment-processes. We claim that substance-processes and moment-processes presuppose each other. Surely a moment-process depends on a substance-process, on the other hand we may assume that a substance-process needs an extension which includes a moment-process.

Finally, there is a category of processes whose boundaries are situations and which satisfy certain principles of coherence and continuity. We call these entities *situoids*; they are the most complex integrated wholes of the world, and they have the highest degree of independence. As it turns out, each of the considered entities (including processes) is embedded into a suitable situoid. A *situoid* is, intuitively, a part of the world that is a coherent and comprehensible whole and does not need other entities in order to exist. Every situoid has a temporal extent and is framed by a topoid. An example of a situoid is *John's kissing of Mary* in a certain environment which contains the substances 'John' and 'Mary' and a relational moment 'kiss' connecting them. Taken in isolation, however, these entities do not yet form a situoid; we have to add a certain environment consisting of further entities and a location to get a comprehensible whole: John and Mary may be sitting on a bench or walking through a park. The notion of being a coherent and comprehensible whole is formally elucidated in terms of an *association relation* between situoids and certain universals. The relation $ass(s, u)$ expresses that the universal u is associated to the situoid s.

How are situoids related to time and space? We use here two relations $chron(s, x)$, and $top(s, z)$, where x is the chronoid framing the situoid s and z is the topoid framing s. The topoid framing a situoid is a fiat object (i.e. given by convention); it can be understood – in a sense – as defined by a local coordinate system. But also the boundaries of the framing chronoid are conventional. Note, that the relation $chron(s, x)$ coincides with $prt(s, x)$ if the situoid is considered as a process; the relations $prs(s, x)$ and $top(s, x)$ are different. The following relation is satisfied:

$$\forall sxy(prs(s, x) \wedge top(s, y) \rightarrow x \leq y)$$

Every temporal part of a situoid is itself a situoid. The temporal parts of a situoid s are determined by the full projection of s onto a parts of the framing chronoid c of s. Boundaries (including inner, fiat boundaries) of situoids are projections to time-boundaries. We assume that projections of situoids to time boundaries are situations. In every situation occurs a substance, and we say that an endurant e is a constituent of a situoid S iff there is a time-boundary t of S such that the projection that is a situation containing e.

Situoids have a rich structure which can be analysed by using some further notions. A *substantial layer* P of the situoid S is a 'portion' of S satisfying the following conditions:

(a) P is a connected process,
(b) P and S are framed by the same chronoid,
(c) every boundary of P contains a substance,
(d) Recall that for a connected process P and t a time-boundary of the chronoid c which frames P, $B(P, t)$ denotes the boundary of P at t. For all time-boundaries p, q of S holds: if a is a substance which is contained in $B(P, p)$ and b is a substance which is contained in $B(S, q)$, $p < q$, and a, b are ontically connected, i.e. $ontic(a, b)$, then b is contained in $B(P, q)$ too.

The notion of a *moment-layer* of a situoid is introduced in similar fashion.

Situoids can be extended in two ways. Let S, T be two situoids; we say that T is a *temporal extension* of S, if there is an initial segment c of the chronoid of T such that the projection of T onto c equals S. We say that T is a *substantial extension* of S if S is a substantial layer of T. Both kinds of extensions can be combined to the more general notion of a *substantial-temporal extension*. The whole reality can be – in a sense – understood as a web of situoids which are connected by substantial-temporal extensions. The notion of an extension can be relativized to situations. Since there cannot be temporal extensions of situations an extension T of the situation S is always a substantial extension. As an example consider a fixed single substance a which occurs in situation S. Every extension of S is determined by adding further monadic or relationary moments to S to the the intrinsic properties of a. A moment-bundle which is unified by the substance a is called *saturated* if no extension of S adds new moments. Is there an extension T of S such that every substance a in T unifies a saturated bundle of moments?

Configuroids. A *configuroid* c in the situoid S is defined as the projection of that substantial layer of S onto a chronoid which is a part of the time-frame of S. In particular, every substantial layer of S is itself a configuroid of S. Obviously every configuroid is a coherent process. But not every coherent process is a configuroid of a situoid because not every process satisfies the substantiality condition.

Occurrents and Situoids. We postulate as a basic axiom that every occurrent is – roughly speaking – a 'portion' of a situoid, and we say that every occurent is embedded in a situoid. Furthermore, we defend the position that processes should be analysed and classified in the framework of situoids. Also, situoids may be used as ontological entities representing contexts. A rigorous typology of processes in the framework of situoids is an important future project. Occurrents may be classified with respect to different dimensions, among them we mention the *temporal structure* and the *granularity* of a occurrent. We conclude this section with an outline of some classification principles.

Temporal structure of occurrents. Let o be an occurrent, then o is embedded in a situoid S. Let y be the temporal projection of o, i.e. it holds $prt(o, y)$. Occurrent o may be classified with respect to the type of the temporal structure y.

(1) Let o be a history and $TB(o)$ the class of time-boundaries which are associated with the constituents of o; then $prt(o, TB(o))$. The history o is said to be dense if $TB(o)$ contains a dense subset. Otherwise o is called discrete. There is a complete classification of all order types of linear orderings which are associated with countable sets $TB(o)$ (Erdös 1964). It is a practical question which of these order types are of use in applications.
(2) A process p is *disconnected* if the projection of p onto time is not connected. We assume that the temporal projection of a disconnected process does not contain isolated time-boundaries. Then the temporal projection of p, denoted by $TP(p)$, is a class of chronoids. There is a natural linear ordering between the chronoids $i, j \in TP(p)$, denoted by $i < j$. The temporal structure of p may be classified with respect to the order type of the system $(TP(p), <)$. Note, that this ordering can be dense.
(3) A occurrent p is said to be *hybrid* it its temporal projection contains chronoids and isolated time-boundaries.

The practical relevance of these distinctions may differ for different applications.

Granularity of Processes. Let p be a connected process, and c the chronoid which frames p. In many cases it is useful to have coarser granularity of p. This can be made precise by using partitions of the the framing chronoid c. A partition of c is a set of chronoids satisfying the following conditions: any two different chronoids $i, j \in Part(c)$. do not overlap and every time-boundary of c is time-boundary (including inner boundaries) of suitable $i \in Part(c)$. We restrict in the following on such partitions $Part(c)$ which are finite or has order type ω. Now we assume a set $PU = \{u(1), \ldots, u(k)\}$ of processual universals. We say that $Part(c)$ is a *PU-partition* of p if every $i \in Par(c)$ instantiates one of the universals from PU. By using suitable partitions of p and collections PU coarser processes may be abstracted from p. This idea of *PU-partitions* of processes has to be tested on practical applications. Here we will use the ideas of Becher *et al.* (2000).

6. CONCLUSION AND FUTURE RESEARCH

One of the aims of the group *Onto-Med* is the application of the GOL-ontology (called GFO) in the field of medical science, but also in other domains. GOL is intended to provide a formal framework for building, representing and evaluating domain-specific ontologies. For this purpose *Onto-Med* is elaborating a *methodology of ontological reduction* (Heller *et al.* 2003a). One of the computer-based applications (called *Onto-Builder*) is the development and implementation of Software tools to support the standardization and reusability of terms in the field of clinical trials (Heller *et al.* 2003d).

The basic categories and basic relations of GOL will be characterized – in the spririt of the axiomatic deductive method – by a family Ax(GFO) of axiomatic systems. By adding new categories and relations from the field of medicine GOL will be extended to GOL-Med and GOL-CTrials. The axioms and categories of GOL-CTrials for example refer to the class of all clinical trials.

Another area of application is the ontological foundation of conceptual modelling. First examples of applying GOL to UML (Unified Modelling Language) are demonstrated by Guizzardi *et al.* in (2002a), (2002b).

REFERENCES

Aristotle: 1984. *The Complete Works*, translated and edited by J. Barnes, Princeton University Press, Princeton.

Allen, J. and P. J. Hayes: 1989. 'Moments and Points in an Interval-Based Temporal Logic', *Computational Intelligence* **5**, 225–238.

Barwise, J. and J. Perry: 1983. *Situations and Attitudes*. Bradford Books, The MIT Press.

Becher, G., F. Clerin-Debart and P. Enjalbert: 2000. 'A Qualitative Model for Time Granularity', *Computational Intelligence*, pp. 137–168.

Borgo, S., N. Guarino and C. Masolo: 1996. 'A Pointless Theory of Space Based On Strong Connection and Congruence', in *Principles of Knowledge Representation and Reasoning (KR96)*, Boston, MA. pp. 220–229.

Booch, G., J. Rumbauch and I. Jacobson: 1999. *The Unified Modeling Language*. Addison Wesley.

Brentano, F.: 1976. *Philosophische Untersuchungen zu Raum, Zeit und Kontinuum*. Hamburg: Felix Meiner Verlag. English translation as *Philosophical Investigations on Space, Time and Continuum*. London: Croom Helm, 1988.

Chisholm, R. M.: 1983. 'Boundaries as Dependent Particulars', *Grazer Philosophische Studien* **20**, 87–96.

Degen, W., H. Heller and H. Herre: 2001a. 'Contributions for the Ontological Foundation of Knowledge Modelling', *Report Nr. 02. Institut für Informatik*, Universität Leipzig.

Degen, W., H. Heller, H. Herre and B. Smith: 2001b. 'GOL: A General Ontological Language', *Formal Ontology in Information Systems*. Collected Papers from the Second International Conference. New York, pp.34–46.

Degen, W., H. Heller and H. Herre: 2002a. 'GOL: A Framework for Building and Representing Ontologies', in B. Heller, H. Herre and B. Smith (eds.), *Ontological Spring* (A Reader in Formal and Applied Ontology), IFOMIS Reports 1 Leipzig, pp. 182-203, 2002.

Degen, W., B. Heller and H. Herre: 2003b. GOL-Manual (Version 1.0α) *OntoMed-Report-3*.

Erdös, P. and R. Rado: 1964. 'A Partition Calculus in Set Theory', *Bull. Amer. Math. Society* **15**, 495–503.

Genesereth, M. R. and R. E. Fikes: 1992. Knowledge Interchange Format, Version 3.0, Reference Manual, Logic Group Report Logic-92-1, Computer Science Department, Stanford University, Stanford, California.

Gruber, T. R.: 1995. 'Towards Principles for the Design of Ontologies Used for Knowledge Sharing', *International Journal of Human and Computer Studies* **43**, 907–928.

Guarino, N.: 1998. 'Formal Ontology and Information Systems', In N. Guarino (ed.), *Formal Ontology in Information Systems*, Proc. of FOIS' 98, Trento, Italy, Amsterdam, IOS-Press.

Guarino, N.: 1998. 'Some Ontological Principles for Designing Upper Level Lexical Resources',. *First Conference on Language Resources and Evaluation.*

Guarino,N. and C. Welty: 2002. 'Evaluating Ontological Decisions with OntoClean', *Communications of the ACM* **45**(2), 61-65.

Guizzardi, G., H. Herre and G. Wagner: 2002a. 'On the General Ontological Foundation of Conceptual Modelling', in S. Spaccapeitra *et al.* (eds.), 2002. Conceptual Modelling, ER,Tampere, Proceedings of the 21st International Conference on Conceptual Modelling, vol. 2503, LNCS, 65-78. Springer Verlag.

Guizzardi, G., H. Herre and G. Wagner: 2002b. 'Towards Ontological Foundations for UML Conceptual Models', in Meersmann *et al.* (eds.), International Conference ODBASE Irvine, California. LNCS vol. 2519, 1100–1117, Springer Verlag.

Heller, B. and H. Herre: 2003a. 'Formal Ontology and Principles of GOL', *OntoMed-Report-1*.

Heller, B., H. Herre and M. Löffler: 2003b. 'Research Projects of Onto-Med', *OntoMed-Report-2*.

Heller, B., H. Herre and M. Löffler: 2003c. 'Formal Specifications in GOL – A Case Study in the Field of Clinical Trials', *OntoMed-Report-6*.

Heller, B., K. Kühn and K. Lippoldt: 2003d. 'OntoBuilder', *OntoMed-Report-5*.

Inwagen, P. V.: 2000. 'Temporal Parts and Identity Across Time', *The Monist* **89**, 437-459.

Johansson, I.: 1989. *Ontological Investigations*. Routledge.

Lewis, D. K.: 1983. *Collected Papers*. Vol. I. Oxford: Blackwell.

Musen, M.: 1992. 'Dimensions of Knowledge Sharing and Reuse', in *Computer and Biomedical Research*, vol. 25, pp. 435–467.

La Podevin, R.: 2000. 'Continuants and Continuity', *The Monist* **83**, 381–398.

Russel, S. and P. Norvig: 1995. *Artificial Intelligence*. Prentice-Hall, Inc.

Seibt, J.: 1997. 'Existence in Time: From Substance to Process', in J. Faye *et. al.* (eds.), *Perspectives in Time*, pp. 143–182.

Seibt, J.: 2001. Formal Process Ontology. *Formal Ontology in Information Systems.* Collected Papers from the Second International Conference. New York, pp. 333-345.

Seibt, J.: 2003. 'Process and Particulars', in M. Weber (ed.), *Process Metaphysics*, Philosophia Verlag, forthcoming.

Sider, T.: 2001. *Four-Dimensionalism. An Ontology of Persistence and Time*. Oxford: Clarendon Press.

Simons, P.: 1994. 'Particulars in Particular Clothing: Three Trope Theories of Substance', *Philosophy and Phenomenological Research* **LIV**(3), 553–575.

Simons, P.: 1998. 'Farewell to Substance: A differentiated leave-taking', *Ratio (new series)*, pp. 235–252.

Simons, P.: 2000. 'How to Exist Without Temporal Parts', *Monist*, pp. 419–436.

Smith, B. and A. Varzi: 2000. 'Bona Fide and Fiat Boundaries', *Philosophy and Phenomenological Research* **60**, 401–420.

Sowa, J. K.: 2000. *Knowledge Representation*. Brooks/Cole.

Varzi, A. C.: 1996. 'Parts, Wholes, and Part-Whole Relations: The Prospects of Mereotopology', *Data and Knowledge Engineering* **20**, 259–286.

Varzi, A. C.: 1997. 'Boundaries, Continuity, and Contact', *Nous* **31**, 26–58.

Welty, A. C. and D. A. Ferucci: 1999. 'Instances and Classes in Software Engineering, Artificial Intelligence', *Artificial Intelligence*.

SVEND ØSTERGAARD

THE CONCEPTUALIZATION OF PROCESSES

ABSTRACT. There are various sources of the human conceptual system that pertain to causation. According to the realism of René Thom the attention network is attuned to existing patterns of singularities in space/time. According to cognitive linguistics the conceptual system is determined by the neural wiring and the embodied experience of the cognizer. Our concepts do therefore not necessarily reflect objective properties of space and time. In this paper I discuss these two positions and their relation. Following Len Talmy, I present a comparison between how causation is conceived in language and how it is conceived in science. Finally, the notion of *agency* and its relation to a basic causative sequence is discussed in more detail.

INTRODUCTION

In this paper I will present some elements of our conceptualization of occurrences or processes. Following Thom (1972) I will argue from specific examples that our conceptualization of events is motivated by patterns of singularities in space/time. However, we cannot assume that there is a one-to-one relation between how dynamics is represented linguistically and how it is manifested in the physical space. In fact, even the notion of a 'dynamics' or a 'process' is a cognitive construction that does not necessarily reflect any objective partitioning in 'observer-independent' space and time. The 'realism' of René Thom must therefore be combined with the general methodological strategy of cognitive linguistics, especially as implemented in the work of Talmy (2000e). According to the general stance of cognitive linguistics, a dynamic scheme is a construction of the mind which serves interaction with the environment, and to the extent it is coded by language it also serves communication. But in general, when we activate a schematic understanding via language we cannot assume that the scheme represents objective properties of the event to which we refer.

By conceptualization I refer to any process leading to a mental representation of dynamics in space/time in a human agent. That is, whenever an observer 'understands' an event in space and time then this understanding is the result of a conceptualization.[1] Whether conceptualization exists independently of language or whether it is imputed by language is of course

J. Seibt (ed.), Process theories: Crossdisciplinary studies in dynamic categories, 77–96.

not a settled question, but there is evidence to suggest the former – for instance, the observation that already at the age of three months infants employ apparently innate causal schemes through which they 'understand' events in space, cf. Spelke (1995). The claim that conceptualization is language independent can also derive support from the 'embodiment' hypothesis – cf. Lakoff et al. (1999) – in which body and action schemes are considered as the basis for language structure. In this tradition language is viewed as a code and independently of the specific form this code takes, its meaning is not to represent a content but to activate dynamic schemes by means of which language users interpret events, acts, etc. Language is thus a privileged medium to get insights into how the mind chunks reality into conceptual units. It seems, however, that no single organizing principle can tell us the whole story. There is no 'ism' – realism, mentalism, pragmatism, etc. – that provides us with the ultimate answer. In the following I will briefly present three aspects of the human conceptualization of dynamic processes.

1. SINGULARITIES IN SPACE /TIME

1.1.

Conceptualization serves the function of enabling the organism effectively to cope with the external environment, and in this respect it is of crucial importance to be able to detect qualitative changes in space and time. For instance, a sound that breaks the silence, an oblong object whose position changes from vertical to horizontal (in falling), the sun appearing above the horizon, a crack appearing in some object (sign of instability), etc. The 'points' in space/time in which there is a qualitative change I shall call *singular points*, in contrast to *regular points*, where no such change happens.[2] The distribution of the singular points is dependent on an observer's *attention*. The part of the attention network which is not responsible for willful attention – the part which Déscartes called passive attention – is attracted by changes in space that provide new information for the observer. Ideally, we can then imagine that space/time is divided into open, regular areas from which the perceptual input corresponds with perceiver's expectations, whose mind is therefore, while perceiving these areas, in a *structurally stable state*. These regular areas are separated by singular points representing a high load of information and therefore putting the cognitive system in a *structurally unstable state*.

It is of course a problem to determine what counts as 'information' for any given observer – which items create structurally unstable states in

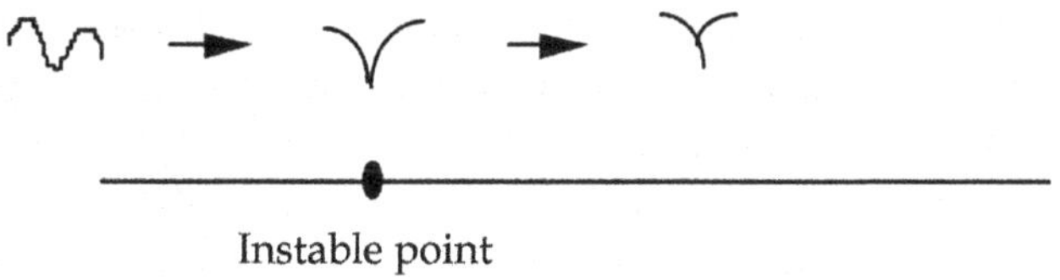

Figure 1.

the perceiver? In Catastrophe Theory this issue has been addressed with a topological formalism; cf. Thom (1972, 1980). It is not possible to go into the technical details here, but allow me to illustrate this with a simple example. Imagine a camel passing an observer. When the camel is next to the observer he/she will see two humps separated by the minimum of a smooth curve. When the camel has passed by and is in front of the observer, he/she will see one of the humps overlap the other. But there is exactly one point on the camel's trajectory when the observer neither perceives the overlapping humps nor the separating minimum. This is a structurally unstable point since a small perturbation (of the camel or the perceiver) will change the perspective point to one of the two above-mentioned cases. The instable point – a singularity – separates the trajectory of the camel into two open, regular areas where the quality of the perceptual input remains the same throughout. No point in the regular areas gives any new information; this is only provided by the singular point; cf. Figure 1.[3]

It is natural to assume – even though this claim has not been proven – that the cognitive system has a built-in capacity for paying attention to the singular points, whereas the regular points are in the background or completely unnoticed.

1.2.

Catastrophe Theory is a mathematical theory for dynamic systems which has the following properties. The system can be represented by an internal space of huge dimensions in which the dynamics quickly develops into a stable state. Whether and how a system reaches one of its stable states is determined by a few external parameters.[4] For example, consider the perceptual/attention network of the human brain. The internal space consists of the neuronal activity patterns which are inaccessible to the observer but are determined by the external space/time representing the system's environment. In a portion of space/time in which there is no qualitative change one can ideally consider the neural activity as stabilized in a fixed state. As another example, consider the thermodynamic state of matter. Here the external parameters are *pressure* and *temperature*, and the internal space can be identified with the position and velocity of the molecules.

In the classical cases one can prove that the external space is partitioned into a discrete set of open areas in which the state of the system is structurally stable, meaning that small perturbations of the external parameters do not change the state of the system. The boundaries of the open sets define the so-called *catastrophe set*; here the system is structurally unstable since a perturbation of the external parameters might change the system qualitatively. Any change of state thus corresponds to a trajectory in the external space which crosses the catastrophe set. For instance, if liquid changes into gas this must be the result of a rise in temperature or a decrease in pressure. If attention changes non-willfully this might be the result of a sufficiently salient change in the external space.

In the last case it is of course impossible to get an idea of what the catastrophe set might look like since the internal dynamics is unknown and individual. It depends on how 'observant' the observer is. However, certain general properties of the conceptual system should, according to René Thom, 'map' onto some ideal partitioning of the space/time of the experienced environment. Human observers are attuned to the environment in such a way that their concepts directly reflect recurring patterns of discontinuous changes. Since language is a coding of the conceptual system, it should also be possible to classify certain types of verb phrases in view of such recurring patterns. Consider, for instance, the verb *fall*. In its prototypical meaning this verb might refer to a structurally stable process taking place between two singularities in space/time. Consider an ideal one-dimensional object standing perpendicularly on a surface. This is a structurally stable state. If the angle between the object and the surface begins to change then the onset of the change is an information-laden moment in time which grasps the attention of an observer. This is the first singularity. The second is when the object hits the ground and now lies in a horizontal position. This is also information laden since it defines new constraints in the environment. The vertical position, the process of falling, and the horizontal position constitute three structurally stable states separated by the two catastrophic points. In order to get a linguistic representation of this process it must appear in a recurrent pattern. The verb form is thus not a coding of any specific instance of the process but of the pattern as such.

To fall from one position to another is an instance of an *accomplishment* in Vendler's classification; cf. Vendler (1967). The description of a fall is to some extent representative for accomplishments in general. Especially for accomplishments that refer to events in physical space, since they are by definition true of a specific time interval, i.e., the reference of such an accomplishment is a portion of time bounded by two singularities. Compare classical examples for accomplishments such as *draw a circle*, where the process is bounded by the point of contact and the point of separation between the pencil and the paper. These singularities – point of contact and point of separation – are from an ideal point of view also present in *paint a picture*, *build a house*, etc. Since the attention network is attracted by the singular points, it seems natural that the category of processes bounded by two discontinuities should make up an 'action type' associated with a certain class of verb phrases. One can generalize this remark further: since human attention is attracted by observable discontinuities in external space, those discontinuities that show up again and again in human experience and thereby constitute patterns of experiences are likely to attract the verbal system. For instance, it is unlikely for any language to have a specific verb which means to heat water from 30 to 40 degrees, since the bounding points are not experientially salient, whereas verbs like *melt*, *boil*, etc. do refer to observable phase changes that appear again and again.

1.3.

René Thom argues for a realism of the sort one finds in Aristotle: events in space and time have a morphology that the human mind can appropriate. There is no doubt that the recognition of information-laden changes in the environment, which recur in stable patterns, plays a role for how humans conceive of occurrences. However, it is only at a very abstract level that one can classify verbs by means of singularities in space, probably at the same level of generality as Vendler's classification. To say that a process fills a bounded portion of time separated by two singularities does not say anything about the force-dynamic structure of the process. Or rather, it does not say anything about how humans impute a force-dynamic scheme onto the observed event.

To illustrate this point further let us take another look at the case of *fall*. Consider the following specific situation: a bucket is falling down from a scaffolding. In a catastrophe-theoretic framework this process has exactly the same structure as in the previous example of falling: a process of downward motion separated by two singularities. But in this case the force-dynamic scheme is different from the previous one. This is clear from the fact that the linguistic representation of the event must include the preposition *from*: – *the bucket fell down from* – or imply it contextually. If we look at the event independently of the linguistic phrasing, any observer's understanding of it will imply a series of micro-details of which the main ones are: 1) prior to the starting of the downward motion, the scaffolding supports the bucket and its position on the scaffolding is stable. 2) A disequilibrium causes the support to give way; this is the onset of the downward motion. 3) The falling object has force-dynamic implications, so that one has to avoid being hit by it. 4) A new stable state appears when the bucket hits the ground.

We can moreover assume that the observer's attention is attracted by the onset of the falling and by the spot on the ground where the bucket is likely to hit, whereas the details of the downward motion probably escape attention. We might further assume that the mind, from specific experiences like this, is able to extract a pattern. In other words, there is an entrenched, experientially based scheme in the mind of the observer. This scheme is abstract, independent of any figurative content, and for the case in question it consists of the points 1) to 4) above: *an equilibrium between an object and a supporting surface is broken; the support gives way, followed by a downward motion until the position of the object is stabilized at a new level.*

In English a code for this scheme is provided for by the verb form *fall from*. On its own the verb*fall* seems to focus on the downward motion and to leave the details concerning the starting point of the motion unspecified. This is an inversion between the foreground and background relative to the experiential situation in which it is the discontinuous changes that catch the attention of an observer, whereas the details of the process are unnoticed. The preposition *from*, on the other hand, focuses on the starting position on the supporting surface, i.e., it profiles 1) and 2) above.[5]

Whenever a person observes a process that on a sufficiently abstract level can be characterized by the above scheme, the verb form *fall from* might be evoked; for instance, in reporting the event. Conversely, on hearing the form *fall from* the hearer will be likely to evoke the scheme as part of his/her understanding of the reported event. This means that the scheme can be extended to other domains than the physical one. In particular, it means that processes whose ontology is unknown can be imputed a schematic meaning by using the form *fall from*. For instance, *he fell from power* evokes an initial situation in which the subject of the sentence is placed in an imaginary space (power space) that supports him and counteracts antagonistic forces in the same space. The balance is then broken, the imaginary space breaks down, and a downward motion is initiated.[6] The sentence profiles or highlights only the disruption of the balance and omits the stabilization on another level.

If only the verb *fall* is used, there will be a slightly different schematic meaning. For instance, in *her spirits fell* there is no explicit or implicit reference to an initial space as the starting point of the process. Instead, the sentence has an initial intra-psychic steady-state conflict between a tendency towards depression and an exertion that counteracts this tendency in its scope of meaning. The onset of the psychic process is then a disturbance of this equilibrium. Note that the two sentences do not explicitly refer to the force-dynamic implications brought forth here, they are the consequences of the schematic meaning evoked by the verbal forms *fall* and *fall from*.

2. ATTENTION

2.1.

It was mentioned above that for any external space/time within an observer's point of view, the discontinuities in space/time are determined by changes that attract the attention of the observer. It was also assumed that singularities appearing in a pattern, independently of single observers, play a crucial role for linguistic coding, such as for instance in the verbal system. It is clear, however, that there is no one-to-one relation between the explicit linguistic representation and the discontinuities that are salient in the experiential situation. For instance, one can imagine that in a given process there is one discontinuity that has a high load of information and therefore is more salient than others. If this is the case we encounter the phenomenon Talmy called a *window of attention*; cf. (2000e, pp. 265–279), which means that language might prefer to represent the most salient singularity explicitly and leave out the rest of the process. The omitted part is outside the focus of attention but still evoked as an integral aspect of the process. We have already seen examples of this with *fall from*, which focuses on the initial part and brackets the final position of the object.

For further illustration consider the verb *reach*. In one of its meanings this verb foregrounds the final point of contact resulting from a backgrounded bounded process of motion, i.e., the meaning is the foregrounded boundary of a backgrounded process, as in *he reached the summit. Reach the summit* is a classical example of an *achievement*; another of Vendler's categories. By its very definition an achievement profiles a specific instant. If this instant is the boundary of a previous process, then this process has to be part of the base meaning of the verb phrase.[7]

As in the case of accomplishments, the profile (background-foreground relation) of the verb phrase does not say much about the schematic meaning of the verb, or about the force-dynamic structure implied by the linguistic utterance. Why should one only pay attention to the final singularity? One reason could be that the history of the process is only of slight interest as long as the goal is achieved. In *he reached the train* we refer to an event in which the means by which he reached the train is unimportant compared to the fact that he reached it. Similarly, if one football team wins over another, then what happens previous to the referee's final whistle is of no importance for the result, although it may be important for the report on the match. This last example is prototypical for a whole class of verb phrases referring to situations in which the process is a non-uniform interaction between two agents or between an agent and a pseudo-agent, and where the outcome of the process is therefore undecided. In that case the final result represents, of course, a high load of information and will therefore attract the attention of any observer or any speaker reporting on the event. This might be the ontological background for the high frequency of verb phrases of the achievement type in language.

The force-dynamic pattern, found in the case of *reach* discussed above, consists in a protagonist with a goal and a resistance working against the protagonist. The verb phrases in question are those that direct attention to the point in time when the resistance recedes and the protagonist achieves his/her goal. However, the previous force-dynamic pattern is a necessary basis for understanding the achievement so it is also evoked by the phrase. This is clear in *he reached the summit*, where the summit is conceived as a resistance, but even in *he reached the train* the speaker seems to imply that the counterfactual state of not reaching the train was a possibility, and that the train was only reached with difficulty.

The force dynamic pattern described above can take many forms and it can be activated by different linguistic material other than the verb, most typically by a preposition. Consider the verb *break*. In the prototypical case this verb refers to an achievement of the punctual kind, cf. *the falling radio broke the vase*. Although the process is punctual we can still conceive it as a force-dynamic interaction in which the radio overcomes the resistance of the vase. In one of Talmy's examples, *the heat broke the guitar*, our understanding of 'heat' makes us conceive the breaking as the punctual termination of an extended process in which the resistance of the guitar gives way to the working of the heat. We might consider this as a general scheme projected onto many phenomena: *a force-dynamic process with a resistance that gives way, but where the observable result of the process appears in an instant*. This scheme can be activated by a preposition, as in *the sun broke through*, where *through* seems to impose a resistance prior to and receding at the instant the sun appears. However, for the observer only the appearance of the sun is accessible. The previous dynamic process is a pure fiction.

2.2.

Extended cases of the achievement kind are verb phrases in which only the final event is highlighted in a chain of events. There are two properties of the attention system that seem to account for this fact. Firstly, our attention works retrospectively in general, which means that we pay attention to the effect before we estimate the cause. Secondly, given an observed asymmetry in the environment, the human mind seems implicitly to assume that there is a causal story leading to that asymmetry. For this reason it is enough to mention the last event – causing the asymmetry – in order to evoke an implicit causal chain leading to it. The linguistic coding will often be in the form of an agent plus the ultimate event, as in *John broke the window*, where an entire causal chain of events is omitted: John taking a rock with his hands, then lifting it, swinging it with his arm, and propelling it through the air, etc. Compare Talmy (2000e, pp. 271–273).

Another trace of this feature of the attention system is the asymmetry between *before* and *after*; cf. Talmy (2000b, p. 321). As noted by Talmy one would rather say *the car exploded after he pressed the button than he pressed the button before the car exploded*. The reason is that the order of the sentences reflects the temporal order of our attention, which moves from the resulting events, from occurring discontinuities in space/time – a sound breaking the silence, a moving object, etc. – towards their causes. We thus see that a non-linguistic aspect of our interaction with the environment is reflected in the syntactic structure of our language.

3. FICTION

3.1.

By *fiction* I am referring to the existence of entities, actions, and morphology in human conceptualization that do not have any correlation in the external space/time. Fiction is ubiquitous in language. A well-known case is fictive motion; cf. Talmy (2000a). For instance, in *the light from the window across the room was hitting me square in the eyes*, we have fictive motion but also fictive action. The light is conceptualized here as an agent that affects a patient at a distance by throwing something at it. We might consider this as an experiential pattern, entrenched in the human mind as a dynamic scheme, and as such it can be projected onto events with the appropriate structure, irrespective of whether there is any actual throwing or not. For instance, in the example the window is conceived as the point of departure of the cause whereas the effect is in the eyes. Since there is a distance between the window and the eyes the cause-effect relation has to be conceived according to a scheme with this structure. The projection mentioned here is in fact a special case of a much more general metaphoric extension, which is called the *events are actions* metaphor in Lakoff/Turner

(1989). This metaphor expresses a general tendency to conceive a determining factor in an event as an agent. In the example the light is such a determining factor.

3.2.

As a combination of the realism of René Thom and the embodied realism of Lakoff allow me to propose the following.[8] How does the human mind get to a conceptualization of an event in space/time? Firstly, it requires the ability to extract a pattern out of the observations of recurrent discontinuities in space. Secondly, it requires the ability to bind the observed changes into a force-dynamic scene, i.e., the observer has to interpret the observations according to their dynamic implications. Those two aspects cannot be separated and together they constitute what is known as a *dynamic scheme*. Is the dynamic scheme something inherent to the observed phenomenon or is it something that the human mind projects onto the scene? I doubt that one can answer this question simply by choosing either Lakoff's embodiment approach or Thom's realism and applying it *tout court* to all dynamic schemes. In the embodiment tradition a dynamic scheme is essentially identical to a pattern of motor activity related to different kinds of actions. The body as a source of energy flow is thus considered the prototypical form of a causal relation. A body scheme like *throwing* can of course be projected onto pure physical events, such as the light affecting my eyes. In this way the hitting becomes a fiction but the causality is not a fiction. There are two observed discontinuities in space, the drawing of the curtain and the effect on the eyes, and there is an experienced causal relation between the two. The speaker might have said: *The light affected me badly because the curtain was drawn*. The embodied scheme is thus not necessary to express the causal relation, but it is a convenient way to express the same content in a more vivid imagery, and also in a simpler format.

In the works of René Thom a dynamic scheme is an aspect of real-world dynamics entrenched in the human mind through perception or experience. Dynamics in space/time is not chaotic but has a *form* that can be appropriated by the mind. Bodily activity has of course a dynamic form, but not all dynamics relate directly to motor activity. For instance, if a heavy falling object hits a fragile object the latter is likely to go to pieces. This is a dynamic scheme that can be experienced bodily but its work in a given context is not dependent on motor activity, although the two can work in conjunction, such as when we use a hammer to crush a shell. But even if we say that the dynamic schemes are aspects of the pheno-physical world we have to acknowledge their fictive status in some cases. We saw this in *the sun broke through the dark*, where 'the dark' is conceived as an antagonist showing resistance to 'the sun.' This probably reflects the phenomenology of the observer. To state this sentence, an observer must have been waiting in the dark prior to the appearance of the sun. The force-dynamic conflict between the sun and the dark is therefore a projection of an internal conflict in the observer between what he/she Expects – it to be light – and what he/she perceives – it is dark. As opposed to the 'hitting light' the projected scheme is thus not motivated by the form of the real-world causality but by the form of an intra-psychic dynamics.

4. DYNAMICS IN LANGUAGE VERSUS DYNAMICS IN PHYSICS

4.1.

An exhaustive analysis of how humans conceptualize dynamic processes would at least have to take real-world dynamics into consideration in addition to the special embodied perspective of humans. Furthermore, one would have to understand the metaphoric projections of dynamic schemes into domains where the ontology of the processes are unknown, as well as the specific viewpoint of the observer reporting on an event. It is thus not an easy task to determine whether a dynamic scheme is a fictive construction or not. In a special sense it is always fictive since it represents the phenomenon in a simplified manner: *the hurricane overturned the shed* seems at first to be an objective report, but on closer examination we see that the only information we get is that the hurricane is a (fictive) agent for the reported event.

We might therefore reflect on whether it is possible to obtain any 'objective' knowledge of physical processes or whether the notion of a dynamic scheme pertains merely to the schematization and conceptualization underpinning language. For instance, Lakoff makes the radical claim that anything we know in science is a direct extension via various kinds of metaphoric mappings of our basic cognition of space/time; see for instance Lakoff et al. (2000). This also goes for scientific physics. However, it should be clear from what has been presented so far that our conceptualization of dynamics as it is revealed in language is strongly at variance with how it is conceived in physics. Let me mention here the main examples of this incongruity; further discussion of this issue can be found in Talmy (2000c, pp. 456–458).

4.1.1. *Force*

It is well known that according to classical physics there is no inherent force in an object moving with constant velocity. If a billiard ball hits another billiard ball, no force is transmitted, only momentum of velocity. This is contrary to an ordinary understanding of the situation, according to which a ball can 'hit with great force', and the faster it goes, the greater its force. We see that cognition is more concordant with Aristotelian physics, in which a moving object contains impetus – an expression of its force. One of the reasons is that in cognition *force* is observer dependent in the sense that anything which can affect the observer has a force. In physics, on the contrary, the observer is left out so if there is no change in the state of the object then there is no resultant force either.

4.1.2. *Antagonist/agonist structure*

The terminology antagonist/agonist stems from Talmy (2000c) and refers to a conceived conflict between two agents. In the example *the sun broke through the fog*, the sun is the agonist and the fog the antagonist. However, this kind of distinction is totally absent in physics. One of the key words in scientific physics is symmetry and there is therefore no physical principle for differentiating between two equivalent objects. This is contrary to human conceptualization and social organization, in which asymmetric relations are pervasive.[9]

4.1.3. *Blocking*

As a consequence of 4.3, *blocking* is not an operative concept in physics because it refers to a steady-state dynamic interaction between two entities in which one of them – the blocking one – is conceived as stronger. However, if one of a pair of opposing forces is stronger than the other, then there must be a resultant force, which – cf. 4.2 – means that the system would accelerate in the direction of the stronger force. In human cognition, on the contrary, *stronger* refers to an agent's possibility of preventing an event from happening.

4.1.4. *Spatiotemporal wholes*

Language seems to partition space and time into wholes that have no meaning from the point of view of (traditional) physics. Take a notion like a *wave*. In human-scale phenomenology a wave is an important and recognizable bounded portion of space/time; especially if you are a surfer. In physics, on the other hand, we are dealing with a continuum of water molecules interacting at a finer scale of magnitude, where the principles for chunking are not phenomenologically relevant. In general, human conceptualization chunks space/time into discrete units that are easy to manipulate and that serve interaction with the physical and social space on a human scale. There is no reason to believe that this operation of chunking should correspond to anything one will find when examining the phenomena at a micro level. In *John sneezed the napkin off the table* it is enough to refer to an act – John sneezed – and a discrete transition – off the table – in order to convey the idea of a causal event, without having to mention causality, let alone specify the micro-level interaction between the air and the napkin.

4.1.5. *Autonomous events*

From the point of view of physics any occurrence is part of an unbroken causal continuum. That is, one will not find an isolated event that exists entirely outside of causality. Language, on the other hand, has specific constructions – called 'absolute construals' in Langacker (1991) – for representing events in isolation, such as for example *the lamp toppled and the guitar broke.*[10] Even if one adds a location, as in *the ball rolled across the green*, the event is still autonomous in the sense that the scope of the sentence does not include any conception of causality. The notion of an autonomous event – from Talmy (2000d) – is just a special case of a more general condition for representing causality in language. Although a sentence like *the falling tile killed the philosopher* expresses causality, the causal chain itself is partitioned out of a causal continuum; for instance, how the tile began to fall is outside the represented scope. This aspect of linguistic representation is a consequence of how the attention network works in general. Firstly, attention is drawn to the singularities that convey the most information, and they might gain prominence over the rest of the causal sequence: given that the philosopher got killed, mentioning how the tile began to fall was not immediately of interest. Secondly, a general condition for the cognitive system is that it has a restricted scope of attention, it not being possible to follow the causal chains indefinitely.

4.1.6. *Agency*

In scientific physics there are of course no agents. But we have seen that in human conceptualization it seems to be the norm that a determining factor in an event is also conceived as the agent. As mentioned earlier this is called the *events are actions* metaphor in Lakoff and Turner (1989); cf. the fictive hitting above. Another typical example is *time healed my wound*, where time is conceived as the determining factor in the event of healing, which is therefore construed as an act with time as agent. Even in expressions where the metaphoric meaning is less conspicuous, such as *the wind blew the napkin off the table*, we tend to consider the wind as an agent, but there are of course no physical principles that allow us to consider 'the wind' as a unit which acts specifically in regard to the napkin.

4.2.

Given the difference between how dynamics is conceived in cognition and in scientific physics, I shall briefly return to the question of the origin of this difference. The short version of the story is that it is due to mathematical representations of the dynamic processes. The invention of the decimal system in the 16th century made it possible to calculate the analytic expression of a number and then to compare analytic models with empirical findings. The invention of differential calculus made it possible to predict the development of a system given some initial conditions. The possibility of representing *force, velocity*, etc. as geometric entities made it possible to perform quantitative operations on these concepts, but now as geometric entities. This tendency has increased in the sense that in modern physics the physical concepts are equally mathematical concepts. The question of whether physics is closer to a true representation of forces than human cognition is therefore connected to the status of mathematics in its relation to the phenomenal world. I will not delve further into this problem since it is outside the scope of this paper.

5. AGENCY

5.1.

Let us finally look more closely at the notion of 'agency' and at its relation to a basic causative sequence. The following is a slight revision of Talmy (2000d, pp. 480–490). At first one might think that the most basic scheme for causality is a tripartite structure with a prior stable state, a discrete state transition, and a subsequent stable state; but on closer examination one can refine this a bit. In the example *the falling radio broke the vase* we see that such a causative consists of two autonomous events: a radio that is falling and a vase that breaks. This example has a structure that is general and that Talmy defines as a *basic causative event*. In the following I will deviate slightly from Talmy inasmuch as I define two different types of autonomous events. One type, denoted E_t, consists of a figure F_t and a translation in space G_t. In the prototypical version E_t is an event in which a figure-like entity moves in space. But to preserve the generality of this description we also include the cases in which F_t is a source and G_t is an emanation from F_t. For instance, in *the heat broke the vase* we implicitly

$$E_t = (F_t, G_t) \longrightarrow E_s = (F_s, G_s)$$

Physical causation

Figure 2.

assume that there is a source from which the heat emanates although this figurative part is backgrounded.[11]

Now consider another type of autonomous event called E_s. It also consists of a figure-like configuration, denoted F_s and a change in F_s's internal symmetry properties, which I name G_s.[12] This is the representation of events in which an object undergoes some kind of (internal) change. But any change of a figure is a change of its symmetry properties, either by the appearance or disappearance of an asymmetry. For instance, a vase that breaks, an indentation made in a bin, and an object put into motion are events that all can be described by the appearance of new asymmetries. Any causal chain therefore creates an asymmetry in relation to some figure-like entity. Conversely, the observation of any asymmetry in the environment causes the observer to infer some causal story leading to the observed asymmetry; cf. Leyton (1992). The last part is of course a consequence of the first: since causality leads to asymmetry it is an advantage for the cognitive system to pay attention to asymmetries in the environment.[13]

The general claim is now that in any basic causative event the causal factor is an event of the E_t-type and the effect is an event of the E_s-type. This comes about in the following way: G_t can be identified with a pattern of trajectories in space/time. If the trajectory of F_s intersects with G_t then G_s results. We can now refine Leyton's claim: whenever humans conceptualize something as an instance of causation it is based on the observation of an asymmetry, i.e., G_s is observed, and it is inferred that the causal factor has been an entity moving in space, i.e., G_t is inferred. For instance, if we observe an indentation in a bin we infer that it is the effect of some entity at some moment being moved towards the bin. Implicitly we reconstruct a basic causative event; cf. Figure 2.

This is not the place to go into details about the linguistic import of Figure 2, but one can mention that many verbs evoke as a minimum the schematic structure presented in Figure 2. For instance, *cut* evokes the idea of an instrument being moved relative to a figure on which it causes an asymmetry – a scratch or a hole is an asymmetry on a surface. Specifications of Figure 2 can be represented by specific verbs or by the addition of adverbial elements. For instance, in *cut through* the adverb specifies some topological properties of the figure, namely, that G_t has a two-dimensional structure – this is probably general for all meanings of *cut* – and the effect,

G_s, consists of a volume being divided into two. A verb like *pierce* also implies Figure 2 and in addition we get the information that G_t has a linear structure that continues throughout a volume F_s, thereby yielding the asymmetry G_s.

Let us now turn to the question of agency. Agency is so pervasive in human reality that it is difficult to imagine a causal chain that lies completely outside the scope of agency.[14] This means that linguistic constructions based only on Figure 2, such as in *the knife cut through the cheese or the hammer broke the glass*, do have a human agent in their scope of predication. If something happens accidentally or incidentally, then *accidentally* and *incidentally* evoke agent causality as counterfactual, exemplified by *he cut himself (accidentally) on the sharp knife*. But even in this case we can say that there is an agent for the necessary motion of a body part, i.e., there is an agent for E_t but not for E_s. This is probably general in the sense that in most cases there is an agent for the moving entity whether this is a body part or not, and if there is no agent then the moving entity is fictively ascribed some agent-like status; cf. *the wind has overturned a tree*.

Let us look at *Floyd broke the glass*. It is part of the scope of this sentence that there has been some body part moving prior to the breaking of the glass. To simplify matters we can then assume that G_t is the trajectory of the moving arm. Now if the glass broke as a result of a spasmodic movement we would not be able to say that Floyd was the agent of the act, so the motion of the arm must result from an act of volition. As noted by Talmy (2000d, p. 512): humans conceptualize volition as "the immediate and prior causing event to a body part's motion".[15]

Voluntary body motion is not enough though, as demonstrated in Talmy (2000d, pp. 510–515) with examples like *Floyd broke the glass in hitting it with his hand* and *Floyd broke the glass in pressing too hard on it*. These sentences show that the closed class forms *in* and *too* imply volition in relation to the motion of the body part, but they also imply the absence of intention concerning the final result. This means that the subject of the act has two relations to the physical chain: one of volition in regard to the body part motion and one of intention in regard to the final result, cf. Figure 3, where *S* denotes the subject.

In the sentence *Floyd broke the glass by hitting it with his hand* the closed-class element by implies that all relations in Figure 3 are activated. Between the mental event of intending to break the glass and the event of moving the body part there is a non-physically manifested relation which can be termed *mental causation*. Of course, there is an underlying neural causation but the terms used in the diagram are those that are operative in the human conceptualization of a basic causative event with an agent: the

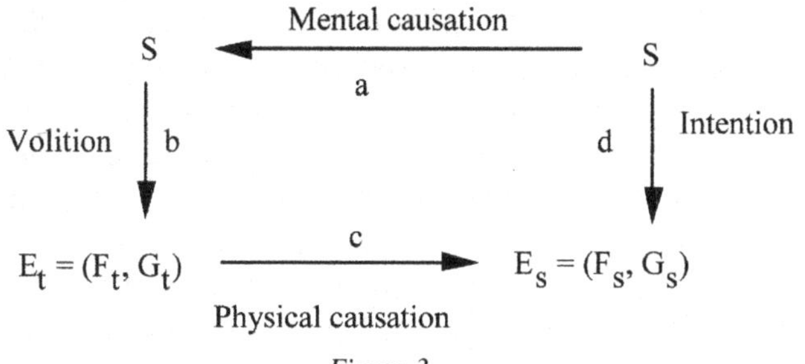

Figure 3.

physical causation (that of breaking), the volition implied in moving body parts, the intention of a human agent, and the mental process of letting one's intentions lead to voluntary movements of body parts.

The reality of the relation labeled a in the figure can be justified by an old thought experiment put forward by J. R. Searle: *X* intends to kill *Y* in order to get revenge, so he buys a gun and drives to *Y*'s place to kill him. On the way he hits *Y* in a crossing and kills him. According to Searle this does not satisfy the requirements of a revenge, exactly because the voluntary motion which *does* kill *Y* is not causally related to the intention of killing. In this example the relations *b*, *c*, *d* in Figure 3 are satisfied but *a* is omitted.

One can now extend Figure 3 to Figure 4.

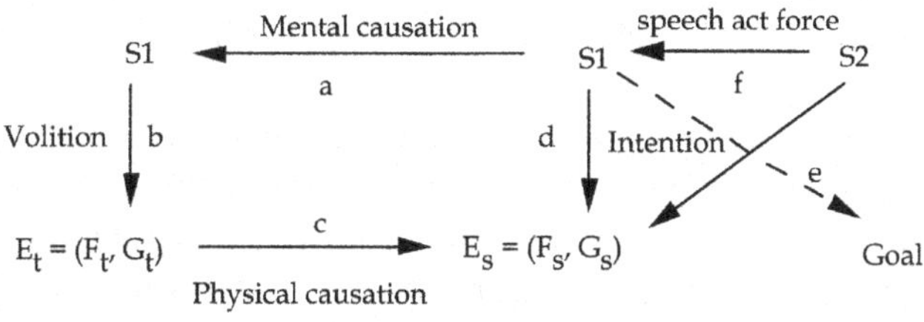

Figure 4.

Firstly, there is the relation labeled *e*, which signifies that the intention of *S* to realize the effect G_s can be justified by something that is not directly part of the physical causal chain. This is manifested by the form *in order to*; in fact, I used this form above when stating that *X* intended to kill *Y in order to* get revenge. I thereby signify that *X*'s killing of *Y* is not exactly the same as the revenge. The latter is a non-physically manifested reality which can be achieved through the killing.

Secondly, I have added another subject, *S*2, who is supposed to exert some speech-act force on *S*1 to realize G_s. There are three main linguistic manifestations of this force. *S2 urged S1 to do* G_s focuses only on the *f* relation in Figure 4; we do not know anything about the outcome of the urging. In *S2 persuaded S1 to do* G_s there is a focus on the f and d relations but the sentence does not state that $S1$ performs the necessary body part motion, because one can say something like *he persuaded her to leave but in the end she decided to stay*. Finally, we have *S2 got S1 to do* G_S. In this case there is a realization of the physical causal sequence, but one could perhaps say that this is a realization that goes via f, a, b, and c, since what motivates the mental causation is not so much $S1$'s intention but rather the speech-act force motivated by $S2$'s intention. *Get to* normally implies some reluctance on behalf of the acting agent.

6. Conclusion

The purpose of this paper has been to show that when speaking of a dynamic process, especially when causality is implied, it is necessary to distinguish between the phenomenological level at which human conceptualization is directed, and the level of physical reality at which the science of physics has directed its investigation. The obvious reason is that humans represent the phenomenal world in a very simplified schematic form which does away with all metric particularities of the physical process. And perhaps a deeper reason for the split between human conceptualization and physics also lies here, namely, that in physics the processes are represented by geometric spaces where it (in principle) is possible to make exact calculations on the differential equations that determine the development of the system.

Certain modern trends in physics work with alternative models in which there is a separation between the interactions occurring at the micro level and the macro- organization resulting from this interaction; cf. Petitot (1985, 1989). In Catastrophe Theory the micro-dynamics constitutes an internal space which causes a partitioning of the external space/time into discrete chunks. The external space can then be considered as a metaphor for how humans carve reality into chunks based on an unknown physical substrate. However, it is only a metaphor since the main motivations for human conceptualization do not have physical correlates. One of them is the notion of 'agency'.

The schema in Figure 3 does not show the factual parts of a causal event, but parts of our conceptualization of that event. By analyzing such conceptualizations one can contribute to systematizing and explaining certain patterns in the verbal system in English (and cognate languages). For instance, verbs that can take part in an absolute construal such as break - *the vase broke* – will always move the 'window of attention' onto G_S, whereas verbs that cannot be construed absolutely will direct the attention towards G_t – *John kicked the dog out of the room.* In the last case G_s has to be profiled by an adverb or a prepositional phrase, whereas in the first case G_t can be gapped.

NOTES

[1] This process is basically neural. Conceptualization is an activity of the brain responding to environmental constraints.

[2] Here I am referring to a mathematical representation of space/time. A singular point will then correspond to some region of interest in the speaker's reality-space.

[3] One can say that the topological qualities of perception remains the same, whereas the metric properties change, but it is a general feature of the cognitive system that it is more sensitive to topological than metric changes.

[4] By *state* I am referring human phenomenology, i.e., a *state* is something that is accessible for a human observer.

[5] It is probably a general function of adverbs to specify the force-dynamic point of departure for the process referred to by the verb.

[6] *Downward* is of course metaphoric - as is the meaning of fall – following the general pattern: up is control and down is out of control.

[7] There is of course the special degenerated case where the event is punctual.

[8] The two realisms do differ in the sense that for René Thom the scheme reflects objective structures in real space/time, whereas for Lakoff and cognitive linguistics in general it is a projection of the mind.

[9] In fact, asymmetric relations are probably essential for biological organization.

[10] In the absolute construal the object of a transitive sentence takes the subject's position. *John broke the window* vs. *The window broke.*

[11] The emanation is also gapped, except that it might be part of our schematic understanding of heat.

[12] Asymmetries are scars on the surface of the moon, chips on vases, graffiti on subway trains, etc. Leyton (1992, p. 7) presents the following abstract definition: *asymmetry is the memory that processes leave on objects*and the dual version: *symmetry is the absence of process-memory.*

[13] A singularity also marks an asymmetry, so there is a family resemblance between Leyton's claim and René Thom's theory of singular points.

[14] This is probably the main difference between causality as conceptualized by humans and causality as it is conceived in physics.

[15] As in the physical case there is a micro-level of dynamic interactions that plays no role in the way the human mind conceives the situation at the macro-level. In this case we have an extended chain of neural and muscular events which are comprised by the notion of voluntary movement. This level of neural and muscular activity belongs to the internal space in the catastrophe-theoretic approach to meaning.

Lakoff, G. and M. Turner: 1989, *More Than Cool Reason*, Chicago: University of Chicago Press.

Lakoff, G. and M. Johnson: 1999, *Philosophy in the Flesh*, New York: Basic Books.

Lakoff, G. and R.E. Núñez: 2000, *Where Mathematics Comes From*, New York: Basic Books.

Langacker, R.W.: 1991, *Foundations of Cognitive Grammar*, Vol. II, Stanford: Stanford University Press.

Leyton, M.: 1992, *Symmetry Causality Mind*, Cambridge MA: MIT Press.

Petitot, J.: 1985, *Morphogenèse du sens*, Paris: Presses Universitaire de France.

Petitot, J.: 1989, 'Hypothèse localiste, modèles morpho-dynamiques et théorie cognitive; remarques sur une note de 1975', *Semiotica* **77**(1-3), 65–119.

Spelke, E.S., A. Phillips, and A.L. Woodward: 1995, 'Infants' knowledge of object motion and human action', in Sperber, Premack, and Premack (eds.), *Causal Cognition*, Clarendon Press, pp. 44–78.

Talmy, L.: 2000a, 'Fictive Motion in Language and "Ception"', in *Toward a Cognitive Semantics*, Cambridge, MA: MIT Press, pp. 99–176.

Talmy, L.: 2000b, 'Figure and Ground in Language', in *Toward a Cognitive Semantics*, Cambridge, MA: MIT Press, pp. 311–344.

Talmy, L.: 2000c, 'Force Dynamics in Language and Cognition', in *Toward a Cognitive Semantics*, Cambridge, MA: MIT Press, pp. 409–470.

Talmy, L.: 2000d, 'The Semantics of Causation', in *Toward a Cognitive Semantics*, Cambridge MA: MIT Press, pp. 471–550.

Talmy, L.: 2000e, *Toward a Cognitive Semantics*, Cambridge MA: MIT Press.

Thom, R.: 1972, *Stabilité structurelle et Morphogenèse*, New York: Benjamin.

Thom, R.: 1980, *Modèles mathématiques de la Morphogenèse*, Paris: Christian Bourgois.

Vendler, Z.: 1967, *Linguistics in Philosophy*, Cornell University Press.

ERIK CHRISTENSEN

OVERT AND HIDDEN PROCESSES IN 20TH CENTURY MUSIC

ABSTRACT. For the purpose of contributing to a clarification of the term "process", different kinds of musical processes are investigated: A rule-determined phase shifting process in Steve Reich's *Piano Phase* (1966), a model for an indeterminate composition process in John Cage's *Variations II* (1961), a number of evolution processes in György Ligeti's *In zart fliessender Bewegung* (1976), and a generative process of fractal nature in Per Nørgård's *Second Symphony* (1970). In conclusion I propose that six process categories should be included in a typology of processes: Rule-determined, goal-directed and indeterminate transformation processes, and rule-determined, goal-directed and indeterminate generative processes.

The term 'process' is fairly indeterminate in its meaning. The purpose of the present paper is to draw attention on different kinds of musical processes which may serve as examples of distinct process categories, thus contributing to a clarification of the term.

Musical processes are experimental devices for composers. For the purpose of investigating unknown or relatively unknown musical and perceptual possibilities, a composer may invent the conditions for the realization of a certain process, consisting of a basic material and a number of rules or goals. The unfolding of the intended process will then provide the composer with a musical result which he may find satisfying or uninteresting, predictable or surprising, dull or promising. Subsequently, the composer has a new chance to adjust or refine the process in order to pursue the desired result.

I will focus on four works by composers who have in very different fashions applied the investigation of musical processes in the evolution of their creative work, the American composers Steve Reich and John Cage, and the Europeans György Ligeti and Per Nørgård.

J. Seibt (ed.), Process theories: Crossdisciplinary studies in dynamic categories, 97–117.

For a period of five years, 1965–1970, Steve Reich concentrated his work on the invention of musical processes, and presented his viewpoints in the manifesto-like essay 'Music as a Gradual Process' (1968, printed 1974)

I am interested in perceptible processes. I want to be able to hear the process happening throughout the sounding music.

To facilitate closely detailed listening, the musical process should happen extremely gradually.

Performing and listening to a gradual musical process resembles: pulling back a swing, releasing it, and observing it gradually come to rest; turning over an hour glass and watching the sand slowly run through to the bottom; placing your feet in the sand by the ocean's edge and watching, feeling, and listening to the waves gradually bury them.

Though I may have the pleasure of discovering musical processes and composing the musical material to run through them, once the process is set up and loaded, it runs by itself (Reich 1974, p. 9)

The distinctive thing about musical processes is that they determine all the note-to-note details and the overall form simultaneously. One can't improvise in a musical process – the concepts are mutually exclusive. (1974, p. 11)

Steve Reich's first process compositions were based on tape loops. After having assembled some tape collages of recorded sounds, he got the idea of experimenting with tape loops from his fellow composer Terry Riley. Reich experienced that when he played identical tape loops on two tape recorders, repeating the same sounds over and over again, the machines would gradually come out of synchronization with each other, producing unforeseen rhythmic patterns. He then gained control of this sounding phenomenon by starting two tape loops in unison and letting them slowly slip out of phase, and produced two tape compositions based on the phase-shifting repetitions of a few words of recorded speech, *It's gonna Rain* (1965) and *Come Out* (1966). Reich was struck by the resulting rich variety of rhythms and sound transformations, and described his experience:

As I listened to this gradual phase shifting process, I began to realize that it was an extraordinary form of musical structure. This process struck me as a way of going through a number of relationships between two identities without ever having any transitions. It was a seamless, continuous, uninterrupted musical process. (Reich 1974, p. 50)

Steve Reich's next step was to transfer the phase shifting process to live instrumental performance. He staged the process by recording a tape loop of a short repeated melodic pattern played on the piano, and subsequently trying to play the same pattern live against the tape loop, changing his tempo gradually. This experiment turned out to be successful and satisfying, and formed the basis of Reich's composition for two live keyboards, *Piano Phase* (1966).

The process of *Piano Phase* is defined by a basic material and a few rules. The material consists of a mere five different pitches, e b d and f# c#, distributed in a twelve-note pattern (Figure 1, bar 1).

Figure 1. Steve Reich: piano phase, bars 1-6.

The rules are:

(1) Both performers play the pattern over and over again. One performer starts, the other fades in in unison (bars 1–2).
(2) The first performer keeps a constant tempo. The other performer gradually increases his tempo, until he is one note ahead of the first performer (bar 3)
(3) After playing in synchronization for a while, the second performer again begins increasing his tempo, and the phase shifting process starts again (bars 3–4)

In the first part of Piano Phase, this procedure is repeated twelve times.

CD reference 1: Reich

Reich was fascinated by the surprising effects:

The use of hidden structural devices in music never appealed to me. Even when all the cards are on the table and everyone hears what is gradually happening in a musical process, there are still enough mysteries to satisfy all. These mysteries are the impersonal, unintended, psycho-acoustic by-products of the intended process. These might include sub-melodies heard within repeated melodic patterns, stereophonic effects due to listener location, slight irregularities in performance, harmonics, difference tones, etc. (Reich 1974, p.10)

The auditory effects of the phase-shifting process in 'Piano Phase' have been described in an article by Paul Epstein:

Where, then, are the mysteries ? They are in fact numerous and stem in part from the fact that while the process is continuous, our perception of it is not. The listener is presented with a rich array of possibilities out of which he/she may construct an experience of the piece. (...)
The phasing process begins with a movement away from unison. Although continuous, it is heard in several distinct stages. At first the impression is of increasing resonance, a change in acoustic quality only. At the next stage one begins to hear the voices separate: echo replaces resonance. At a certain point the irrational division of the beat caused by the echo presents a dizzying rhythmic complexity. When the voices are nearly 180 degrees, or one half beat, out of phase, a doubling of the tempo is perceived; one has a momentary sense of stability, of a simplification of the irrational rhythmic relationship heard previously. This stage is very brief and is one of those events that seem to occur suddenly. The out-of phase quality quickly returns and lasts until the new phase locks in. (Epstein 1986, pp. 497–499)

CD reference 2: Reich

Epstein concludes that the listener's task in experiencing process music is also one of discovery, and that the listener may focus his attention in different ways by adopting different listening strategies.

Reich's phase shifting pieces are examples of determinate processes, unfolding from a clearly defined initial material and a set of transformation rules. By introducing this kind of rigorously structured repetitive music, Reich was revolting against two prevailing trends on the New York music scene in the 1960's, the complexity of serial music and the indeterminate music of John Cage. Reich's alternative can be denoted as 'determinate simplicity'.

JOHN CAGE: *VARIATIONS II* (1961)

In 1958, Cage had introduced his ideas of indeterminacy in Darmstadt, Germany, in his lecture entitled 'Composition as Process', printed in *Silence* (1961). This lecture is a manifesto from a radical period in Cage's life, when he had dismissed the ideas and methods underlying earlier works such as numerical structures, considered improvisation, unambiguous notation and preconceived form. Here, Cage presents his ideas of 'non-intention':

> This is a lecture on composition which is indeterminate with respect to its performance. That composition is necessarily experimental. An experimental action is one the outcome of which is not forseen. (Cage 1961, p. 39)

> The early works have beginnings, middles, and endings. The later ones do not. They begin anywhere, last any length of time, and involve more or fewer instruments and players. They are therefore not preconceived objects, and to approach them as objects is to utterly miss occasions for experience... (1961, p. 31)

> ... constant activity may occur having no dominance of will in it. Neither as syntax nor structure, but analogous to the sum of nature, it will have arisen purposelessly. (1961, p. 53)

Cage aims at music which is unforeseen and purposeless, music which lets the sounds be themselves, free from the intentions and expressions of the composer's mind. Models for the realization of this kind of music are found in Cage's *Variations I* (1958) and *Variations II* (1961). The *Variations* are not scores, but 'compositional tools' which present procedures for creating scores or successions of musical events. Any trace of preconceived composition has disappeared, and the work exists solely as a model for a process (Pritchett 1993, p. 126).

The materials of *Variations II* consist of a set of transparent sheets with dots and lines, to be arranged in any order and interpreted as musical parameters by the performers. This is the beginning of Cage's instructions for *Variations II:*

> Six transparent sheets having single straight lines. Five having points. The sheets are to be superimposed partially or wholly separated on a suitable surface. Drop perpendiculars by means of any rule obtaining readings thereby for (1) frequency, (2) amplitude, (3) timbre, (4) duration, (5) point of occurrence in an established period of time, (6) structure of event (number of sounds making up an aggregate or constellation). A single use of all sheets yields thirty determinations.

In order to illuminate the process defined by Cage, I have produced two short and simple versions for piano of *Variations II*. I have added the following rules for the realization: (a) The established period of time is 45 seconds, divided in three parts of 15 seconds. (b) Each composition consists of five musical events. (c) Each event is defined by Cage's parameters 1, 2, 3, 4 and 5. (No. 6 is omitted for the sake of simplicity). (d) Each of the parameters may adopt three different values.

The practical procedure is the following: (1) Five transparent sheets having straight lines and five having dots are superimposed on the glass plate of a copying machine, and a copy is taken. (2) On the copy, the lines

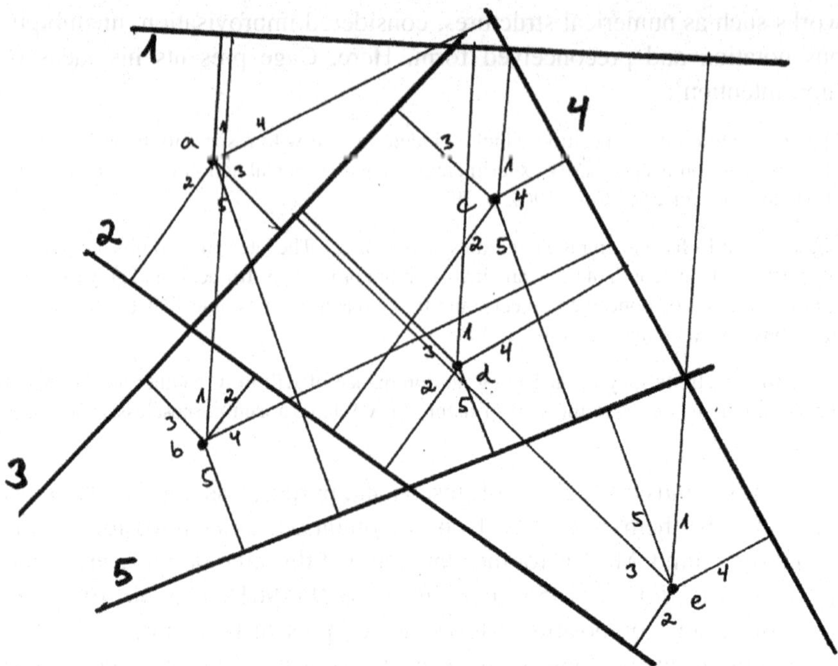

Figure 2a. John Cage: *Variations II*, Version 1.

	1. Pitch	2. Volume	3. Timbre	4. Duration	5. Occurrence
Event					
a	middle	middle	keyboard	long	last part
b	high	low	strings	long	middle
	middle	high	strings	short	middle
d	high	high	strings	short	middle
e	high	low	knocking	short	middle

Figure 2b. John Cage: *Variations II*, Version 1: Properties of five musical events resulting from the application of Cage's instructions and a number of additional rules.

are numbered 1 2 3 4 5, each representing a musical parameter, and the dots are marked a b c d e, each representing a musical event. (3) From each dot, perpendiculars are drawn to each line, and marked 1 2 3 4 5 (Figure 2a).

(4) The length of each perpendicular is estimated as 'short', 'middle' or 'long', and these three categories are interpreted as parameter values for: (1) Frequency (pitch): low/middle/high. (2) Amplitude (volume): low/middle/high. (3) Timbre: keyboard/strings/knocking. (4) Duration: short/middle/long. (5) Point of occurrence within the period of time: first part/middle/last part. This reading of the perpendicular lines results in the definition of five musical events shown in Figure 2b.

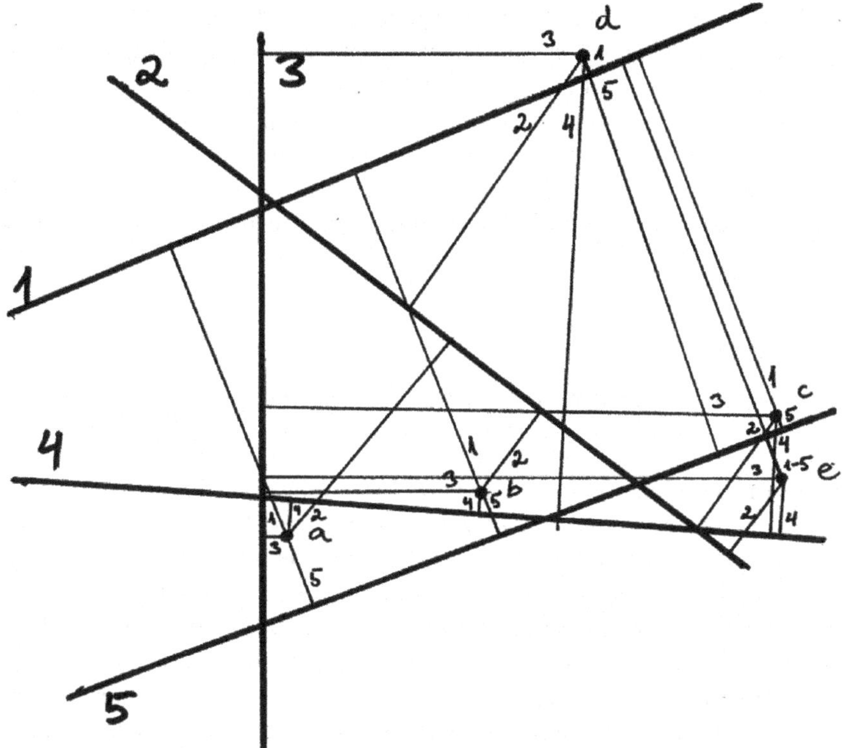

Figure 3a. John Cage: *Variations II*, Version 2.

	1. Pitch	2. Volume	3. Timbre	4. Duration	5. Occurrence
Event					
a	high	middle	keyboard	short	first part
b	high	low	strings	short	first part
c	high	middle	knocking	short	first part
d	low	middle	strings	long	last part
e	low	low	knocking	short	last part

Figure 3b. John Cage: *Variations II*, Version 2: Properties of five musical events.

The performer enjoys considerable liberty in the realization of the events. In this version, no events occur in the first part (15 seconds of silence). In the middle part, the strings are struck, plucked, or scraped 3 times in high, middle, and high pitch range, and the wood of the piano is knocked once. In the final part, a long sound (tone or cluster) is played at the middle of the keyboard. For the realization of a performance, the events may be notated in a performance score, applying appropriate symbols.

Figures 3a and 3b display another version of *Variations II* and the properties of the resulting musical events.

In this version, three different events occur in the first part: keyboard, knocking and strings in the high register. The middle part is silent, and the piece ends with a short knock and a long string sound.

These are two examples out of an infinite multiplicity of musical events and structures which may result from the superposition of transparent sheets having lines and dots. Cage's *Variations II* is a model for indeterminate processes, designed to produce unforeseen results from random inputs.

CD reference 3: Cage

GYÖRGY LIGETI: *IN ZART FLIESSENDER BEWEGUNG* (1976)

In his indeterminate works, John Cage has arranged his composing means so that he has no knowledge of what may happen. The composer György Ligeti works in the opposite manner. He relates that his music takes shape in his imagination as a sonorous form, and that he is able to listen to the piece from beginning to end in his mind (Ligeti 1971). His next step, then, is to invent constructions which can transform his inner vision into a web of musical relations and connections. These constructions will often be conceived in the form of carefully designed musical processes. One striking example is Ligeti's piece for two pianos, entitled *In zart fliessender Bewegung* (In a Gently Flowing Movement), which is the last of the three pieces for two pianos *Monument, Selbstporträt, Bewegung* (1976). This piece is a meticulously woven web of arpeggiated piano tones, often moving in canons or mirror canons, an incessant flow of rise and fall, fluctuation and acceleration, expansion and contraction. The beginning of the score is shown in Figure 4.

In order to render the ongoing processes in this piece intelligible, I have visualized the musical flow in the first 53 bars of Ligeti's piece in a graph (Figure 5). In the graph, the vertical axis represents pitch, the divisions in equal steps corresponding to the chromatic scale. The horizontal axis represents time, each unit corresponding to one quarter note in the score. Four quarter notes make up one bar of music. The bars are numbered in accordance with the score.

Each vertical column in the graph shows all tones (between 6 and 12) played by both pianists within the quarter-note time unit. Each short tone of the piano arpeggios is plotted in the graph as a black dot. A tone which occurs twice within the time unit is marked as a white dot, and a tone which occurs three times is marked as a white dot with a point. Accented short notes are marked as small triangles, strong accents (sforzando) as large

Figure 4. György Ligeti: In zart fliessender Bewegung, bars 1–8.

triangles. Longer notes are drawn as squares or lines. The graph displays the temporal and spatial evolution of the music.

CD reference 4: Ligeti

A detailed analytical description of this music has been provided by Herman Sabbe (1987, pp. 37–48). With reference to Sabbe's analysis, the following musical processes can be pointed out:

1. *Expansion and contraction of the pitch space*: Expansion in bars 1–14, contraction in bars 15–22. Renewed expansion from bar 25, reaching the upper and lower limits of audible pitch in bars 51–53.
2. *Increase and decrease of interval size*: The musical intervals are enlarged and diminished in accordance with the overall expansion and contraction throughout bars 1–36. Then follows a gradual diminishing of intervals, as the music moves towards the extremes in bars 51–53. These processes are heard as changes and fluctuations of harmonic color.
3. *A hollowing out of the total soundspace* begins in bar 39, to be followed by a filling of the void by a succession of slow chords in bars 49–63. (The chords are not shown in the graph)
4. *Increase and decrease of tempo*: From the beginning until bar 48, a stepwise increase of the number of tones per time unit takes place, perceived as increasing tempo. This is reinforced by a gradual accel-

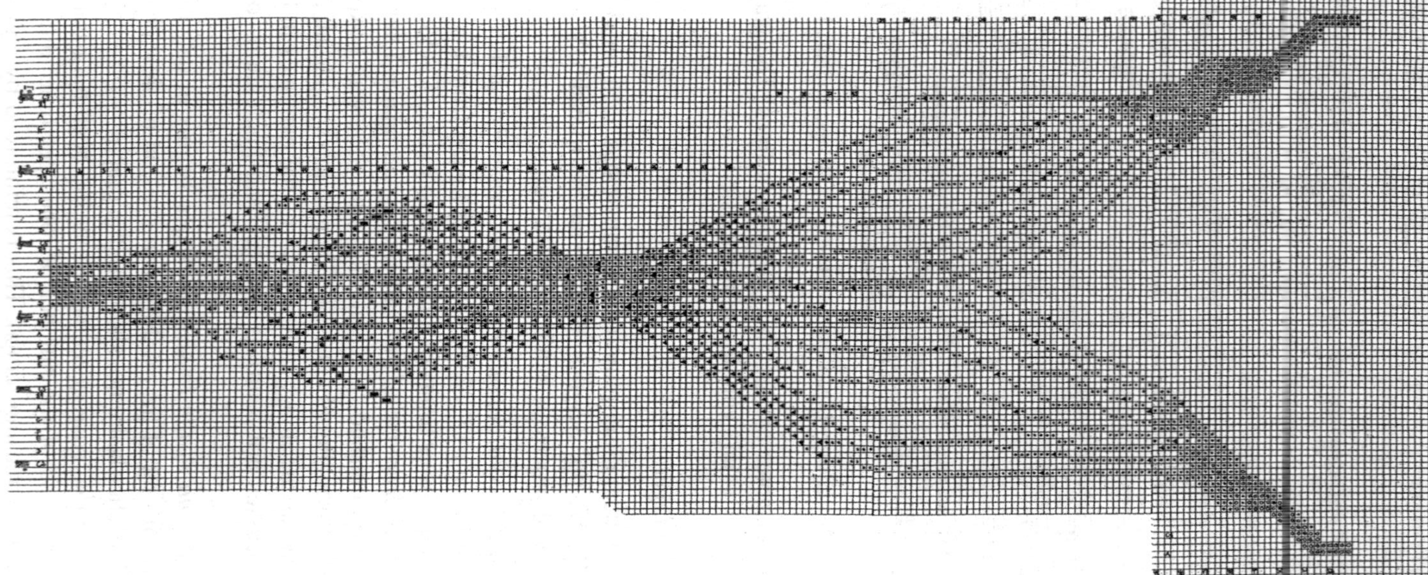

Figure 5. Graph of the temporal and spatial evolution in Ligeti's *In zart fliessender Bewegung*.

erando in both instruments in bars 31–48, leading to a culmination in the trill in bar 52, whereupon a sudden shift to slow tempo takes place.

5. *Increase and decrease of volume*: A gradual crescendo follows the accelerando from bar 31, culminating in bars 47–48, then cut off by a subito pp in bar 49.
6. *Emergence of patterns of accents and sforzato accents*: In the flow of fast tones, certain accented tones stand out like clear points on a blurred background. The moderate accents articulate the overall process by accentuating the emergence of new tones in the flow, as well as the outlines of expansion and contraction. The succession of strong sforzato accents, beginning in bar 21, constitute independent patterns of melody, which display a canonic structure (Febel 1978).

Altogether, the unfolding of the interrelated processes in Ligeti's *In zart fliessender Bewegung* represents the audible realization of the composer's vision of a musical form. As a means of attaining a preconceived compositional goal, they can be characterized as goal-directed processes. The rules or principles guiding these processes are not meticulously defined, but clearly observable as predominant tendencies in the evolution of the process.

PER NØRGÅRD'S *INFINITY SERIES*

My final example of musical processes is a generative process of fractal nature which creates multiplicity out of an initial unity. When the Danish composer Per Nørgård was investigating principles of organic growth in the 1960's, he discovered a tone series which possesses unique properties of self-similarities, repetitions and symmetries. Nørgård named it the 'infinity series', and continued to explore its properties for several decades. The infinity series is generated by mirroring an initial interval symmetrically downwards and upwards, thus producing new intervals which, by repetition of the mirroring procedure, form an endless succession of further intervals. The generation of the series is explained in Figure 6.

The unfolding of the first 64 tones of the series, displayed in Figure 7, shows a remarkable emergence of self-similar figures, symmetries and interrelations. The series generates a repeated succession of figures, which can be described as M- and W- shapes separated by rising axes: M/M/W/M. The interval ranges of the M- and W- shapes correspond two by two, and numerous other intervallic relationships between the figures can be observed.

Generation of Nørgård's infinity series

a) Two tones are given, 1 and 2.
Tone 1 is the beginning of a lower voice of odd-numbered tones, drawn in black. Tone 2 is the beginning of a higher voice of even-numbered tones, drawn in white.

New tones are generated by the projection of a previous interval; each interval is projected twice. In the lower voice, the interval is projected in the opposite direction. In the higher voice, the interval is projected in the original direction.

b) Generation of tones 3 and 4
The interval 1-2 is a rising semitone. In the lower voice, it is projected as a falling semitone, producing tone 3. In the higher voice, it is projected as a rising semitone, producing tone 4.

c) Generation of tones 5 and 6
The new interval 2-3 is a falling whole tone. In the lower voice, it is projected as a rising whole tone, producing tone 5. In the higher voice, it is projected as a falling whole tone, producing tone 6.

d) Generation of tones 7 and 8
The new interval 3-4 is a rising minor third. In the lower voice, it is projected as a falling minor third, producing tone 7. In the higher voice, it is projected a a rising minor third, producing tone 8.

e) Generation of tones 9 and 10
The new interval 4-5 is a falling semitone. Projections produce tones 9 and 10.

f) Generation of tones 11 and 12
Produced by projections of the new interval 5-6, a falling semitone.

g) Successive generations
Produced by successive projections of new intervals.

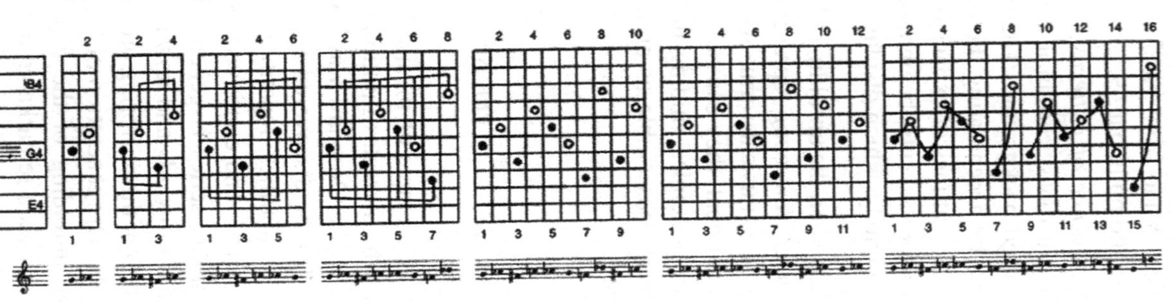

Figure 6. Generation of Nørgård's *Infinity Series*. Reproduced from Christensen (1996, Vol. II, p. 59).

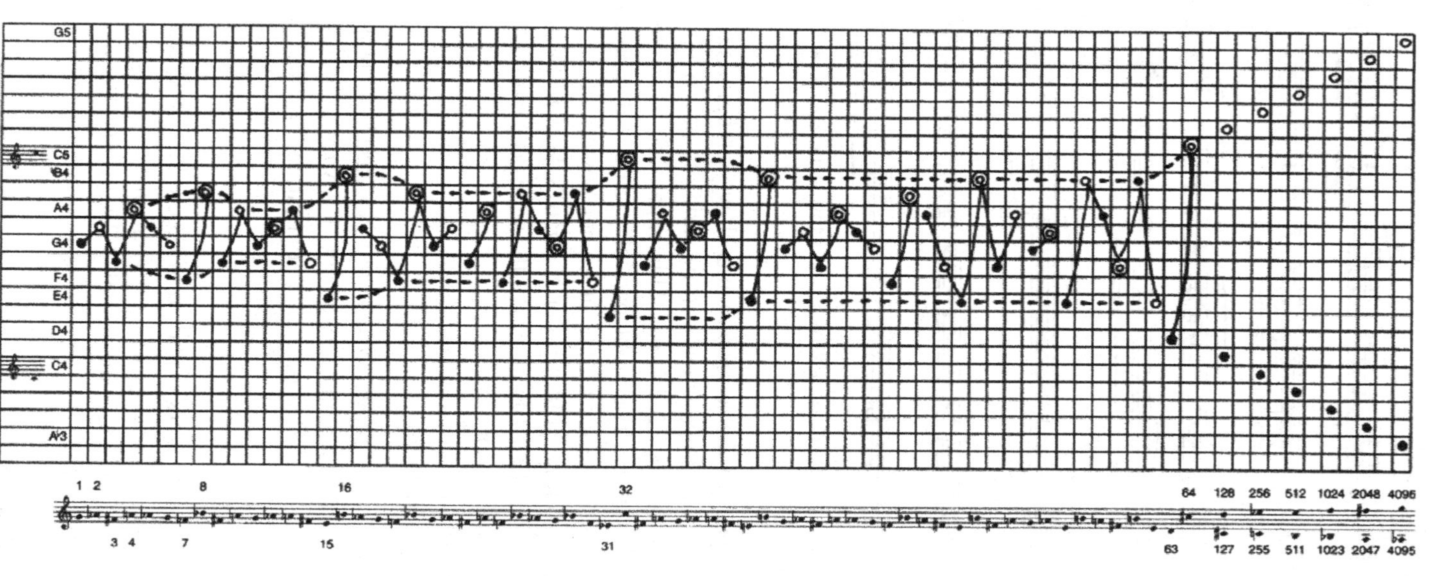

Figure 7. Unfolding and expansion of the *Infinity Series*. Reproduced from Christensen (1996, Vol. II, p. 58).

Figure 8. *Voyage into the Golden Screen*, 2nd Movement, bars 1–4. Reproduced from Kullberg (1996, p. 79).

The pitch space of the series expands slowly and gradually, adding one step up and one step down whenever the series comes to a power of 2, that is, at tones 2, 4, 8, 16, 32, 64, etc. This expansion is continued infinitely and ever more slowly. The right part of Figure 7 displays the intervals generated when the series arrives at higher powers of 2: 128, 256, 512, 1024, 2048, up to tone 4096. In Per Nørgård's *Second Symphony* (1970), the unfolding of the first 4096 tones of the series is heard in different layers of varied instrumentation.

CD reference 5: Nørgård

A few years before the *Second Symphony*, Nørgård had composed a shorter version of the same music in a movement of *Voyage into the Golden Screen* (1967), encompassing 1024 tones in a transparent instrumentation.

CD reference 6: Nørgård

The beginning of this work is shown in Figure 8.

In the first bars of *Voyage into the Golden Screen*, some characteristic self-similarities of the series are revealed: Flutes play the basic series (Figure 7). Oboes play tones 4, 8, 12, 16, ... which make up a transposition of the series (Figure 9). Clarinets play tones 2, 4, 6, 8, ... forming another transposition of the series (Figure 10). Horns play tones 1, 5, 9,

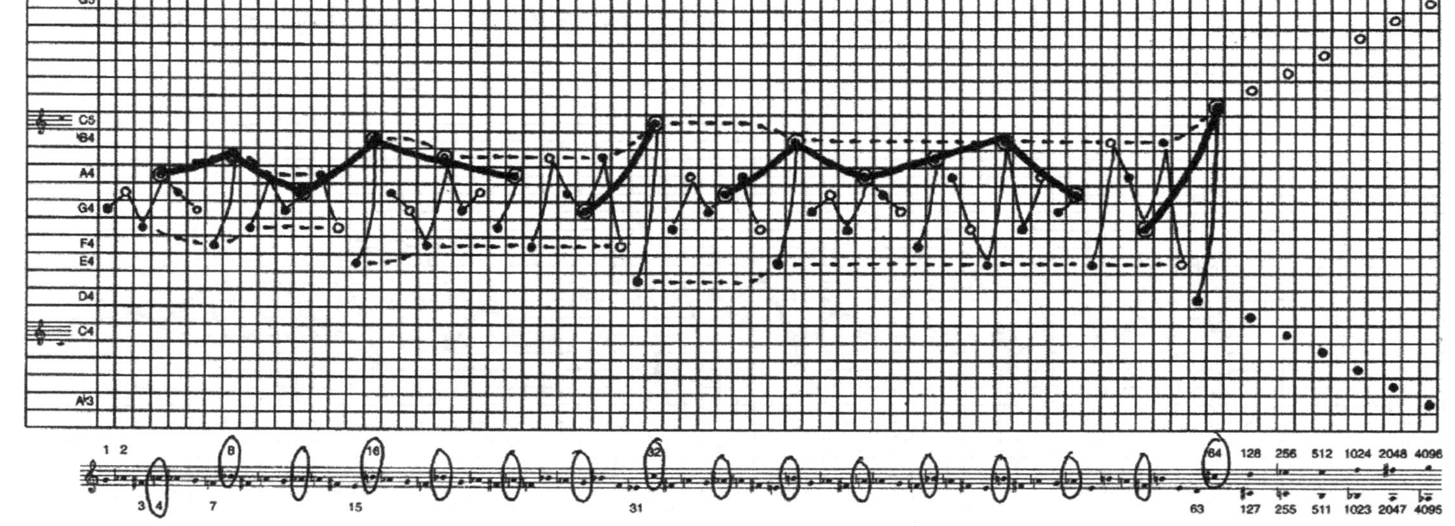

Figure 9. Tones 4, 8, 12, 16, . . . constitute a transposition of the series.

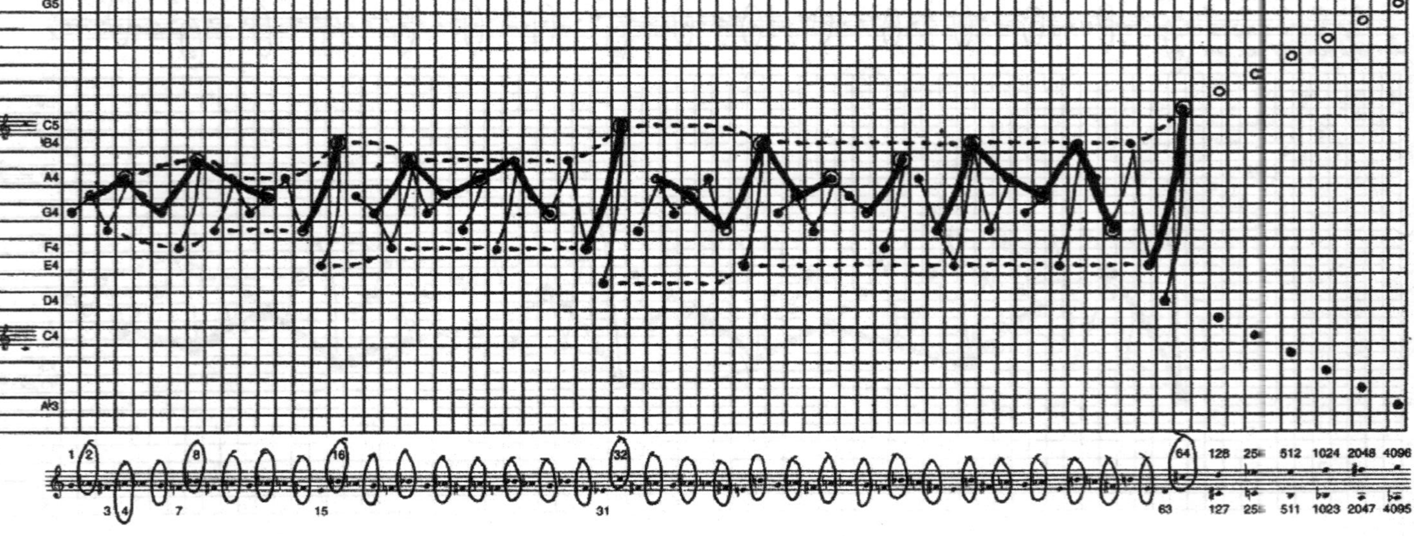

Figure 10. Tones 2, 4, 6, 8, ... constitute another transposition of the series.

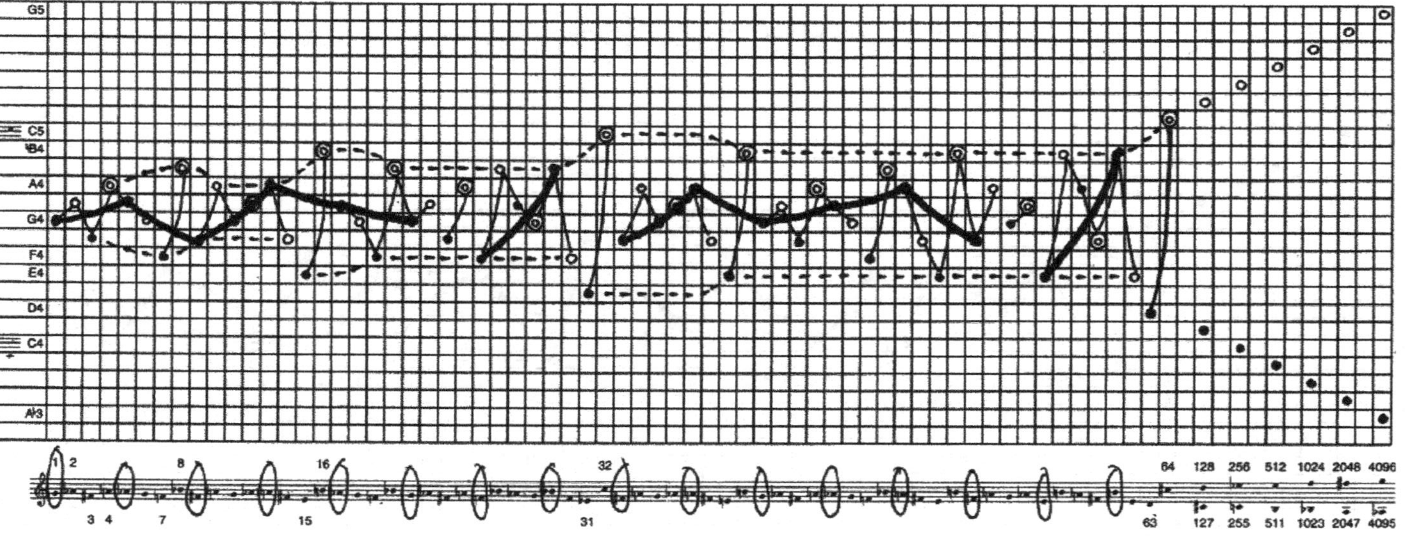

Figure 11. Tones 1, 5, 9, 13, . . . constitute a replica of the series.

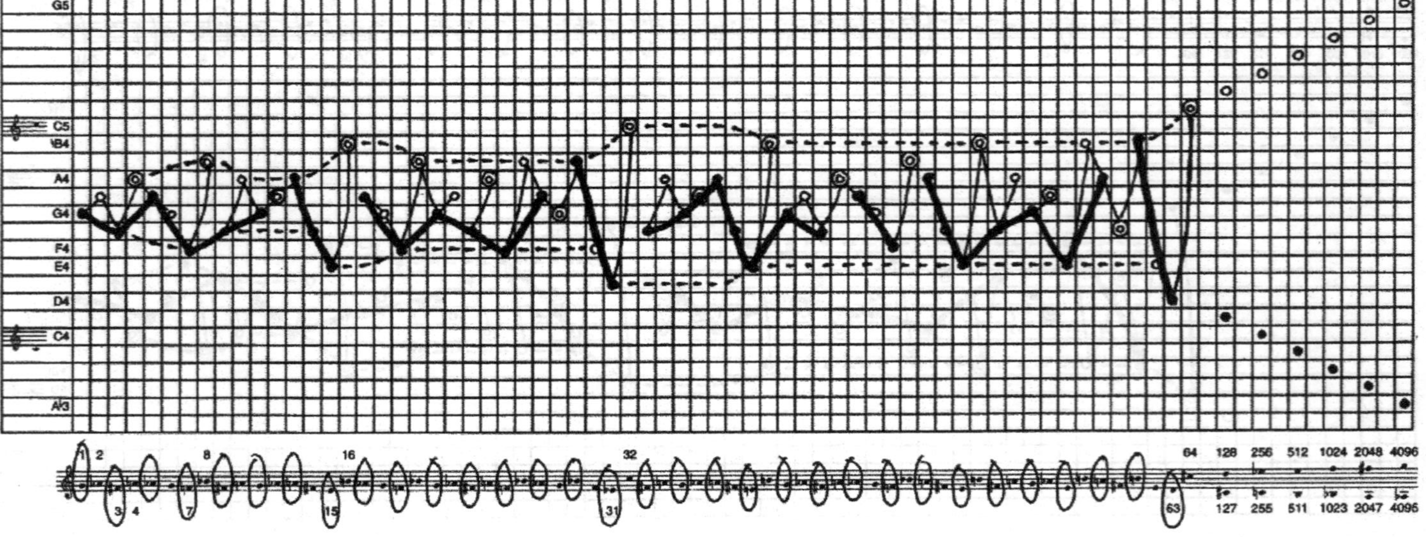

Figure 12. Tones 1, 3, 5, 7, 9, . . . constitute a mirror image of the series.

1 (-2) 3 (-1) (-1) (-2) 5 (-4) 3 (-2) 1 1 (-3) (-2) 7
3 (-1) (-2) 5 (-3) 1 (-2) 3 (-4) 5 (-2) (-1) 3 (-5) (-2) 9 .... etc.

Figure 13. The beginning of the infinity series represented by numbers .

transformation process	rule-determined goal-directed indeterminate
generative process	rule-determined goal-directed indeterminate

Figure 14. Six process categories.

13, ... constituting a four times slower replica of the basic series (Figure 11). In these first measures and throughout the work, the infinity series is omnipresent in different self-similar transpositions and different tempo layers. By the mere presence of the series, time is stretched, compressed and stratified. For example, tones 16, 32, 48, 64, 80, 96, ... will form a new, slower transposition. A further characteristic property of the series is its self-mirroring, as seen in the shapes constituted by tones 1, 3, 5, 7, 9, ... (Figure 12) or tones 1, 9, 17, 25, 33, ...

The infinity series, discovered in the 1960's, is an early example of a generative fractal process, a process which reproduces and mirrors its own shapes and structures infinitely in different orders of magnitude. From the emerging shapes of this series, it is possible to extract innumerable melodies and melodic fragments. Per Nørgård has included selected features of the series in a large number of his compositions, and he has extended the principles of the series to the generation of rhythms and rhythm patterns (Beyer 1996)

In the above examples, the unfolding of the series has been demonstrated within the frame of an equidistant scale. The succession of intervals, as seen in Figure 7, can be read as a number series which displays characteristic recurrences (Figure 13).

Read as a succession of scale steps upwards and downwards, this number series can be applied to any kind of scale, equidistant or non-equidistant, such as diatonic or pentatonic scales, quarter-tone scales, the overtone series etc., producing different musical shapes of the infinity series. A discussion of the mathematical properties of the series is provided by Thor A. Bak (2002).

In conclusion, let me suggest a modest contribution to the typology of processes. With reference to the process examples discussed above, six possible categories of processes can be proposed, shown in Figure 14.

The four musical examples represent four of these six categories. Steve Reich's *Piano Phase* represents a rule-determined transformation process, and György Ligeti's *In zart fliessender Bewegung* can be described as a goal-directed transformation process. Per Nørgård's Infinity series is a model for a rule-determined generative process, and John Cage's *Variations II* is a model for an indeterminate generative process.

ACKNOWLEDGEMENTS

I am grateful to the Danish pianist Erik Skjoldan for his introduction to the practical application of Cage's instructions.

REFERENCES

Anderson, Julian: 1996, in A. Beyer (ed.), *Perception and Deception*, pp. 147–166.

Bak, Thor A.: 2002, in Jensen, Hansen and Nielsen (eds.), *Per Nørgårds uendelighed-srække* (Per Nørgård's Infinity Series), pp. 207–216.

Beyer, Anders (ed.): 1996, *The Music of Per Nørgård*, Scolar Press.

Cage, John: 1961, *Silence*, Wesleyan University Press.

Carlsen, Jan Pilgaard: 1999, *Rum i Pelle Gudmundsen-Holmgreen: Udstillingsbilleder (1968) med György Ligeti: Monument (1976) og Steve Reich: Piano Phase (1967) som indforende eksempler* (Space in Pelle Gudmundsen-Holmgreen: Pictures at an Exhibition (1968) with György Ligeti: Monument (1976) and Steve Reich: Piano Phase (1967) as Introductory Examples), manuscript, Copenhagen University.

Christensen, Erik: 1993, 'Zwischen Chaos und Ordnung. Per Nørgård in Gespräch', (Between Chaos and Order. Interview with Per Nørgård), in *MusikTexte* Vol. 50, pp. 31–36. Cologne.

Christensen, Erik: 1996, *The Musical Timespace*, Vols. I and II, Aalborg University Press.

Christensen, Erik: 2002, in Jensen, Hansen and Nielsen (eds.), *Min slutning er min begyndelse. Interview med Per Nørgård* (My End is My Beginning. Interview with Per Nørgård), pp. 286-0298.

DeLio, Thomas: 1980-1981, *John Cage's Variations II – The Morphology of a Global Structure*, in *Perspectives of New Music*, Vol. XIX, pp. 351–371.

Epstein, Paul: 1986, 'Pattern Structure and Process in Steve Reich's Piano Phase, *The Musical Quarterly* **72/4**, 494–502.

Febel, Reinhard: 1978, *Musik für 2 Klaviere seit 1950 als Spiegel der Kompositionstechnik*, Herrenberg.

Gefors, Hans: 1996, in A. Beyer (ed.), *Make Change Your Choice!*, pp. 35-55.

Jensen, Jørgen I., Ivan Hansen, and Tage Nielsen (eds.), *Mangfoldighedsmusik. Omkring Per Nørgård* (Multiplicity Music. Around Per Nørgård), Gyldendal, Copenhagen.

Kostelanetz, Richard (ed.): 1971, *John Cage*, London: Allen Lane, The Penguin Press.

Kullberg, Erling: 1996, in A. Beyer (ed.), *Beyond Infinity*, pp. 71–93.

Ligeti, György: 1971, 'Fragen und Antworten von mir selbst', *Melos* **12**, 509–516.

Pritchett, James: 1993, *The Music of John Cage*, Cambridge University Press.

Rasmussen, Karl Aage: 1996, in A. Beyer (ed.), *Connections and Interspaces. Per Nørgård's Thinking*, pp. 57–70.

Reich, Steve: 1974, in Reich (ed.), *Music as a Gradual Process, 1968*, pp. 9–11.

Reich, Steve: 1974, in Reich (ed.), *Notes on Compositions, 1965–1974*, pp. 49–71.

Reich, Steve: 1974, *Writings about Music*, Halifax: Nova Scotia Series.

Reich, Steve: 2002, *Writings on Music, 1965–2000*, Oxford University Press.

Revill, David: 1992, *The Roaring Silence. John Cage: A Life*, London: Bloomsbury.

Potter, Keith: 2000, *Four Musical Minimalists*, Cambridge University Press.

Sabbe, Herman: 1987, *György Ligeti*, Musik-Konzepte Heft 53, Schott, Mainz.

Schwarz, K. Robert: Steve Reich: 'Music as a Gradual Process I–II', Part I in: *Perspectives of New Music*, Vol. XIX (1980–1981), pp. 373–392. Part II in: *Perspectives of New Music*, Vol. XX (1981–1982), pp. 225–286.

Schwarz, K. Robert: 1996, *Minimalists*, London: Phaidon.

CD REFERENCES – RECOMMENDED LISTENING

Steve Reich: *Piano Phase.* Wergo – WER6630-2, Track 3: 0'00–2'15.

Steve Reich: *Piano Phase.* Track 3: 2'00–3'20.

György Ligeti: *In zart fliessender Bewegung.* SONY – SK 62397, Track 15.

John Cage: *Variations II.* The described versions are not recorded. Other versions of *Variations II* are available on Etcetera – KTC 2016 and Hat Hut – HATARTCD 6146.

Per Nøgård: *Second Symphony.* Point – PCD 5070, Track 1: 0'56–2'06.

Per Nøgård: *Voyage into the Golden Screen.* Dacapo – DCCD 9001-B, Track 12: 0'58–1'44.

PART II: APPLICATIONS OF PROCESS-BASED THEORIES

MARK H. BICKHARD

PROCESS AND EMERGENCE: NORMATIVE FUNCTION AND REPRESENTATION

ABSTRACT. Kim's argument appears to render causally efficacious emergence impossible: Hume's argument appears to render normative emergence impossible, and, in its general form, it precludes any emergence at all. I argue that both of these barriers can be overcome, and, in fact, that they each constitute reductions of their respective underlying presuppositions. In particular, causally efficacious ontological emergence can be modeled, but only within a process metaphysics, thus avoiding Kim's argument, and making use of non-abbreviatory forms of definition, thus avoiding Hume's argument. I illustrate these points with models of the emergent nature of normative function and of representation.

1. BACKGROUND

Tensions between naturalism and normativity are of ancient provenance. We can find them, for example, in Plato and Aristotle's analogy between perception and the impression left by a signet ring in wax: Wax impressions are factual; How do they acquire the normativity of representational content? How could they represent falsely?

With Descartes, such tensions become expressed in a fundamental metaphysical split between two kinds of substances, one of the factual, non-normative world, and one of the mental, normative (and intensional) world. Some, such as Hobbes, attempted to account for the world only in terms of the factual realm, and Hume argued that the normative could not be recovered from strictly factual, empirical, grounds – 'ought' could not be derived from 'is'.

This diremption between fact and norm has been generally accepted since Hume, sometimes yielding an anti-naturalism, such as with Kant and Frege, and sometimes yielding an anti-normative naturalism, as with Quine. In any case, we seem to be faced with a small set of unattractive alternatives: (1) an anti-naturalistic dualism of fact and norm, (2) attempting to account for the world with a pan-normative idealism, (3) a rejection of normativity yielding an identification of naturalism and physicalism. Kant introduced the two realm, fact and norm, framework in reaction to Hume,[1] and logical positivism was the last failed attempt at making good

J. Seibt (ed.), Process theories: Crossdisciplinary studies in dynamic categories, 121–155.

on this approach. Idealisms are not prominent in today's scene, but remain a temptation, even if hidden, such as in some versions of contemporary linguistic idealism. The austere rejection of normativity in favor of a strictly factual world has become the dominant contemporary view since Quine, though it is seldom realized how deeply this fails to account, *scientifically* account, for normative, mental, phenomena.[2]

There is a fourth possibility: naturalistic emergence. If norms were emergent from non-normative phenomena, that could unify the factual and normative world, thus transcending the trilemma. But ontological emergence encounters serious problems, so serious that they have been taken to be fatal by many, if not most. Nevertheless, I argue that emergence is the required dissolution of this aporia, but that an acceptable model of emergence itself requires fundamental shifts elsewhere. In particular, it requires a shift from a substance or particle metaphysics to a process metaphysics.

1.1. *Process and science*

This shift has strong historical support. Every science has passed through a phase in which it considered its basic subject matter to be some sort of substance or structure. Fire was identified with phlogiston; heat with caloric; and life with vital fluid. Every science has passed beyond that phase, recognizing its subject matter as being some sort of process: combustion in the case of fire; random thermal motion in the case of heat; and certain kinds of far from thermodynamic equilibrium systems in the case of life.

The exception to this historical pattern are sciences and philosophies of mind. Mind is still approached from within a substance and structure framework of background presuppositions. This is well illustrated with the case of representation: perceptual representations are construed, for example, as consisting of transduced encodings of the light in the retina, but this process of 'transduction', and how it could yield normative representations, is just as mysterious in this technologically updated version of wax impressions as it was in the original. The account, that is, is still caught in the strictly factual, and cannot account for normativity.[3]

2. CHALLENGES TO EMERGENCE

I will address and critique two fundamental challenges to emergence, one metaphysical and one logical. These challenges, I argue, are fundamental and valid, but unsound. In fact, diagnosing them yields two basic false assumptions which, when corrected, point the way toward a legitimate approach to emergence. Within this metaphysical and logical framework, then, I address two primary forms of normative emergence, function and representation.

2.1. *Metaphysics: Particles and process*

New substances cannot emerge. Only combinations or organizations are possible. Furthermore, if all is substance, or, in its contemporary atomistic form, if all is particles, then all causal power is resident in that basic substance or particle level. In particular, there is no emergent causal power.

Kim (1989, 1990, 1991, 1992a, b, 1993a, b, 1997) has developed these basic points into a subtle and sophisticated argument against emergence. In effect, his arguments pose a dilemma: either naturalism is false, or genuine emergence does not exist:

- If higher level phenomena are not supervenient on lower levels, then we have some sort of dualism and naturalism is false.
- If higher level phenomena *are* supervenient, then all causality is resident in the lowest level supervenience base of fundamental particles, whatever they may turn out to be. In particular, no genuine higher level causal powers can be emergent. All causality is located in the fundamental particles.[4]

In this view, higher level causal regularities are just the working out of the causal dance of the particles within whatever configuration they have with each other. Higher level organization, which is the usual purported locus for emergent causal power, is merely the stage on which the basic particles engage in their causal interactions. Therefore, all higher level phenomena are causally epiphenomenal, and causally efficacious emergence does not occur.[5]

The crucial center of this argument depends on the fact that particles participate in organization, but do not themselves have organization. Thus, the presumed locus of causal power, in this framework, is something that has no organization. Consequently, organization is not a legitimate locus of causal power. The emergence assumption that new causal power can emerge in new organization would require breaking the monopoly of causal power that is held by things that have no organization.

There is, however, a strong rejoinder to this argument: there are no particles. First, a pure particle metaphysics has serious coherence problems because dimension zero particles would have zero probability of ever encountering each other. Worse (for a particle model), however, is that our best contemporary physics argues that there are no particles (Brown and Harré 1998; Cao 1999; Davies, 1984; Huggett, 2000; Saunders and Brown 1991; Weinberg, 1977, 1995, 1996, 2000). Instead, everything is quantum fields. What appear as particle interactions are instead quantized oscillatory field processes, and this quantization is akin to the quantized number of waves in a guitar string. There are no guitar sound particles.

But quantum fields are processes, and processes are inherently organized;[6] a point process is an incoherent notion. If all is process, then all causal power is resident in process organizations. Everything that has causal power is organized, and has the particular causal power that it does by virtue of, among other things, its organization. Organization cannot be delegitimated as a potential locus of causal power without eliminating causality from the world.

Organization, then, is a legitimate locus of causal power. Different organization, including at higher levels of organization, can have different, novel, *emergent* causal power. The possibility of emergence is ubiquitous in new organizations of process. In effect, since it is clear that emergence has occurred, Kim's argument is a reductio of substance and particle metaphysics. Conversely, acceptable models of emergence must be framed within a process metaphysics (Bickhard 2000).

2.2. *Logical: No 'ought' from 'is'*

The second challenge I will address is a logical one. It derives from Hume's argument that norms cannot be derived from facts, that 'ought' cannot be derived from 'is'. The form of the argument is that it "seems altogether inconceivable" that *ought* could be deduced from *is*. There are two aspects to the argument:[7]

1. an assumption that facts are the proper beginning of any such deduction or derivation, consistent with the empiricism that Hume is entertaining here, and
2. that the only valid form of introduction of new terms in a derivation is by abbreviatory definition, in which a new term abbreviates some clause or phrase consisting of already available terms.

The empiricist assumption in this case is about the presumed empiricist origin of representational or semantic content: it must come from the senses, and, therefore, be factual. In particular, at least in contemporary versions, it is not legitimate to begin with normative terms when attempting to account for normativity. I will set this point aside temporarily, and focus on the second issue, that of validly introducing new terms.

The structure of this part of the argument is that, if all terms in a conclusion are validly introduced, then, in principle, all terms could be back-substituted through their definitions, eventually converting the conclusion into an equivalent conclusion that used only terms from the original premises. But those terms, by assumption, are all factual, not normative, and, therefore, any valid conclusion will be strictly factual, and not normative. In its general form, this argument precludes the introduction of anything fundamentally new: valid derivations do not go beyond whatever is available in the premises with respect to their basic terms.

This general conclusion precludes any form of emergence. Nothing new can come from what we start with, only new relations, whether logical or physical. This is the logical analogue of an underlying ontological commitment typical of substance or particle metaphysics: new substances cannot emerge from old, only new blends or structures. And even the restriction to factual premises reflects this substance-ontological commitment: substances motivate empiricist notions of perception and representation, and substances are themselves not normative.

2.3. *On legitimate definition: Abbreviatory and implicit*

The argument, however, is unsound. The false assumption is that the only legitimate form of definition is abbreviatory definition. If all acceptable definitions are abbreviations for constructions using already available terms, then the backtranslation argument at the core of the Humean argument is itself acceptable: the backtranslations are merely unpacking the abbreviations.

But there is an alternative form of acceptable definition that does not support such backtranslation, and, therefore, the mere existence of such an alternative renders the Humean argument unsound. The alternative is implicit definition. In model theory, a set of formal sentences *implicitly defines* the class of models that would satisfy those sentences. That is, the set of formal sentences implicitly defines the class of translations of the (non-logical) terms that yield a consistent interpretation of the overall set of sentences (Chang and Keisler 1990; Keisler 1977; Kneale and Kneale 1986). In geometry, for example, a sentence of "Two *X*s determine a *Y*" might be interpreted with *X*s as points and *Y*s as lines. In this case, it might also be interpreted in the reverse manner, with *X*s as lines and *Y*s as points (in which lines determine their point of intersection, and parallel lines determine the point at infinity). Implicit definition is not restricted to formal languages (Hale and Wright 2000), though it is perhaps easiest to convey what it does in that setting.

The fundamental point, however, is that implicit definition is a legitimate form of definition (relatively common, in fact, once one learns to recognize it – almost nothing is rendered or is renderable in terms of a sense data reduction) that does not support the backtranslation argument. Hume's argument, then, is unsound, and the block against emergence in general, and normative emergence in particular is removed.

I also note that, if implicit definition is an acceptable form of definition, an acceptable provider of meaning, then the basic empiricist stance that all representational content must derive from the senses is itself refuted.[8]

2.4. *Emergence and normative emergence*

Emergence, then – causally efficacious emergence – is not defeated either by Kim's argument or by Hume's argument. But avoiding Kim's argument requires taking process seriously, requires, ultimately, a process metaphysics, and avoiding Hume's argument requires recognizing the power of implicit definition. Definitions of emergent phenomena cannot be given as abbreviations of base level phenomena. Dispensing with Hume's argument clears the way not only for emergence in general, but also for the possibility of normative emergence, the original focus of the argument.

3. THE EMERGENT NATURE OF NORMATIVE FUNCTION

To *clear the way* for the possibility of models of emergence, however, is not to *provide* any such models. That requires additional development, and risks further errors. I will begin the presentation of a model of the emergence of normative phenomena with normative biological function, the sense in which it is the function of the heart to pump blood, and that it is *dysfunctional*, a normative notion, for a heart to not do so or to do so badly. The dominant model in the literature for normative function is the etiological model. So, I preface the outlining of the proposed model with a brief exposition of the etiological model and of why it is not itself already acceptable.

3.1. *Etiological models of function*

The central intuition of the etiological approach to biological, and, thus, normative, function is that an organ, a heart, say, has the function of pumping blood because its ancestral hearts were selected for having that causal consequence. In particular, it is because of those evolutionary ancestral selections that this heart under consideration exists at all, and it is with respect to those selections that the function of the heart is to pump blood and not some other of its consequences, such as contributing mass to the organism or filling space or producing heart beat sounds (Millikan 1984, 1993).

This model presents as a naturalistic model of the emergence of functional normativity in evolution. If successful, it would constitute a refutation by counterexample of both Kim's and Hume's arguments; it is a fundamentally important argument, therefore, for these reasons as well as for its relevance to biology and the philosophy of biology. Unfortunately, it is not successful.

In etiological models, the having of a (proper) function is constituted in the having of the right evolutionary history – the having of the right etiology. That is, function is constituted in having the right history. But this implies that function is not constituted in the current state of the system, because two different systems could have the same current state even though they had different histories.

This point has been recognized, though not in these terms, in the etiological function literature. One example is the science fiction thought experiment in which a lion is supposed to pop into existence in the corner of the room – just from molecules in the air coming together in the right ways – that is, by assumption, molecule by molecule identical to the lion in the zoo (Millikan, 1984, 1993). The organs of the lion in the zoo have the right history, and, therefore, they have functions. But the lion that just popped into existence has no selection history at all, and, therefore, its organs do not have functions. This seems strongly counter-intuitive. But we are all too familiar with our intuitions being wrong – witness quantum mechanics and quantum field theory – so this science fiction violation of intuition might well be worth the naturalization of function that is provided by the etiological model for real organisms.

The two lions example, however, does more than draw attention to a counterintuitive consequence of the etiological account of function. It illustrates the point that etiological function is not constituted in current system state: the two lions, by assumption, have the same state. But only current state can be causally efficacious. The two lions will, because they have the same current state, have the same causal properties, but one has functions and the other does not. Etiological function is causally epiphenomenal (Bickhard 1993, 2002; Christensen and Bickhard 2002). Etiological function does not succeed in providing a naturalization of function.[9,10]

This causal epiphenomenality is a window into a number of serious problems with etiological function (Bickhard 1993; Christensen and Bickhard 2002, in preparation), but it does by itself suffice to demonstrate that, if we are seeking a model of the naturalistic emergence of causally efficacious normative function, this is not it.

3.2. *A dynamical model of the emergence of normative function*

I will outline a dynamical model of the emergence of normative function. Function, in this model, will be emergent in the dynamic organization of the system, and, therefore, will be constituted in the current state of the system – and, therefore, will be causally efficacious.[11]

One basic problem in naturalizing normative function – and in naturalizing any kind of normativity – is that normativity inherently involves an asymmetric distinction between the normatively good and the normatively bad. In the case of function, this is the distinction between function and dysfunction. In the case of representation, it is the distinction between correct or true, and incorrect or false. And so on. But the laws of physics, in general, do not manifest such asymmetries. They do not seem to provide the grounds for such asymmetric distinctions.[12]

There is an exception, however, and it is in terms of this exception in physics that I propose to model emergent normativity, the emergence of normative function, in particular. The exception is thermodynamics. The asymmetry is between energy well stable systems and far-from-equilibrium systems.

In particular, some patterns of process are fleeting. The fall of a leaf from a tree may take a few seconds, but then it is over. Some process organizations, however, can be stable, and perhaps persistent for very long periods of time. An atom may last for billions of years, so long as the ambient energy is not too great. An atom is an example of an energy well stability: it is a process organization that will remain stable so long as above-threshold energy does not impinge on the system. Among other consequences, if such a system is isolated, and goes to thermodynamic equilibrium, it remains stable.

This is in contrast to far-from-equilibrium systems. Because they are far-from-equilibrium, they cannot persist without explicit intervention, without explicit interaction with their environment, in order to maintain their far from thermodynamic equilibrium condition. If such a system is isolated, it goes to equilibrium and ceases to exist as whatever far-from-equilibrium system it began as. The fundamental asymmetry to which I appeal, then, is that between stability with no intervention and stability that is dependent on intervention.

Experimentally, a far-from-equilibrium system may owe whatever stability it has to strictly external interventions, as when a chemical bath is maintained far-from-equilibrium with pumps pumping various chemicals into the bath. Such systems can exhibit interesting and important properties, such as self- organization (Nicolis and Prigogine, 1977).

For current purposes, however, a different class of far-from-equilibrium systems is central. Some far-from-equilibrium systems make contributions to the maintenance of their own far-from-equilibrium conditions. A canonical example is a candle flame. A candle flame maintains above combustion threshold temperature; it melts wax so that it percolates up the wick; it vaporizes wax in the wick into fuel; in standard atmospheric and gravitational conditions, it induces convection, which brings in fresh oxygen and gets rid of waste. A candle flame exhibits *self-maintenance* in several ways (Bickhard 1993, 2002).

This is the core for the emergence of function: a contribution to the maintenance of the far-from-equilibrium conditions of a far-from-equilibrium system is functional, it *serves a function*, for the stability, the persistence, of that system. This is a model of function as usefulness, rather than as (evolutionary) *design* (Christensen and Bickhard in preparation).

This model of function is of a causally efficacious property: the persistence or cessation of the far-from-equilibrium process makes a causal difference to the world. It is a normative property, in that such a contribution can be positive or negative, adequate or inadequate. It is a relational property: the heart of a parasite is functional for the parasite, but is dysfunctional for the host.

There is an important contrast with etiological models here not only in the specifics of the model of function, but also in the broader explicatory strategy involved. Etiological models of function focus on the property of *having a function*. Some *part* of the organism has a function insofar as it has the right selection history.[13] The notion of *serving a function*, insofar as it is considered at all, is derivative in such models. Something serves a function insofar as it accomplishes the function that it has.

The dynamic model just outlined turns this explicatory dependency on its head. The primary notion is that of serving a function, and all others will be derivative from that. But this point issues a promissory note to in fact account for *having* a function in terms of *serving* a function.

The key to this derivation is the relation of *functional presupposition*. A part *has a function* insofar as the rest of the system functionally presupposes that it is serving that function. Functional presupposition, in turn, refers to the sense in which the organization of a system may presuppose that some part of the system has a particular (set of) consequence(s), that it serves certain functions, because those are the conditions under which the rest of the system can continue to be functional for the system. That is, functional presupposition is a kind of functional dependence that traces dependencies from the overall system down through structures and organizations in terms of their dependencies on each other for their being successfully functional themselves. In other words, if the serving of a particular function in the system is dependent on some part serving some function of its own, then that part is *presupposed* as serving that function, and, therefore, as *having* the function of serving that function.[14]

This inversion of the explicatory relationship between serving and having a function is not only quite different from etiological approaches, it offers its own advantages as well – over and above the naturalization of the notion of function per se. For example, the serving of a function by something that does not have that function makes little sense in the etiological view, and serious contortions must be undertaken in order to avoid this diremption. There is no such difficulty for the dynamic model. So, for example, the fact that the legs serve the function of helping blood circulation on long airplane flights, even though they do not have that function (certainly not as a proper function), is a perfectly natural point to make on this model (Christensen and Bickhard 2002). In the etiological view, it's not coherent. Similarly, because function in this model is not restricted to the case of a biological part having a biological function and derivations from that, but instead takes serving a function as its broadest category, it can address, for example, artifactual functionality (both of serving and of having a function, e.g., a coat in cold weather) directly as well as being derived from purposes (Millikan 1984).

I submit, then, that the dynamic model of emergent normative function does succeed in naturalizing function. It is normative in a sense relative to particular far-from-equilibrium systems (there is no God's eye view notion of normativity to be had here). It is causally efficacious: it makes a difference to the world whether or not a particular far-from-equilibrium system persists or ceases and goes to equilibrium. It supports the biologically crucial notion of having a function. And, in fact, it provides a much richer way of analyzing function and functional relationships than the part-focused, and the dichotomous "have or don't have", functional framework of etiological approaches (Christensen and Bickhard 2002, in preparation), as witnessed here with the example of the functional contribution to blood circulation that can be made by the legs on long flights. Normative function, however, is just the bottom of a long hierarchy of normative emergences. All of mind and mental and social phenomena are fundamentally normative, and they all emerge in a hierarchy with biological functional normativity at its base. Some other locations and levels in the hierarchy include representation, perception, memory, learning, emotions, sociality, language, values, rationality, and ethics. I will not be able to address most of these in this paper, but will address in some detail the naturalistic emergent nature of representation.

Representation still resists naturalism. Lest there be some confusion about this point, I will preface the interactive model of representation with a discussion of the fact that representation does still resist naturalization in the current literature. This discussion will proceed first by examining particular models, in particular, those of Millikan, Dretske, Fodor, and Cummins, and then with a more general critique of dominant assumptions about the nature of representation.

4.1. *Millikan and etiological approaches to representation*

A first demonstration of the failure of the etiological approach to capture a naturalistic model of representation is already mostly done: the demonstration of the causal epiphenomenality of etiological function. Etiological representation is a function performed by particular systems, and, therefore, inherits the epiphenomenality of the general etiological approach to function. This point suffices to refute etiological representation, but it does not stand alone.

Just as etiological function is constituted in the past, so also is etiological representational content. The content is constituted in the particulars of the selection history. One consequence is that the contents of an organism's representations are not accessible to that organism. As for function, only current state is accessible. But representational error is constituted as the misapplication of content to the present situation. So error is not constituted in current state, and is not accessible to the organism. That is, system detectable error is impossible.[15] Therefore, error guided behavior and learning are impossible.

More deeply, comparison of content with current situation requires not only accessibility of content, it also requires representing the current situation. But this is the original problem of representation. Checking a representation, then, is circular: it can only be checked against itself.[16] This is the classic radical sceptical argument (Greco 2000; Rescher 1980), and, as indicated below, it applies to multiple contemporary attempts to model representation.

4.2. *Dretske*

There are many differences between Dretske's model and Millikan's model, but, for current purposes, they are also very similar. In particular, Dretske's model is also an etiological model, but one in which the relevant history is a learning history rather than an evolutionary history (Dretske 1988). Nevertheless, content is constituted in the past, and, therefore, Dretske's representation is epiphenomenal.[17]

Similarly, because content is constituted in the past history of the organism, representational error is not constituted in current state, and, therefore, is not causally efficacious. Consequently, error guided behavior and learning are not possible. And the circularity of having to compare the (inaccessible) content with the current represented situation also recurs. Representing the current situation for the purposes of comparison and checking is the original problem of representation yet again.

There is an additional problem that should be pointed out. Dretske renders representation in terms of its usefulness in explanations of the system processes. A key sentence reads: "*C* is recruited as a cause of *M* because of what it indicates about *F*, the conditions on which the success of *M* depends".(Dretske 1988, p. 101). *C* is a mental state. It constitutes a representation in virtue of its having been recruited as a cause of some behavior *M*, and its having been so recruited because of what it, '*C*', indicates about the conditions on which the success of *M* depends (*F*). So, *C* indicates the success conditions for *M*, and, consequently, indicates the success of *M* in virtue of indicating those success conditions. *C* is a *representation* of those conditions *F* because of its having been recruited (via learning) as a cause of *M* in virtue of its indicating *F*, and it was so recruited because those conditions *F* (indicated by *C*) are the success conditions for *M*.

Note that the 'because' in this sentence cannot be a causal relation. *C* indicating the conditions *F* is a relation to the environment, and, although *C* may be causally or functionally accessible to the organism, that indicating relationship between *C* and *F* is not so accessible, and, therefore, cannot itself be a cause of anything in the organism. This is, however, not a problem for Dretske's intended reading. He intends 'because' to be an explanatory relationship. We, as observers and analyzers of the organism, can explain why *C* is recruited as a cause of *M*: it is so recruited because it indicates the success conditions for *M*.

Dretske, in other words, is not attempting to model representation per se in an organism, but, instead, is attempting to model the legitimate ascription of representation to an organism. However, unless representation has a strictly social or linguistic ontology, such as, perhaps, money or marriage, this will not do. Clearly it does not have such an ascriptive nature: if it did, then, again, representational error guided behavior and learning would not be possible, and they clearly are possible, and not just in humans. Furthermore, if we inquire about the representations involved in making such ascriptions, we encounter either a circularity or infinite regress of ascriptions of representations that constitute ascriptions of representations, and so on.[18]

4.3. *Fodor*

Fodor's model of representation is a version of an information semantics approach. In such approaches, semantic content is purported to be constituted in the carrying of appropriate information about the object of representation. Information is rendered in strictly factual terms as some kind of special correspondence relation between representation and represented, such as mathematical information (statistical covariation), causal, or, for Fodor, lawful or nomological correspondence (Fodor, 1975, 1986, 1987, 1990a, b, 1991, 1998).

One problem that such models face is the error problem. If the special correspondence exists, then the representation exists and it is correct. If the special correspondence does not exist, then the representation does not exist. The correspondence either exists or it does not; there is no third possibility. But there is a third condition that must be modeled: the representation exists and it is incorrect (Millikan, 1984). There has been a minor industry in the last decades attempting to solve this problem, without success.

Fodor's attempted solution rests on a notion of asymmetric dependency. False instances of what would otherwise be legitimate correspondences are asymmetrically dependent on correct instances in the sense that the false instances would not occur if the correct instances did, but that dependency is not reciprocated: the correct correspondence instances could very well occur even if the false ones never did. The intuition is that error is parasitic on success, and that the asymmetry of the dependency relation captures that.

But a quick counterexample begins the demonstration that this will not do. Consider a neural transmitter docking on a receptor molecule in a receiving neuron. There is all the information, causality, nomologicalness, and subsequent biological activity here that anyone could want. Now consider a poison molecule that mimics the transmitter. There is a dependency between the possibility of the poison docking in the same receptor molecule as the transmitter, and the transmitter itself docking in that receptor molecule – and it is not reciprocated, it is asymmetric. Yet we have at best a functional error here. There is no representation at all (Bickhard 1993; Levine and Bickhard 1999).

Furthermore, the asymmetric relations among classes of counterfactuals that constitute asymmetric dependency (what could be the case) cannot be modeled in terms of current system state. Therefore, representation content, on this model, cannot be constituted in current system state, and, in the by now familiar manner, Fodor's model is epiphenomenal.

Similarly, error is epiphenomenal, because content is inaccessible, so error guided behavior and learning are not possible. And comparison with current, represented, situation encounters the original problem of representation. A circularity. In practice, in all purported examples, for Fodor and for others, this circularity is avoided by making such comparisons only between an ascribed content to some organism state and the observer's representations of what 'really' is the case in the situation. In other words, it is only such an external observer who is in a position to figure out the relevant histories or counterfactuals in order to determine the purported content involved, and to independently represent the current organism environmental situation to be able to compare the content to the represented situation. Only the external observer, therefore, can determine if error has (hypothetically) occurred – but this is not error detectable by the system.

4.4. *Cummins*

Cummins (1996) introduces an important distinction between a representation and the target to which the representation is applied. Error occurs when the representation is applied to a target that it does not fit. Content, in this model, is structure, and representational 'fit' is the relationship of structural isomorphism.

A distinction like that between target and representation is roughly the correct way to account for error. If representational content is determined by the object that is currently being represented, whether via informational, causal, or lawful correspondences, then error becomes at best extremely difficult to account for.

Nevertheless, there are problems. First, contrary to what Cummins assumes, there is no fact of the matter about what the structure is in a physical system. This point is obscured in Cummins' discussion by always considering either mathematical structures (in which the structure is determined by the mathematical definition), or by considering physical examples in which the relevant structure seems so strongly intuitive that the question doesn't arise. Consider one of these examples: a toy car designed to run mazes. The wheels in this car are steered by a peg that runs in a slot in a card that is inserted in the car. If the shifts in the slot are isomorphic with the required turns of the car, then it will succeed in running the maze. And, different cards with different slot patterns will run different mazes.

But if instead of a peg there is a read-head that reads the domains of magnetization along the edges of the slot, the "structure" will be totally different. Structure, in other words, is a matter of read-out, and that is a functional matter. Structure, then, is not constituted in the purported bearer of that structure – the card with the slot in this case – but, at best, in the relationship between the functional read-out process and that bearer.[19]

Furthermore, if the goal is to hit the side of the maze at a certain point, instead of running through and exiting the maze, then the card is no longer a correct 'representation'. The correctness of the representation, in other words, is dependent on the normativity of the goal involved in the action, and that normativity is not inherent in the system, but only in the observer or user or designer. The relevant 'structure' then is functional in both the sense of influencing system process and in the sense of the normativity of the actions involved, and, therefore, the normativity of the representations involved. Neither sense is naturalized here.

4.5. *Encoding models of representation*

At least since Plato and Aristotle's analogy between perception and a signet ring pressed into wax, models of representation have assumed that representation is some form of encoding, an encoding correspondence between the representation and the represented. The issue has been to model what sort of correspondence would constitute such an encoding correspondence. Many possibilities have been considered: the special encoding constituting relationship is informational, causal, nomological, structural isomorphism, the locus of historical evolutionary or learning selections, and so on. I have called this general assumption about the nature of representation *encodingism* (Bickhard 1987).

Encodings clearly exist. In Morse code, for example, '...' encodes '*S*', or, if the conventional nature of the relationship in Morse code is disturbing, we might consider the sense in which, say, this neutrino count encodes properties of the fusion process in the sun. The issue is whether or not encodings can capture the nature of all representation, mental representation in particular.

Unfortunately, encodingism, in whatever form, encounters myriads of problems, some of ancient provenance, some discovered relatively recently. Already mentioned is the problem accounting for the possibility of representational error: If the special correspondence exists, then the representation exists and is correct, while, if it doesn't, then the representation doesn't exist. There is no third option for modeling the possibility of the representation existing, but being incorrect. Also mentioned is the problem of system detectable error. Not all organisms are capable of it, and even humans are thus capable not all of the time, but error guided behavior and learning do occur, and these require system detectable representational error. So, any model that makes such system detection impossible is thereby refuted. I have also pointed out that this error detection problem encounters the circularity and regress of the classic radical skeptical argument: To check a representation against that which it purports to represent requires epistemic access to what it purports to represent, but that is available only via the representation to be checked. No independent check is possible.

But there are many other problems. One is that there are too many of the candidate special correspondences, and most all of them (at least) are not representational. Every instance of a causally related pair of events in the universe is an instance of an informational, a causal, and a nomological correspondence. Any physical system is subject to having point-to-point and relation-to-relation correspondences defined between it and any other physical thing. That is, anything can be shown to be in isomorphism with anything else with an 'appropriate' definition of the correspondence mappings. Still further, if some activity in someone's occipital lobe is in correspondence with the table in front of that person, then it is also in correspondence with the biochemical activities in the retina, the light activities in the space directly in front of the eyes, the quantum activities in the surface of the table, the table one minute ago, the table yesterday, the logging and chemical extraction of the materials out of which the table is made, the stellar processes that produced the atoms out of which the table is made, and so on back to the Big Bang. Which of all of these instances of the special correspondence is the representational one, and how does the organism accomplish figuring out which one it is and what is on the represent***ed*** end of that special one?

Jean Piaget had an argument against such models, called the copy argument. If our representations of the world are in some sense copies of the world, then we would have to already know about the world in order to construct our copies of it (Piaget 1970). This is, in effect, the constructive side of the skeptical problem of not being able to check our representations. It is another manifestation of the circularity of encoding models.

Yet another problem has been called the incoherence problem (Bickhard 1993; Bickhard and Terveen 1995). Genuine encodings, such as Morse code, borrow their representational content from whatever they encode. Genuine encoding is a kind of stand-in relationship: '. . .' stands-in for '*S*' and serves a function because '. . .' can be sent over telegraph wires while '*S*' cannot. So, as long as there is something that an encoding can stand-in for, there is nothing illegitimate about it as an encoding. The problem is that, while such encoding stand-in or definition relationships can iterate multiply – '*X*' in terms of '*Y*' and '*Y*' in terms of '*Z*', and so on – there must be a level of grounding encodings in terms of which all others are defined, or out of which all others are constructed. If we consider some element of this purported ground, say 'X', and ask how it manages to have any representational content, and if we assume that all representations are encodings, then there are only two possibilities: either we can define 'X' in terms of (some) other encodings, in which case it is not at the grounding level, contrary to assumption, or we define 'X' stands-in for 'X', 'X' represents whatever it is that 'X' represents. But this does not provide 'X' with any content at all, and, therefore, fails to make 'X' an encoding representation at all. Grounding encodings in encodings is not possible, yet this is presumed by all encodingisms. Encodingism is incoherent.

One insight into the problems of encoding models is that the relationship between an encoding representation and its representational content is *external*. Internal and external relationships were a major factor in, for example, Green and Bradley's Idealisms of the 19th century, but they are also part of what Russell reacted so violently against in those Idealisms. With Quine's austere ontology, eschewing as much as possible intension, modality, essence, normativity, and so on, internal and external relations are seldom discussed. An internal relation is one that is essential to one or more of the relata. It is an essentialism of relations, not just of properties. An arc of a circle, for example, could not be that arc of that circle unless it were related to the point that is the center of that circle. That relationship to that point is internal to that arc. An external relationship is one that might or might not exist, without the relata changing in any way. This book might or might not be above the table, and both the book and the table per se are indifferent to the existence or non-existence of the relationship. Quine, in effect, banished internal relations in favor of all relations being external.[20]

But if a representation is externally related to its content, then that representation, whatever it is, could be just the same even if it did not have that content (Bickhard in press). Consequently, the representation has any content at all only because it is known and held to have that content by some agent who knows the encoding relationship. In fact, since the encoding relationship is external, that relationship exists only insofar as it and the relata are themselves known, and are known as having that relationship. But this is precisely the point that underlies the infamous regress of interpreters that is involved in rendering encodings in terms of encodings. In fact, there are two complementary regresses: one that attempts to provide content to a representation in understanding it, and one that attempts to provide content to a representation in defining it in the first place (Bickhard and Richie 1983).

Fundamentally, such models are models of representation in terms of factual relationships of some sort or another. But content is a normative property, and Hume's argument blocks getting anything normative out of strictly factual grounds. I have argued that both Hume's argument and Kim's argument are unsound, and, therefore, their conclusions can be avoided, but doing so requires modeling normative emergence, and none of the candidates on offer has succeeded in doing that. Encodings borrow content; they do not emergently generate it.

The problem, then, is deep. It will not do to push it off onto evolution, claiming that all grounding encodings are innate, and all further representations are defined in terms of them (Fodor 1981). There is no model of how evolution could transcend Hume either, and, if Fodor were to provide one, there is no argument forbidding whatever that process or relationship might be from being realized in learning and development in individual organisms. There may be innate supports for some kinds of learning and development, such as language, but innatism per se is not a solution to the problem of representational content.[21] If we were required to already have representation in order to get representation – more generally, to have *anything* X in order to get X – then representation (X) would not be possible. It could not have emerged in cosmology or evolution. In fact, it is precisely the inability to model emergence in general, and, therefore, the emergence of any particular X, that yields conclusions such as that we must already have something X in order to get X. Emergence is the transcending alternative.

5. THE NATURALISTIC EMERGENCE OF REPRESENTATION: THE INTERACTIVE MODEL

Let us turn then to the emergence of representation. The emergence of normative function in the sense of serving a function has been modeled in terms of contributions to the self-maintenance of far-from-equilibrium systems, and having a function in terms of the functional presuppositions involved in the organized functioning of an organism (or other relevant system). But self-maintenance is a(n emergent) property that is relative to a range of environments. A candle flame's self-maintaining processes will not succeed if there is no oxygen, if the energy flow away from the flame is too great, or if it is running out of candle.

5.1. *Recursive self-maintenance*

The candle flame has no options, but other systems do. A bacterium, for example, might swim so long as it is swimming up a sugar gradient, but tumble if it finds itself swimming down a sugar gradient (Campbell 1974,

1990). The swimming is self-maintaining so long as it is oriented toward higher sugar concentrations, but it is *not* self-maintaining if it is oriented toward lower sugar gradients. Conversely with tumbling. So, swimming is self-maintenant under some conditions and not under others, and the bacterium can detect the difference in the conditions and switch its activities accordingly; it can select between a pair of possible interactive processes that which would be appropriate for current (orientation) conditions.

This is an ability to maintain the property of being self-maintenant in the face of variations in relevant conditions that determine what will be self-maintenant and what will not. It is, in other words, a self-maintenance of self-maintenance (in the face of variation), a *recursive self-maintenance* (Bickhard 1993).

Recursive self-maintenance requires some means of differentiating environments, two or more possible kinds of interactions, and a switching capability that relates the differentiations to the selections of interactive possibilities. Bacteria, for example, may do a front-end to back-end comparison of sugar concentration, or a time-delay comparison of sugar concentration, while swimming to differentiate orientation up a sugar gradient from down a sugar gradient. The subsystems in the bacterium that engage in such differentiation and switching and swimming or tumbling must be relatively constant on the time scale of such detection-swim or tumble. This is unlike a candle flame, for example, in which the openness and environmental interchange of all parts of the flame occur on roughly the same time scale. Recursive self-maintenance, in contrast, requires some processes in the system to be functioning at much slower time scales than others; it requires that the system contain *infrastructure* that can engage in the relevant detections, switchings, and interactions (Bickhard, 2000b).

This difference in temporal scale, in turn, implies that the processes that engage in and maintain the different scales of process in the system must similarly be differentiated in their temporal scales. This is the origin of metabolism. Metabolism is the group of self-maintenance processes that support infrastructure (Moreno and Ruiz-Mirazo 1999).

Recursive self-maintenance requires infrastructure to engage in differentiation, switching, and interactions. There are both informational (switching) and work (swimming) aspects. This suggests the possibility of a self-maintenant system that engages in work to maintain itself, but does the same thing all the time – there is no switching among alternative possibilities, there are no informational aspects. An example might be a sulfur metabolizing bacterium at a deep sea vent that does nothing but metabolize sulfur continuously. Such a work-but-not-information self-maintaining system is the simplest version of what might be called *autonomy*: the ability to do something (work) that is functional (contributes to self-maintenance) for the system (Bickhard 2000b; Christensen 1996; Christensen and Bickhard 2002; Christensen and Hooker 1998, 2000). In general, I will use the term autonomy as an umbrella term for this and more complex forms of self-maintenance.[22]

5.2. *Representation and interaction selection*

A system's dynamics can functionally presuppose various conditions, including other functional contributions, to be the case. With respect to the selection of one among an alternative set of interactions with an environment, this involves a functional presupposition that the current environment is in fact an appropriate environment for that interaction. For the selection of swimming, there is a presupposition that the orientation is up a sugar gradient.

That presupposition can be false. The differentiations, for example, may not be fine enough to distinguish between sugar and saccharin: a bacterium will swim up a saccharin gradient just as readily as a sugar gradient, but, under those conditions, swimming is *not* self-maintenant, it is not a contribution to autonomy, it is not functional. The functional presupposition is *false*, so the interaction is *dysfunctional*.

This emergence of the normativity of truth value, true or false, out of the normativity of function, functional or dysfunctional, via functional presupposition is, I claim, the point of emergence of representation out of pragmatics. This is the most primitive form of representation: in effect, a representation of the current environment as being appropriate for the selected (inter)action.

For the bacterium, there is a direct triggering relationship between the differentiations of environments and the corresponding interactions. For more complex systems, a more complex kind of relationship is required. A frog, for example, may have many possible interactions available in a given particular environment, perhaps tongue flicking and eating indicated by a particular visual scan of a fly, or jumping into the water to avoid a hawk indicated by the visual scan of a shadow. Multiple possibilities must be indicated in some way and then selected among, rather than there being a direct trigger. The selection processes will generally be with respect to goals and other normative phenomena, and are deeply interesting and important.[23] My focus here, however, is on the indications.

The central point is that, just as the *triggering* of an interaction presupposes that the environment is appropriate for that interaction, so also the *indication* of the appropriateness of an interaction is an indication that the environment is appropriate for that interaction. Such an indication functionally presupposes that the success conditions for that interaction hold in this environment. Such an indication, then, is a functional relationship that has a truth value in virtue of its presuppositions. It is a representation.

5.3. *Properties of interactive representation*

Note first that those functional presuppositions are about the environment: only if the environment in fact is appropriate could the interaction succeed. The interactive representation has representational truth value. Second, that truth value can be false. The environmental conditions may not obtain. The frog could have an indication of the potentiality for tongue flicking and eating, but the visual scan was of a speck of dirt or a pebble, not a fly. Third, if the truth value is false, the organism has a chance of detecting that. If the interaction is initiated, and it doesn't follow the anticipated path, then the presuppositions of the indication were false, and they were falsified, *for the organism*. System detectable error is possible; error guided behavior and learning are possible. The radical sceptical argument is avoided: the representation constituting indications or anticipations are of future interactive potentialities, and those interactions are accessible for the organism, and, thus, their potentialities – their appropriateness, the truth of their presupposed success conditions – are 'checkable'.[24]

The functionally presupposed properties of the environment are internally related to the indications of the potentialities of the interactions. They could not be those interactions without having those success conditions. There is, therefore, no problem of interpreters, either to provide or assign content or to 'understand' or translate content. The content is intrinsic, internal, to the interaction indications.

That content, however, is implicit rather than explicit. The organism does not have any direct representation of what those success conditions are.[25] This implicitness of content is quite different from the explicitness that is required for encoding representations. It is the internal relatedness and the implicitness of interactive content that avoids the circularities that encoding models inevitably encounter: if content must be explicit and must be provided or assigned in order for the state to be a representation at all, then it must already be representationally available to be so provided or assigned, but the representation to which it is to be assigned was purported to be the source of epistemic access, so that content could *not* be available. The purported existence of the representation presupposes that the representation already exists.

Among other consequences,[26] the emergence of interactive representation is not an in-principle problem. Any recursively self-maintaining system with the right kind of interaction selection processes will have at least primitive representations, and the internal functional construction of new such process (should such internal construction be possible, as in learning) can inherently involve emergent representation.[27] Whatever innate scaffolding there may or may not be for various kinds of learning and development, there is no in-principle necessity that all grounding representations be innately provided.

Finally, note that interactive representation can only emerge, can only exist, in autonomous agents, systems that functionally interact with their environments. It is not possible for passive input processors, such as a computer. The locus of representational emergence is action, not consciousness. In this, the interactive model is akin to Pragmatism (Joas, 1993), though the representational model per se is closer to Peirce's model of meaning than it is to his model of representation.

5.4. *What about input processing?*

This point, however, raises the question of how to account for the 'input processing' that is known to occur in the nervous system. Human perception, for example, seems to involve the 'encoding' of properties of the light into line and frequency codes that carry that information to the central nervous system. In fact, the interactive model accounts for such phenomena in a very direct way.

An interactive system must be sensitive to the environment. It must differentiate appropriate environments for interactions in order to successfully indicate the potentialities for particular interactions. System interactions will proceed in accordance both with internal system organization *and* with the environment being interacted with. A visual scan depends on the scanning process and on what is being scanned. The internal outcome or flow of such an interaction, then, differentiates those environments that yield that outcome or flow from those that don't. It is important, however, that it does so without representing anything in particular about those environments. Such differentiations serve as the basis for setting up consequent indications of interactive potentiality.

Passive such differentiations – 'interactions' with no outputs – can occur as well. They are not as powerful as full interactive differentiations, but they are differentiations nevertheless. The internal outcome of a passive 'differentiator' will differentiate just as much as the internal outcome of a full interaction.

But passive differentiators are input processors. They are the paradigm of classic correspondence models of representation. In such models, they are presumed to represent whatever it is that they differentiate, as in sensory 'encoding' (Carlson 2000). It is certainly the case that such input processing generates internal states and processes that are correlated with environmental conditions, and, in that sense, are in informational, causal, and perhaps nomological correspondence with the differentiated environmental conditions. But standard models assume that such correspondences constitute representations of whatever the correspondence is with, and then are unable to account for virtually any further properties of representation or avoid the multiple perplexities of encodingism. The interactive model needs such internal environmental differentiating conditions just as much as do standard models, but the interactive model makes no assumption that such signals constitute representations. They are, instead, aspects of control processes enabling the organism to set up interactive indications appropriately to the environment, and it is those indications that have representational content and truth value, not the differentiating processes that control the setting up of those indications. The frog doesn't represent either flies or pebbles, but, instead, represents the opportunity for tongue flicking and eating (Bickhard 1993).

5.5. *More complex representation*

Thus far, interactive representation doesn't look much like paradigmatic representation. Representation of interactive possibilities seems different from representation of, say, physical objects, or abstract objects. How are such kinds of representation accounted for?

To address these questions, the resources available to interactive representation must be elaborated further. Thus far the discussion has remained limited to differentiations that either directly trigger an interaction, or directly set up one or more indications of interaction potentiality. But those differentiation processes and their associated indications are themselves potential in the organism even when they are not active. That is, there will be differentiation-indication relationships that are resident in the organism even when not immediately active.[28] There is a visual-scan-of-a-flying-speck to the-possibility-of-tongue-flicking-and-eating connection in the frog even when there are no flies or other specks around. Furthermore, those differentiation-indication relationships can branch – a single differentiation could yield indications of multiple possible indications. Still further, they can iterate: the successful completion of an indicated interaction could be the differentiating condition for indications of still further interactions. Interaction indications, then, can branch and iterate and connect into potentially vast and complex webs. This is one of the primary resources for accounting for more complex representation.

To illustrate, consider a child's toy block. It offers multiple possibilities of manipulation and visual scans, some of which are contingent on others, such as when the block needs to be rotated to recover a particular visual scan. In general, the interactions afforded by the block form a web that is internally completely reachable: any point in it is reachable from any other, perhaps with appropriate intermediary interactions, such as rotations. Further, this reachable sub-web in the child's overall interactive environment remains invariantly available under a large range of other possible interactions. The child could drop the block, leave it and walk away, put it in the toy box, and so on, and the internally reachable pattern of interaction potentialities remains invariant, though with changing intermediary interactions required to recover it directly. That is, the child may have to return to the room in which the block was left, or open the toy box, and so on, in order to gain immediate access to this special patterns of interactive possibilities. The pattern does not, however, remain invariant under all possibilities. Burning the block or crushing it, for example, will eliminate that pattern of possibilities.

An internally reachable pattern of interactive possibilities that remains invariant under a class of interactions such as locomotion, physical movement, enclosing such as in a toy box, and so on is the model for the interactive representation of a small manipulable object. Because both models are action based, I have here been able to simply borrow Piaget's model of object representation and translate it into the terms of the interactive theory (Piaget 1954).[29]

What about representations of abstractions? The interactive model might be able to capture representation of objects, but what could the interactive realm be for electrons or numbers? Again the model is roughly Piagetian, though in this case the similarities are a little rougher than for objects. Consider an interactive strategy of "Do *X* 3 times before giving up and trying something else" where *X* is some task strategy. This is simply an iterative count on control flow that can be useful in some circumstances. Note, however, that it instantiates the ordinal number '3'. If there were a second level of interactive representation interacting with and representing the first level in the same sense in which the first level interacts with the environment, then that property of '3' could itself be differentiated and anticipated – represented – in the second level. Similarly, a second level of interactive representation might have properties that would be useful to represent, and could be represented from a third level, and so on. This is a model of a hierarchy of levels of interactive 'knowing', in which the representations at a higher level are generated via 'reflective abstraction' from organizations and processes at lower levels.

As mentioned, this is a roughly Piagetian model, though both the levels and the reflective abstraction diverge in some crucial respects (Campbell and Bickhard 1986). I cannot fill out any details of these models here; the basic point is that the interactive model encounters no aporia regarding more complex kinds of representation. Instead, it offers a rich array of resources of various kinds of patterns of interactive indications and of abstractions into higher levels of representation.

6. EPISTEMIC CONTACT AND CONTENT

Interactive representation, then, offers the possibility of addressing representational and cognitive phenomena in general. It captures the fundamental normative character of representation, yet avoids the myriads of fatal problems of encodingist models. But the interactive model also forces changes in ubiquitous assumptions about representation and cognition. One interrelated set of these changes illustrates further some of the crucial issues involved in the theoretical shift from correspondence or encoding models to the interactive model.

In the interactive model, differentiation of the environment constitutes *contact* with that environment. It is on the basis of that contact that anticipations of interactive potentiality can be set up.[30] Those anticipations involve functional presuppositions that have truth value, and can be (fallibly) discovered to be false. Those presuppositions constitute content (Bickhard, 1980, 1993; Bickhard and Terveen, 1995).

For example, the frog's anticipations of being able to flick its tongue and eat will be false, and falsified if the frog does actually flick its tongue in the relevant way, if the visual scan was in fact of a pebble. The frog's content is of the (presupposed) opportunity for tongue-flicking and eating, not of the pebble at all. The visual scan of the pebble is the contact; the opportunity for tongue-flicking and eating is the content.

Standard approaches to representation construe passive differentiations, as in sensory processing, as constituting encoded representations of that which is differentiated. This is a conflation of contact and content. The content is assumed to be that which the contact was with. That is, content is assumed to equal, or be constituted in, contact. This is yet another perspective on the origin of the multiple aporias and circularities involved in encodingist models.

The interactive model, then, distinguishes between two properties that are identified in standard models.[31] Furthermore, it is the conflated *combination* of the contact and content that is called representation in these models. With contact and content distinguished in the interactive model, then, which should be called representation?

Contact as representation is the classical position, but this implies that representation has no truth value, plus myriads of additional problems, such as too many correspondences. *Content* as determinative of representation captures the primary character of representation, truth value. That is, it captures the fundamental normative character of representation, that which a representation is *supposed* to represent. So, I have made what is in part an arbitrary semantic decision, though well motivated in the sense outlined, to reserve 'representation' for that which carries truth value, not that which makes contact with the environment.[32]

7. THREE DESIDERATA FOR A MODEL OF REPRESENTATION

The interactive model satisfies three fundamental desiderata for a model of representation. The relationship between representation and content is an internal relationship, thus avoiding the necessity of importing or assigning content. The relationship is accessible to the system itself, so the representation qua representation can be functional in the system's interactions. And the relationship is inherently normative for the system, emergently normative, thus avoiding the need to import normativity from an observer or designer or user or analyzer or explainer (etc.) of the system. Such internal normativity also makes possible an account of system detectable error, thus error guided behavior and learning, and, therefore, avoids the radical skeptical argument that would make these impossible. There are other desiderata, but there is no other model available that satisfies even these three.

8. SO WHAT?

There are many legitimate questions at this point, but one is 'So what?'. What difference does the model make? Perhaps it avoids multiple problems, but does it leave everything else unchanged? Another is: Is it adequate to frame all aspects of cognition and other forms of mentality, or will some further consideration force still further changes back at this foundational level?

These questions cannot be fully answered except by elaborations of the basic model to address many other mental phenomena, to see if it is adequate, and to find out if further changes are required. But many of those elaboration tasks have, in fact, already been undertaken. As examples of what difference the interactive model makes, consider the possibility that it eliminates the frame problems, motivates and permits much richer models of scaffolding in child development (and permanent scaffolding in, for example, human sociality), and permits a fully dynamic model of cognitive processes in which representation flits in and out of emergent existence in exquisitely context sensitive ways – perhaps to be held if striking, salient, or possibly useful, perhaps to simply disappear if not – rather than modeling cognition as operations on inert symbolic representations, or even as the dynamics of trained connectionist vector correspondences.[33] The model does *not* leave everything else unchanged.

With respect to adequacy issues, the core interactive model of representation has been elaborated and extended to address many additional phenomena, and the underlying process metaphysics has been extended even beyond representational issues per se, such as to emotions and psychopathology. The backbone of the information processing perspective, for example, in which perceptual encoded inputs are processed in cognitive systems, perhaps to be re-encoded into language utterances, fails completely – none of the presumed steps of encoding are possible – and the entire framework, as well as the presumed steps within it, must be replaced, and has been replaced. As for representation per se, the resulting models are familiar in what they address, but not always familiar in how they address them.[34] Again, the model does not leave everything else unchanged.

9. CONCLUSION

Emergence, including normative emergence, has occurred. All of our familiar world has emerged since the Big Bang. But emergence, and especially normative emergence, is not possible in a naturalistic framework that accepts a substance and particle metaphysics. Such metaphysical frameworks, therefore, are refuted (so long as naturalism is accepted).

Process metaphysics together with the logical and semantic power of implicit definition renders (accounts of) emergence possible. The asymmetry of normativity, I argue, emerges from a fundamental asymmetry of thermodynamics: energy well stable systems do not require maintenance, while stable far-from-equilibrium system do require maintenance.

Both function and representation can be accounted for with models of self-maintenant and recursively self-maintenant systems – of autonomous systems. Crucial to this account are the internally related functional presuppositions of functional dynamics in a system, especially those of interaction indication and anticipation and their presuppositions about the environment.

The interactive model of representation is just the beginning of the need to re-address all phenomena of the mind free of substance metaphysical blinders and of corresponding encodingist assumptions about representation. That is, to re-address mental phenomena in genuinely dynamic, process terms. Adopting a process perspective does not automatically provide correct models of mental phenomena; it 'simply' clears multiple barriers and *aporiae* out of the way for the attempted construction of such models. Mental states do not exist, any more than do flame states – both are processes.

NOTES

1 Though both Aristotle and Descartes postulated two fundamental realms as well, substance and form for Aristotle and two kinds of substance for Descartes.

2 I once saw a prominent psychologist reject a question about the normativity of representation as being 'mystical'.

3 There are several closely interrelated issues here that are collapsed together. I will argue in the following that a shift to a process metaphysics is required in order to, among other things, account for causally efficacious emergence, an account of emergence is required in order to, among other things, account for normative emergence, and an account of normative emergence is required in order to, among other things, account for representation. Nevertheless, a substance metaphysics makes process problematic, emergence impossible, and normativity, including representational normativity, inexplicable. A substance framework, then, collapses all of these issues into one 'antithetical' realm split off from substance.

4 There is an additional crucial assumption of the causal closure of the (micro-)physical realm. This assumption fits physical and physicalistic intuitions, and rules out, for example, British emergentist postulates of higher level causal laws that apply only to higher level structures and organizations, and, therefore, do not come into play unless and until those higher level patterns are instantiated. One of the British emergentists core examples was chemical valence, and the approach faded when quantum mechanics succeeded in explaining valence phenomena (Stephan, 1992).

5 In his (1998) Kim develops a less reductive model, and even endorses a kind of emergence, but the differences from his earlier work turn on a change in definition of supervenience, not on anything more fundamental, so his model is vulnerable to exactly the same arguments, using his old definition of supervenience, that he has so successfully deployed in previous publications. In particular, he does not avoid the problems demonstrated in his earlier arguments concerning the epiphenomenality of higher levels of organization (Campbell and Bickhard in preparation).

6 The crucial point here is that processes are distributed in space and time, unlike dimensionless point particles. In the case of fields, this is inherent in their mathematical formulation in terms of differential equations: such equations are not definable on discrete point sets.

7 A great deal has been interpreted into "seems altogether inconceivable", which is the limit of Hume's actual argument (Hume 1978, Book III, Part I, Section I, pp. 469–470), and

the interpretations are not without their own controversies. I will not venture in Humean scholarship here, but will stay with the interpretations that seem to have had the strongest historical influence since Hume.

[8] Beth's theorem in model theory (Chang and Keisler 1990), which proves that, under certain conditions, implicit and explicit definition are of equal power, is at times taken to justify ignoring implicit definition (Doyle 1985). But implicit definition is, in fact, more powerful than explicit definition under other conditions (Dawar et al. 1995; Hella et al. 1994; Koaitis 1990), and has never been found to be less powerful. Explicit definition is not a substitute. Implicit definition cannot be safely ignored, and it does refute this fundamental assumption in the argument attributed to Hume.

[9] It is certainly legitimate to appeal to distant causes, including in the past, and it might seem that that is all that is at issue here. But the problem is that distant causes must have their effects via temporal trajectories through current states. So, if differing distant causes result in identical current states, as in the case of the two lions, then those differences in history do not constitute causally efficacious differences. Because those differences in this case constitute the difference between having a function and not having a function, function, in this model, is not causally efficacious.

[10] With regard to normativity per se, there is also a normative inconsistency involved. The selection histories in an etiological approach are selections for properties that are useful to the organism, and usefulness is already a normative notion. For this and other problems, see (Christensen and Bickhard 2002, in preparation).

[11] A number of arguments have been offered for the conclusion that the kind of model of function to be presented here is not possible. Several of these are addressed and refuted in Christensen and Bickhard (2002).

[12] The asymmetry at issue here is the sense in which 'good' is 'preferred' over 'bad'. Physical laws, in general, can provide the basis for making distinctions – between one position and another, or one direction or another, or one velocity or another, and so on – but the laws themselves do not provide any basis for either side of such distinctions being picked out relative to the other side of the distinction. The laws are invariant, symmetric, relative to such distinctions. The exception to this symmetry of physical laws is in thermodynamics.

[13] I ignore special issues concerning 'proper functions' here.

[14] There is an issue here of what constitutes a relevant *part*, how it is constituted, how it is differentiated, and so on. For example, is this damaged kidney still a kidney in the relevant sense of having the function of filtering blood? What about this mass of scar tissue located where a kidney was once located? How are such cases differentiated? This is an issue that is central for etiological model – they take *a part as having a function* as the central focus of explication – but it is not addressed. It's complexity seems not to be appreciated. The notions of infrastructure and functional presupposition provide an approach to the issue of "What is a part?", but I will not develop it here.

[15] For example, discovering that I am wrong about there being a cow in front of me – it's a horse on a dark night instead – requires that I have some functional access to its being a cow that is being represented as being in front of me in the first place, access to my own representational contents. If the content of my own representations is not functionally accessible to me, then such a check and discovery of error is impossible for me, by *any* means.

[16] The most natural rejoinder intuition here is that we don't check a representation against the current situation, we check representational consequences against later situations. The model that I advocate, in fact, makes good on this intuition, but it cannot stand as stated:

(1) that later check is just as subject to the skeptical argument as the first, and (2) it is not the correctness of the representation that would be checked in such a case, but the accuracy of one or more presumed or inferred consequences.

[17] It should be noted that, for etiological models, both the having of *any* content at all is constituted in having some history of the right kind, and the having of some particular content is constituted in having a history of some much more restricted particular kind. In neither case is it constituted in current state of the system, and, so, the epiphenomenality follows for both the general and the particular cases.

[18] This point holds for all models of representation that focus strictly on ascriptions, e.g., Clark (1997).

[19] Furthermore, all that such a read process can do is to influence the functional flow in the system itself, and such influences can always in principle be built directly into the functional organization of the system (Bickhard 1980, 1982).

[20] Like Bradley and others, I intend the notion of internal relations to make a modal claim, and do not intend to combine it with a reductive thesis (cf. Armstrong 1989; Castaneda 1975; Denkel 1997; Von Wachter 1998).

[21] Chomsky's innatism for language is of a somewhat different form from Fodor's, and its problems are even worse (Bickhard 1995; Bickhard and Campbell 1992; Campbell and Bickhard 1992).

[22] This is a notion of autonomy that is profoundly consistent with the Aristotelian notion: "Autonomous entities rely on themselves both for the realization of their capacities and for their persistence" p. 213. "An organism's activity is much more than an expression of what it is; it is also the means by which the organism preserves itself from deterioration" p. 219. 'Self-maintenance is the preservation that results from an organism's self-directed behavior" p. 227. "Living organisms are ... autonomous self-preserving systems"(Gill 1989, p. 241).

[23] Explorations in this direction develop various kinds of motivation, including emergent motivations, such as curiosity and esthetic motivation, and values (Bickhard 2000c, in press-b).

[24] It is important to note that the indications of the possibility of future interactions *constitute* representations in virtue of their presuppositions about the environment, but they are not themselves *constituted out* of representations. In particular, they are (properties of) *functional* organizations in the overall system, and are functionally accessible to and useable by the system. Most importantly, they are not accessed by the system by being represented to the system: that would constitute a circularity in the model. They can be accessed as representations *for* the system, but they are not represented to the system, except perhaps from some higher level of reflection in the system (should the species at issue be capable of reflection), but such reflection is not a part of the overall model that I will develop here.

[25] In complex cognitive organisms such as humans, later elaborations and theories may fill in some degree of explicitness – that it is a child's toy block, for example, that is providing all these manipulation and visual scanning and chewing, etc. opportunities.

[26] One consequence is that implicit content can be unbounded relative to explicit content: it can require lists of explicit contents of unbounded length to capture one implicit content. For example, "the chair in front of me" (such deixis can be one form of implicitness) could apply to an unbounded number of particular chairs. Attempting to make explicit all of what is involved in an implicit content is one source of the frame problems (Bickhard 2001; Bickhard and Terveen 1995).

[27] It is not only learning or development that can generate emergent new representation. The ongoing dynamics of a cognitive system can emergently create and dissolve representational organizations of processes like froth on a wave. This fluid dynamics of representation emergent in cognition is drastically different from standard assumptions of cognition as operations on static, inert, and relatively long lasting representational elements (Bickhard 2000d, 2002).

[28] At this point, note that, if we attempt to construe the 'because' in Dretske's sentence (Dretske 1988, p. 101) as a causal relation, instead of as an explanatory 'because' as Dretske does, we are forced to alter the sentence in order for such causality to be possible. In particular, we could have "C is recruited as a cause of M because it indicates the success of M". These properties and relationships *are* accessible to the organism. In fact, they are the only properties and relationships that are causally real even in Dretske's original version. Note now, however, that C, in indicating the success of M, is *functionally presupposing* F, the success conditions of M. A genuinely causal reinterpretation of Dretske's model yields a kind of desiccated version of the interactive model.

[29] This action, pragmatic, framework is a powerful similarity between Piaget's model and the interactive model. There are also, however, many strong differences (Bickhard 1988; Bickhard and Campbell 1989; Campbell and Bickhard 1986).

[30] Such anticipations can be based on indications, as discussed earlier, or on more sophisticated processes such as *microgenesis* (Bickhard 2000c; Bickhard and Campbell 1996). Microgenesis is also more likely than discrete indications to be realized in brain processes. The crucial point is that there be future oriented anticipations, paths of interaction that are *prepared for* – and in that sense anticipated – as potentials in the current environment, that involve presuppositions about the environment. It is these presuppositions that can be false, and, therefore, true, and, therefore, that can be representations.

[31] Though the contact-content distinction is roughly Cummins' (1996) target-representation distinction.

[32] One advantage, among many, is that content is untied from direct contact with the environment, removing perplexities about, for example, counterfactual representation, fictional representation, musings, and so on. If representation is determined by what it is in correspondence with, these pose awkward problems.

[33] See, for example, Bickhard (2001, in press).

[34] See, for example, for perception (Bickhard and Richie 1983), learning (Bickhard, 1998; Bickhard and Campbell 1996), motivation and emotions (Bickhard 2000c, in press-b), consciousness (Bickhard 2000c), sociality (Bickhard 1992, 1992b, in press-c), language (Bickhard 1980, 1987, 1992, 1995, 1998; Bickhard and Campbell 1992; Bickhard and Terveen 1995), rationality (Bickhard 2002b).

REFERENCES

Armstrong, D.M.: 1989, *Universals. An Opinionated Introduction*, Boulder: Westview.

Bickhard, M.H.: 1980, *Cognition, Convention, and Communication*, New York: Praeger Publishers.

Bickhard, M.H.: 1982, 'Automata Theory, Artificial Intelligence, and Genetic Epistemology', *Revue Internationale de Philosophie* **36**(142–143), 549–566.

Bickhard, M.H.: 1987, 'The Social Nature of the Functional Nature of Language', in Maya Hickmann (ed.), *Social and Functional Approaches to Language and Thought*, Academic Press, New York, pp. 39–65.

Bickhard, M.H.: 1988, 'Piaget on Variation and Selection Models: Structuralism, Logical Necessity, and Interactivism.', *Human Development* **31**, 274–312.

Bickhard, M.H.: 1992, 'How Does the Environment Affect the Person?' , in L.T. Winegar and J. Valsiner (eds.) , *Children's Development within Social Contexts: Metatheory and Theory*, Erlbaum, pp. 63–92.

Bickhard, M.H.: 1992b, 'Scaffolding and Self Scaffolding: Central Aspects of Development', in L.T. Winegar and J. Valsiner (eds.), *Children's Development within Social Contexts: Research and Methodology*, Hillsdale, NJ: Erlbaum, pp. 33–52.

Bickhard, M.H.: 1993, 'Representational Content in Humans and Machines', *Journal of Experimental and Theoretical Artificial Intelligence* **5**, 285–333.

Bickhard, M.H.: 1995, 'Intrinsic Constraints on Language: Grammar and Hermeneutics', *Journal of Pragmatics* **23**, 541–554.

Bickhard, M.H.: 1998, 'Levels of Representationality', *Journal of Experimental and Theoretical Artificial Intelligence* **10**(2), 179–215.

Bickhard, M.H.: 2000, 'Emergence', in P.B. Andersen, C. Emmeche, N.O. Finnemann, and P.V. Christiansen (eds.), *Downward Causation*, Aarhus, Denmark: University of Aarhus Press, pp. 322–348.

Bickhard, M.H.: 2000b, 'Autonomy, Function, and Representation', *Communication and Cognition – Artificial Intelligence* **17**(3–4), 111–131.

Bickhard, M.H.: 2000c, 'Motivation and Emotion: An Interactive Process Model', in R.D. Ellis and N. Newton (eds.) *The Caldron of Consciousness*, J. Benjamins, pp. 161–178.

Bickhard, M.H.: 2000d, 'Dynamic Representing and Representational Dynamics', in E. Dietrich and A. Markman (eds.) , *Cognitive Dynamics: Conceptual and Representational Change in Humans and Machines*, Hillsdale, NJ: Erlbaum, pp. 31–50.

Bickhard, M.H.: 2001, 'Why Children Don't Have to Solve the Frame Problems: Cognitive Representations are not Encodings', *Developmental Review* **21**, 224–262.

Bickhard, M.H.: 2002, 'The Biological Emergence of Representation', in T. Brown and L. Smith (ed.), *Emergence and Reduction: Proceedings of the 29th Annual Symposium of the Jean Piaget Society*, Hillsdale, NJ: Erlbaum, pp. 105–131.

Bickhard, M.H.: 2002b, 'Critical Principles: On the Negative Side of Rationality', *New Ideas in Psychology* **20**, 1–34.

Bickhard, M.H.: in press, 'The Dynamic Emergence of Representation', i n H. Clapin, P. Staines, and P. Slezak (eds.), *Representation in Mind: New Approaches to Mental Representation.*, New York: Praeger Publishers.

Bickhard, M.H.: in press-b, 'An Integration of Motivation and Cognition', in Les Smith and Colin Rogers (eds.), *British Journal of Educational Psychology*.

Bickhard, M.H.: in press-c, 'The Social Ontology of Persons', in J. Carpendale and U. Mueller (eds.), *Social Interaction and the Development of Knowledge*, Hillsdale, NJ: Erlbaum.

Bickhard, M.H. and R.L. Campbell: 1989, Interactivism and Genetic Epistemology, *Archives de Psychologie* **57**(221), 99–121.

Bickhard, M.H. and R.L. Campbell: 1992, 'Some Foundational Questions Concerning Language Studies: With a Focus on Categorial Grammars and Model Theoretic Possible Worlds Semantics', *Journal of Pragmatics* **17**(5/6), 401–433.

Bickhard, M.H. and R.L. Campbell: 1996, 'Topologies of Learning and Development', *New Ideas in Psychology* **14**(2), 111–156.

Bickhard, M.H. and D.M. Richie: 1983, *On the Nature of Representation: A Case Study of James Gibson's Theory of Perception*, New York: Praeger Publishers.

Bickhard, M.H. and L. Terveen: 1995, *Foundational Issues in Artificial Intelligence and Cognitive Science: Impasse and Solution*, Amsterdam: Elsevier Scientific.

Brown, H.R. and R. Harré: 1988, *Philosophical Foundations of Quantum Field Theory*, Oxford: Oxford University Press.

Campbell, D.T.: 1974, 'Evolutionary Epistemology', n P.A. Schilpp (ed.), *The Philosophy of Karl Popper*, LaSalle, IL: Open Court, pp. 413–463.

Campbell, D.T.: 1990, 'Levels of Organization, Downward Causation, and the Selection-Theory Approach to Evolutionary Epistemology', in G. Greenberg and E. Tobach (eds.), *Theories of the Evolution of Knowing*, Hillsdale, NJ: Erlbaum, pp. 1–17.

Campbell, R.J. and M.H. Bickhard: in preparation, 'Physicalism: Particulars and Configurations'.

Campbell, R.L. and M.H. Bickhard: 1986, *Knowing Levels and Developmental Stages*, Contributions to Human Development. Basel, Switzerland: Karger.

Campbell, R.L. and M.H. Bickhard: 1992, 'Clearing the Ground: Foundational Questions Once Again', *Journal of Pragmatics* **17**(5/6), 557–602.

Cao, T.Y.: 1999, 'Introduction: Conceptual Issues in Quantum Field Theory', in T.Y. Cao (ed.), *Conceptual Foundations of Quantum Field Theory*, Cambridge: Cambridge University Press, pp. 1–27.

Carlson, N.R.: 2000, *Physiology of Behavior*, 7th edn, Boston: Allyn and Bacon.

Castaneda, H.N.: 1975, 'Relations and Identity of Propositions', *Philosophical Studies* **28**, 237–244.

Chang, C.C. and H.J. Keisler: 1990, *Model Theory*, Amsterdam: North Holland.

Christensen, W.D.: 1996, 'A Complex Systems Theory of Teleology', *Biology and Philosophy* **11**, 301–320.

Christensen, W.D. and M.H. Bickhard: 2002, 'The Process Dynamics of Normative Function', *Monist* **85**(1), 3–28.

Christensen, W.D. and M.H. Bickhard, in preparation, 'Function as Design versus Function as Usefulness'.

Christensen, W.D. and C.A. Hooker: 2000, 'An Interactivist-Constructivist Approach to Intelligence: Self-Directed Anticipative Learning', *Philosophical Psychology* **13**(1), 5–45.

Christensen, W. D. and Hooker, C. A.: 1998, 'Autonomous Systems and Self-Directed Heuristic Policies: Toward New Foundations for Intelligent Systems', in B. Hayes, R. Heath, A. Heathcote, and C.A. Hooker (eds.), *Proceedings of the Fourth Australian Cognitive Science Conference*, Newcastle, Australia.

Clark, A.: 1997, *Being There*, Cambridge, MA: MIT/Bradford.

Cummins, R.: 1996, *Representations, Targets, and Attitudes*, Cambridge, MA: MIT.

Davies, P.C.W.: 1984, 'Particles Do Not Exist', in S.M. Christensen (ed.), *Quantum Theory of Gravity*, Adam Hilger, pp. 66–77.

Dawar, A., L. Hella, and Ph.G. Kolaitis: 1995, 'Implicit Definability and Infinitary Logic in Finite Model Theory', *Proceedings of the 22nd International Colloquium on Automata, Languages, and Programming*, ICALP 95, Szeged, Hungary, July 10–11, 1995, New York: Springer-Verlag, pp. 621–635.

Denkel, A.: 1997, 'On the Compresence of Tropes', *Philosophy and Phenomenological Research* **57**(3), 599–606.

Doyle, J.: 1985, 'Circumscription and Implicit Definability', *Journal of Automated Reasoning* **1**, 391–405.

Dretske, F.I.: 1988, *Explaining Behavior*, Cambridge, MA: MIT Press.

Fodor, J.A.: 1975, *The Language of Thought*, New York: Crowell.

Fodor, J.A.: 1981, The Present Status of the Innateness Controversy', in J. Fodor (ed.), *RePresentations*, Cambridge, MA: MIT Press, pp. 257–316.

Fodor, J.A.: 1986, 'Why Paramecia don't have Mental Representations', in P.A. French, T.E. Uehling, and H.K. Wettstein (eds.) , *Midwest Studies in Philosophy X: Studies in the Philosophy of Mind*, Minneapolis, MN: University of Minnesota Press, pp. 3–23.

Fodor, J.A.: 1987, *Psychosemantics*, Cambridge, MA: MIT Press.

Fodor, J.A.: 1990, *A Theory of Content*, Cambridge, MA: MIT Press.

Fodor, J.A.: 1990, 'Information and Representation', in P.P. Hanson (ed.), *Information, Language, and Cognition*, Vancouver: University of British Columbia Press, pp. 175–190.

Fodor, J.A.: 1991, 'Replies', in B. Loewer and G. Rey (eds.), *Meaning in Mind: Fodor and His Critics*, Oxford: Blackwell, pp. 255–319.

Fodor, J.A.: 1998, *Concepts: Where Cognitive Science Went Wrong*, Oxford.

Gill, M.-L.: 1989, *Aristotle on Substance*, Princeton: Princeton University Press.

Greco, J.: 2000, *Putting Skeptics in Their Place*, Cambridge: Cambridge University Press.

Hale, B. and Wright, C.: 2000, 'Implicit Definition and the A Priori', in P. Boghossian and C. Peacocke (eds.), *New Essays on the A Priori*, Oxford: Oxford University Press, pp. 286–319.

Hella, L., P.G. Kolaitis, and K. Luosto: 1994, 'How to Define a Linear Order on Finite Models', *Proceedings: Symposium on Logic in Computer Science, Paris, France, July 4–7*, Los Alamitos, CA.: IEEE Computer Society Press. Huggett, N.: 2000, 'Philosophical Foundations of Quantum Field Theory', *The British Journal for the Philosophy of Science* **51**(Suppl.), 617–637.

Hume, D.: 1978, *A Treatise of Human Nature*, Index by L.A. Selby-Bigge; Notes by P.H. Nidditch, Oxford: Oxford University Press.

Joas, H.: 1993, 'American Pragmatism and German Thought: A History of Misunderstandings.', in H. Joas ed., *Pragmatism and Social Theory*, Chicago: University of Chicago Press, pp. 94–121.

Keisler, H.J.: 1977, 'Fundamentals of Model Theory', in J. Barwise (ed.), *Mathematical Logic*, Amsterdam: North Holland.

Kim, J.: 1989, The Myth of Nonreductive Materialism', *Proceedings and Addresses of the American Philosophical Association* **63**, 31–47.

Kim, J.: 1990, 'Supervenience as a Philosophical Concept', *Metaphilosophy* **21**(1–2), 1–27.

Kim, J.: 1991, 'Epiphenomenal and Supervenient Causation', in D.M. Rosenthal (ed.) , *The Nature of Mind*, Oxford: Oxford University Press, pp. 257–265.

Kim, J.: 1992a, 'Downward Causation in Emergentism and Non-Reductive Physicalism', in A. Beckermann, H. Flohr, and J. Kim (eds.), *Emergence or Reduction? Essays on the Prospects of Nonreductive Physicalism*, Berlin: Walter de Gruyter, pp. 19–138.

Kim, J.: 1992b, 'Multiple Realization and the Metaphysics of Reduction', *Philosophy and Phenomenological Research* **52**, 1–26.

Kim, J.: 1993a, *Supervenience and Mind*, Cambridge: Cambridge University Press.

Kim, J.: 1993b, 'The Non-Reductivist's Troubles with Mental Causation', in J. Heil and A. Mele (eds.) , *Mental Causation*, Oxford: Oxford University Press, pp. 189–210.

Kim, J.: 1997, 'What is the Problem of Mental Causation?', in M.L.D. Chiara, K. Doets, D. Mundici, and J. van Benthem (eds.) *Structures and Norms in Science*, Dordrecht: Kluwer Academic Publishers, pp. 319–329.

Kim, J.: 1998, *Mind in a Physical World*, Cambridge, MA: MIT Press.

Kneale, W. and M. Kneale: 1986, *The Development of Logic*, Oxford: Clarendon Press.

Kolaitis, Ph.G.: 1990, 'Implicit Definability on Finite Structures and Unambiguous Computations', in *Proc. 5th IEEE LICS*, pp. 168–180.

Levine, A. and M.H. Bickhard: 1999, 'Concepts: Where Fodor Went Wrong', *Philosophical Psychology* **12**(1), 5–23.

Millikan, R.G.: 1984, *Language, Thought, and Other Biological Categories*, Cambridge, MA: MIT Press.

Millikan, R.G.: 1993, *White Queen Psychology and Other Essays for Alice.*, Cambridge, MA: MIT Press.

Moreno, A. and K. Ruiz-Mirazo: 1999, 'Metabolism and the Problem of its Universalization', *BioSystems* **49**, 45–61.

Nicolis, G. and I. Prigogine: 1977, *Self-Organization in Nonequilibrium Systems*, New York: Wiley.

Piaget, J.: 1954, *The Construction of Reality in the Child*, New York: Basic.

Piaget, J.: 1970, *Genetic Epistemology*, New York: Columbia.

Rescher, N.: 1980, *Scepticism*, Totowa, NJ: Rowman and Littlefield.

Saunders, S. and H.R. Brown: 1991, *The Philosophy of Vacuum*, Oxford: Oxford University Press.

Stephan, A.: 1992, Emergence – A Systematic View on its Historical Facets', in A. Beckermann, H. Flohr, and J. Kim (eds.), *Emergence or Reduction? Essays on the Prospects of Nonreductive Physicalism*, Berlin: Walter de Gruyter, pp. 25–48.

Von Wachter, D.: 1998, 'On Doing without Relations', *Erkenntnis* **48**(2/3), 355–358.

Weinberg, S.: 1977, 'The Search for Unity, Notes for a History of Quantum Field Theory', *Daedalus* **106**(4), 17–35.

Weinberg, S.: 1995, *The Quantum Theory of Fields. Vol. 1. Foundations*, Cambridge: Cambridge University Press.

Weinberg, S.: 1996, *The Quantum Theory of Fields. Vol. II. Modern Applications.*, Cambridge: Cambridge University Press.

Weinberg, S.: 2000, *The Quantum Theory of Fields: Vol. III. Supersymmetry*, Cambridge: Cambridge University Press.

WAYNE CHRISTENSEN

SELF-DIRECTEDNESS: A PROCESS APPROACH TO COGNITION

ABSTRACT. Standard approaches to cognition emphasise structures (representations and rules) much more than processes, in part because this appears to be necessary to capture the normative features of cognition. However the resultant models are inflexible and face the problem of computational intractability. I argue that the ability of real world cognition to cope with complexity results from deep and subtle coupling between cognitive and non-cognitive processes. In order to capture this, theories of cognition must shift from a structural rule-defined conception of cognition to a thoroughgoing embedded process approach.

1. INTRODUCTION

Despite the fact that processes predominate in the natural world, thoroughgoing process theories are rare in science and philosophy; for many reasons structures tend to take the focus of attention. In this paper I will argue that the emphasis on structures has led to deep problems in cognitive science, and that contemporary evidence favours a systematically process-oriented approach. Developing such an approach poses enormous challenges because it requires us to rethink the central issues for understanding cognition, but the payoff is a more integrated perspective that helps reframe traditional problems and opens up a more productive relationship between high-level theory and the various empirical sciences of cognition.

2. COGNITION SANDWICHED BETWEEN STRUCTURES

Standard approaches to cognition emphasize structures far more than processes. In particular, the following structural assumptions are typical in cognitive science:[1]

- Representations are structures (symbols) that have the function of corresponding to features of the world.
- Cognitive processes are operations on representations.

J. Seibt (ed.), Process theories: Crossdisciplinary studies in dynamic categories, 157–175.

- Cognitive architectures are structures that specify possible cognitive processes.
- The world has a structural ontology consisting of objects, properties and categories.

It has been very difficult to think about cognition in any other way. Cognitive processes are intuitively thought of as transformations between perception, ideas and action. As such, it is clear that they can vary in how well they work, with accurate perception and reliable judgement of consequences being clearly important for cognitive success. These factors in turn seem to essentially require that cognitive processes satisfy certain kinds of relations: truth and the principles of valid inference being primary candidates. These are then taken to be the normative standards for cognition.

In explaining normative standards it is natural to think that there must be an encompassing cognitive system that defines possible meaningful relationships and rules for cognitive processes (e.g., the principles of rationality, the computational architecture of classical artificial intelligence, or Chomsky's universal grammar), and basic structures that serve as *meaning bearers*. In a complementary way, the world is thought of as consisting of objects with defining properties that interact in a law-like manner and are organised into hierarchical categories, such as natural kinds.[2] Cognitive processes must be rationally structured so that they can represent these relationships.

So, once we have bought into the compelling idea that cognition is systematic we are then forced from cognitive processes upwards to general structures that define the systematicity of the processes, and downwards to the specific structural types that participate in the processes. Cognition therefore seems to be necessarily sandwiched between structures below and above. And when it works *well* – that is, it is rational – the structural articulation of cognition isomorphically mirrors the structured relationships in the world.

3. COGNITIVISM AND THE PROBLEM OF FLEXIBILITY

Cognitivism is the modern elaboration of these intuitions, and normativity enforces its structuralist orientation with an iron grip: cognition can only be rational if it is rule-governed. However there is a fundamental tension between the idea that cognition is rule-governed and another important property of intelligence: *flexibility*. Complex cognition is highly context sensitive in at least two interrelated ways. Firstly, the interpretation of meaning is shaped by background knowledge and knowledge of the context. When dining in a restaurant, for example, small cues like placing the knife and fork together on a half finished plate can indicate both that you want the plate removed and that you were not happy with the meal, so reducing the waiter's expectations concerning the tip. Secondly, the control of action can be made sensitive to many simultaneous factors in a situation. Whether you order the salmon can depend on a combination of vital factors, including the way it is prepared, the sauce, the selection of white wines, which desert you would prefer, the evening meal you had the previous day, whether the restaurant has a reputation for sea food, and the counter-temptation of the lamb cutlet. I shall generically refer to this ability to take many factors into account when deciding action as *situational awareness*, and argue that it is a crucial factor in most complex cognitive activities because it facilitates flexible, context-sensitive action.

At first glance situational awareness seems to be just a matter of accurately representing the world in the manner envisioned by the traditional normative account of cognition described above. Yet, curiously, cognitivist artificial intelligence (AI) – which has carried the rule-governed approach to its logical conclusion by modelling cognitive processes as formal symbolic processes such as algorithms and heuristic searches of problem spaces – has faced apparently intractable difficulties in trying to understand situational awareness. One key difficulty concerns the fact that something like situational awareness is required for cognition at all. Early AI approaches attempted to model reasoning abilities that only employed general principles in order to solve problems, such as Newell and Simon's General Problem Solver program (Newell and Simon 1972). However, it became apparent that most human problem solving involves 'knowledge-based' rather than 'general' reasoning. The AI approach to modelling so-called knowledge-based reasoning has been to program AI agents with large databases encoding 'common sense' and 'domain specific' information. Such a database is assumed to constitute a representation of a problem domain, and the problem solving process proceeds by searching the space in some way. However this strategy has led directly to further problems, in particular computational intractability due to combinatorial explosion.

Combinatorial explosion occurs because the computational task of updating perception and calculating the effects of changes in the world grows exponentially as the complexity of the agent's representation of the world is increased. The well known 'frame problem' is related to this and concerns the issue of representing a dynamic environment. To keep computation manageable it has been assumed that the agent should restrict its calculation of states of the world to only those properties that are changed by an event. Thus, to use a standard example, when a robot calculates the effects that result when it moves into a room, it should *not* have to compute the fact that the colour of the room will remain the same. Without such an assumption it would seem that for *each* event that occurs the agent would need to search its complete knowledge base and recalculate the entire state of the world. The trouble with implementing the idea lies in finding a way of pre-specifying what will change and what won't for any given event. One strategy has been to build in a default assumption that properties persist, unless there is an explicit rule to the contrary. However this leads either to implausibly simplistic assumptions about the world or an inordinate number of rules: most properties change in at least some circumstances.

Indeed, whilst the motivation for such a strategy seems straightforward enough within the computational framework in which it arose, in the light of more recent research emphasising cognition in realistic situations it looks rather dubious. To assume prevailing stasis as a way of achieving computational tractability is effectively an attempt to represent a dynamic world by assuming that it is not very dynamic. Clearly not everything changes all the time, or cognition would be impossible. But on the other hand the real-world problem contexts that demand intelligence are *characteristically* highly dynamic. Predominantly static environments are not ones that require or would give rise to intelligence. The crux of the issue can be seen more clearly if we contrast the frame problem with the human ability to be sensitive to relevance. A human performing a skilled activity at a high standard is able to focus quickly on *just* the relevant factors in a situation, and to quickly to shift attention as changes in the situation make new factors relevant. Attempting to restrict calculation to actual change in the environment is, per se, at best a very crude approximation to the ability to detect relevance: situational factors can change and yet not be relevant to the agent given the agent's goals, whilst a change in goals can make new factors relevant to the agent that have not in themselves changed. So the frame problem as understood in AI is really just the tip of the iceberg, so far as the problem of effective cognitive functioning in a complex environment is concerned.

Two principle strategies for solving the frame problem have been pursued in AI: (1) finding ontologies for knowledge representation that aptly express relationships in the world, and (2) finding effective search heuristics. An important thing to note is that these are general computational strategies. As such, there is something essentially misconceived about them: in environments complex enough to require intelligence the information that is relevant to action varies in a highly context-dependent way.

Knowing what is relevant is therefore inherently something that cannot be achieved via a general computational strategy. Situation-specific learning must be involved. This is supported by the empirical fact that in highly novel situations people have great difficulty in effectively differentiating between relevant and irrelevant information.

A possible response to this might be to propose a layered solution that involves first solving the representational ontology and heuristic computational issues, and then implementing learning processes that refine the agent's knowledge about particular situations. The difficulty is that, no matter how the representational ontology is constructed, realistic cognitive situations have enough significant variables to present astronomical numbers of possible states. And since AI learning techniques rely on searching the represented possibility space, learning itself becomes intractable. In other words there is simply no way around the problem of combinatorial explosion. It might be protested that this simply can't be true, since we have an existence proof that cognition is possible in situations of at least the complexity that humans face; namely, human cognition itself. But this fails to consider the possibility that human cognition might operate according to very different principles to those assumed by cognitivist AI.

To give some flesh to this prospect we can consider briefly some of the key tenets of alternative paradigms, in particular pragmatism and situated cognition research. Perhaps the most basic principle of situated cognition is that an adaptive agent doesn't have to represent the world internally in order to successfully interact with it. Brooks (1991) argued that a critical problem with the cognitivist AI approach was the fact that all information used to control behaviour had to first pass through a central representational module. Robots built according to this principle became paralysed as they attempted to update perception and calculate what action to take. He advocated an alternative design paradigm known as behaviour-based robotics, which entirely dispensed with internal representation of the world and instead relied on well-tuned actions triggered by environmental stimuli.

However, although Brooks originally proposed that behaviour-based robotics could serve as a model for intelligence generally, it soon became apparent that it doesn't scale up very well. One of the basic reasons is very simple: some kinds of tasks require complex sequences of actions, and it is difficult or impossible to find external cues that will evoke the actions in the right order.[3] Consequently, in these circumstances some form of internal control of action sequencing is required. This brings us back to the issue of representation, and one stream of research in robotics has sought to develop hybrid systems that combine elements of the behaviour- based and classical approaches in an effort to gain the benefits of each. However this would seem to have only limited prospects, since the original problems with AI have not been solved.

The pragmatist tradition suggests the basis for an alternative route. AI robots became bogged down because to do anything they require extremely detailed, and hence computationally expensive, representational models of the world. Behaviour-based robots gained real-time effectiveness by dispensing with representations, but at the cost of being unable to perform tasks that require internal ordering of action. In contrast, pragmatism does not eschew representation, but it shares with behaviour-based robotics the idea that the world is not represented in cognition as a collection of objects with intrinsic properties and law-like behaviours. Rather, representation emerges out of the control of action. The agent learns the properties of objects by interacting with them, and is particularly focussed on learning those factors that affect action success.[4] This is very different to the cognitivist approach to representation, and the differences have important implications for understanding the way in which real cognitive agents are able to manage complex relationships. For cognitivism representation is *prior* to action, whereas for pragmatism it emerges *out of* action and is learned in an action- relevant way. Thus, the interaction process acts as a filter helping to determine the relevance of the information that goes into the *formation* of representations.

The nature of the representations on this interpretation is also inherently action-relevant. The base representational format for cognitivism is explicit description, whereas for pragmatist approaches it is some form of control function. J. J. Gibson's theory of affordances, for example, claims that we perceive the world in terms of opportunities to interact with it (Gibson 1977). Empirical evidence provides support for this: both infants and adults perceive objects and spatial relations in relation to their body size and orientation, and respond to stimuli prospectively in terms of the actions they may need to perform (e.g., Bertenthal and Clifton 1997). Control-based representations can be much more efficient than explicit description because they are able to rely on implicit regularities, for example through implicit predication (Bickhard, this volume), and can exploit distributed sensorimotor and affective information.

Clearly humans do *acquire* knowledge in a form that makes possible explicit description, referred to as declarative knowledge. But it is learned in a rich sensorimotor context and there is evidence that deeper non-declarative forms of knowledge are required in order for declarative knowledge to appropriately influence action (e.g., Bechara et al. 1997). Research in cognitive linguistics also suggests that conceptualisation is deeply shaped by embodiment (e.g., Lakoff 1987). Moreover a further advantage of control-based representation is that it is easy to understand how the burden of control can be distributed between the agent and the environment, and redistributed through learning. Research on human cognition in real world contexts has shown that humans can simplify the complexity of the cognitive problems they must solve during a complex task by using environmental organisation – such as the layout of the navigation deck on a warship or an aeroplane cockpit – as part of the problem solving process (Hutchins 1995a, b).

Situated cognition research thus suggests that there are at least two complementary ways in which real cognitive agents avoid the computational complexity problems that afflict AI, or at least transform them into a tractable form.[5] Firstly, in many cases they avoid the need for representing the environment by allowing the environment itself to organise their behaviour. Secondly, where representation is required, it is constructed from interaction in a way that is inherently related to action, and is guided by learning processes that help build in relevance. At the most general level, then, we can see that situated cognition gains an advantage by treating cognitive processes as deeply coupled to non-cognitive processes. In particular, it treats the organisation of cognition as being in part the result of interaction, which means that understanding how cognition can focus on the relevant aspects of interaction is much less problematic.

On the other hand, from a traditional perspective the price is high. Formal computation provides a framework that appears to satisfy the key normative requirements that a theory of cognition must satisfy, by showing how cognitive processes can be both representational and rule-governed. On the pragmatist account it appears that the formation of representations must be a non-cognitive process, since *ex hypothesi* the process is not itself mediated by representations and rule-governed cognitive processes. Moreover, since the properties of the pragmatist representations lie beyond the scope of cognitive norms, it would appear that the behaviour of these representations in cognitive processes must also be cognitively ill-defined. It is therefore not at all clear how normative cognitive standards could be applied to pragmatist cognition. I will examine this problem in more detail in the next section, but first I will consider one final strategy for rescuing cognitivism because it highlights just how important the issue of flexibility is.

Another approach to the problem of combinatorial explosion that remains within the classical framework is to try to bypass it by compartmentalising cognition into modules, with each module being responsible only for some specialised cognitive domain. Computational complexity is reduced because the cognitive module is equipped with just a limited set of domain specific knowledge. It is also possible to use representational formats for each domain that facilitate the type of computations performed in the domain (e.g., visual or auditory perception), further speeding cognition. And it has been further argued that evolution should have furnished us with a suite of cognitive modules apt for solving the problems we characteristically face.

However there are a number of problems with the so- called 'massive modularity' principle as a solution for cognitive science, one of the more striking being that it abandons, or at least greatly reduces, any attempt to capture the generality or flexibility of cognition. Instead, cognitive problems are claimed to fall within well-defined domains for which we are equipped with pre-specified solutions. It is certainly true that human intelligence shows considerable domain specific organisation: a chef may be able to make highly intelligent choices in a kitchen, but this will not translate in any direct way to an ability to be an effective platoon commander in combat. But it is also true that in many normal situations we effortlessly blend information from a variety of types of sources, and are able to call up various types of information as it becomes relevant. A simple shopping trip to the supermarket, for instance, calls on numerical knowledge, cooking knowledge, experience with the products available, ability to deal with parking lots, queues and checkout personnel, and so on.

One possibility which might seem plausible is that, although successfully completing the overall shopping trip requires the input of knowledge from a variety of sources, it can be broken down into a series of sub-problems, such as parking, finding the right entrance, selecting the products, paying at the cash-register, each of which could be managed by a domain-specific computational module (spatial, numerical, social, etc.). But whilst there are some situations that might be treated in a strictly modular way, modularity cannot be a general solution to effective behaviour in realistic situations. Many of the phases of the shopping process involve some varying *mixture* of kinds of knowledge, with a high degree of interdependency between background factors across domains.

For example, having invited a friend to dinner whom I want to impress, I plan to cook my most effective dish. Unfortunately the supermarket doesn't have the required ingredients, and nor does it have the ingredients for my second most effective dish. The butcher next door has some high quality steak, but I'm not entirely confident about making an appropriate sauce, and including a decent bottle of red wine will push the total cost up to the point where I will be without money for the last two days before payday. A Thai green curry with beer or a Riesling will be much cheaper, but might not create the intimate social atmosphere I'm aiming for. A simple gorgonzola gnocchi accompanied by an aged Semillon could set a more appropriate mood. But the steak will have more class. Can I make the sauce? Can I borrow money from a friend 'till payday? I'm strongly tempted to gamble on the sauce, but I also know that almost all of my friends are even more likely to have spent all their money than I am, and the only one who isn't is out of town. Reluctantly, I toss the packet of gnocchi into my shopping basket and head for the cheese aisle.

Fodor (2000) argues that abduction (inference to the best explanation) is an important feature of human reasoning, and consequently the massive modularity thesis cannot be right. Nor, indeed, can any variant of computational cognitivism. The thrust of his argument is that abduction requires calling on information from many different domains for a given inference. The domains therefore *can't* be strongly modular in the way that the massive modularity thesis supposes, and cognitive science has no theory of how knowledge can be organised in the global ways required for abduction. My supermarket scenario is designed to suggest that this argument is essentially correct. But it is also too limited: the problems concern flexibility in intelligence generally, not just abduction, especially if abduction is interpreted to be a special, epistemically demanding cognitive process, perhaps distinctive to science as Carruthers (forthcoming) proposes. To the contrary, the ability to call on many different forms of knowledge fluidly and appropriately as the situation unfolds is a feature of most human practical reasoning.

Insofar as cognition does show domain organisation, extensive evidence in developmental and cognitive neuroscience suggests that these domains are permeable: their inner workings are accessible to each other and to higher executive processes that are not domain specific. A process as important to cognition as learning the properties of an object, for instance, requires the coordination of information from different modalities, such as visual shape and tactile shape. Learning about objects thus requires integrating information across modalities and through time. Moreover it also appears to be the case that the principles governing the specialisation of cognitive organisation in the higher brain areas are not primarily geared towards separation into complete problem types, which is what is required for domain specific modularity in the sense assumed by the massive modularity thesis. Rather, they seem to concern *aspects* of complex behaviour – numerical, social, culinary, and so on – that are *combined* to produce situationally appropriate action. For example, individuals with brain lesions may lose the ability to recognise tools but not animals, or vice versa, suggesting that specialised regions process these forms of knowledge.

Nevertheless there is no barrier in normal cognition to thinking about both animals and tools at the same time or combining the information in problem solving, and there is evidence that retrieval of different types of semantic information is managed by executive systems in the prefrontal cortex (Thompson Schill 1997). Moreover, a study by Tranel et al. (1997) has shown that the category related specialisation can be explained in terms of perceptual and use-related factors, suggesting that the specialisation is the result of learning rather than innate.

4. COGNITION ENMESHED IN PROCESSES

Cognitivist AI has struggled to explain how agents could act intelligently in complex environments because on the one hand this requires extensive situational awareness, and on the other hand such awareness seems to immediately produce computational intractability. In the previous section I argued that pragmatist approaches might provide a way out of the problem because they treat cognition as being intimately coupled with other processes, thus helping to explain how cognition could be tuned to the relevant aspects of interaction. Cognitivism hasn't looked for such coupling, and it is inherently difficult for it to explain.

The reason for this lies in the fact that formal systems are essentially closed. If cognition is formal computation there are only two ways in which a non- cognitive process can be related to a cognitive process: as an implementation, or as an input that is first represented and then enters the computational stream. Any other kind of influence can only interfere with cognition, since (a) an influence that results in the production of an undefined symbol (or symbols) cannot be treated as meaningful by the computational rules that transform symbolic statements into other statements, (b) an influence that manifests as a change in the rules that define the computations that the system can perform will result in false, invalid or meaningless output, and (c) any causal effect that does not change the symbol properties or computational rules (such as painting the computer green) will have no computational influence at all. The whole point of the development of formal systems theory has been to eliminate the need for subjective (i.e., not formally defined) interpretation in proofs by regimenting the inferential steps as simple mechanical rules (Seig 1999). But this makes it apparent why any formalist theory of cognition will face the problem of computational intractability: all coupling between the cognitive system and the world must be mediated by representations, and realistic cognitive situations have an extremely large number of potentially relevant factors. To be able to represent the inputs and solutions to the problems that may arise the system has to have an extremely large representational space, but then it faces the problem that all the cognitive processes it must perform involve searching that space.

4.1. *Reciprocal coupling between low and high order cognition*

At this point it is worth examining the issue of the formation of higher order cognitive organisation, including concepts, models etc. The ability to be sensitive to relevance is connected to the ability to form higher order cognitive organisation because it involves, at least in part, grouping into like and unlike, important and unimportant. The basic principle by which cognitivism explains higher cognitive organisation is composition: simple representational constituents are combined to form complex expressions. However there is evidence that suggests some important forms of cognition do not follow principles of compositionality. Traditional theories of speech perception have favoured building block models in which auditory perception is processed for phonetic categorization and phonemes are assembled into words and higher grammatical structures. But recent evidence has shown that phoneme perception is strongly sensitive to higher cognitive influence. That is, the perception of a sound pattern as a phoneme is greatly influenced by the linguistic context and the expectations of the listener (Remez 2000).

On this basis, Remez argues that the reification of linguistic entities as 'things' that are assembled is mistaken, and that a 'system of differences' approach better explains the nature of information processing in speech perception. Though Remez doesn't quite describe it like this, the 'system of differences' idea treats the information processing ability of the system as an emergent and variable property affected by learning. The problem of reducing the ambiguity in a perceptual signal is achieved by the imposition of constraints restricting the possible interpretations, and these constraints can be assembled in higher cognition and used to shape early perception. For example, without contextual information, to interpret a group of sounds as a word an English speaking listener has 100 000 possible words to choose from. With sentence and pragmatic information about speaker intentions the possibilities can be much reduced, sometimes to the point where no sound signal at all is required in order to guess the word – e.g., "A stitch in time saves..." (ibid., p. 108). Thus, increases in the higher knowledge that an individual can bring to bear in a situation help to improve the sharpness of the perceptual differentiations the individual can make. Neisser (1997) points to the same principle as a crucial factor in the accuracy of memory recall.

The principle of convergent constraints provides a way for high order cognition to improve low order cognitive processes such as auditory perception and word recall. Goldstone et al. (2000) demonstrate an even stronger top-down effect in visual perception (one that is implicit in Remez' discussion) whereby early perceptual feature recognition is learned, and the learning is guided by high order conceptual knowledge. They cite neurological evidence showing that cortical areas involved with early perceptual processes can be flexibly and context sensitively tuned by training, and behavioural evidence that in areas as diverse as radiology and beer tasting experts make different perceptual differentiations compared to novices (pp. 193–194). Experimentally they show two related effects in perceptual learning: dimension sensitization and unitization. Dimension sensitization involves paying increasing attention to stimulus dimensions (e.g., colour, shape, etc.) that are important for classifications. These attention biases can occur in early perception, where they are the result of perceptual learning, or late perception, where they are based on strategic judgment concerning the situation. Indeed, the prior learning of the individual can affect the way in which a whole is decomposed into parts. Unitization involves grouping together a complex of features into a single functional whole. For instance the letter *A* may initially be perceived as a group of oriented lines, but after extended learning come to be perceived as a functional unit (ibid., p. 215).

Collectively this evidence suggests that, rather than being combinatorial, cognition – at least in these cases – involves mutual influence between sub-featural, featural and higher order cognition. Rather than being stuck with a fixed set of representational primitives, cognition is able to construct new representational abilities based on higher order task demands and conceptual learning. Furthermore, to a significant extent the robustness and precision of cognition depends, not on stable representational atoms and algorithms that are built into the fundamental architecture, but on flexible cognitive processes that construct higher order contextual information which then serves to refine lower order processes in a reciprocal and open ended cycle of adjustment.

The reciprocal modulation between low and high order cognition indicates that there is a significant amount of plasticity in cognition, and that this plasticity is regulated by functional activity. Evidence from neuroscience concerning activity-dependent plasticity in neurons and neural systems provides some explanation for how this is possible. Hebbian learning is a simple example: the synaptic connection strength between two neurons increases if their activity is correlated, and decreases if it is uncorrelated. But there are also many other kinds of activity-dependent plasticity, ranging from changes in dendritic morphology based on activity in the immediate vicinity (Quartz and Sejnowski 1997), regulation of plasticity in a cortical region by reward systems (Schultz 2000), through to global synaptic scaling mechanisms (Desai et al. 2002). The sheer extent of activity-dependent plasticity in the brain suggests that a great deal of cognitive functional organisation is shaped by the activities performed by the individual (e.g., Stiles 2000). Direct evidence for this comes from neuroimaging studies showing that musicians have functional and anatomical brain differences related to music processing (Münte et al. 2002).

Such activity dependent plasticity helps explain how the reciprocal connection between low and high order cognition is possible, but it is very problematic for cognitivism for several reasons. It violates the 'virtual machine' hypothesis, which is a straightforward corollary of the assumption that cognition is formal computation, and from a methodological point of view is used as a reason to ignore neural phenomena as being merely a matter of implementation. But activity-dependent plasticity criss-crosses the supposed boundary between cognition and implementation, changing lower level processes based on higher functional processes, and in turn changing the nature of the higher processes. Activity-dependent plasticity also violates the cognitivist structure-content distinction, which ensures that cognitive processes are only sensitive to syntax and are hence algorithmically implementable. More subtly, the structure-content distinction reflects the normative idea that cognition does – or should – operate according to general rules that employ placeholders for representations of particulars. However the separation between rules and content also makes the system inflexible: precisely because of this, the system can't modify the rules it uses. In contrast, the *content* of what a person sees can be used to change the information processing used to make visual discriminations.

Thus, although the presence of an interconnection between content and structure violates some deep assumptions about computation, it has important advantages for learning: it allows the functional organisation of the system to be tuned by the nature of its activities. There are a large variety of neural processes that show activity-dependent plasticity, and this provides an extremely large number of degrees of freedom for the functional organisation of cognition. These degrees of freedom are selectively locked down or released by regulative signals operating with a range of spatial, temporal and pathway specific characteristics. Stepping back from the details, the overall regulative loop from activity to architectural change can build properties of the organism and environment into the functional organisation of the neural system. The selective regulation ensures that the various types of plasticity are only expressed in restricted circumstances. This helps to minimise the potential complexity the cognitive system must deal with at any given time by pre-tuning it to its context, whilst preserving a background capacity to adjust to different functional contexts.

4.2. *The process context of intelligent action*

Just as the structuralist rule-based conception of cognition is flawed, so too is the cognitivist world ontology, especially as it relates to intelligent action. Cognitivism treats the world as an agent-independent, hierarchically structured collection of objects. But from the point of view of intelligent action the world is an agent-related complex of interdependent processes. Of course, objects are an important part of an agent's environment, and object tracking, categorisation and modelling is a significant part of cognition. But objects are of derivative, not primary importance, and tracking process relationships presents different cognitive demands to that of recognising objects per se.

The situations that form the context for intelligent action are much *more* than just assemblages of objects because intelligent agents are biological organisms, and as such fundamentally constituted in and shaped by processes.[6] The most basic biological process is the reproduction cycle, but interconnected processes define virtually all aspects of life. For humans these include the sleep-wake cycle, the hunger-satiation cycle, the working day, the budgetary cycle between paydays, the dynamics of social relations, and so on. These processes are multidimensional, often continuously valued, and interdependent. And it is the process interdependencies that are the source of the constraints that differentiate better and worse action. Metabolism is the fundamental process that makes action possible for a biological organism, and also determines one of the primary goals for organismic behaviour: acquire food. For many humans activity in the workplace that is useful, or apparently useful, to the larger processes of the employer organisation is the means for acquiring food and other resources. But there is more to human life than sustaining metabolism. Human goals are such that quality of life requires taking part in other processes, including recreational, social and intellectual, and these can compete with work and basic metabolism for the individual's resources. Thus, any given situation within which cognitive problems occur – the workplace, the supermarket, dinner with a friend, etc. – exists within a complex network of process relationships. A single action in a particular situation can have ramifications for many of these relationships, so the challenge for the individual is to find the best way of balancing the interdependencies.

Cognition is thus enmeshed in processes, and it is this that makes flexibility such a central feature of intelligence. As factors in a situation shift, action needs to be reorganised. After spending five minutes in the queue at the checkout I remember that my flatmate is going to give me money for the telephone bill tonight, and I'm fairly sure that the final demand from the telephone company won't arrive before payday. This means that I can put off actually paying the bill until then, and since I can live off the telephone money, I can afford the steak. I leave the queue, return the gnocchi and gorgonzola to the shelves, and go looking for a good bottle of red.

5. SELF-DIRECTEDNESS, NORMS AND FLEXIBILITY

A great deal of recent cognitive science has abandoned the strict formalist framework of cognitivist AI, but has failed to supply any alternative normative conception of cognition. For a philosopher steeped in the traditional rationality project this trend amounts to a descent into the maelstrom.[7] In Section 3 I raised the objection against pragmatism that it does not appear to be normatively evaluable by traditional standards. Unless an alternative theory of cognitive norms can be supplied this is a critical problem because there is no account of the distinction between better and worse cognition, and any theory that cannot explain this distinction is at the very least seriously incomplete.

Pragmatist approaches do appeal to a normative standard: the ability to achieve effective interaction. The trouble is that per se this appears to be a merely functional rather than cognitive norm. Traditionalists are right to want distinctively cognitive norms, because it is only with such norms that cognition can be effectively distinguished from other kinds of functional processes. But they are wrong to look for an entirely sui generis theory of such norms. Cognitive processes have evolved as specialisations of deeper biological processes, and even in their highest order form they are intimately connected to other kinds of processes. It follows that whilst cognition may have some distinctive normative attributes, its norms are entwined with functional biological norms.

Elsewhere in collaboration I have developed a theory of self-directed agents that characterises the evolutionary emergence of cognition as a capacity deeply integrated with other biological processes. The central idea is very similar to Romanes' claim that the evolution of intelligence is marked by a transition from reflexive behaviour to more flexible action under higher 'mental' control (Romanes 1904). The account of self-directedness focuses on the ability to integrate multiple factors in action production, permitting action to be sensitive to a broader range of conditions and hence to be able to succeed in complex variable situations where reflex action cannot be relied on (Christensen and Hooker 2000, 2001). There is not the space to present the details of this theory; for the current purposes I shall focus on a few key aspects of it.

The first point to make is that although animal cognition research has traditionally focused on stimulus-response learning, there is evidence that more complex integrative abilities are phylogenetically surprisingly ancient. Invertebrates such as locusts and honeybees possess integrative neural systems and are capable of multidimensional tradeoffs and abstract learning (Raubenheimer 1999; Menzel and Giurfa 2001). Self-directedness arises from integration involving a number of key interrelated organism processes: interaction in a complex environment, affect, motor control, and development. Complex context-sensitive action requires coordination between these processes. Foraging problems can require the capacity for time and space awareness, and choice involving multidimensional tradeoffs. Affective processes must detect conditions relevant to the animal's viability, including states such as hunger, tiredness and the presence of things in the environment that should be approached or avoided. Complex motor control requires coordination between multiple sensorimotor maps, dynamic control and predictive modelling. Development involves building and maintaining systems of sensorimotor coordination and the learning of skills. The elaboration and integration of all these processes has been achieved in brain evolution through the addition of new structures that refine and coordinate the function of pre-existing structures (Arbib et al. 1998).

With regard to learning, the capacity for memory and anticipation plays an important role in facilitating higher cognitive processes such as goal and concept construction. However a distinctive claim of the theory of self-directedness is that this learning is highly relational, and primarily builds knowledge of situational interdependencies rather than context-independent representations. Representation is involved in this learning, but higher cognition is not representationally *driven*. Rather, cognition involves multiple processes, including semantic and non-semantic, conscious and non-conscious, which work in coordination. Learning helps refine the coordination, balancing low order implicit processes with higher order control to achieve effective situational activity. Situational awareness develops through experience as the agent learns to represent and attend to the key factors that must be controlled, but just as importantly the agent learns *not* to attend to factors that are irrelevant to the agent's goals.

This account turns the cognitivist structural sandwich inside out. Complex interacting micro, meso and macro processes allow high order cognitive organization to form and dissolve. Rules in cognition only appear as embedded constructs: as low order behavioural stimulus-response relations or as codified high order strategic and policy-level reasoning. Moreover it is very plausible that all cognitive rules should be thought of as systematically defeasible. There are no convincing general rules for epistemology or ethics, for instance, and the reason for this is, arguably, directly related to the extremely complex, multidimensional conditions of human existence. In seeking generality and rigour philosophers have fetishized mathematico-deductive argument and largely ignored practical knowledge. But reasoning processes modelled on mathematico- deductive argument are inherently unsuited for real time, real world cognition.

6. Conclusion

The position of cognitivism as the theoretical centre of cognitive science has been eroded by a growing stream of counter evidence, new non-symbolic and non-computational modelling techniques, and some spectacular efforts to launch rival paradigms. However the in-principle cognitivist arguments to the effect that only cognitivism can capture core cognitive properties, and hence must ultimately be correct, have not received a great deal of direct challenge. This paper has attempted to mount a counter-argument that cognitivism misframes the core cognitive issues by focusing on representations and rules rather than the adaptive embeddedness and flexibility of cognition, and that it is unable to capture these properties. In order to understand these aspects of cognition we must develop process-based theories that directly model the complex coupling between cognition and deeper biological processes.

Acknowledgements

During the period of writing this paper I have been supported as a Postdoctoral Research Fellow at the Konrad Lorenz Institute for Evolution and Cognition. The theory of self-directedness was originally developed in collaboration with Cliff Hooker, and the ideas presented here have also benefited greatly from many discussions with Mark Bickhard.

Notes

1 Or at least have been until recently – see below.

2 For an extended discussion and critique of this assumption see Lakoff 1987.

3 See Bryson 2000, Brooks 1997.

[4] For a contemporary theory of representation in the pragmatist tradition see Bickhard 1993; this volume.

[5] I will suggest some more in the next section.

[6] When artificial intelligent agents become possible it is likely that they will need to replicate many of the features of biological agents and will be just as process-bound

[7] It might be objected that the traditional normative project is prescriptive rather than descriptive (in contrast to empirical cognitive science), but if cognition isn't – even in its ideal form – a well-defined representational process then the traditional norms simply don't apply, even prescriptively. If the problem were that real cognition is a sloppy representational-computational process then a prescriptive program to eliminate the slop is viable. But the problem being raised here is that *rigorous*representational computation theory is inherently unable to explain the properties of cognition. Ipso facto, the norms associated with representational computation cannot be directly applied to real cognition. At the very least they must be supplemented or revised in some way.

REFERENCES

Arbib, M., P. Érdi, and J. Szentagothai: 1998, *Neural Organization: Structure, Function and Dynamics*, Cambridge, MA: MIT Press.

Bechara, A., H. Damasio, D. Tranel, and A. Damasio: 1997, 'Deciding Advantageously before Knowing the Advantageous Strategy', *Science* **275**, 1293–1294.

Bertenthal, B. I. and R. Clifton: 1997, 'Perception and Action', in D. Kuhn and R. Siegler (eds.), *Handbook of Child Psychology*, New York: Wiley.

Bickhard, M. H.: 1993, 'Representational Content in Humans and Machines', *Journal of Experimental and Theoretical Artificial Intelligence* **5**, 285–333.

Brooks, R. A.: 1991, 'Intelligence Without Representation', *Artificial Intelligence* **47**, 139–159.

Bryson, J. J.: 2000, 'Cross-Paradigm Analysis of Autonomous Agent Architecture', *Journal of Experimental and Theoretical Artificial Intelligence* **12**, 165–190.

Carruthers, P.: forthcoming, 'Distinctively Human Thinking: Modular Precursors and Components', in P. Carruthers, S. Laurence, and S. Stich (eds.), *The Structure of the Innate Mind.*

Christensen, W. D. and C. A. Hooker: 2000, 'An Interactivist-Constructivist Approach to Intelligence: Self-directed Anticipative Learning', *Philosophical Psychology* **13**, 5–45.

Christensen, W. D. and C. A. Hooker: 2001, 'Self-directed Agents', in J. S. McIntosh (ed.), *Naturalism, Evolution & Intentionality*, Calgary: University of Calgary Press, pp. 19–52.

Christensen, W. D. and M. H. Bickhard: 2002, 'The Process Dynamics of Normative Function', *Monist* **85**, 3–28.

Desai, N. S., R. H. Cudmore, S. B. Nelson, and G. G. Turrigiano: 2002, 'Critical Periods for Experience-dependent Synaptic Scaling in Visual Cortex', *Nature Neuroscience* **5**, 783–789.

Fodor, J.: 2000, *The Mind Doesn't Work That Way: The Scope and Limits of Computational Psychology*, Cambridge, MA: Bradford Books/MIT Press.

Gibson, J. J.: 1977, 'The Theory of Affordances', in R. Shaw and J. Bransford (eds.), *Perceiving, Acting and Knowing*, New York: Wiley, pp. 67–82.

Goldstone, R. L., M. Steyvers, J. S. Smith, and Kersten, A.: 2000, 'Interactions between Perceptual and Conceptual Learning', in E. Dietrich and A. B. Markman (eds.), *Cog-*

nitive Dynamics: Conceptual and Representational Change in Humans and Machines, Mahweh, NJ: Lawrence Erlbaum, pp. 191–228.

Holyoak, K. J. and J. E. Hummel: 2000, 'The Proper Treatment of Symbols in a Connectionist Architecture', in E. Dietrich and A. B. Markman (eds.), *Cognitive Dynamics: Conceptual and Representational Change in Humans and Machines*, Mahwah, NJ: Lawrence Erlbaum, pp. 229–264.

Hutchins, E.: 1995a, *Cognition in the Wild*, Cambridge, MA: MIT Press.

Hutchins, E.: 1995b, 'How a Cockpit Remembers its Speeds', *Cognitive Science* **19**, 265–288.

Lakoff, G.: 1987, *Women, Fire and Dangerous Things*, Chicago: University of Chicago Press.

Menzel, R. and M. Giurfa: 2001, 'Cognitive Architecture of a Minibrain: The Honeybee', *Trends in Cognitive Sciences* **5**, 62–71.

Münte, Thomas F., F. Altenmüller, and L. Jäncke: 2002, 'The Musician's Brain as a Model of Neuroplasticity', *Nature Reviews Neuroscience* **3**, 473–478.

Neisser, U.:1997, 'The Ecological Study of Memory', *Philosophical Transactions of the Royal Society of London B: Biological Sciences* **352**, 1697–1701.

Newell, A. and H. A. Simon: 1972, *Human Problem Solving*, Englewood Cliffs, NJ: Prentice-Hall.

Payton, D.: 1990, 'Internalized Plans: A Representation for Action Resources', in P. Maes (ed.), *Robotics and Autonomous Systems, Special Issue on Designing Autonomous Agents: Theory and Practice from Biology to Engineering and Back*, pp. 89–104.

Quartz, S. R. and T. J. Sejnowski: 1997, 'The Neural Basis of Cognitive Development: A Constructivist Manifesto', *Behavioural and Brain Sciences* **20**, 537–596.

Raubenheimer, D. and S. Simpson: 1999, 'Integrating Nutrition: A Geometrical Approach', *Entomologia Experimentalis et Applicata* **99**, 67–82.

Remez, R. E.: 2000, 'Speech Spoken and Represented', in E. Dietrich and A. B. Markman (eds.), *Cognitive Dynamics: Conceptual and Representational Change in Humans and Machines*, Mahwah, NJ: Lawrence Erlbaum, pp. 93–115.

Romanes, G.: 1904, *Animal Intelligence*, London: Kegan Paul, Trench, Trubner & Co.

Schultz, W.: 2000, 'Multiple Reward Signals in the Brain', *Nature Reviews Neuroscience* **1**, 199–207.

Sieg, W.: 1999, 'Formal Systems, Properties of', in R. A. Wilson and F. Keil (eds.), *The MIT Encyclopedia of the Cognitive Sciences*, Cambridge, MA: Bradford Books.

Stiles, J.: 2000, 'Neural Plasticity and Cognitive Development', *Developmental Neuropsychology* **18**, 237–272.

Thompson-Schill, S. L., M. D'Esposito, G. K. Aguirre, and M. J. Farah: 1997, 'Role of Left Inferior Prefrontal Cortex in Retrieval of Semantic Knowledge: A Reevaluation', *PNAS* **94**, 14792–14797.

Tranel, D., C. G. Logan, R. J. Frank, and A. R. Damasio: 1997, 'Explaining Category-related Effects in the Retrieval of Conceptual and Lexical Knowledge for Concrete Entities: Operationalization and Analysis of Factors', *Neuropsychologia* **35**, 1329–1339.

Wilson, R. A. and F. Keil: 1999, *The MIT Encyclopedia of the Cognitive Sciences*, Cambridge, MA: Bradford Books.

MICHAEL FORTESCUE

THE PATTERN AND PROCESS OF LANGUAGE IN USE: A TEST CASE

ABSTRACT. This paper is concerned with the kind of non-linear causation that lies behind the production and comprehension of speech in discourse, where multiple 'input' data typically act in concert towards a determinate output. To this end Whitehead's philosophy of Process - in particular his theory of 'prehensions' – is applied to the analysis of pragmatic implication and inference in a short literary excerpt, which involves the most complex kind of prehension, the 'intuitive judgment'. This leads to a number of conclusions concerning the way in which patterns of linguistic behaviour can be evoked out of mere potentiality by contextually embedded prehensions that constrain and modulate the expression of intended discourse acts. Some of these constraints are attributable to the nature of focal consciousness itself. An extension of the Whiteheadian notion of 'subjective aim' (usually understood to preside over a single 'actual occasion') will be necessitated, since in discourse this can clearly both be suspended and divided into subordinate sub-aims while the overall aim, still unfulfilled, is inherited from occasion to occasion.

1. WHITEHEAD AND THE FLOW OF DISCOURSE

The uniting purpose of the interdisciplinary workshop at Sandbjerg Castle in June 2002 was, as I understand it, to investigate common ground for describing non-linear forms of causality, in particular as manifest in the behaviour of self-organizing systems, i.e., organisms. I believe that Whitehead's philosophy of Process provides such a common ground to investigate the explanatory power of process-based descriptive frameworks. It is at all events a prime candidate for a process-based framework that can be applied across disciplines to describe the behavior of self-organizing systems. In Fortescue (2001) I attempted to apply Whitehead's basic notions to language and its cognitive underpinnings. I took as my starting point the Whiteheadian notion of the 'prehension' as the basic unit of process in its broadest sense (including language processing). Prehensions – the integration of new data by an organism according to a unifying 'subjective aim' – are what constitute the 'concrescence' of the minimal ontological unit of reality, the 'actual occasion'. The term 'actual occasion', note, refers to reality as seen 'from within' rather than to an external 'event' or the like. Actual occasions are like Leibnizian monads with open

J. Seibt (ed.), Process theories: Crossdisciplinary studies in dynamic categories, 177–218.

windows to the past. They are 'quanta of experience' that come into being out of the data of the relevant past on the crest of the present, work themselves out according to their internal dynamism, and perish into data for subsequent occasions. The prehensions of which they are made up cover perceptions as well as the accessing of concepts.[1] Although Whitehead's alternately calls them 'feelings' this should not be understood as implying emotional response, which is rather a matter of certain kinds of 'subjective form' which a prehension can assume. Prehensions match felt input with the felt content of memory/experience. A self-organizing system, an 'organism' in Whitehead's terminology, persists through time as a quanticized sequence of 'actual occasions', it is a spatio-temporal 'nexus' (or 'fact of relatedness between actual occasions'). Societies are a type of nexus wherein actual occasions involve each other, i.e., contain information about each other by mutual prehensions, and enjoy a certain type of internal coherence (serving a common 'purpose') that ensures the causal continuity from occasion to occasion within it. A human subject is a purely temporally organized 'personally ordered nexus'. In a nut-shell, the human mind/brain can be seen as a society-of-societies embedded in still broader social societies, and organized hierarchically in such a way that higher level prehensional activity is maximized, founded on and arising out of the physical prehensions of neurons, which in turn arises out of organized nex_s of molecules, etc.. In this paper I use the terms 'actual occasion' and 'concrescence' (or 'concrescent occasion') more or less interchangeably as the basic unit of (cognitive) process; the former emphasizes the entity as a unity, the latter its dynamic nature.

The 'actual occasion' contrasts with the notion of 'eternal object' (or 'concept'), which includes the patterns of linguistic potentiality that define the human language capacity. 'Eternal' should not be taken too literally and certainly without any associations of thinghood ('repeatable' is better, since this also covers fuzzy prototype categories).[2] The development that is the concrescence of an 'actual occasion' has the architecture of a parallel distributed flow of information, with multiple, partially indeterminate 'input' factors calling forth a variety of increasingly complex prehension types according to their immediate relevance to the occasion's developmental goal (its 'subjective aim') on the way to a determinate resolution or 'satisfaction' (some action, either physical or purely mental). From this perspective, 'pattern' – for instance grammatical, lexical or pragmatic patterning – reflects the developmental potential of the organism, while 'process' provides the actual *realia* in which that pattern is manifest, for instance in individual acts of expression on the part of speakers. The two are inextricably intertwined and interdependent.

However intuitively appealing such general cognitive and philosophical speculations are, it may be objected that it behoves us as practitioners of the human sciences to go further and show that this general description of self-organizing systems of information transfer can be instantiated in a formal model of human cognition and language processing. I shall not in fact be attempting a full-blown formalization – for reasons which will be elaborated below. I shall attempt rather to show in the following that the language-specific constraints that define human communicative behaviour (the patterns behind the processes) can be profitably approached via Whitehead's basic taxonomy of prehension types and their successively more complex combinations. This undertaking was not carried out by Whitehead himself, and the hints he gave in this direction need to be fleshed out with what has been unearthed by the cognitive sciences in the years intervening since his death, as adumbrated in my book. In the present paper I shall be focusing primarily on the application of these key Whiteheadian notions to the analysis of pragmatic implication and inference, as expressed in and through language, which involves the most complex kinds of 'intellectual' prehensions. This will lead to some conclusions concerning the way in which patterns of linguistic behaviour can be evoked out of mere potentiality by contextually embedded prehensions to constrain and modulate the expression of the intended speech act.[3] Amongst these will be certain constraints on the form of linguistic expression attributable to the nature of focal consciousness - a matter that has come more and more to the forefront in Cognitive Linguistics in recent years thanks to the pioneering work of Wallace Chafe.

In the present framework this is a matter of the span of the unitary concrescent occasion that determines the 'pulse' of consciousness. It will be found that an extension – or redefinition – of the Whiteheadian 'subjective aim' (usually understood as presiding over a single concrescent occasion) is necessitated, since in discourse this can clearly both be suspended (embedded in broader intentional structures) and divided into subordinate sub-aims while the overall aim is inherited, still unfulfilled, from one occasion to the next.

2. WHO DONE IT – AND WHY?

Consider the following passage from a contemporary who-done-it. Although this is a literary rendition of natural discourse it serves my purpose well since the context of the processes of inference involved is of a familiar and reasonably clear type (and not too far from what might actually occur in a 'real' criminal investigation). It combines a degree of conventionality of expression with reflexes of on-line reasoning, two aspects of language-in-use that have always been difficult to partition out systematically. Superintendent Cheever is here questioning Reverend Charles Harris and his wife Caroline about the recent movements of their daughter Meg, whose remains they have just identified in the nearby mortuary. The parents are both grief-stricken though the husband – a stern churchman – does not show it as openly as his wife does; another officer takes notes. Cheever has just explained that the police have found almost no personal effects in Meg's flat and presume that she had been in the process of moving out of it before she disappeared (the Leo mentioned is the young man she was running away with after he had apparently jilted her best friend, and with whom she was found dead):

> 'She was going on holiday with Leo,' said Charles quietly, 'but she didn't say anything about moving out of her flat. Not to me anyway.' He looked irresolutely at his wife.
>
> 'She didn't tell you anything, Charles, because she knew you'd disapprove.' Caroline mopped her red-rimmed eyes. 'She had an abortion ten years ago. She didn't tell you about that either, did she? And why not? Because you'd have ruined her life for her.' She crumpled the handkerchief between her palms. 'Well, it's ruined anyway, but it might not have been if she'd been able to talk to you as a father instead of a priest. Everything had to be kept secret in case you preached at her.'
>
> Her husband stared at her, the planes of his face bleached white with shock. 'I didn't know,' he murmured. 'I'm sorry.'
>
> 'Of course you're sorry. Now,' she added bitterly. 'I'm sorry, too. Sorry for her, sorry for the baby, sorry for me. I'd like to have been a grandmother.' Her voice broke into a sob. 'It's such a waste. It's all such waste.' She turned to the Superintendent. 'We have a son, but he's never wanted to marry. He wanted to be ordained like his father.' Her eyes filled again. 'It's such a terrible waste.'
>
> Cheever waited while she fought for control. 'You implied that you knew Meg was moving out of her flat, Mrs Harris,' he said at last. 'Could you tell us about that? Where was she going?'
>
> 'To live with Leo. He had a house. It made more sense for her to move in with him.'
>
> (Minette Walters 1995: 172)

Imagine now that we are privy to what is going on in superintendent Cheever's mind here. What conclusions is he drawing and in what order? Why does he proceed as he does and how does he interpret the words of the couple he is questioning in the light of his on-going investigation? The crucial question to be focussed on is this: how can one explain the implicature recognized by the superintendent that Mrs Harris knew that their daughter had moved out of her flat? What chain of actual mental processes might have brought him to that conclusion? There is of course a wide literature on the subject of 'on line' implicatures of this kind, founded on the cooperative principles and maxims of Grice (1975) and continued most recently in Levinson's distinction between Q-, I- and M- implicatures (Levinson 2000).[4] But this tradition tends to concentrate – by necessity, given the complexities involved – on short, relatively isolated exchanges, on what could be called 'local' implicatures. One might well wonder whether the principles and generalizations adduced by this pragmatic tradition are sufficiently developed to explain what is going on in more extended situations such as the passage above. Consider all the cumulative background information necessary for Cheever to conclude that since Meg had not told her father anything her mother must have known that she *had* moved (the implicature of *She didn't tell you anything, Charles, because she knew you'd disapprove*). He certainly has to believe that she is following (at least ostensibly) the general principles of cooperation, for example making a statement that is relevant to the situation – his investigation. But that is not all that is going on: the reason behind her utterance here obviously has something to do with built up resentment that is sparked off by her husband's bland answer (perhaps specifically prompted by the word 'anyway' in Charles' preceding utterance). The relationship between Caroline and Charles and between both of them and their daughter, separately, are highly relevant to the interpretation, and as of yet Cheever has glimpsed no more than we have of these matters. If he forms a tentative hypothesis as to Caroline's implied meaning already at this point, it is going to be further reinforced when he hears the utterance containing the word 'either' a little later – and if it is *not* yet manifest to him it must surely be so by the time he hears that utterance. There is obviously a certain amount of indeterminacy here: we can never know at exactly what point the implication struck him, or how certain he was of it when it did. Yet we, as readers, actually have enough information provided by the passage as given to know that Cheever must at some point have recognized this implicature, since his further actions presuppose it – even though he will refer to it as a mere implication, not as something confirmed or asserted directly. Is this out of politeness and tact, or is it because he really is unsure of its truth?

To account for what is (minimally) going on 'in Cheever's head' here we need at very least something like a computer-interpretable 'means-ends' diagram in the manner of Leech (1983) which can reflect the complex nature of Cheever's purposes. This involves various layers of 'suspension', most notably of his expectation of a clear answer to his (indirect) question as to whether Meg had moved out of her flat, which is embedded within his broader purpose of elucidating the circumstances around the murder. It is difficult, however, to capture on a diagram the flexibility applied by rational human beings to attaining specific ends in concrete situations with a variety of alternative strategies at their disposal, constantly reassessed and adjusted 'on line'. Cheever, for example, is presumably entertaining at this point various competing, tentative hypotheses as to what these circumstances were and as to the veracity of Charles' and Caroline's statements, and new intuitions concerning these matters (evoked by input from without or by his own internal thought processes) may slightly alter the nature of his aim under way, or at least alter his strategy for attaining it. And what about the role of conventionalized – or at least expected – 'turns' of the sort conversation analysts would discern here? Certainly there is the expectation of a more precise answer to Cheever's original question than that supplied by Charles, but this is not actually forthcoming, and the exchange between husband and wife veers off in quite another direction, although the expectation does not vanish. How can things like the conventions of discourse, Gricean maxims and 'means-ends' diagrams be related to the Whiteheadian 'concrescence'? As suggested, Cheever may not have grasped the implicature until after he had thought over the whole exchange for a while, and we do not know how much – and with what degree of conviction – he has penetrated the background relationship between the three characters at this point, so if we are to apply a Whiteheadian analysis in terms of successive actual occasions here how are we to draw a boundary around the coherent, unitary processes involved? What is the causal sequence here? To which 'occasion' in Cheever's 'life-line' is the implicature in question to be assigned, or should we give up right away and claim mere indeterminacy?

Without denying the indeterminacy involved, I believe we can indeed divide this dialogue up objectively into a sequence of actual occasions – albeit a sequence containing overlapping and suspended 'subjective aims'. After all, actual occasions are defined in terms of multiple, largely indeterminate input and determinate, expressive output, and, as I have proposed elsewhere (Fortescue 2001, p. 34), if one can pinpoint the sequence of units of *expression* into which no further input (introducing new information from without) can be seen as relevant, one is at least half way to determining how to break up such an exchange into self-contained concrescent units. Of course, distinguishing between concrescent occasions of which the expression is a sequence of words and those of which the expression is a change to Cheever's internal model of the circumstances of the crime[5] is going to be more tentative, but we shall see below that one can follow this through logically a good deal of the way.

The Whiteheadian conceptual apparatus supplies us moreover with the necessary categories to describe the *types* of cognitive processes inferable behind the reasoning displayed in this passage, namely in his theory of prehensions, of which the most complex kind, the 'intuitive judgment', is particularly relevant. These prehensions are not evoked according to strict linear chains of causality, but according to their graded relevance to the subjective aim of the occasion with respect to potential patterns discernable within the activated embedding context (or 'nexus'). This is constrained both by the initial situation pertaining to the concrescence and by the cognitive patterns that can be brought to bear on the solution of its aim. The result is an integration of the initial 'data' with the initial aim in a maximally useful or informative way compatible with the overall model of the circumstances. The first 'stage' in this overall process is when the initial data (the sum total of all the physical prehensions acting as input to the concrescence) is narrowed down by 'negative' prehensions to the 'objective data', which is just that part of the initial data that is highlighted or activated as relevant to the subjective aim. It is here that a refinement of the Whiteheadian notion of 'subjective aim' is called for, however. I shall assume that the complex aims felt by a subject in a 'personally ordered nexus' can be multi-layered – from global to local – and can be inherited from occasion to occasion within the life-line of that personal nexus. Longer-term layers of intentionality are eminently suspendable, whereas more local ones must be satisfied (as far as this is possible at all) within a single occasion, e.g., by the analysis and integration of spoken or read words into an accreting mental model by the extraction of all relevant 'meaning'. What is 'left over' or suspended from a longer-term aim is passed on as part of the objectified 'expression' of the individual occasion, understood as not just spoken words and gestures but also as adjustment to the internal model of the on-going exchange on the part of the speaker him or herself. The mechanism for this kind of inheritance is already in place in the form of the 'hybrid prehension'.[6]

3. A CLOSER LOOK THROUGH THE MAGNIFYING GLASS

In the following 'first pass' analysis I shall concentrate on three key utterance sequences from the passage, namely (A), (B) and (C), repeated below. As a first approximation (that suggested above) I shall treat each sentence-length utterance as corresponding to a single Whiteheadian occasion (i.e., one concrescence). This approach (somewhat adjusted) will be justified as the paper proceeds.

A) Charles: *She was going on holiday with Leo, but she didn't say anything about moving out of her flat. Not to me anyway.*

Charles' immediate goal here is ostensibly to answer a request for information (actually the statement of an assumption calling for confirmation) just made by Cheever; this is the aspect of his broader intentional stance – to help the police with their enquiry – which is in focus right now. There are further factors of still more distant background intentionality and behavioural potentiality that have a bearing upon his response, but they can be ignored. We can assume that Charles thought – like most people – that Meg had not been moving out permanently but was just going on holiday with her friend.[7] Note that he starts with the most important new information relevant to the implied question (at odds with the assumption made by the police and supplying the reason why her flat was in fact empty), then adds a comment explicitly contrasting this information with the – as far as he knew - erroneous assumption whereby the question was framed. It is embedded in a constatation that Meg had not informed him of what would have constituted grounds for this assumption (directly telling him she was moving out - the implication is that she would have done so if she were). This further highlights the foregrounded information by contrasting it with the erroneous supposition and hence renders the satisfaction of his subjective aim – here expressed in words – more intense and informative. The prehensions needed to formulate this response would all have been evoked in order to match Cheever's expectation of a relevant and sufficient answer to his – Charles' – own internal model of the events being inquired about.

More interesting is the final qualifying remark he adds, which cannot be said to be an integral part of the satisfaction of his initial aim and therefore should be treated as belonging to a separate – but 'chained' – actual occasion: it reflects further mental activity of self-editing. This is accompanied by a look towards his wife, as if he knows that she knows that there were things that Meg would tell her but not him - he may be forestalling (by an anticipatory prehension) a possible objection or correction on Caroline's part. He has realized at all events that he has made too strong a claim, which may not strictly have been true (Grice's maxim of quality could be evoked here). The local sub-aim (still falling under the superordinate aim of answering Cheever's question, presumably as truthfully as possible) is to correct a possible misinterpretation – or overinterpretation – of his own words. It is formulated by just pronouncing the 'new' information which is to be interpreted against the background of his immediately preceding words. Not only is the data to this 'new' occasion different from that of the first statement (it includes that statement as input), it also involves at least one additional 'imaginative' prehension,[8] namely his recognition that his words leave open the possibility that she told someone else, in particular

Caroline. (He must surely know that Meg was more likely to confide in her than in him – a hint of background information not yet clear to either Cheever or the reader.)

At this point let us shift to Cheever's perspective: he is expecting something like a direct yes/no answer to the question he has implied (this is the conversational 'rule of the game' of turn-taking in an interrogational setting). Charles' response (up until the final *caveat*) could be taken as providing just that: he contrasts what he (thinks he) knows with what the police assume, thus supplying the quantity and quality of information they both understand that Cheever wants. His additional *caveat* – which may have been innocent and/or honest but may also have been motivated in part by peevishness ('no one ever tells me anything') – causes Cheever to realize that he has not got such a simple and complete answer after all. Is Charles washing his hands of Meg's actions? He obviously knows she 'went on holiday' with the unsuitable 'friend' Leo, but mentions this with suspicious casualness, given that he is a High Church clergyman. And why the sideways glance at his wife here? Note that a part of Cheever's constant professional (but also quintessentially human) intensionality is to ask why, when, who, why, etc., all the time. So he is 'primed' now to straighten this out in his mind, to find out why Charles added his *caveat* (this is his 'adjusted' subjective aim – or local sub-aim – at this point), but he is forestalled by Caroline:

B) Caroline: *She didn't tell you anything, Charles, because she knew you'd disapprove.*

Her subjective aim here (again within an overall intentional stance geared to answering Cheever's questions) is to express her immediate reaction to Charles' words. It is an objection involving both factual correction (Charles' answer to the question put by Cheever clashes with her own knowledge and potential answer) and a more spontaneous, heated accusation aimed at Charles. Instead of simply correcting his reply by stating that Meg had indeed told *her* she was moving out, she makes a more general statement about Meg's attitude to her father (fear of his disapproval), breaking the maxim of quantity – under emotional rather than communicative pressure – but not that of relation, since it provides a *reason* for Meg not telling him 'anything'. (She exaggerates of course, in the manner of resentful spouses!) Observe that she only *implies* that Meg was moving out (contrary to Charles' belief); the immediate implicature is that Meg did not tell Charles about moving out (this is a token instance of the general accusation she makes), and only more indirectly that she did tell somebody else (probably herself), which in turn she would presumably only have done if she had really been moving out. If she had been following conversational protocol to the letter (which she might well have were Charles not present) she would have given a more direct and clear answer, but the subjective intensity of the resentment Charles' answer has sparked off in her overrides the ordinary cooperative principle of just filling in the objective facts required of her. Instead, the three-sided exchange is diverted into an elaboration of her incrimination: *She had an abortion ten years ago. She didn't tell you about that either, did she? And why not? Because you'd have ruined her life for her*, etc., plus Charles' pathetic response to this.

The key prehensions in the concrescence leading up to (B) – over and above Caroline's recognition of the literal meaning of the words – are of certain patterns in and behind her husband's response (as potential motivation to them); these she is all too familiar with and obviously does not like. As regards the form of her complex utterance, notice that the main foregrounded statement links back to and comments on Charles' utterance (as 'topic'), which it parallels with variation, followed by a causal clause. His name is interjected at a suitable syntactic break between clauses, redundantly indicating the addressee of her utterances. (It no doubt functions also to underscore an attitude of indignant blame of the type 'it's you I'm talking to!')

During all of this, through several successive exchanges between husband and wife, Cheever remains silent, observing them: his overall subjective aim is not affected, but there are new bits of relevant evidence (the couple's words and actions playing out in front of him) that must be integrated into his accreting model of the circumstances of the case. Thus, for instance, the occurrence of the word 'either' in Caroline's next utterance but one strengthens quite considerably the likelihood of Cheever's hypothesis as to the implication of her words (another instance of Meg's withholding embarrassing information from Charles that Caroline reveals she did know about). The matter of the abortion may also prove relevant later on. Surely this supports an analysis into several successive 'occasions' along Cheever's life-line, since successive changes result in his internal model corresponding roughly to the sentence-length occasions behind Charles' and Caroline's utterances here, each containing one essential new unit of information. In real natural speech, intonation would also be relevant to such 'packaging' decisions. This is the line I shall follow at all events, as an approximation to the 'truth' of the matter.

Throughout the ensuing elaboration by Caroline, then, Cheever is probably wondering why she is making an issue of the point, and this in turn will probably be relevant to the answer to his original question that he is still waiting for. One way or another, he is able to extract the implication – by a 'suspended judgment'[9] – that Meg had told Caroline that she was moving out. This would explain why she chose to foreground the 'fact' that Meg did not ever confide in Charles (for the reason she gives). On the other hand, the resentment towards her husband that his coldly objective words have sparked off in her would explain (as a contributory – if tacit – ingredient behind her outburst) why it is she nevertheless fails to give a full account of what Meg did tell her, as if she is overcome by her own pent-up negative emotions towards Charles. Cheever's inference is simply the 'best fit' with all the facts at this point – plus his general background knowledge. It 'straightens out' the wrinkle in the information his question has elicited and modifies his immediate subjective aim in such a way that he can now return to the original aim at eliciting more about the circumstances of Meg leaving her flat. Cheever's sub-aim at ascertaining 'why is she saying all this?' has been satisfied as well as circumstances and time allow. Of course there is no guarantee that Caroline's apparent bitterness towards her husband's sanctimonious blindness is particularly relevant to solving the crime, but it does cast useful light on the nature of their relationship and that of their daughter to them. As a good investigator Cheever is continually formulating and modifying on-going hypotheses and suspicions as regards the motives and general veracity of the various 'suspects'. He is always on the look-out for inconsistencies disturbing the 'normal' causal flow of events, as we all are in normal waking life (but in his case this is finely honed by professional training). His next utterance is formulated after a thoughtful pause:

C) Cheever: *You implied that you knew Meg was moving out of her flat, Mrs Harris. Could you tell us about that? Where was she going?*

The subjective aim behind this short sequence of interrelated utterances is that of obtaining confirmation of his inference (that Meg had moved out) and then – by eliciting a more detailed answer to his original question – continuing with the same overall line of inquiry he has not yet finished with and whose satisfaction is still suspended. Note that the concrescence resulting in these words cannot start until all the relevant data is 'in' – the essential ingredient is the suspended judgment formulated above, although the time elapsing before he speaks could well have included other purely internal thought processes (consisting of several actual occasions), but it is not necessary to known their content – or even if there were any – in order to understand the causal flow of the narrative here. Possibly they were 'low grade' occasions (not involving reflective consciousness) as he repeatedly probed for feelings of discrepancy between the inference with all that he knew about the case up to that point and did not affect any significant changes to his model of it. When he does speak, he does not need to repeat the 'point' of his further questions – that is given by the shared context.

What is highlighted as the core around which his utterances crystallize, stripping the potential input down to an economical minimum for comprehension, is the nature of the new information he wants to elicit over and above the partial (and indirect) answer he has received so far. Confirming the implication is the most pressing business at this point (though it is only preparatory to the 'directive' act that follows), and this is what his first utterance is aimed at. Note that it refers to Caroline's *knowing* a proposition X (that Meg was moving out of her flat), not only the truth of X. Here he adds an additional implicature of his own, namely that Meg had in fact moved out of her flat – or why would Caroline have implied she knew this? (Suggesting otherwise would be to accuse her of lying.) Actually his hypothesis is so strong in his mind now that he states the implication as a fact, going on to ask for more details about it and not allowing her time to retract it (a clever police technique!). In asking her subsequently to tell him 'more about that' he does not expect to be told about why Caroline knew about X, but more about X itself (accepted as true). The correct understanding of the reference of anaphoric 'that' depends on the additional implicature. Cheever's actual choice of words and his organization of the information to be conveyed into sentence-length chunks is constrained by various factors apart from the decision to take the most important element first (the confirmation of the inference), namely the lexical, grammatical and discourse conventions of spoken English.

His next sentence-length utterance (*Could you tell us about that?*), to which the first statement is in speech-act terms simply a lead-in, is in the typical guise of a polite (indirect) yes/no question under a suitable prosodic contour. It crystallizes out in conformity with the immediate subjective aim (itself in harmony with the suspended superordinate one) as a request making conventional use of the formula 'could you + P', where P refers to a Searlian preparatory condition for a directive speech act. The lexical core of the proposition is the verbal complex 'tell about', which brings with it its own argument structure requirements and indicates the action required of the addressee. But immediately he elaborates upon his own request, sensing that it is perhaps not quite specific enough: his intention might not be fully recognized by the distraught – and possibly evasive – woman before him. Note that the breaking up of his utterance here into two sentences is also partly determined by the information packaging preferences of spoken English – a yes/no question followed by an information ('wh') question is normally not combinable into a single sentence. The additional wh-question homes in on one specific aspect of Cheever's more general request for relevant information. So, on the assumption that the third sentence was not part of the originally planned realization of the immediate aim behind the first two, but arose through further input of a self-monitoring nature, there are two successive occasions here, their expression closely linked by the shared (inherited) overall subjective aim and chained together in their execution as lead in, main illocutionary point, and more specific elaboration of that point, respectively. One could also treat them as a single occasion if the potential for the last sentence were solely determined by the subjective aim associated with the production of the first one, with conceptual reversions within the unitary occasion rather than further input (here the expression of the first two sentences) determining the outcome. There is no way of being entirely certain, since all we have to go by is the reported external action. The first solution seems intuitively to have the edge, that is all. Note that in the decision to treat the first two sentences as a single occasion we have an exception to the rule-of-thumb of equating sentences to actual occasions mentioned earlier. The criterion for such decisions must lie in the subjective aims linking the discourse acts involved, whether they are reasonably analysable as divided or combined.

We can now diagram what we have analysed so far. On the left of the diagram can be seen Cheever's layered model of the on-going situation in its broader setting (in 'episodic memory', if you will), with his ongoing attempt at reconstruction of the crime near the centre and his knowledge of the characters suspected to have been involved overlapping around it. Within these circles are in turn represented the inter-relationships among them, in so far as this is known to Cheever at the outset. The bold circle contains the actual dialogue situation, sustained by Cheever's immediate perception, and this has been 'blown up' in Cheever's working memory (the bold intermediate 'lens' shape). Successive changes to his on-going model (from top to bottom) are shown by arrows back and forth from the dynamic 'funnel' of successive occasions along Cheever's life-line on the right as new information is abstracted or added via the perceptually highlighted 'lens' of working memory.[10] The broken arrowheads (A_2, A_3 and A_6) represent direct changes to the model in episodic memory that are not mediated by verbal acts. Each successive occasion (subscripted O) is associated with a subjective aim (subscripted S) and an action *vis-à-vis* the model (subscripted A). Note that occasion O_3 is really an abbreviation of several chained internally-expressed occasions, into which at some stage Charles' responses to Caroline (as registered by Cheever) are integrated. Within each occasion specific prehensions are activated according to the 'fit' of the (complex) subjective aim to the state of the model at that moment. The chained succession of occasions from S_1 to S_6 reflects the suspension of S_1 on its way towards final satisfaction – note the 'resettings' underway evoked by Caroline's intervention at O_3 and by Cheever's recognition of the implication of her words at O_4. At the end S_1 is indeed satisfied (Cheever now knows Meg had definitely moved out), but the broader aim of solving the crime (S_0) remains unfulfilled of course. Observe how in the final state of the model the '?' at the point where Cheever's representation of Meg and that of the circumstances of the crime overlap has been replaced with the information that she had indeed moved out. The small double-headed arrows within the individual representations indicate the additional information Cheever now has about the mutual relationship between Meg and her parents.

4. A STILL CLOSER LOOK

Now I shall proceed to a more detailed Whiteheadian analysis – in terms of specific prehension types – of the input to and output from Cheever's final utterance sequence (C). This can, as implied above, be broken down into a minimum of three actual occasions, each with their afferent and efferent aspects, one purely 'internally' objectified occasion plus two 'externally' objectified ones. This is not a distinction Whitehead himself made explicit, but seems essential if one is to account for the overlapping and suspended subjective aims and sub-aims that are so typical for actual conversation (I shall return to this below). The internally objectified one corresponds to Cheever's integration of the implicature of Caroline's utterance (B) (the essential 'new' piece of data he must have assimilated) into his on-going model, and the two externally expressed ones, as analysed above, consist of – in all – three sentence-length utterances. The first two of these are linked by a common subjective aim, and the third one, which is also causally linked to the preceding ones, only less tightly so, has its own subjective aim, modified (if only slightly) by further self-monitoring on Cheever's part.

First, let us consider the logical order of prehensions within the purely internally objectified occasion that led to Cheever's grasping of (B) and in turn supplied the essential new input to the externally objectified occasions producing (C). These phases correspond to Sherburne's characterization of the successive phases of the Whiteheadian concrescence (with the latter's 'comparative feelings' split into two stages – Sherburne 1966, p. 40), namely:

i) *Conformal feelings* ('initial responsive phase'): physical prehensions extracting the 'objective data' from the initial data of Caroline's utterance according to the initial aim, i.e., recognizing the holistic speech string under a unifying intonational contour and initiating the abstrac-

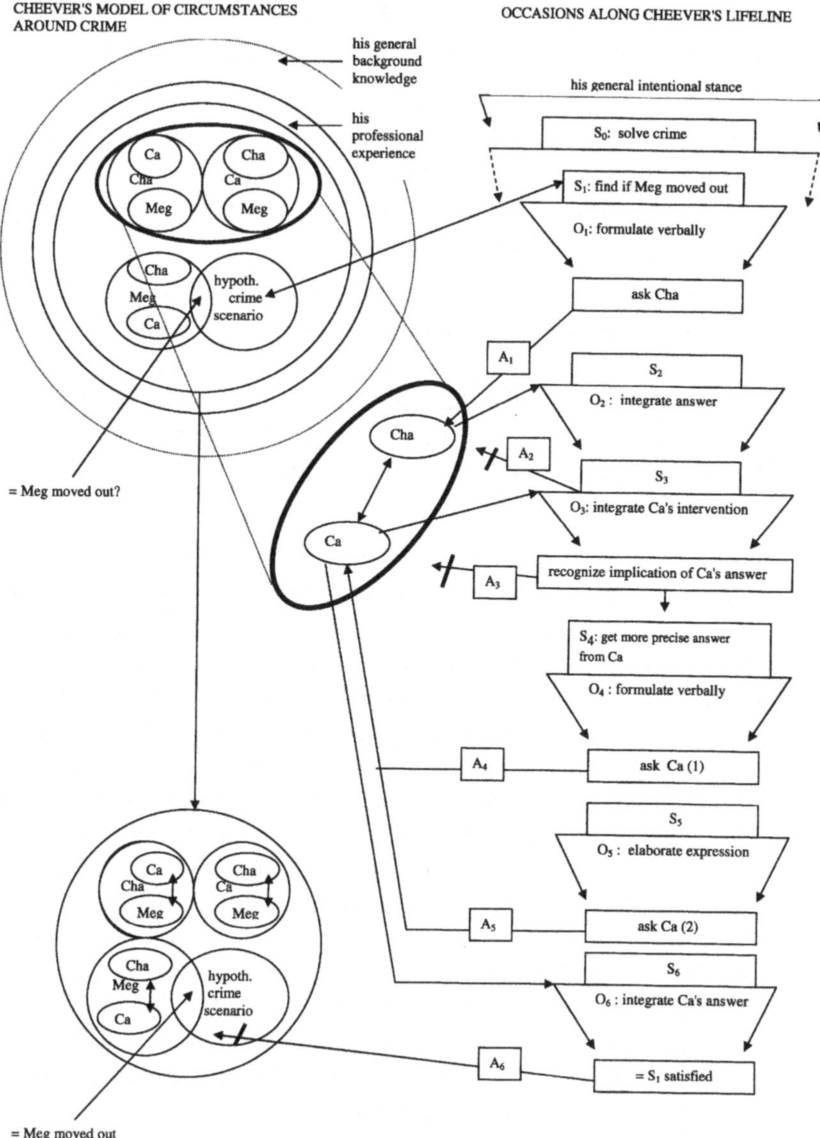

Figure 1.

tion of phonological units from it. (But this may include reference forward to the corresponding concepts/eternal objects under (ii), where conceptual prehensions provide top-down expectations.)

ii) *Conceptual feelings*: grasping via conceptual valuation of the physical input the meaning of the ingredient morphemes in (B), i.e., by linking perceptual prehensions of the ingredient words to the concepts causally related to them in memory. (But this may include perception in the mode of 'symbolic reference' of the propositional structure under (iii).)

iii) *Simple comparative feelings*: recognition of the propositional structure prompted by the words of (B) and their mutual ordering via an 'imaginative' prehension. (But this can only be fully determinate when the context is grasped as under (iv).)

iv) *Complex comparative (intellectual) feelings*: here the suspended intuitive judgment (a comparison of comparisons) involving the embeddedness of the proposition in its overall nexus (i.e., Cheever's model of the on-going dialogue plus the associated circumstances of the crime); the implicature behind (B) is recognized and potential non-relevant interpretations are eliminated or down-graded.

This results in the 'satisfaction' of the occasion, whereby the information abstracted is felt by Cheever to represent the best (intensest, richest) fit possible with his on-going model of the situation surrounding the crime and with his general background knowledge and expectations and is therefore integrated into the model. The 'decision' closing the occasion is the change to Cheever's on-going model that recognition of the implicature effects.[11] As regards the subjective aim behind this occasion, we must distinguish (as Whitehead did) between the vague initial aim, its satisfaction not yet worked out in terms of detailed, propositionally organized planning, and the successive clarification of that aim as the 'subjective aim proper', involving imaginative/anticipatory propositional prehensions evoked at phase (iii) as a precise goal ('lure') for the satisfaction of the occasion. But just as there appears to be overlapping between the phases sketched above, the elaboration of the subjective aim is a continuous matter. Already at phase (ii) one can imagine conceptual prehensions specifying the particular concepts that will be integrated into the propositional lure in the next phase. The initial aim was to understand what Caroline is saying and implying in (B), and the subjective aim proper simply spells this out more and more specifically in the context; it does not 'change' *vis-à-vis* the initial aim other than by rendering it sharper by relevant contrasts and eliminations on the way to its satisfaction. The subjective aim as a whole is conformal with Cheever's more superordinate long-term aim of solving the case and also with the intermediate one of trying to ascertain Meg's movements prior to the crime by questioning her parents. Both of these are 'suspended' (unfulfilled) during the present occasion, a matter I shall return to below.

The corresponding approximate sequence of prehensions behind the first of the two externally expressed occasions objectified in Cheever's words in (C) can be analysed as follows:

i) *Conformal*: recognizing (by direct physical feeling of the 'updated' model) discrepancies/gaps remaining that need filling in in order to obtain closer approximation to satisfaction of the suspended subjective aim, i.e., information which needs to be elicited by further verbal action. (Recognizing which specific kind of act is required – and how to organize it – must wait until the ensuing stages.)
ii) *Conceptual*: choosing a maximal positively weighted strategy for attaining this end, i.e., by activating a suitable discourse act type (a species of eternal object) conformal with the aim. (But this depends in part on the propositional content, as in (iii).)
iii) *Simple comparative*: formulating a (complex) proposition corresponding to the discourse act chosen (and conventionally linked to its realization) and crystallizing around the choice of predicates and associated arguments and satellites. (Reversions allow for a degree of novelty here as regards choice of words and constructions, etc., but this depends in part on discourse factors in phase (iv).)
iv) *Complex comparative*: self-monitoring and adjusting of potential expression according to politeness and other conversational principles (as graded by relevance to the subjective aim and constrained by maxims of relevance, manner, quality and quantity, etc.); this includes anticipatory prehensions of Caroline's probable reaction to his words (thereby relating the discourse act to the contextual setting).

This will result in Cheever's utterance of (C), in conformity with his aim of eliciting an answer to his still open question regarding Meg's movements prior to her disappearance. (He still needs to know more precisely why there was no trace of her presence in the flat.) Note that although this is 'externally' objectified in verbal form, it *also* effects an addition/adjustment to Cheever's own internal model – he obviously has to keep track of his own sequence of questions as well as the responses they elicit. This 'decision' results in the passing on of the now transformed data to automatic objectification by the vocal organs of the two linked utterances (under an appropriate prosodic contour), the declarative one providing the grounds for the ensuing interrogative one. They are built up around the contentive words and their grammatical argument structures (corresponding to the prehended 'logical subject' and 'predicative pattern' of the proposition) that have been drawn into the concrescence as potential propositional content. These were activated as corresponding best to the objective data to be expressed. The production of the information-seeking question itself is suspended while the first utterance crystallizes into a well-formed transitive sentence of English. (The direct wh-question that follows, clarifying the more exact kind of answer required, belongs to a subsequent occasion, as analysed above.) This sentence has three levels of embedded propositions that correspond to the embedding of Caroline's preceding utterances in their overall nexus (as interpreted by Cheever) – the structure is iconic of Cheever's construal of this embedding. I shall return below to some language-specific factors (reflecting deeper cognitive constraints) that may lie behind this organization.

Only when all of this is 'sent off' for production is the immediate subjective aim satisfied and the 'reset' subject (the next occasion on Cheever's life-line) can take in new data. The more superordinate subjective aim ('solving the crime') remains unfulfilled, of course, and it is this – 'inherited' along Cheever's life-line – that holds the causal sequence on course, prompting him to elicit further information for integration into his accreting model of the circumstances of the crime. At the outset of the concrescence the immediate subjective aim is still the vague overall one of elucidating the circumstances around the crime, plus the subordinate 'perlocutionary' sub-aim of eliciting information from Meg's parents about her movements prior to the murder. It becomes gradually more precise and focussed during the four phases as sketched, resulting in the specific question asking Caroline to elaborate on Meg's moving out of the flat. Cheever presumably feels that this emerging strategy is the best means of getting at that information at this point. Its actual expression is realized, as we have seen, by a sequence of a pre-act followed by an indirect question as to the preparatory conditions pertaining for providing the kind of answer required. Note that the concrescence as a whole makes use of both *ad hoc* rationality and convention on the way towards its objective expression.

In the above account you will have noted a certain uneasiness – if not skepticism – as regards the internal division of actual occasions into a precise sequence of four phases of concrescence. It looks as though continual anticipatory feed-forward to more complex stages of the concrescence as well as self-monitoring feedback to earlier, less differentiated stages must be allowed to occur within a single occasion in order to account for the causality of normal conversational discourse. Could one not just as easily say that the relevant patterns for more complex prehensions are present from the outset *in ovo* in the objective data – for example the contextual information that may prove relevant in determining the choice of logical subject or predicate later in the concrescence of an utterance? We need to clarify what is 'there in the data', explicit and available through straightforward conceptual valuation, and what has to be accessed by more complex prehensions within the single concrescent occasion (including tacit implications that necessitate conceptual reversions to access and anticipatory propositional prehensions of likely consequences of the intended verbal action). To that end let us consider more closely the multiple ingredients in the data to the occasion objectified by (C) that must have been at least semi-active in Cheever's mind at the point just prior to its utterance. These (constituting the 'objective data') would have included:

- The propositional content of Caroline's utterance (B) (and its following elaboration), from which Cheever inferred that Meg had told her she was moving out.
- Cheever's graded hypotheses (however vague or tentative) as to what might have happened to Meg and why.
- His assessment (based on the subjective form of adversion or aversion adhering to his prehensions of their words and actions) of the general veracity of Charles' and Caroline's words and how much they might be holding back.
- His understanding from the preceding exchange of at least something of the nature of the relationship between Charles and Caroline and Meg and of their attitudes towards each other.
- The professional context of his questioning and knowledge of conduct generally befitting such situations.
- His recognition that the present line of questioning must be handled tactfully (given Caroline's incomplete answer and heated reaction to his previous promptings).
- His knowledge of conventional ways of leading up to a tactful information question.
- His knowledge of the English language in general.

From this the subjective aim of getting further useful information from the couple further highlights those explicit conceptual valuations relevant to forming higher-order propositional prehensions that may heighten its specificity. It also teases out implicit conceptual 'reversions' that contribute to the integration of the data at hand with known eternal objects on the way towards maximal intensity of 'satisfaction'.[12] These eternal objects include quite complex patterns, such as the conventional strategy of introducing a request for information by a pre-act justifying it – as reflected in Cheever's statement of the inference which forms the basis of his actual question that follows. Also the general conversational maxims that constrain his words to no more nor less than necessary (given the degree of tact called for) are relevant eternal objects. None of these can be said to be explicit in the objective data to the occasion. Nor are the complex propositions prehended via anticipatory feed-forward (or the 'hybrid' prehensions of the possible motives behind Caroline's behaviour). Yet they are all part and parcel of the 'weighing' of relevant alternatives here. So conceptual reversions – the feelings of relevant alternatives introducing novelty into cognitive activity – must surely be allowed to arise at any point, not just at the phase of conceptual feelings. We can go further and state that reversions are what are responsible for teasing out pragmatic dimensions to verbal discourse not explicitly contained in the words employed. For this to work we have only to allow a more fluid succession of intra-concrescential phases, i.e., interpret it by and large logically as opposed to temporally, something definitely in Whitehead's spirit. In fact he himself makes it quite clear that the 'genetic passage' from phase to phase should not be understood as taking place in physical time. The causal flow is from occasion to occasion; within each occasion everything 'happens at once'.

5. AN EXCURSION INTO GENERAL MATTERS

A paradox may appear to have arisen in the preceding account, in so far as an actual occasion (of the sort relevant to human cognition) is said to have a duration of 'a second or two' (cf. Fortescue 2001, p. 34, citing modern psycholinguistic findings as well as Whitehead himself). The problem of the internal 'succession' of phases of a single concrescence is, however, circumvented by the cognitive hypothesis of parallel processing that I am assuming. The 'duration' in question can be taken as the length of a single 'focus of consciousness'. It is more important to recognize that successive concrescences *amongst themselves* are subject to the suspension and mutual overlap/embedding of complex 'inherited' subjective aims, for it is not immediately apparent how the Whiteheadian apparatus should handle this kind of suspension across occasions.[13] On my cognitive interpretation such long-term aims and purposes are inherent in the nature of the mind/brain as a self-organizing organism. Whole receding layers of intentionality come to bear on present decisions, from vaguely felt 'static' background propensities (in many overlapping or independent domains) to precise conscious intentions (dynamic processes in focus). The driving force at all levels is the urge to maintain 'intensely living' homeostasis. The overall aim of any occasion is self-creation, the progression from inherited feeling into future action as data for further occasions along its life-line (at least within the personally ordered nexus). This is manifest in the constant search for answers to internal 'why, how, what, who, when' questions, etc., as the organism confronts novel situations (as well as in externalized self-expression). In other words, we must distinguish between 'immediate' (or local) subjective aim and (levels of) superordinate, long-term subjective aim, receding into general human potentiality for action.

The specific problem of how to deal with subjective aims that both remain constant during a single actual occasion and yet change as the result of successive prehensions within it is resolved by two factors. Firstly, by distinguishing between the subjective aim and the subjective *form* of a concrescence, where the latter does indeed by necessity change with each successive prehension. Secondly, by recognizing that overall subjective aims can be broken down, as suggested, into unitary sub-aims chained together by the inherited 'suspension' of the superordinate aim. This does not undermine the original Whiteheadian notion, since the subjective aim for Whitehead typically becomes both simplified and elaborated during successive phases of a single concrescence even though there is just a single initial aim that dominates the self-formation of the subject of every occasion. This may seem like splitting hairs, but the possibility of confusion arises with certain statements in Process and Reality such as that on p. 69, where the subjective aim is stated to be not divisible according to successive prehensions within a single concrescence. Note, however, that this does not preclude the same vague initial aim becoming more and more refined and determinate as the unitary concrescence proceeds. Each concrescence *produces* its own subject, by integrations.[14] Perhaps one should say that an overall or long-term subjective aim can be changed (e.g., added to, made more specific), or even change tack altogether between occasions according to new data, whereas an immediate, local aim (a sub-aim) does not change as such but is elaborated *en route* within an occasion until it becomes objectified for passing on to subsequent ones. A non-satisfied long-term aim is inherited not directly from the satisfaction of an occasion (as objectified data for the next), but rather – via a hybrid prehension on the part of the subsequent occasion – from the eternal object(s) defining that occasion's overall subjective aim within the relevant, enduring nexus. Such on-going chainings of hybrid prehensions characterizes the 'personal nexus'. Note that Cheever's model of the circumstances of the case as he listens to utterance (B) and to Caroline's ensuing elaboration of it is continually being adjusted as he grasps the import of her words, so the 'internally objectified' concrescence whereby Cheever comes to understand the implication of (B), as analysed above, is an abstraction. Its immediate aim (at understanding the ultimate reasons behind her utterance) could in reality have been suspended while piecemeal adjustments were 'tried out' on the model as each successive new utterance by Caroline was taken in and analysed – i.e., in a whole series of subsidiary 'internally objectified' occasions prior to the externalized one that produced his own next utterance. The notion of 'immediate' (or local) aim is obviously a relative one here.

The subjective *form* of an actual occasion, as initially determined at the beginning of a complex concrescence, can and indeed *must* change. This form is, if you like, the way the actual occasion feels to its own subject. It reflects how individual prehensions within the concrescence are valued (up or down) *vis-à-vis* the subjective aim; *which* successive prehensions are evoked is determined by the 'mental pole' (conceptual) activity of the occasion as it 'creates' itself by fulfilling its aim. Subjective forms are subject to the category of 'subjective harmony', whereby the subjective form of successive conceptual prehensions are adjusted in accordance with the prevailing subjective aim, and this constrains its range of variation within a single concrescent occasion. I shall return to how this may be expressed in language below. Consciousness is the subjective form of (certain) complex prehensions, and conscious self-monitoring in the production of an utterance-in-context is one relevant possibility here (though the degree or intensity of consciousness can of course vary). More specifically, consciousness is the subjective form of the mental pole activity of actual occasions associated with comparative (propositional) prehensions only at the later, supplementary phases of concrescences. It is consequently quanticized and pulse-like within on-going processes guided by long-term or rationally 'chained' aims. It may affect in turn the linguistic expression of information organization along the lines discussed by Chafe (1994). Basic generalizations recognized here include 'at most one chunk of new information per focus of consciousness' (and 'at least one chunk of old/given information'). This constrains – critically for my present purpose – how long a stretch of propositional material is typically expressed by one concrescent occasion. The immediate contents of a speaker's focus of consciousness will (at least co-)determine how much information is selected and organized for expression – typically one clause-length chunk corresponding to a single focus of new information. This organization of information into 'chunks' must allow, of course, for some degree of competition and variation of expression.

This brings us back to the distinction I have introduced between internally and externally objectified occasions, the former resulting in changes purely to the subject's own internal model, the latter also causing a change in that of some other subject (typically via an utterance aimed at a hearer). In Fortescue (2001) these were treated as compressed into single afferent-efferent units rather than as chains of alternating afferent and efferent occasions along the same personal life-line. This was a kind of descriptive shorthand that was justified by emphasizing the afferent and efferent aspects of *all* actual occasions. The distinction I am now making only concerns the nature of the 'decision' – the objectification – closing the concrescent realizations of different kinds of subjective aims. As regards the linkage between chains of occasions of the two types, note that there is both an 'eternal object' and a 'physical inheritance' side to the matter: a combination of recognized expectations of conventional discourse act 'turns', etc., and the usual workings of subjective aims as inherited from occasion to occasion (reflecting general rationality at the mental pole of successive occasions). This inheritance is mediated by hybrid prehensions of the subject's own conceptual prehensions – or those of others - earlier in the causal sequence, as defined above. The output of an internally objectified concrescence in a dialogue situation is typically the input (as objectified data) to a following externally objectified one by the 'same subject' (a subsequent occasion along the same life-line), i.e., to an utterance whose formulation has been affected by preceding mental activity. This contrasts with purely internal thought, where successive internally objectified occasions are chained. A linked succession of occasions of either kind (or an alternating mixture) constitutes in effect a 'personally ordered nexus'. This is a complex enduring – but changing – structure consisting of numerous ingredient 'enduring objects' (objects that are repeated in time throughout sequences of occasions, e.g., knowledge structures/models, etc.) that constitute the organism's memory. Over – or through – these a temporal chain of 'presiding occasions' (the on-going focus of mental activity, in so far as there is any) wends its aim-driven way, all supported by the organism's physical frame.[15] On my cognitive interpretation of Whitehead I shall continue to maintain the basic Whiteheadian notion of the unitary afferent-efferent concrescent occasion, exactly one corresponding to each causal change to some internal model of the relevant personal nexus, although some occasions must be seen to 'stop short' of the phase of comparative (potentially conscious) prehensions. Some of them are 'satisfied' with internal changes, others only with external actions, such as speaking.

It remains to say a few words about the nature of the 'internal models' that I have been taking for granted. As we have seen, Cheever's accreting model of the circumstances he is investigating consists of representations (however abstract and derivative) of several mutually embedded and – as far as possible – harmonized 'worlds', for example his own view of Charles' and Caroline's perspectives on the matters in question, plus at least one tentatively entertained scenario for the crime itself. It is a 'feeling structure' of 'enduring objects' constituted by objectified feelings/prehensions of earlier occasions in his life-line and accessed via matching feelings.[16] Part of it is 'heightened' through continual perceptual feedback, anchoring and stabilizing the whole, namely the actual on-going discourse situation where Charles and Caroline and Cheever are talking together in the same room at the same time. It is in turn embedded in a broader cultural/experiential background nexus that fades off into 'common knowledge' – a presupposed pool of potentiality assumed to be roughly the same for all human beings. The accreting model constitutes the core of the overall context necessary for understanding the meaning of the successive utterances of the passage (and for assessing their relevance). Within this overall complex, Cheever's sketchy and tentative representation of Charles' world further contains links to overlapping nex_s concerning his relation to his wife and to Meg, and that of Caroline's world in turn contains information on her relationship to Charles and Meg. Access to their worlds (by Cheever) is facilitated by their physical presence before him, maintaining in potential focus the relevant enduring models of them in his memory. Note that the phrasing of Caroline's and Charles' utterances contain a number of linguistic 'space builders' – e.g., the verbs 'say about' and 'know' that signal a particular 'intentional stance' to their propositional complements. These help him manouevre through the complex structure, adjusting and adding to it as new information is absorbed for integration.

6. BACK TO THE CASE: LINGUISTIC ASPECTS

These general points out of the way, let us turn to some specifically linguistic aspects of the exchanges in the example passage. The key notion relevant for linguistic behaviour in the Whiteheadian scheme of things is 'symbolic reference'. This is our usual mixed mode of perception linking the precise imagery of 'presentational immediacy' with the vague but feeling-laden depth of 'causal efficacy'. In the case of the perception and understanding of a word (phonological or written), for example, its form is accessed via perceptual prehensions in the first of these two simple modes, and its content by physical prehensions in the second mode, that of causal efficacy. (This includes the feel of the word's potential for use in specific contexts.) In the case of a whole sentence-length utterance more complex (propositional) prehensions are necessary, but they are again in the mixed mode of symbolic reference, matching higher level perceptual patterns (here morphosyntactic) to their felt 'meaning' within the relevant nexus. This is essentially a matter of recognizing familiar eternal objects (both of form and – via conventional and/or iconic association – of content) finding ingression in the physical data. One side-effect of this mixing of modes is the enhancement of prototype effects, since the fit from expression to world is never perfect, only 'best possible in the circumstances'.

One needs to be careful with the term 'meaning' here, however – it can be understood in two ways according to perspective. From the Pattern view of language (emphasizing the mode of presentational immediacy) it is a matter of potential 'eternal objects', whereas from the Process view (emphasizing causal efficacy) it is a matter of the feelings induced in the subject of an actual occasion by the prehension of a sign (typically underdetermined until context is taken into account). It is perhaps preferable from the Whiteheadian perspective to limit the use of the word 'meaning' to the latter, dynamic sense, reserving the term 'eternal object' (i.e., 'concept' or 'conceptual structure') for the static, purely potential patterns to which the 'meaning' of words point through association. In fact Whitehead himself described symbolic reference in terms of the relationship between the (perceptual) species of 'symbol' and the species of 'meaning', where the former refers to the perceptual mode of presentational immediacy and the latter to that of causal efficacy (Whitehead 1978, p. 181). Between the two is a 'faintly relevant nexus' (i.e., between the word token and some specific contents of memory). Note that the distinction here is not quite the same as the more traditional one drawn between (conceptual) meaning ('Sinn') and indicative denotation ('Bedeutung'), since the prehensions afforded by a sign are not to be equated with referring or denoting alone. (The latter kind of act draws upon these 'affordances' rather, particularly in the form of indicative prehensions.)[17] In reality the one cannot be entirely abstracted from the other, as is generally the case with the relationship between pattern and process in Whitehead: patterns presuppose processes – the real *quanta* of reality, actual occasions – but the latter cannot help but manifest pattern. The eternal objects constituting the grammar and the meanings of the words of English here represent a system (or interrelated set of systems) of rather precise structural contrasts with a potential linkage to vaguer – and highly context-sensitive – feelings of causal efficacy.[18]

At all events, expression and comprehension are both dynamic for Whitehead. What is essential to grasp is that 'meaning' in the Whiteheadian scheme of things – though he did not spell this out himself in detail – must cover not just conceptual meaning (in terms of logical subjects, predicates and quantifying/ qualifying expressions), but also relational meaning (as expressed in syntactic structure, including subordinate clause type), subjective-form-expressing meaning (both discourse-functional and epistemic/evidential), and intentional meaning (as reflected in discourse/speech act choice). *All* of these are from the Pattern perspective eternal objects (as are 'nexus types', the contexts of situation relevant to meaning), in so far as they are repeatable and communicable – they represent the *sine qua non* of communication, in fact. This is why I hesitate to replace outright the problematical term 'eternal object' by 'concept'.

As a concrete example of the accessing and activating of lexical items, consider the predicate 'go' in the last sentence of (C). Cheever could 'read' (or rather 'feel') this off the internal model that he is all the time trying to add details to and reorganize in the most efficient way conformal with his understanding of truth, likelihood, etc. This is so because the felt meaning of that verb (its causal efficacy) matches and potentially bridges a gap in his certain knowledge of the circumstances of Meg's disappearance: since she is known to have left her flat she must have gone *somewhere* else, though the precise goal is unknown. Cheever prehends the potential relevance of the concept of 'going' and the best fitting associated phonological word to that part of the model. Her 'going' somewhere specific is certainly worth asking about at this point. Note that the general verb 'go' is chosen since nothing is known at that point of the specific means or manner of her leaving the flat. At any one point in the concrescence of an utterance only those eternal objects that contribute to maximal intensity of its satisfaction are 'summoned out of potentiality', and here the meaning of the word 'go' can provide such a contribution. The category of 'subjective intensity' ensures that the prehensions that are integrated in a concrescent occasion are those that match the aim 'in virtue of near-identity', supported by relevant contrast-intensifying reversions (Whitehead 1978, p. 279). In linguistic terms, those that ensure maximal fit between form (linguistic means available) and function ('felt' meaning to be expressed). There may of course be only an approximate form-function match. In general one may say that precise meanings may be communicated, despite the loose, prototypically organized nature of conceptual aggregates associated with lexical items, thanks to the convergence in a given aggregate of eternal objects of various types so that only a part of it is highlighted/activated in a given context.

It must be borne in mind that the activation of 'function words' (e.g., definite articles or certain prepositions) is of a somewhat different nature from that involving 'content' words: it is not their intrinsic conceptual or referential meaning but rather their necessary association – like inflectional and agreement phenomena – with specific grammatical constructions (including individual lexical frames) that draws them willy nilly into the organization of the utterance. 'Discourse function' words (like 'anyway' or 'well' in our passage) are different again, since their activation involves the matching of the 'subjective *form*' of the occasion with its expression.[19] More generally, not all sentences are generated or understood 'from scratch', i.e., from individual words or morphemes – much of human speech is tinkered together 'on line' from pre-existing 'chunks' and constructions of varying degrees of abstractness. Sometimes these are fixed idioms, sometimes variations can be made in their exact formulation. Consider Caroline's outburst *It's such a waste*. This is in itself a standard locution in such 'tragic' situations, but is nevertheless open to variation, as in her immediate repetition: *It's all such a waste*, and incrementally a little later: *It's all such a terrible waste*. We can easily imagine that the idiomatic chunk was 'summoned' as a pre-formed whole to match the sentiment she wanted to express, but that the subjective aim behind this utterance was not wholly satisfied by it. Under the pressure of her pent-up frustration she repeats the cliché, giving it additional emotional intensity by broadening the subject to 'all' (the whole complicated situation), and then, when that is still not enough to vent her feelings, by adding the adjective 'terrible' to emphasize the extent of the tragedy still further. That she can do this suggests that she has access to the 'chunk' both as a unit and as a simple (and variable) syntactic template.

Hierarchically further along in the elaboration of the concrescence is the organization of the data into propositional structures (eternal objects of a syntactic type).[20] Here the amorphous input (felt in the mode of causal efficacy) that is ultimately to be expressed in an utterance can begin to take on a determinate shape as the relevant options available to the organization of 'information structure' are harnessed through suitable prehensions activated by the subjective aim in context. These options would appear to be inherently hierarchically organized, in the sense that the choice of one type of overall utterance type (e.g., a question) precludes certain options available to another (e.g., the subject-first order of declarative utterances). The same applies to choices of constructional configurations at intermediate levels of organization, such as the choice of the passive construction in English, which precludes starting a clause with the agent as subject. At the superordinate 'discourse' level principles such as Prague School 'functional dynamism', Givón's 'most pressing material first' (Givón 1988), Lambrecht's '(wholly new) sentence focus' vs. 'constituent focus' organization (Lambrecht 1994), and Hawkins' 'Early Immediate Constituent' principle (Hawkins 1994) may come to bear, all of which have a 'trickle-down' effect on subsequent, lower-level choices.[21] These are typically linked to higher-order discourse act choices of the kind discussed by Hannay under the rubric of 'discourse management' (Hannay 1991), and may all be seen as falling out unconsciously from more general principles of economy and clarity (often with no 'on line' choices involved at all, since they have been historically conventionalized within the language concerned). Consider Caroline's final reply to Cheever at the end of the passage: *To live with Harry. He had a house.* She organized the first sentence fragment of her reply this way because the 'missing' matrix clause has already been supplied by Cheever's preceding question and the fragment supplies the required new information. The factors behind the formulation of the second (full) sentence are more complex, since it supplies a supplementary *reason* for the initial part (in isolation the sentence as it stands would be rather unusual), but it clearly represents a single new item of information predicated of a given logical subject (the topic Harry). A presentational packaging of this information as if it were all new (e.g., 'There was a house belonging to Harry') would have been quite unwarranted.

Of course all this concerns primarily the focussed/foregrounded material already selected for expression by the subjective aim; presupposed background material (given information) is either suppressed at the outset by negative prehensions in the 'conformal' phase of the concrescence, or referred to by special devices like anaphoric pronouns, as needs and usage dictate. What is presupposed and unexpressed – left beneath the surface of the overt expression – is highly context-dependent and requires on-going assessment of what constitutes 'common knowledge' to the discourse participants involved. Think of what was presupposed and left implicit behind Caroline's utterance in the middle of the quoted passage: *I'd like to have been a grandmother.* She expects the logical link between Meg's abortion, her feeling sorry for the unborn baby, and the current statement to be understood by Charles and Cheever – also the following leap to her mention of her son not wanting to marry. Within the Whiteheadian framework the mutual compatibility of such utterances is something that can be directly felt (e.g., by Cheever as he fits them to his accreting model of the circumstances of the case) and does not need propositional spell-out of every intermediate link.

We can now return to the uncertainty expressed earlier about the breakdown of the concrescence into exactly four 'phases', and see how this may or may not relate to specifically linguistic processing. If the production or analysis of a linguistic utterance proceeds, as has been suggested, in a largely parallel manner, does talking of the successive phases of such a process even make sense? Apart from the final dispatching of a bundle of commands to the speech articulators, can anything really be said about the order of prehensions in the passage from complex and largely indeterminate input to determinate output, given the degree of flexibility of on-line planning (including restarts and reformulations, not to mention the idiosyncrasies of the internal models that the individual speaker/hearer brings to bear on comprehension)? One way of relating such processes to the four phases would be in terms of 'modules' – one might suggest, for example, that each successive concrescent phase in the production or comprehension of an utterance corresponds to treatment by a particular module of the brain, abstracting away from whatever parallel processing actually goes on among modules. Thus – to take the case of comprehension – phonological processing involving physical prehensions in the early conformal phase of concrescence might continue in one area (in the left temporal lobe) at the same time as the extraction of meaningful units (morphemes) via conceptual prehensions of the associated 'eternal objects' from the speech chain is initiated in adjacent 'semantic memory'. The extraction of a propositional structure may in turn be initiated in yet another area (say in the vicinity of Broca's area) before morphemic analysis is complete, while the abstraction of further contextually determined implicatures (involving intuitive judgments and other 'intellectual' prehensions) might in turn be initiated before syntactic processing is complete (e.g., in certain cortical areas of the non-dominant hemisphere).[22] Still other kinds of prehensions *en route* might be related to the parietal (quasi-spatial) 'logic' area, e.g., in interpreting prepositions. All of this would be under the general control of an overall complex subjective aim sustained through time in the frontal cortex with the aid of sub-cortical 'echo' loops (allowing both continual feed-back and anticipatory feed-forward). Certainly psycholinguistic research suggests that relatively rapid and automatic 'sub-goals' can be carried out while superordinate 'frontal' production or comprehension processes are suspended, and that top-down processing (from subjective aim) and bottom-up processing (from physical data) typically proceed in parallel. Each such 'module' presumably has its own long-term potentiality (manifesting as rules, patterns, conventions, default expectations) over which particular types of dynamic process operate – input–output pairings, if you will. Just as in the case of the more general schema of concrescences, the fact that one module is superordinate to – or requires input from – some other module refers only to the initial order of activation and does not preclude the overlap of *completion* of that activity.

So overlapping phases (reflected in the activation of specific functional brain regions) seems to provide a possible answer to the problem of linear organization here (one supported by accumulating psycholinguistic evidence). But this in turn raises the question of how, within the Whiteheadian framework, one is to handle the passing on of information from one 'module' to the next. Are these intermediate stages best seen as unitary concrescences in their own right, with definable constraints on input and output types, or do they correspond to individual prehensions chained together to fulfil a unitary higher-level aim? Clearly the latter, it seems to me. It is, after all, at the highest organizational level that teleological processes such as the production and comprehension of utterances is relevant, namely that of the mind/brain as a whole. The activity invoked in populations of neurons at lower levels of organization is subordinate to the overall, potentially conscious intentionality of the individual organism or 'personally organized nexus'. The key notion here is that of the unitary, sustained subjective aim that overarches and 'licences' all such chained 'sub-programs'.

7. FOR THE RECORD

This is all very well for the general adjustment of the Whiteheadian framework to account for linguistic behaviour, but what, one may ask, about formalization? Is that not what the scientific endeavour is all about – to make succinct, repeatable generalizations? But is that really always possible with a teleological, human science such as linguistics? Is it even desirable? I am certainly not about to propose that some kind of quantum-like wave function formalism indicating probabilities rather than 'rules' and 'laws', for example, should replace the linear formulae of syntactic theory, even though it is clear that the latter cannot help us very much with the representation of the 'chaotic' cascading of language-as-process. Where I think the Whiteheadian perspective can be of some practical use to linguistics (as opposed to philosophy, where it needs no justification) is in focussing upon the systematic correlation of Whitehead's typology of prehension types to specific tasks and phases of language processing, a matter of the cognitive underpinnings of language. In particular, this approach can in principle show (a) how linguistic patterns (grammar, conventions of use, etc.) reflect hierarchically organized combinations of intentional processes realizing specific types of contextually embedded subjective aims (discourse acts), and (b) how, conversely, the processes behind the production or comprehension of utterances-in-context might summon (via specific types and sequences of prehensions) only the linguistic patterns relevant to overt expression of the focally activated aspects of the nexus concerned. Since a degree of creativity is involved in all language use this cannot be a fully determinate matter and is therefore not fully formalizeable. Why should one expect actual language use to be describable in terms of predictive, absolute rules in the manner of the natural sciences, as opposed to general principles applying or not to specific situations according to complex intentional factors? From the Whiteheadian perspective linguistic ‘rules’ belong to the purely potential realm of eternal objects, to Pattern rather than Process. They are norms guiding process, no more, no less, and can be broken at will for specific ends. What can be demanded of the present approach (if it is to be preferred over others) is that it should display a certain economy of description, and this it provides not by forcing concise and potentially misleading formulae upon the analysis but by limiting the descriptive apparatus to a manageable yet adequate array of prehension types.

Means-ends diagrams in the manner of Leach, taxonomies of the more or less conventional patterns available for expressing specific speech acts *à la* Searle, and the informal generalizations of Conversation Analysis, represent useful analytical tools already well established within Pragmatics, but they need to be embedded in a richer theory of complex intentions if they are to be integrated more tightly into a broad Cognitive Science. The goal of extracting constraints on the form of linguistic productions from a typology of introspectively accessible prehensional processes is neither utopian nor unscientific – the Whiteheadian theory of prehensions, nexus types and concrescent processes as a whole is subtle and complex, and promises to throw new light on such matters as complex propositional embeddedness and the relationship of consciousness to information packaging. What would such Whiteheadian ‘soft’ constraints look like? Bear in mind that we are looking for generalizations about ‘eternal objects’ (those finding ‘ingression’ in the grammar of English and the conventionalized aspects of the pragmatics of its use) that can be motivated by process considerations, i.e., in terms of prehensions, subjective aims, and the embeddedness of utterances in complex nex_s. Since our perspective is cognitive we are talking here about language-specific grammar as 'internalized’ by the individual participants, not about typological meta-statements across grammatical patterning in a range of languages.

Consider Caroline’s words from the cited passage (they follow her rebuke to Charles: *You’d have ruined her life for her* – i.e., if Meg had told him about the abortion.):

D) *Well, it’s ruined anyway, but it might not have been if she’d been able to talk to you as a father instead of a priest. Everything had to be kept secret in case you preached at her.*

We can use this to explain – with the aid of the notion of ‘prehension’ – certain constraints on anaphora and referential explicitness.

The use of the pronoun ‘it’ in the first sentence of (D) is sanctioned by its well-known general function of stimulating in the hearer a search for the most accessible candidate referent in the immediate, ‘activated’ context (here ‘her life’). For ‘context’ one could read ‘nexus’ – the whole on-going situation involving the three participants. Note that there is no trouble with such 'instructional' meanings in the Whiteheadian framework: eternal objects include potential action types such as the one this grammatical morpheme evokes. The relevant constraint here concerns the choice between this item and the semantically equivalent full noun phrase 'her life'. This grammatically quite acceptable possibility is repressed (by negative prehension) because it is redundant (non-efficient) for the fulfilment of the subjective aim, given the near-repetition of the propositional frame from the preceding sentence. Caroline could nevertheless have chosen to express the full noun phrase - for example if she had felt it was worth the effort to repeat it in order to render the new utterance more independent (a matter of matching output with intended 'rhetorical strength').

The use of 'she' in the next sentence and of 'her' in the last sentence are somewhat different. Here the general principle again indicates that the pronouns must refer to the most relevant 'activated' female referent (namely Meg), but Caroline's options are now more restricted: the use of 'Meg' in place of 'she' would be unusual, given the uniqueness of possible reference (although not entirely impossible, especially if preceded by a lengthy pause). 'Her' in the last sentence, however, could simply not be replaced by 'Meg' at all, since even without contrastive stress it would suggest a contrast (say with another female); there is no such contrast possible, given the nexus concerned. This has nothing to do with grammatical acceptability as such, but rather with the unambiguous status of Meg as 'topic' at this point – both extra-linguistically (within the activated portion of the nexus), and as reinforced by the syntactically subordinate status of the clause in which 'her' occurs. One could certainly construct an approximate 'rule' here, but it would still have to refer to discourse matters such as topic-tracking and context type. The motivation for such a constraint is at all events not at all rule-like: it is the need to ensure correct interpretation of one's referential intentions – as efficiently as possible given the need to avoid ambiguity.

One could further invoke Levinson's I- and M- heuristics (mentioned earlier), the former being reflected in the preference for pronouns in local coreference conditions and the latter leading to expectation of some special reason for nevertheless using a full noun phrase when such conditions apply (cf. Levinson 2000, p. 267ff). Such broadly applicable principles are part of the permanent background potentiality behind all discourse - they only become active when a potential mismatch is sensed, e.g., when a decision has to be made weighing alternative expressive strategies, but their effects can also be grammaticalized and/or semanticized, that is, become absorbed into the grammatical and lexical conventions of a language.

Both grammatical configurations and heuristic principles of language use are - as norms that can function as constraints on the expression and interpretation of subjective aims – species of 'eternal object', so this causes no particular problem within the Whiteheadian framework. They both reflect the interaction of subjective aim and eternal objects, the former 'perspectivizing' the latter (by relevance) and in turn constrained by them. What the grammar of English contains at this point is a set of relevant options and discourse context types where the one or the other of such options is preferred. These options can be manifest cognitively as a feeling of relative degree of 'fit'.

In so far as this is structurally determined, it can in principle be stated in terms of propositional organization, a matter I have addressed elsewhere in Whiteheadian terms (Fortescue 2001, p. 46 and 81ff.). In particular I have argued that a Whiteheadian proposition may well be expressed by more than one clause, according to the exigencies of discourse management. Let it suffice here to suggest that there is a limit on the hierarchical complexity of what can be grasped within a single focus of consciousness by a single Whiteheadian 'intuitive judgment'. This, it will be recalled, binds an entire proposition (of whatever internal complexity) to an indicative prehension of the relevant nexus from which it derives, providing it with the possibility of a truth value and thus allowing integration into the on-going internal models of speaker/hearers. Here we can discern a reason for the natural disposition of speakers to express 'just one new chunk of information per utterance', referred to above, but what is added to the model is a unitary 'new' comment that is attached to some specific nexus (or part of a nexus) as its 'topic'.[23] The natural unit of such 'chunking' is the proposition, which has its roots in perception.

In excerpt (D) we can see how competition between, on the one hand, this long-term principle of 'just one new chunk of information per utterance' and, on the other, the possibility of linking the expression of several propositions together in a complex sentence structure when they share one overall subjective aim, can potentially lead to conflict (and thus indeterminacy in the analysis). Nevertheless, we can use the logic of the relationship between the evolving subjective aim and an ongoing model to make principled decisions as to what is most likely to be going on in such cases. Thus we can see competition between syntactic 'packaging principles' of the type adumbrated above and the topic-tracking principles discussed earlier in this section. This competition arises when the latter principles extend across several interrelated worlds or nex_s. How exactly do they interact? Well, there are constraints on the depth and type of embedding that come to bear on the choice between anaphoric or full lexical referring phrases when tracking topics across complex structures, and these in turn appear to reflect in part the well-known constraints on 'extractability', which have been variously analysed in terms of purely structural 'island constraints' or of the scope of functional 'focus' or 'topic' categories (cf. Fortescue 2001, p. 72, where suggestions are made as to how these can be 'translated' into Whiteheadian terms). One relevant constraint might go as follows. The reason why Caroline did not say *Everything had to be kept secret in case you preached at Meg* (rather than *at her*, as she did in fact put it) could be that referring in a proposition (whether simple or complex) to the same, non-contrastive, non-ambiguous referent by a full noun phrase rather than by a suitable pronoun when that referent has already been introduced or reinforced as the topic of the on-going discourse in an immediately preceding proposition is only sanctioned if the immediate subjective aims of the discourse acts behind the utterance of the two successive propositions do not stand in a relationship of inclusion within the same actual occasion. This would rule out 'Meg' here – but not, as a possibility, in the preceding sentence with 'she' as subject of the subordinate clause, where the topic was *not* introduced in the superordinate clause, and where the nearest preceding reference to her was in the expression of a separate actual occasion with its own distinct – though related – subjective aim.[24] This is not circular, since the scope of the individual occasion has been defined independently.

8. CONCLUSION: NON-LINEAR PROCESSING AGAIN

It is time to return to the central question of non-linear causality. It should be clear by now that this is also central to Whitehead's philosophy of Process. This framework accommodates both Aristotelian 'final causes' (the subjective aim *within* occasions) and 'efficient causes' (the 'objectification' of one occasion for following ones). The many elements of a concrescence 'become one' and in turn become one of many elements in the data for a subsequent actual occasion. But it is especially final causation that has been highlighted in the present application to linguistic behaviour, the most distinctively human of activities, since the teleology of self-organizing 'occasions' lies at its heart. The emergentist, relativistic (and non-reductionist) cast to Whitehead's thinking is what renders it particularly relevant to modern Cognitive Science today. Applying it to linguistics was something Whitehead himself did not progress very far with, but I hope I have shown how his basic ideas find a natural arena for deployment there. It may well be that any attempt to develop a precise formalism in this connection is unlikely to give us much mileage, given the degree of novelty and freedom permeating linguistic behaviour. Of course Whitehead was much concerned with both mathematics and the natural sciences earlier in his career, but it is his later, holistic philosophy (often branded as 'metaphysical') that is of interest here. His analysis of the specific stages of emergence of (potentially conscious) mentality from physical processes has a direct bearing on the constraints human cognition brings to bear on linguistic processing. Through the general framework of his philosophy we can directly address the question of the nature of non-linear causality in teleologically organized beings. For Whitehead, non-linear causation is self-causation. In this he is in harmony with much recent thinking that goes under the rubric of 'complex adaptive systems theory' (cf. Gell-Mann 1994, p.16ff). Essentially, this is the attempt to ensure that a 'top-down', intentional approach to the investigation of complex organisms comes to mesh smoothly with a 'bottom-up' one (reductionist only in the positive sense of 'emergentist') by taking into account all the 'accidental' factors – the additional degree of informationally rich 'chaos' – involved in the emergence of the more from the less complex.

At the centre of my application of Whitehead's ideas to linguistics is the notion of the 'prehension'. All prehensions (whether physical, conceptual or hybrid) are acts of information transfer, and their mutual ordering within a unitary concrescent process is, as we have seen, logical rather than linear in the strictly temporal sense. The non-linear relationship between the subjective aim of an actual occasion and its satisfaction is determined by those prehensions that enhance what is most relevant for that satisfaction in the semi-activated nexus concerned – in the case of the production of an utterance, those prehensions that 'summon' relevant linguistic signs and principles for linearizing the intended expression. The result is linear, the productive (causal) factors are not. The only adjustment to the Whiteheadian framework that is called for here is, as I have suggested, allowance for longer-term subjective aims being suspended across successive occasions while immediate sub-aims are satisfied in a piecemeal fashion. This can be effected by the reabsorption of unfulfilled or only partially fulfilled aims back into the nexus pattern sustaining successive actual occasions within it, whereby they become purely latent again until that part of the nexus becomes active in a new occasion arising out of it. In terms of Whitehead's original system this is, as I have suggested, presumably a matter of the inheritance, via hybrid prehensions, of already objectified data *indirectly* reflecting earlier subjective aims. Such prehensions continually recreate the focal life-line of the 'personally organized' nexus. Recall that it is the perspective of the actual occasion on eternal objects (as determined by its embedding in a particular nexus) that determines the subjective aim of that occasion, so if the nexus is changed, the future aim of occasions within it will also be changed accordingly. Immediate subjective aims need to be seen not only as heuristic and creative and limited to a single focus of (potential) consciousness, but also as embedded in receding layers of more and more stable 'background' intentionality within the personally ordered nexus. The dynamic, changing nature of immediate subjective aims (constantly negotiating between reality and possibility) is surely the major stumbling block to modeling teleological behaviour in computer simulations to date, despite the advent of more sophisticated techniques of parallel distributed processing.

Luckily we do not have to wait for bottom-up computer modelling to 'catch up' in order to pursue our understanding of the overall organization of cognitive processes. The essential thing, it seems to me, is to uphold the complementary perspectives of Pattern (in terms, say, of eternal objects, conceptual 'repeatables' including norms, expectations and socially sanctioned rules) and Process (in terms, say, of individual actual occasions, and the prehensions and subjective aims of which they consist). It is important to understand that the former is both the source and the result of the latter in self-organizing systems. In Whitehead's universe, process is complemented by potentiality, which is in itself dynamic. Pattern only exists in so far as it is borne by actual occasions (processes). This amounts to a cognitive analogue of modern Superstring Theory within physics (cf. Gell-Mann 1994, p. 27ff).[25] What is foregrounded in on-line language use – as in perception – is dynamic, transitive action and what remains in the background is the more static context, largely presupposed and out of focal awareness. Process is in general the passing on of feeling into expression, and feeling (or prehension) is the reception of expression (Whitehead 1968, p. 23). The human organism is composed of entities mutually expressing and feeling themselves. Bohr (1948) spoke of the complementarity between feeling and thought - one could perhaps extend this in the present context to the complementarity pertaining between implicit cognition (i.e., 'feeling' in the sense of purely internally objectified prehensions) and explicit verbal expression. The total meaning of utterances involves both, namely the 'on-line' implicatures of discourse as well as the explicitly coded semantics of the language concerned. Language as such is the systematization of expression – it renders past experience more vivid, precise and 'manipulable'. In doing so it emphasizes the accidental and presupposes the necessary (Whitehead 1968, p. 101), for we are in general most aware of (and interested in) what differs from our expectations. Language, although the neural underpinnings from which it emerges may for the time being remain obscure, is constrained by – and reflects – the nature of consciousness, and it is here that Whiteheadian philosophy and linguistics have something to offer each other. Perhaps together they may even coming a step closer to solving the deeper mystery. I rest my case.

NOTES

[1] The main types of prehension relevant for language are the following: physical, conceptual, hybrid, indicative, propositional, negative, and anticipatory. These will be explained below, as relevant, though a general characterization can be found gathered in Fortescue (2001, p. 14ff.). Note that 'symbolic reference' (the core of Whitehead's approach to language) is itself a complex process.

[2] I sometimes use the terms 'eternal object' and 'concept' as equivalent, though strictly speaking the former covers all 'forms of definiteness', e.g., grammatical patterns, strategies and plans of action, types of emotional reaction, the phonological or orthographical shape of words, as well as the abstracted 'sense' of lexical items (anything recognizably a 'type' rather than a 'token'). Whitehead himself only distinguished between eternal objects of the 'objective' and of the 'subjective' species, the former being 'elements in the definiteness of objectified nex_s' (including mathematical 'forms'), while the latter are 'elements in the definiteness of the subjective form of a feeling' (i.e., involve the perspective of a subject). It is the latter type that is particularly relevant for language.

[3] Or more accurately 'discourse act' (cf. Leech 1983, p. 59), since we will be concerned with complex intentions, i.e., perlocutions that involve both speakers and hearers, not just illocutions in the manner of Searle (1969), isolated from the broader discourse context. The role of convention in such acts cannot be ignored, but the position I shall take is that conventionality is a matter of degree, with some parts of pragmatics – different in different languages – being conventionalized and indeed grammaticalized, while for others basic rational and social principles of a Gricean nature must be appealed to. In Whiteheadian terms, this means that certain relevant 'eternal objects' – namely patterns of conversational interaction – are potentially accessible to the speaker of a language in the form of specific rule-like expectations, whereas others have to be reached by conceptual 'reversions' from analogy to known patterns or by the application of more basic general principles (themselves eternal objects) to new situations. The answer to the question of how much of pragmatics is a matter of conventional pattern must thus be left indeterminate – the extremes of integration into the coded categories of the grammar on the one hand, and of indirect hinting on the other are clear enough, but for the grey zone in the middle we can only talk of individual variation and likelihood of interpretation in context.

[4] These reflect respectively the heuristics whereby the hearer infers that 'what you do not say is not (the case)', as when the weaker of a scalar pair like 'some/all' implies that the stronger does not apply; that whereby 'what is expressed simply is stereotypically exemplified' can be inferred (Grice's 'say no more than you have to'), as in bridging inferences; and that whereby 'what is said in an abnormal way is not normal' can be inferred, as in the interpretation of overly elaborated information. All three refer to norms or expectations and the implications of breaking with them.

[5] This model is a nexus of the kind Whitehead calls an 'enduring object'. As a realist he does not distinguish in principle between 'objects' in the objective world and their mental 'images' (however abstractly transformed) – both are species of enduring objects, i.e., assemblies of causally interrelated actual occasions constituted by prehensions either

physical or conceptual or both. In the case of physical objects this is a simple matter of the causal chaining of physical prehensions which ensures relative persistence from occasion to occasion through time, whereas in the case of mental 'images' (actually 'feeling' images, not necessarily isomorphic with any particular sensory modality) it is a matter of the prehension by a subject of the accretion in memory of previous actual occasions objectified for it as potential data (via 'causal efficacy').

6 Hybrid prehensions are of another subject's 'pure' (conceptual or physical) or 'impure' (propositional) prehensions. That 'other subject' can be either an earlier occasion in the same personally ordered nexus or another entity altogether providing 'data' to that nexus from the past. They provide developmental continuity with an individual's own past self and allow empathy with the motives of other entities in one's memory of the past. This kind of prehension is thus the means of building up an internalized Theory of Mind. It allows hearers to work out the probable intentions between speakers' utterances not by impossibly open-ended 'reverse' deduction, but by abduction involving the mutual coordination of situational and referential nex_s and relevant eternal objects (including discourse act types, pragmatic principles, etc.), i.e., answering the question 'What would have made *me* say that in these circumstances?' This instinctive stance is reflected in the fact that people by and large share the same 'feel' for what is perceptually salient in a given situation, irrespective of linguistic expression. Among the various means available (besides 'propositional content') that help us home in on the intentional complex behind an utterance there is also prosodic packaging, which reflects the subjective form of an utterance, including affect and emphasis. Patterns in this 'parallel' channel have their own (at least partially) conventionalized interpretations that can contribute to anchoring the utterance in the 'context of situation' by eliminating potential but irrelevant implications.

7 Actually, 'going on holiday' has the potential flavour of a euphemism here: from what we know of Meg's actual involvement with (boyfriend) Leo and of the prudishness of her parents (especially her father), it is pretty certain that this 'holiday' would not have involved separate rooms. Charles may simply be being naive here, but Cheever and the reader are not required to be so. Note that the meaning of the word 'holiday' cannot simply be severed from its total context – the latter delimits the expectations aroused by the word, highlighting just certain aspects of the constellation of affordances associated with it. This constellation is already reduced by the use of the word within an idiomatic phrasal construction and is further reduced by the broader 'context of situation'.

8 An imaginative prehension is a type of propositional prehension that 'predicates an eternal object of an actual occasion (or nexus of occasions) from which it has not been derived'. It is one of the two ingredients in the higher level 'intuitive judgment', the other one being a further 'indicative prehension' of the embedding of the logical subject in its own broader nexus, the basis for truth value assignments. Here, for example the eternal object 'tell (someone)' might be predicated of logical subject 'Meg', but this imaginatively prehended proposition can only take on a truth value (here as a 'suspended' intuitive judgment) when combined at a higher level with a further indicative prehension of the (relevant parts of the) whole nexus or 'historical' situation in which Meg is embedded.

9 Or abduction of the 'most likely hypothesis'. A 'suspended judgment' is not to be confused with a suspended subjective aim. It is a complex species of prehension which occupies a single occasion.

10 This 'feeling lens' corresponds to the focus of consciousness equated with the individual occasion, and thus one needs to envisage it as constantly reactivated from occasion to occasion while its relevance to the satisfaction of subjective aim S_1 (which activated it in the

first place) continues. The absorption and abstraction of episodic memory into Cheever's more permanent general knowledge (the outer 'rings' surrounding the on-going model) is another matter, and can be assumed to take place gradually as experiential 'tokens' accrete into – or add to – existing 'type' memory traces over time.

[11] In this technical sense the 'decision' of a concrescence is its closure (the 'satisfaction' of its subjective aim) in a determinate action, namely the 'objectification' of the occasion as data for some subsequent occasion.

[12] Conceptual valuation merely relates a physical feeling to the corresponding concept (eternal object) in memory, whereas conceptual reversions (in both internally objectified and externally objectified occasion types) introduce novelty since they recognize partial correspondences of data with eternal objects, so prototype effects and analogy may be involved. It relates new data to the best approximation to a fit from previous experience, and can tease out tacit relations (e.g., propositional ones) between previously objectified data that have not been directly prehended before. A reversion produces conceptual feelings that are 'partially identical to and partially diverse from' the eternal objects forming the primary data (i.e., explicit in the objective data to the occasion).

[13] The background 'gradedness' of long-term intentionality is handled by Whitehead in cosmological/ theological terms, but he leaves open an alternative analysis in terms of 'conceptual reversions' (if one chooses the latter the former is not necessary, and *vice versa*). He preferred in the end to see the 'initial aim' of a concrescence as deriving from a 'hybrid prehension of God' – one could alternatively say 'from the entire past conceptual experience of the personal nexus concerned'. The subjective aim proper develops by ordering possibilities, subordinating, contrasting, and eliminating alternatives, i.e., working out how best to realize the initial (vague) aim. In the present context one could say that the long-term initial aim behind Cheever's successive utterances is the result of a hybrid prehension of his earlier self's general aim, namely of solving the crime.

[14] A subjective aim arises in the primary (receptive) phase of an occasion as a valuated 'image' – a species of eternal object – of what the subject could become. It is the reason why concrescent occasions can be determinate processes after all. It is what determines the 'balanced intensity' of the integration of competing forces that results in their 'satisfaction'. It selects only those eternal objects – through valuations and reversions – that contribute to that intensity and do not counteract each other (it 'grades' them accordingly). The subjective aim integrates conceptual prehensions with primary physical data and its goal is maximal intensity of feeling (a) in the immediate subject and (b) in the relevant future, anticipatory feelings of which are ingredient in the immediate concrescence (cf. Sherburne 1966, p. 53). This 'relevant future' consists of those elements that are felt by the concrescing subject (by anticipatory prehensions) as real potentiality for its satisfaction.

[15] Note that I hold to a 'wheels-within-wheels' – perhaps fractally organized – interpretation of actual occasions, whereby these may display differing scale (both spatial and temporal) and differing internal complexity according to nexus level (see Whitehead 1978, p.18ff on different 'grades' of actual occasions). Only actual occasions at the level of the personally organized nexus are potentially conscious; their extent and internal structure obviously differ from occasions at neural or still lower nexus levels.

[16] One must distinguish 'enduring objects' carefully from 'eternal objects' here. Generalized schemata (albeit abstracted from experiential tokens) belong to the latter category. Enduring objects of the kind under discussion here could be said to belong to 'episodic' memory (including the experiencer's 'perspective') rather than 'semantic' or (in so far as this is really distinguishable from the latter) 'encyclopedic' memory. They are neverthe-

less built up largely from the prehension of 'type' concepts/eternal objects (in particular contexts or nex_s). The process whereby 'type' schemata are abstracted from (perceptual) tokens (by approximate match or analogy) is called 'transmutation' by Whitehead. Little is known of the discrete location of these two types of information in the mind/brain, but there are intriguing possibilities suggested by, for example, the gradient from posterior to anterior along the dominant temporal lobe that corresponds to increasing specificity of meaning (or decreasing degrees of abstraction) of nouns, from general perceptual type to proper name (cf. Damasio and Damasio 1992, p. 70).

[17] I follow MacWhinney's (1999, p. 216f.) use of the Gibsonian word 'affordance', referring by and large to any sensory-motor sensation (including visual images and kinaesthetic schemas) evoked in the hearer by the word. MacWhinney expressly excludes propositional knowledge from them, however, whereas there is no reason why Whiteheadian propositions could not be 'read off' sensori-motor schemas or associations between such schemas. MacWhinney discusses four different perspective-taking systems in the human brain, namely that of perceptual 'affordances' (in the posterior cortex), that for spatio-temporal orientation (mainly parietal), and those for causal action frames and for social frames (both frontal). These are all involved in the integration of complex mental models and form the basis for imagining performing actions and projecting oneself into objects and situations outside of oneself.

[18] The use of the term 'eternal object' about a specific language like English should not be understood in a Platonic sense as indicating its unchanging status. Obviously languages contain variation and change all the time, and even a single individual's competence – and expertise in applying it – varies over time. The point is simply that it represents a set of conventions facilitating communication on the understanding that the 'code' is to be treated *as if* it were static, i.e., as an accepted norm. It is a pure potentiality abstracted from actual language use (cf. Fortescue 2001, p. 238f. for elaboration of this point).

[19] This in turn can be related to the notion of 'subjectivity', as discussed by Traugott and others in the papers collected in Stein and Wright (1995). This is a matter of the expression of speakers' attitudes and higher-level inter-personal discourse organization through the use of discourse adverbials and connectives, etc. (as well as through the prosodic modulation of token utterances-in-context).

[20] Note that propositional prehensions as such (before being integrated into intellectual prehensions) are instinctive rather than conscious, and this is true of all but the very highest level of linguistic behaviour: the production of propositions without further integration into judgments, etc., is usually opaque to consciousness, and its rules function 'tacitly'. See Fortescue (2001, p. 184) for the argument that children are born with an instinctive ability to recognize propositional relations, as manifest in early perception before language begins to develop.

[21] Such 'processing' principles can generally be translated into Whiteheadian terms. For example, Hawkins' EIC principle begs an explanation in terms of the nature of the propositional prehension, which consists of two parts, the 'indicative feeling' and the 'predicative feeling where the former concerns the arguments – including the agent and/or or 'theme' – of the proposition in the relevant nexus. Since these are prehended in a single act it is natural (in terms of economy and clarity) that their expression should also not be separated by too much subordinate material ('heavy shift' can often alleviate this).

[22] Linearization during production can best be regarded as an on-going process throughout the duration of the objectively expressed occasion, as successive prehensions activate specific lexical and constructional patterns, which may have to compete for integration. (There

may be morphosyntactic incompatibilities between the ingredient morphemes activated.) It does not have to be fully determinate until the 'decision' sending off 'instructions' to the articulators. The linear syntactic framework – the set of constructional clausal 'templates' chosen to realize the complex intention requiring expression – is in itself merely a potential, static pattern that can be successively elaborated and integrated with specific lexical material via mutual adjustment as the concrescence proceeds. This can be visualized as a kind of creative, multi-faceted process of 'crystallizing out'.

[23] An exception is when the higher level aim of the occasion is the discourse act of setting up a new referent or topic. This is expressed in many languages by a special 'presentational' or 'thetic' propositional construction. This often has a bipartite structure (e.g., 'There were some sheep grazing in a field' in English) directly reflecting the linking of an indicative prehension ('there was X') with a propositional one ('X was grazing in a field') in the Whiteheadian 'intuitive judgment'.

[24] It is not enough to say that being the object rather than the subject of the subordinate clause is the reason for the exclusion of the full NP, since 'Meg' would have been acceptable (at least as an alternative) if an utterance by the same speaker (Caroline) with a different, non-overlapping subjective aim, had intervened (e.g., one introducing a different topic, perhaps a friend of Meg's, whose life had been 'ruined' for parallel kinds of reasons). However, the fact that 'her' is object and not subject here is indeed relevant: the options available for referring to the topic-as-subject are wider than for referring to the topic-as-non-subject, and this has presumably something to do with the tight default linkage between subject and topic in English. Referring to the subject by a full NP in such a redundant position might at least serve the function of reinforcing its status as topic, whereas doing the same with an object would be contradictory: to be reinforced in this way a topic in English must – apart from specific conditions of contrast – be in its default position as subject. Note that the factors involved in the choice of reference strategies I suggest are somewhat more structurally specific than just relative distance between successive expressions referring to a topic, as quantified in the well-known studies by Givón (e.g., Givón 1983).

[25] Where process provides the constraints on its own patterning. In this 'bootstrapping' (if not 'bootlacing'!) synthesis of particle physics and quantum mechanics, every fermion (read 'particle') has its own complementary boson (read field or 'wave' quantum), and the minimal units of reality are regarded as miniscule 'loops' rather than point-like particles. Process generates its own pattern, which in turn sustains the process as a kind of 'glue-like' force field. The kernel of the analogy as regards language is the relationship between linguistic process (expressive 'act') and pattern (constraints on expression), whereby every 'minimal unit' (at whatever level of analysis) consists of an indissoluble unity of form and content, the sign (the 'particle') and its potential for meaning (the corresponding 'wave form').

REFERENCES

Bohr, N.: 1948, 'On the Notions of Causality and Complementarity', *Dialectica* **2**, 312–319.

Chafe, W.: 1994, *Discourse, Consciousness, and Time*, Chicago: University of Chicago Press.

Damasio, A. and H. Damasio: 1992, 'Brain and Language', *Scientific American* Sept. 1992, 63–71.

Fortescue, M.: 2001, *Pattern and Process: A Whiteheadian Perspective on Linguistics*, Amsterdam/Philadelphia: John Benjamins.

Gell-Mann, M.: 1994, *The Quark and the Jaguar*, New York: W. H. Freeman and Company.

Givón, T. (ed.). 1983, *Topic Continuity in Discourse: Quantified Cross-Language Studies*, *TSL 3*, Amsterdam: John Benjamins.

Givón, T.: 1988, 'The Pragmatics of Word Order: Predictability, Importance and Attention', in M. Hammond, E. Moravcsik and J. Wirth (eds.), *Studies in Syntactic Typology*, Amsterdam/ Philadelphia: John Benjamins, pp. 243–284.

Grice, H. P.: 1975, 'Logic and Conversation', in P. Cole and J. Morgan (eds.), *Syntax and Semantics, Vol. 3: Speech Acts*, New York: Academic Press, pp. 83–106.

Hannay, M.: 1991, 'Pragmatic Function Assignment and Word Order Variation in a Functional Grammar of English', *Journal of Pragmatics* **16**, 131–155.

Hawkins, J. A.: 1994, *A Performance Theory of Order and Constituency*, Cambridge: Cambridge University Press.

Lambrecht, K.: 1994, *Information Structure and Sentence Form*, Cambridge: Cambridge University Press.

Leech, G.: 1983, *Principles of Pragmatics*, London/New York: Longman.

Levinson, S.: 2000, *Presumptive Meaning: The Theory of Generalized Conversational Implicature*, Cambridge, MA: MIT Press.

MacWhinney, B.: 1999, *The Emergence of Language*, New Jersey: Lawrence Erlbaum.

Searle, J.: 1969, *Speech Acts*, London/New York: Cambridge University Press.

Sherburne, D.: 1966, *A Key to Whitehead's Process and Reality*, Chicago/London: University of Chicago Press.

Stein, D. and S. Wright (eds.): 1995, *Subjectivity and Subjectivisation in Language*, Cambridge: Cambridge University Press.

Walters, M.: 1995, *The Dark Room*, London: Macmillan.

Whitehead, A. N.: 1968 (1938), *Modes of Thought*, New York: The Free Press.

Whitehead, A. N.: 1978 (1929), *Process and Reality*, New York: The Free Press.

BARIS PARKAN

A PROCESS-ONTOLOGICAL ACCOUNT OF WORK

ABSTRACT. This essay offers a process-ontological account of work, addressing two challenges in particular. First, I try to show that even though the phenomenon of work is extremely diverse, all occurrences to which the word 'work' correctly applies – according to the current semantic intuitions of the relevant linguistic community – share the feature of being: *the creation of something of value.* Second, guided by this initial conceptual delineation of the phenomenon, I argue that traditional ontology would face fundamental difficulties in giving an ontological analysis of work, which can be surmounted in a Whiteheadian framework. I take Whitehead's analysis of an actual occasion as a model for an analysis of work that explains how toil (exertion of effort) can result in the accomplishment of a desired end. On this basis of this explanatory model the complex dynamicity that we experience in work as the creation of something of value can be accounted for in the well-defined terms of an ontological theory.

Work is one of the most important aspects of our lives. We spend most of our waking life working – if not to make a living, then to bring about some other desired changes, such as making our surroundings more pleasant, learning something, practicing a skill, or producing and enjoying works of art. That we have to work is a natural consequence of the fact that we, as organic beings, are not self-contained, but dependent on our surroundings. Therefore, it is when we work that we are most fully aware of 'being alive' and of our 'human condition', most aware of who we are, of our capabilities, our environment, and of how we are related to that environment. In other words, work is not only our means of survival, but also a very important dimension of self-experience and experience of the world.

It is therefore surprising that work has not received much attention as a subject of ontological analysis. This may in part be due to the fact that work has a social dimension, which quickly transforms an ontological analysis into a political one. For example, Marx and Hegel have given ontological accounts of work, but these accounts are quickly overshadowed by the political and ideological proclamations intertwined with the ontological analysis.

I see two other and more decisive reasons why work has not received much attention as a subject of ontological analysis, residing in the subject matter itself: (1) The extremely diversified nature of the phenomenon of work makes it very difficult to tackle the concept in a coherent and consist-

J. Seibt (ed.), Process theories: Crossdisciplinary studies in dynamic categories, 219–235.

ent manner. (2) Work is a dynamic phenomenon, and traditional ontology is not well-equipped to deal with dynamic phenomena.

These latter two aspects of work are not insurmountable obstacles, however, but merely challenges to a satisfactory ontological analysis. In this paper I shall present an account of work that attempts to meet them. In the first part of the paper, I shall characterize and classify various kinds of work. In the second part, I shall explain and utilize Alfred N. Whitehead's process-ontological scheme to provide an ontological analysis.

1. THE CONCEPT OF WORK

Let us begin with a simple rehearsal of semantic intuitions. During all the following activities we consider ourselves to be working: cooking, building objects, making music, tilling the soil, studying, negotiating, picking trash, designing an airport, designing a web page, conducting a scientific experiment, giving or taking orders, applying formulas, etc. Further, some activities like dancing, having sex, and making conversation could in some cases be considered work. Some very personal acts and deeds that involve only a few people in an intimate context can also qualify as work. For example, we 'work on' a relationship, or on our friendship, or our family. In this context, writing a letter, making a tape for someone, paying a visit, remembering occasions and buying gifts, trying to communicate, trying to understand and forgive, helping can all be work. Doing a beautiful deed such as saying encouraging words to someone who is in despair perhaps is not work, but has a lot in common with it. On the other hand, the following are not work: sleeping, lying in the sun, eating, drinking beer, playing backgammon, taking a walk, watching the sunset, watching an action movie, some of the reading and music-listening we do, chatting to a friend about this and that, 'hanging out', relaxing, bar-hopping, etc.

Given the extremely diversified nature of all the phenomena to which the word 'work' applies, is there a justification for grouping them all under the same name? Some have argued that it is more appropriate to speak of a family resemblance, claiming that an activity like work, which is "complex, highly variable, and meaning-dependent on the individual",cannot be "made into a simple, invariable activity of universal meaning" (Pence 1978/79, p.310).

Standard dictionaries resort to a disjunctive strategy:

1. [work is an] activity in which one exerts strength or faculties to do or perform something **a.** sustained physical or mental effort to overcome obstacles and achieve an objective or result **b.** the labor, task or duty

that is one's accustomed means of livelihood **c.** a specific task, duty, function, or assignment often being a part or phase of some larger activity.[1]

Dictionaries also draw attention to the diversified semantic field governed by 'work' and a host of 'near synonyms' such as 'labor', 'travail', 'toil', 'drudgery', etc. which reinforces the impression that this field is organized by 'family resemblance' rather than clear cut genus-species relations. 'Labor' involves "great and often strenuous exertion"; 'travail' involves 'pain and suffering' and 'drudgery' is 'dull and irksome'. In *The Human Condition*, Hannah Arendt attempts a more thorough distinction between work and labor. According to Arendt, labor is 'life-sustaining' activity concerned with 'animal life' whereas work is 'world-building' activity concerned with the specifically 'human life'. While work produces objects for use which become part of a human world, labor produces objects for consumption and therefore cannot make a tangible durable contribution to the human artifice.

Another dimension of diversification enters the picture when we consider the social area of the semantic field. Dictionaries remind us of conceptual distinctions between activities engaged in remuneratively in earning one's living. These distinctions are based on various factors, ranging from the personal meaning of the activity for the individual to whether the activity is imposed by someone in authority or not. For example, 'job' is defined as "small miscellaneous piece of work undertaken on order at a stated rate"; 'calling' is defined as "occupation viewed as a vocation or profession"; and so on.[2]

While the position that only family resemblance connects the various instances of work seems well-justified, others have found it easy to locate what all these various instances have in common. Work has been characterized – not fully in earnest, one might suspect - - as "first, altering the position of matter at or near - the earth's surface relatively to other such matter; second, telling other people to do so".[3] More seriously one might take the description of work as 'human activity which takes existing conditions and transforms them so that the new conditions more completely satisfy our needs and desires" (Okrent 1978/79, p. 322).

In fact, the various ways in which work has been conceived throughout history can be reduced to two basic meanings: misery and toil (*molestia*), accomplishment (*opus, opera*) (Lowith 1967, p. 261). Even though Arendt distinguishes *operae* – the mere activity –, from *opus* – the work – and refers to the former as "toil and effort", such a distinction fails to take notice of, or at least obscures, the inherent connection between activity, exertion of effort, toil (*operae*) on the one hand, and the product (*opus*), on the other – i.e., that the exertion of effort is constitutive of the achieved end.

I believe that the best way to approach the topic is by devising an ontological account that is arguably broad enough to cover the instances of all varieties of work yet structured enough to formulate the differences between these varieties.

Later, in the second part of this paper, I will make use of Whitehead's ontology to demonstrate the connection between exertion of effort and the achievement of the objective, thus proving that the various concepts of work can be reduced to the two basic meanings of toil and accomplishment, which can further be shown to be connected to each other. Since the desired end (the objective accomplished) is arguably something of value and the exertion of effort is constitutive of that result, I contend that what characterizes the phenomenon of work in general is the creation of something of value.

1.1. *Classificatory parameter*

In a second step, after an initial exploration of semantic intuitions and various attempts of explicating them, let us now reconstruct the content of our concept of work in a more organized fashion. The point of departure for such an enterprise is the thesis I just introduced, namely, that all the variegated phenomena discussed above can be brought under the heading of 'work', understood as 'the creation of something of value'. This will become clearer if we place these phenomena into a matrix spanned by two descriptive perspectives, answering to the questions of (1) why we work and (2) how we work. These two perspectives correspond to the two components of the definition of work suggested. The first perspective corresponds to the desired end – the value aimed at. The second perspective corresponds to the expenditure of effort.

1.1.1. *Why we work*

In order to bring the first classificatory perspective into view, it is important to draw two distinctions. First, we need to distinguish between the proximate end and the final/ultimate end of a work activity. The ultimate end can be the continuance of life (as in the case of labor) or anything else that is considered an end-in-itself – in other words, anything that contributes to the good life (as in the case of art). The proximate end is the production or performance of anything that is instrumental to another end. This distinction is not absolute, but is rather a matter of degrees, with some proximate ends more removed from the final end than others. Building shelter, for example, more directly aims at life as its end than a job held in a company.

The second important distinction is between the value that work has for the individual engaged in the activity and the value that our work has for others. That for the most part we do not work to satisfy our needs and desires on our own but engage in massive cooperation on a social or even global scale requires us to maintain two different perspectives at once: the point of view of the individual and a bird's eye view of the social system as a whole. A person's motivation to perform a certain work task and the social function of that performance frequently do not correspond to each, the prime examples being tele-marketers or blue-collar workers at a weapons factory. Therefore, we need to break the question of why we work at least into two parts: (a) Why do we, as individuals, engage in a certain work activity? (b) What role does our work have in the system as a whole? Since the distinction between proximate end and final/ultimate end is not absolute, in answering questions (a) and (b), I will refer directly to the final end.

(a) The question "why do we, as individuals, work?" has three answers I think: we work (a1) for direct survival (e.g., fishing, hunting); (a2) for money, which can be for indirect survival (making a living), attainment of other desired objects, and/or for power and status; (a3) to change things for the better (which also includes intrinsic satisfaction from self-expression involved in the change). (a1) and (a3), or (a2) and (a3) are not mutually exclusive; they can overlap.

(b) The second question concerns the role our work plays in the whole. Here, there are two or three possible answers: (b1) for the system's livelihood – which is why it, at once, has a place in the system and is our livelihood, too. (e.g., engineering, irrigation, transportation, insurance, government); (b2) not just for maintaining the system but for making it good – i.e., not just for life, but for the good life.

With respect to the first descriptive perspective, then, we can classify occupations by creating possible combinations of answers (a1–3) and (b1–2). (W1) There are work activities which serve the direct needs of the individual and of the embedding social system. For example, a custodian, a truck-driver, a burger flipper, a factory worker, a tax collector, a civil engineer, accountants, managers, electricians, plumbers, cashiers, reporters, merchants, butchers, policemen all work to make a living while they keep the wheels turning by providing food, housing, security, maintenance, transportation, communication, and other necessary goods and services. (W2) In some cases, while someone does something for intrinsic satisfaction or to make a change for the better, it is conducive for society at large. An architect or an engineer, a doctor, a lawyer, a teacher, a computer programmer, a judge, an advertiser, or a journalist may derive intrinsic satisfaction from their work, while they contribute to providing health care and education, maintaining the justice of the system, transferring information, and so on. (W3) Someone can do her work to make a change for the better while it contributes to the good life (e.g., scientists, philosophers, writers, dancers, actors, directors, comedians, musicians, etc.). (W4) Someone can work to make a living while her work contributes to the good life (e.g., a factory worker producing luxury items, a tour guide, people who work at movie sets, a waiter, a gypsy who sells flowers, etc.) (W5) Finally, there are those cases where the work we do for ourselves has no social significance (e.g., cooking my favorite dish for myself).

1.1.2. *How we work*

In attempting to classify work activities with respect to the second descriptive perspective answering to the question of how we work I will consider the following factors: (1) the presence of a model of the desired end; (2) the types of medium; (3) interaction with other agents; and (4) faculties and skills that are used.

1.1.2.1. *The presence of a model.* While work is by definition teleological and most work is performed under the guidance of a model, some work activities appear to lack this teleological structure. For example, consider the activity of a cashier at a supermarket, a factory worker at an assembly line, or someone flipping burgers at a fast food restaurant. One might thus want to make a distinction between teleological and non-teleological kinds of work.

In fact, the presence or absence of a guiding model in the process of a work activity is one of the main criteria by which Arendt distinguishes between work and labor. According to Arendt, work is performed under the guidance of a model, and it is this model as an end that organizes the means and the process of work. Hence, the process of work is linear, and at the end of the activity, the process disappears into the finished product (Arendt 1958, p. 140). In contrast, since labor is caught in a cycle of production and consumption and therefore leaves no tangible marks behind, the process of labor is cyclical; in labor, the "process consumes the service" (ibid., p. 87). Since labor's productivity is measured in terms of reproducing life and not in terms of the "qualities or the character of the things it produces" (ibid., p. 87), labor does not introduce any novelty into the world; it merely restitutes a certain set of states or a certain process structure.

However, it is misleading to associate non-teleological activities with activities whose end is the continuance of life. A doctor's aim is the continuance of life, but it could not be claimed that she works without the guidance of a model. Even a laboring activity like farming is rationally planned. Babysitting consists of a long series of finite tasks, but there is a model for each task. We therefore need to distinguish, instead, between continuance of life as an end and the continuance of the same activity as the (proximate) end. It is the latter which appears to take place without a model. Division of labor creates this appearance. When the work process is compartmentalized, the vision of the project might get lost to the individual agent, and the proximate end of the activity for the individual becomes filling the 8-hour stretch of time from the moment of punching in to the moment of punching out. But from a bird's eye view of the system, the activity is part of a process that is guided by a model even if, from an individual's point of view, activities that aim at continuance of the same situation may appear to be non-teleological.

1.1.2.2. *The medium.*[4] All work takes place in a medium whether that medium be earth and seeds, test tubes and burners, or tags, text files and pictures. The various types of media can be grouped into four categories: organic nature, inorganic nature, symbols (words, signs, numbers), and sounds and images.

The organic medium can be plants, animals, other people, or ourselves. Work activities that take place in this medium include farming (plants and animals), teaching and medicine (other people), and sports (ourselves). The inorganic medium includes forces of nature (wind, water, etc.) as well as matter. Construction and production of physical objects (houses, factories, cars, clothes, etc.) take place in this medium. Physical objects might also be constructed with the further end of utilizing forces of nature (such as building windmills, dams, canals).[5] Most white collar workers and academicians (accountants, scientists, writers) work in symbols, even though in many instances these symbols are applied to/translated into other types of media. Finally, artists (film directors, musicians, dancers, etc.) work in sounds and images.[6]

1.1.2.3. *Interaction with other agents.* Some types of work such as some arts, crafts, and professions, are performed in solitude. Others are performed in interaction with other people. Of those performed in solitude, some are nonetheless performed for a social end. For example, a translator may work in solitude, but she might be working with and for someone else, such as a pharmaceutical company. The work of a gardener tending her plants, on the other hand, involves no interaction with others.

The interaction involved in non-solitary work activities can be classified into two groups: social and communal. By 'social work', I mean work that involves division of labor (not division of labor in civil society in general, but division of labor in one particular workplace, such as a factory). In such cases, the workers contribute to the same outcome, but without controlling the process through interaction with each other. In communal work, in contrast, the agents work towards the realization of a goal in interaction with each other. The production of a symphony, a play or a movie and teamwork in a soccer game are such examples.

1.1.2.4. *Skilled vs. unskilled work.* A more important distinction among work activities with respect to how we work concerns the use of faculties involved. The first distinction that comes to mind is between physical and mental work. However, this distinction on its own is not very telling since the work of the mind and the body cannot be easily separated from each other. Physical work is not necessarily simpler or more trivial than mental work (e.g., a surgeon performing an operation). Thus, it may be more useful to make a classification in terms of the complexity of the skills involved, such as abstraction, imagination, technical and analytical skills, or dexterity.

In economics, a common problem with the labor theory of value is "the need for an explanation of the wage differentials of different types of labor" (Bowley 1973, p. 96). Since labor theorists of value hold labor to be the measure of value, they must assume that all labor is homogeneous. The higher wage that skilled labor receives is explained by the labor cost of extra training or differences in natural ability (Blaug 1968, p. 116). But this appeal to a 'difference in natural ability' remains entirely obscure. The Whiteheadian analysis of a work process which I shall present now also furnishes an explanation of the distinction between skilled and unskilled work.

2. PROCESS ONTOLOGY

2.1. *Traditional ontology's shortcomings*

The most important reason why I believe that the subject of work has not received much attention as a subject of ontological analysis is because traditional ontology is not as well equipped to deal with dynamic phenomena in general.

Moreover, work is a dynamic phenomenon posing a particular difficulty. Work is not only a continuous progression towards a goal but it has a complex dynamic structure where main and subgoals can be recursively modified and other processes (consumption, enjoyment) are engendered in the process of production. Since traditional ontology takes static entities as basic, it can merely describe a series of states of affairs that gradually represent closer approximations of the desired state, but it cannot make sense of a continuous progression towards that state, let alone the continual modifications of the manifest goings on by feedback loops and other elements of a complex, hierarchical dynamic organization. Traditional ontology can present only snapshots of a linear development where the desired end is conceived of as something static and separate from the process of production and processes connected to it. A process ontological account, on the other hand, can take the end-result as a set of interactive relations (with its past and future), thus allowing for a more inclusive description of work, holding out the prospect to account not only for typical productions like the building of a table, but also highly interactive forms of work such as psychological counselling in group therapy or brain storming sessions of the company's management.

Further, the traditional divisions in ontology between domains of fact and domains of value discount from the outset the very idea of a realization of value. Traditional ontology can describe work as the production of a desired object or state of affairs, but then cannot explain the interaction of different existential types – i.e., how something that is desired but does not yet exist (value) is brought into existence (fact).

2.2. *A Whiteheadian account*

In a Whiteheadian framework, the connection between exertion of effort and the bringing about of something of value can be accounted for rather straightforwardly. Yet, there is one problem in presenting a Whiteheadian account and that has to do with his terminology. Since Whitehead consciously avoids the concepts and conceptual metaphors of traditional ontology, his terminology is highly idiosyncratic. Therefore, our account must be prefaced by an introduction to the ontological framework.

2.2.1. *A Whiteheadian framework*

The framework I shall sketch here is not Whitehead's but Whiteheadian, since in several places I assume without argument certain readings of Whiteheadian terms that are controversial interpretations. For present purposes it suffices to introduce the basic elements of this framework.

2.2.1.1. *A process of concrescence.* In contrast to traditional ontology, Whitehead's ontology takes processes as basic. The primary unit of being in his ontology is what he terms 'an actual entity'. The actual entity, a 'process of concrescence', is a process of many diversities growing into a new unity (Emmet 1932, p. 41). There is much controversy as to whether 'an actual entity' designates a sub-atomic occasion or *any* entity that constitutes a processual unity, such as a tree, a human being or a table. I take the latter position: quite independent of scale, any set of processes can exhibit the kind of organization that characterizes actual entities.

2.2.1.2. *Prehensions.* Every actual entity constitutes itself by incorporating various elements of the environment out of which it is arising into a new unity. The dynamic connection through which any one element becomes appropriated or excluded by the becoming entity is called a 'prehension'. The choice of the term 'prehension' is probably intended to indicate that a prehension is a process that is much like apprehension except that it does not necessarily involve consciousness (Hartshorne 1972, p. 125). It is modeled on an intentional mental act, but is not restricted to mental acts; prehension is 'information transfer' in the most general sense of the term. For instance, we can think of a snail as prehending the rock it is moving on or a piece of paper as prehending the charcoal touching it. It is important to note that a prehension is not "a merely static and mechanical link" (Emmet 1932, p. 87) but a *transaction.* In Whitehead's words, it is "referent to an external world, and in this sense [has] a vector character" (Whitehead 1967, p. 28).

Since the actual entity does not appropriate all the elements of its environment, but leaves some out, Whitehead divides prehensions into two species: positive prehensions and negative prehensions. When an element is negatively prehended, it is excluded from the concrescence in the making, whereas an element that is positively prehended becomes a component in the internal constitution of the becoming entity. A positive prehension is also called 'a feeling'.

2.2.1.3. *Subjective aim and eternal objects.* The actual entity is a dynamically organized complex process whose unity is created in the making, not as an accidental result but as the result of a goal-driven process. Thus, even though an actual entity can be analyzed into its prehensions, Whitehead takes the actual entity and not the prehension as basic, because the reason why an entity appropriates some elements in its environment and leaves others out cannot be accounted for with respect to the prehension itself alone. The explanation for why certain elements are prehended, and others left out, refers to the *subjective aim* of the concrescence. The subjective aim, which is basically a final cause, is Whitehead's principle of individuation for actual entities. Thus, the actual entity has a teleological and self-referential structure – teleological, because it is goal-driven; and self-referential, because it is self-creative.

Since the subjective aim is what triggers the process, however, it itself cannot be entirely in the making. Rather, it is given to the actual entity through what Whitehead calls "*eternal objects*". An eternal object is a potentiality – a notion that is fundamental to process philosophy.

Eternal objects are universals, like Platonic forms, determining the definiteness of the processes in which they are realized. However, unlike Platonic forms, eternal objects are both immanent and transcendent. The realization of an eternal object in an actual entity is not a distortion of a perfect form in a transcendent realm; when an eternal object is realized in an actual entity, the eternal object is actually in that entity. As a potentiality, however, the eternal object is transcendent. Eternal objects are patterns of some defining characteristics, but as such, they can be abstracted from the processes they qualify; "what an eternal object is in itself . . . is comprehensible without reference to some one particular occasion of experience".[7]

Another difference between eternal objects and Platonic forms is that eternal objects are relational. Eternal objects can be simple or complex. Simple eternal objects define the relation between the becoming entity and the objects it prehends; they explain how the subject prehends the object, and in that function, they are what we call 'qualities' (e.g., a color). A complex eternal object is a group of eternal objects, related to each other in a determinate way. Complex eternal objects serve to explain what we call essence or abstractness. In short, eternal objects define potential or actual relations among the actual entities that are in a concrescing entity's environment.

Both eternal objects and the subjective aim are potentialities. The difference between an eternal object and a subjective aim is that while eternal objects are completely general, the subjective aim concerns particular subjects. For that reason, eternal objects are also called 'pure potentialities', while the subjective aim is an 'impure potential'. In other words, while facts are *physically* prehended, eternal objects are *conceptually* prehended, and the subjective aim is a hybrid of physical and conceptual prehensions. For example, while a certain melody is a pure potential, that melody being played through a particular instrument is an impure potential. When thought of in terms of work, it is the difference between a pure possibility that is entertained and a possibility that is adopted as a project in a certain setting.

The subjective aim should be thought of more in terms of a tentative objective than a strict blueprint. Therefore the final form of the process (its subjective form) will differ from its subjective aim. The entity approaches the realization of its goal, not by directly realizing a complex eternal object, but by gradually realizing an increasing number of eternal objects while trying to integrate them with each other and with actual entities. This process is best explicated by bringing into play Whitehead's concept of contrasts.

2.2.1.4. *Contrasts.* In the process of appropriating and integrating various prehensions into its own unity of form, the actual entity forms what Whitehead calls 'contrasts'. By 'contrast', Whitehead in fact means unity. In a contrast, two or more elements are brought together, compared as well as contrasted, and, if they are compatible, synthesized. If they are incompatible one or more of them are left out.

As the process advances towards its goal, more contrasts can be formed among previously formed contrasts, thus yielding structures of greater complexity. As an example, one can here imagine the composition of a piece about a storm. The piece starts with the xylophones representing raindrops. Then the xylophones are contrasted with the tambourine, signaling impending trouble. As the piece advances, more instruments are introduced forming further contrasts. As stated above, during these phases, the subjective aim may be modified. Nevertheless, the concrescing entity prehends its own aim with a certain intensity, an intensity which increases as more contrasts are formed and the process comes closer to realizing its goal.

While the intensity of a process is proportionate to the number of contrasts formed between eternal objects, the number of eternal objects realized are limited by conditions of compatibility. Whitehead calls the conditions for compatibility "the demand for balance".

The intensity of a process is also due to conceptual reversions, which account for creativity and novelty.

2.2.1.5. *Conceptual Reversions.* We have seen that the realization of an eternal object in an actual entity is the actualisation of a possibility. Typically, the realization of a possibility takes place through the prehension of an eternal object that is already realized in a previous actual entity. For example, the blue-gray color in a painting comes from the sky; it is already realized in the sky, and the painter abstracts that color as potentially realizable in another entity – i.e., the painting. But in choosing to actualize a certain potential, the entity has avoided other alternatives. The sky could have been a different shade of blue. In a conceptual reversion, an unrealized possibility that a previous entity has avoided is prehended. This means that a conceptual experience is derived without the physical experience to match it. For example, someone else, contemplating this painting, is able to prehend that different shade of blue that has not been realized in actuality.

In such cases, the question of where potentialities are located becomes especially puzzling. How can we account for the prehension of eternal objects that one has not come in physical contact with? One explanation is that eternal objects are ordered in a certain relevance "independently of their joint exemplification" in reality (Kraus 1979, p. 110) so that the prehension of an eternal object that one comes in physical contact with (via an actual entity it is realized in) may prompt the prehension of a related one. A conceptual reversion brings about the prehension of an as yet unrealised potential when that potential is "partially identical with and partially diverse from" a potentiality that is realized (Whitehead 1967, p. 40). The idea is that if all that one physically came in contact with were vertical and horizontal straight lines, for example, one might still be able to imagine vertical lines that make curves, based on a spectrum of lines ranging from straight vertical on one end of the spectrum to straight horizontal on the other.

2.2.2. *Applying the framework: An ontological analysis of work*

The dynamic, teleological and self-referential structure of an actual entity parallels the complex structure of a process of work, which may contain all or some of the features explained above. In work, one forms, in light of the subjective aim, contrasts between certain data and conceptually prehended alternatives, occasionally eliminating some, performing conceptual reversions to arrive at others, and modifying the subjective aim as necessary along the way.

In applying the Whiteheadian framework, I take the worker in interaction with her medium to be *one* process of concrescence. In other words, work is a relational affair with one relatum: the product in the making. The worker and the materials that she works with are *antecedent* actual entities that are now entering a novel unity with various components of each other. This novel entity in the making is the final product aimed at by the work process. One might here object that, in the end, the worker emerges intact from the work process; thus, the final product cannot be a unity of parts of the agent and the material. It is true that the worker does not necessarily and explicitly lose body parts or mental faculties in the process, though she might, and usually gradually does. As a matter of fact, Karl Marx describes the laboring process as expenditure of human energy and tissue, which is why *animal laborans* needs constant replenishment. But I believe that the wasted tissue is not the worker's main contribution to the emerging product. Rather, in the work process, the worker 'mixes with' her medium, receiving, altering and passing on various elements of it into the future stages of the process. In this process, certain potentialities are actualized. These potentialities in the worker that are getting actualized leave the worker and enter the product; we might thus think of the worker as 'fissioning' and 'fusioning'. What an ontology of work needs to explain is where these potentialities come from and how they are actualized. And of course, the worker does not emerge from the process unchanged. She may be tired, wiser, stronger, more famous, full of sores, emotionally attached to the product of her work, and so on. Thus we are justified in taking the worker and the material in interaction up until the culmination in the final product as one process.

2.2.2.1. *Reconfiguration.* With the notion of a subjective aim, coupled with the theory of prehensions (positive, negative, conceptual and physical) and contrasts, we get the conceptual tools for giving an account of the work process as *reconfiguration* as well as an idea of how this reconfiguration is motivated.

A process of work is motivated when the complex interrelatedness among the entities in our environment is not 'ordered', but merely 'given'. Sometimes we may desire things just as they are given. But at other times, there may be incompatibilities in an interrelatedness that is thus merely given that need to be eliminated for a desired state to be reached. For example, a farmer needs to pull out the weeds that grow around her vegetables, a janitor needs to empty wastebaskets, etc. In some other cases, the merely given may be in need of some additional elements, like a barren land in need of water or a circuit in need of a wire. In yet others, the present elements may need to be rearranged (e.g., consider a carpenter building a chair or a poet writing a poem).

Negative prehensions explain how a process of work eliminates some elements; positive prehensions explain how it incorporates others; and contrasts and conceptual reversions explain how it re-arranges them so that "the mere complexity of givenness which procures incompatibilities is superseded by the complexity of order which procures *contrasts*" (ibid.).

Also, because prehensions are transactions with a vector character, they aid in understanding how one phase of the process is passed on to the next.

Further, a process, both in its initial stage and in its culmination, is related to other actual entities through prehensions. It begins with prehensions of antecedent actual entities, and is, upon culmination, prehended by subsequent entities. Therefore, a Whiteheadian account can demonstrate how other processes are engendered in the process of production. For example, the culmination of one process of work, such as transportation of lumber, may be the beginning of another process, such as manufacturing of furniture; the culmination of individual thought processes may enter a new unity with each other in a group discussion; or the culmination of a product of art may inspire and initiate a new process. These individual processes can also be ongoing within a larger process, since the picture Whitehead paints is one of nature as a network of events which "extend over other events, so that the large scale events are systems of atomic events" (Emmet 1932, p. 79). Thus, we can make sense of highly interactive processes of work, such as the production of a movie or a symphony.

2.2.2.2. *Exertion of effort.* Contrasts and conceptual reversions account for the intensity of a process of concrescence. All these concepts together (intensity, contrasts and conceptual reversions) can also help us understand exertion of effort. To take a simple example, the exertion of effort involved in jogging can be explained in terms of Whitehead's idea that intensity increases as the subject comes closer to realizing its subjective aim. The jogger's effort has an intensity that increases proportionately as she envisions the distance lying ahead of her traversed. Exertion of effort also increases proportionately to intensity with the formation of an increasing number of contrasts. For instance, consider a musician composing a piece, who, as she is writing the music for the piano, is simultaneously trying to hear in her head the drums, thus contrasting the drums with the piano music. In this instance, the exertion of effort required in beholding various musical elements in one's head at once (is the 'glue' that) allows for the formation of contrasts.

If , as argued above, conceptual reversions are possible because eternal objects are ordered in a certain relevance to each other, then prehending unrealized alternatives (i.e., creativity) also requires exertion of effort. Being able to envision a wide spectrum of eternal objects in detachment from reality requires considerable mental effort. For a simple analogy, consider the mental exertion involved in a chess game where the player tries to envision possibilities.

Thus, a Whiteheadian analysis clarifies how exertion of effort is related to reconfiguration leading to desired change in a medium.

3. CONCLUSION

Even though the term 'work' is used in a variety of ways and applied to widely diverse phenomena, I believe that all instances to which the word 'work' applies have a common essence that can be grasped: when we want something, and we *try* to make it happen, we are working. All work is the creation of something of value, whether that value be the continuance of life or the satisfaction of some other desire. The word 'creation' in this characterization refers to the instance of trying, not the product, and I hope I have shown that it is possible to give an ontological analysis of such instances of trying, or, 'exertion of effort'.

To recapitulate, the desired change in a medium is achieved through a reconfiguration of the components of that medium. This reconfiguration can be explained via four or five Whiteheadian ideas: the analysis of an actual occasion into prehensions, subjective aim and eternal objects, conceptual reversions, and contrasts. In this analysis, the subjective aim accounts for the value-oriented teleological structure of a process; prehensions and contrasts allow for a breaking down and re-synthesizing of various elements in the process; and conceptual reversions account for novelty and creativity. Whitehead's theory of prehensions also accounts for the continuity of the process, as well as its relatedness to other processes that precede or supersede it, while contrasts and conceptual reversions also account for the 'intensity' of the process of concrescence.

In a process of work, this intensity is equivalent to the exertion of effort. Thus, the Whiteheadian concepts of prehensions, contrasts, and conceptual reversions allow us to understand the connection between exertion of effort and the bringing about of the desired change, revealing 'exertion of effort' and 'reconfiguration towards attainment of objective' to be two sides of the same coin. It is through exertion of effort that one is able to behold possibilities and thus envision alternatives, discard certain elements and integrate the ones retained while sustaining the balance required for the desired outcome.

NOTES

1 *Webster's Ninth New Collegiate Dictionary.*

2 Ibid.

3 Bertrand Russell, here quoted after Ciulla (2000, p. 22).

4 I prefer the word 'medium' to 'material', because it is less biased with respect to the ontological structure of the material, and includes not only physical objects, but all of or spatial and temporal surroundings.

5 My classifications in terms of the types of medium is developed along the lines of a similar classification in Hegel's *Philosophy of Right.*

6 One important detail that might lead to confusion is the role that the use of tools plays in our interaction with our media. Is the tool to be considered part of the agent or the medium? When the tools used are simple, such as a shovel or a pen, we tend to think of the tool as an extension of the agent. But what about more complicated machinery and

electronic apparatus? We might then need to add machinery and electronic media as a fifth type of medium.

[7] Whitehead, quoted after Hall (1963, p. 104).

REFERENCES

Arendt, Hannah: 1958, *The Human Condition*, Chicago: The University of Chicago Press.

Blaug, Mark: 1968, *Economic Theory in Retrospect*, Homewood, IL: Richard D. Irwin, Inc.

Bowley, Marian: 1973, *Studies in the History of Economic Theory before 1870*, London and Basingstoke: The MacMillan Press Ltd.

Ciulla, Joanne B.: 2000, *The Working Life*, New York: Three Rivers Press.

Emmet, Dorothy M.: 1932, *Whitehead's Philosophy of Organism*, St. Martin's Street, London: MacMillan and Co., Limited.

Hall, Everett: 1963, 'Of What Use are Whitehead's Eternal Objects?', in George L. Kline (ed.), *Alfred North Whitehead: Essays on His Philosophy*, Englewood Cliffs, N.J.: Prentice-Hall Inc., pp. 102–116.

Hartshorne, Charles: 1972, *Whitehead's Philosophy*, Lincoln: University of Nebraska Press.

Hegel, G.W.F.: 1967, *Philosophy of Right*, Oxford/London/New York: Oxford University Press.

Kraus, Elizabeth M.: 1979, *The Metaphysics of Experience*, New York: Fordham University Press.

Lowith, Karl: 1967, *From Hegel to Nietzsche*, Indianapolis/Cambridge: Anchor Books, Doubleday and Company.

Okrent, Mark: 1978–1979, 'Work, Play and Technology', *The Philosophical Forum* **X**(2–4), 321–340.

Pence, Gregory E.: 1978–1979, 'Towards a Theory of Work', *The Philosophical Forum* **X**(2–4), 306–320.

Whitehead, Alfred N.: 1967, *Process and Reality*, New York: The MacMillan Company.

'Work'. Def.: 1990, *Webster's Ninth New Collegiate Dictionary*, Springfield: Merriam-Webster Inc.

PAUL NEEDHAM

CONTINUANTS AND PROCESSES IN MACROSCOPIC CHEMISTRY

ABSTRACT. Chemistry deals with substances and their transformations. School chemistry provides a picture of this in terms of small balls called atoms and ball-and-stick structures called molecules which, despite its crudity, has been taken to justifiably reflect a reductionist conception of macroscopic concepts like the chemical substances and chemical reactions. But with the recent interest in chemistry within the philosophy of science, an extensive and determined criticism has developed of the idea that the macroscopic world has been, or is likely to be, reduced to microscopic theory. From this perspective, it is of interest to see macroscopic ontology treated autonomously. I try to take a first few steps towards spelling this out. It involves recognising entities falling into two broad categories: continuants-things which can have different properties at different times – and processes – things whose temporal parts may have different features, but which themselves stand in contrast to continuants in this respect. The character of each and their interrelations depends on their mereological structure of parts, the exploration of which is one of the prime purposes of the paper.

1. INTRODUCTION

Continuants are things which can have different properties at different times. Or to put it more precisely, they stand in different relations to different times. A continuant might be warm at one time and cold at another, subject to a pressure of 1 atmosphere at one time and to half that at another, and so on. The term "continuant" was introduced by Johnson (1921, pp. 199–202), who contrasted it with "occurrent" – a term denoting what are variously called events or processes. Occurrents are not necessarily instantaneous; even when temporally extended, the contrast holds. A process might slow down, for example, or develop in some unexpected fashion. This has been taken to mean that processes are more like continuants than Johnson thought, and to lessen the contrast. In Dretske's famous example (1967), the party began in the garden, but moved into the house. Although the grammatical construction seems much the same as what we have when describing continuants changing, it is not, however, easy to understand what the corresponding ontological conception of processes would be, and by no means necessary to take the superficial grammatical parallel

J. Seibt (ed.), Process theories: Crossdisciplinary studies in dynamic categories, 237–265.

seriously. The first *part* of the process was quick, later *parts* were slower; the first *part* of the party was in the garden, and later parts indoors. Some authors speak of phases rather than temporal parts. But apart from wanting to reserve the word "phase" for another use, explicitly speaking of the parts of the entities at issue here emphasises their mereological features, which will be of prime concern in this paper.

These two categories have not always been equally welcomed into the philosopher's ontology. Contemporary philosophers complain of a "traditional bias towards substance" (Seibt 2000, p. 242), which, as this terminology suggests, can be traced back to Aristotle. Once change is described as alteration – the possession by a continuant of different properties at different times – then there is no need, so the argument runs, to postulate a distinct ontological category of changings. Understanding occurrents as changings, they seem to be mere manners of speaking, not corresponding to anything in the fundamental ontology. More recently, and especially during the 20th century, occurrents have been taken more seriously by philosophers such as Russell, Whitehead and Quine. There are contemporary philosophers, such as Seibt (2000, 2001) and Simons (2000), who go so far as to claim that there really are no continuants, or that they can be reduced to, in the sense of defined in terms of, occurrents. Some of the arguments for these extreme positions will be briefly considered shortly, but the middle line taken here eschews bias in either direction, largely on grounds of the interdependence of the two notions which will transpire in the course of the descriptive project pursued in the major part of this essay. The primary concern here will be with certain aspects of chemical ontology, distinguishing certain kinds of continuants and certain kinds of processes, and describing how the features of the one are intimately related to those of the other. Chemical substances are taken up first, against which the more elusive category of processes can then be contrasted. Several differences on points of detail with views held by various philosophers, particularly on the description of processes, will emerge. But before getting down to the nitty-gritty, something must be said about the general approach.

2. REDUCTION AND THE MACROSCOPIC PERSPECTIVE

Chemistry deals with entities on both the macro- and the microscopic scale – with substances and the transformation of substances, and with molecules, ions and their transformations. Under the influence of a certain view, encouraged by the teaching of science and adopted by substantial sections of the philosophical community, the macroscopic perspective is considered to be essentially connected with the sphere of observation and phenomenological laws – i.e. observational regularities with no explanatory value of their own. According to this view, claims in the macroscopic realm either have been successfully *reduced* to microtheory, or are outstanding problems in a science whose criterion of success is to solve the reduction problem.

An analogous view in biology, of the reduction of concepts and claims in biology to cell chemistry has been challenged with the rise in interest in the philosophy of biology since the 1970s, and the antireductionist position might now be described as the standard view there. More recently, the reductionist view within the physical sciences has been challenged as philosophical interest in chemistry begins to make an impression in the literature.

The core of reductionism seems to be a thesis about explanation, that only a microscopic explanation would be adequate to explain macroscopic phenomena – hence the label "phenomenological" – and such explanations will be forthcoming in a fully detailed, successful science. Exactly how this thesis is to be made precise has been subject to much discussion, and although the classical formulation of Nagel (1961, Ch. 11) has been extensively criticised, there is no general consensus on an alternative. This makes for a certain ambiguity in anti-reductionist stances like that of Kitcher (1984) over whether the claim is that Nagel's scheme cannot be applied to biology to establish the reductionist's thesis, or the more definite claim that reduction means what Nagel says it means, and by that criterion, biology is definitely not reducible to chemistry. But however the idea is unpacked, reductionism is usually understood to have ontological implications, to the general effect that what there is at the primitive level is confined to the microscopic realm, and macroscopic entities are no more than bundles of microentities.

If, as has been argued, there are good reasons for rejecting the reductionist thesis in chemistry, then there is no need to accept these ontological implications of reduction. I don't propose to go into the reasons for questioning reductionism here (see van Brakel 2000, Ch. 5 for an overview). But I must take issue with some authors who would circumvent the reasonable consequences of rejecting reductionism by claiming that what is denied is *epistemological* reduction while holding on to *ontological* reduction.[1] Facts expressed in the macroscopic vocabulary are, according to ontological reduction, nothing "over and above" those expressed entirely in the basic (microscopic) vocabulary. Since, however, existence claims are formulated in terms of predicates by saying they are instantiated, and according to the antireductionist arguments these predicates are not eliminated, it is the ontological thesis which is called into question. Accepting antireductionist arguments therefore involves a commitment to a macroscopic ontology – if only provisionally in the hope that satisfactory reductionist arguments will be forthcoming. If temperature is not reducible to the average kinetic energy of molecules, then the property of having a temperature (or the more fundamental relations of having the same temperature – being at thermodynamic equilibrium – and being warmer than – having the capacity to heat the other body if brought into diathermal contact) remains unreduced, and there are bodies to which it applies. Note that questioning reductionism in this area is compatible with recognition of both micro- and macroscopic entities. This stance is not to be confused with an older anti-atomist view, even if it does draw on certain ideas from that time. But little will be said here about the difficult problem of how to understand this more complete picture.

Arguments opposing reduction of macroscopic to microscopic realms also undermine an argument for the general reduction of continuants to occurrents mentioned above. Simons defends the thesis that "what minimally makes it true that a given continuant exists at a certain time is not the continuant itself but whatever occurrents are constitutive of its continued existence ... the *vital* occurrents" (2000, p. 70). No persuasive argument is offered for this claim beyond the question-begging statement that "it seems reasonable to suppose the continuant C could not exist at t unless there were some occurrent involving it which has a t-slice" (p. 69). Appeal to the special case of biological organisms' dependence on life-preserving processes of metabolism, reproduction, and so on, hardly motivates an analogous view of stones. Nevertheless, Simons thinks that the challenge of the boring continuant, exemplified by a rock which lies unchanged for centuries, calls for special treatment rather than overturning the general thesis. This he provides by appealing to the chemical binding of the atomic constituents, which is "described in physics by essentially dynamic equations, the Schrödinger wave equations. The rock is in fact teeming with occurrents, ... and they are vital to it" (p. 71). It is a moot point whether stationary, equilibrium-state solutions are appropriately described as dynamic, but the general thesis is that reduction on this philosophical scale is thus made to depend on reduction of macroscopic to microscopic phenomena falling within the purview of science. Apart from objections mentioned above, there is something improper in trying to save a thesis about our everyday conceptions of things by resorting to the unfamiliar terrain of microphysics. If specialists find it difficult to provide a satisfactory account of microscopic ontology, we should be as suspicious of the imposition of familiar concepts in this realm as we are of the portrayal of atomic constituents by the diagrams Dalton drew. At all events, the integration of macroscopic and microscopic viewpoints is beyond the scope of the present paper.

3. SUBSTANCES

Chemistry deals with substances and their transformations. Chemical substances, I take it, are kinds of matter – for example, water, caustic soda, rubber, and so on. The use of the substantive "substance" doesn't imply that these kinds are things. At all events, since it is macroscopic *ontology* that is at issue here, it is important to see that the claim that chemistry deals with substance doesn't imply that there are substances understood in any other way than that there are things which are substances of one kind or another – have the substance properties of being water, caustic soda, etc. Rather than introducing abstract objects like waterhood, named by singular terms, then, substance terms are taken to be predicates, and what exists is the things these predicates apply to. "Water", for example, is a predicate, and more specifically, a mass predicate, applying to entities which, following usage in the literature on mass terms (Cartwright 1970, 1975 and Roeper 1983), I call quantities, and denote by variables π, ρ, σ, ... An important feature of mass predication is the so-called distributive condition:

(1) $\varphi(\pi) \wedge \rho \subseteq \pi. \supset \varphi(\rho).$

where "$\supset$" and "$\wedge$" stand for material implication and conjunction, respectively, φ is a mass predicate and "$\subseteq$" is the part relation of mereology. Standard mereology (cf. Leonard and Goodman 1940) is assumed unless otherwise indicated. Substance predicates are not the only predicates satisfying this condition. But how, exactly, they are to be distinguished from phase properties in physical chemistry – describing states of aggregation such as being solid, liquid or gas – and predicates describing solutions – such as brine, bronze and jade – is an issue calling on details which I don't want to pursue here.[2] Rather, two lines of criticism of the distributive condition (1) will be addressed.

The first derives from the fact that substances participate in chemical reactions, which are processes in which substances are used up in the creation of new substances. Processes will be introduced into the ontology later. For the moment, the point I want to draw from this is that what is one substance at one time may well be another at another time. There is a time factor to be brought into the account, then, with the recognition that what is, say, water at one time may not be at another, so that water is both consumed and produced in various processes. Substance predicates should, accordingly, express relations between quantities and times. Moreover, since times themselves have a mereological structure of parts, the general distributive condition should be written

$$(2) \qquad \varphi(\pi, t) \wedge \rho \subseteq \pi \wedge t' \subseteq t. \supset \varphi(\rho, t').$$

Second, it might be objected against the distributive condition that sufficiently small parts of what is water are not water. What characterises a quantity as water is features such as its state of aggregation (phase) at the quantity's pressure and temperature – e.g. water freezes at 0 °C at normal pressure. Individual molecules cannot be ascribed a phase, a temperature or a pressure. Irreversible thermodynamics is based on the idea of local equilibrium, according to which thermodynamic features such as temperature and pressure are assigned to points in space at instants of time. Having introduced this idea in their textbook, Kondupudi and Prigogine (1998, p. 335) go on to consider how small a volume we can "meaningfully associate" thermodynamic features. Thermodynamic features undergo fluctuations on the microscale, and the limiting condition is that the fluctuations be small in relation to the magnitude of the feature in question. For an ideal gas at normal temperature and pressure, volumes of one cubic micrometer (10^{-6} m) are safely above this limit, and the limiting size is much smaller for liquids and solids. The formalism, which treats temperature, pressure, and so on, as functions of 3-dimensional points and instants, and allows the application of the operators of mathematical analysis, is therefore inconsistent with the credible interpretation. The present approach seeks to follow the interpretation, i.e. what the theory is held to say that is true, rather than the formalism. Accordingly, the variables for quantities considered here don't range over the material occupants of arbitrarily small regions of space. Quantities are whatever can, in principle, bear thermodynamic features, even though, at any given time, they might not actually do so because not at equilibrium. But, in accordance with the assumptions of irreversible thermodynamics, even quantities not at equilibrium and consequently not bearing a temperature have parts which do. Having a temperature means having a degree of warmth, i.e. being *as warm as* something and *warmer than* something else. Taking the warmer than relation as primitive, and writing "π is at least as warm as ρ at t" as $W(\pi, \rho, t)$, an axiom of the kind

$$(3) \qquad \forall\pi\exists\rho\exists\sigma\exists t(\rho \subseteq \pi \wedge W(\rho, \sigma, t))$$

is required. Perhaps even this is too strong, and a modal formulation,

$$(3') \qquad \forall\pi\exists\rho\exists\sigma\exists t(\rho \subseteq \pi \wedge \Diamond W(\rho, \sigma, t)),$$

safer. I won't take a stand on this issue here, but simply note that this would require both time and quantity variables to range over an outer domain on the standard free-logic account of quantified modal logic. The idea of indestructible quantities might be contended in the light of modern microphysics, but accords well with the everyday world of chemistry.

A similar question arises about a lower limit on the time a quantity can sustain a substance property. The formalism of irreversible thermodynamics assumes that thermodynamic properties are assigned to points at *instants*. But just as only claims about finite regions of space are considered true, so too for intervals of time. Perhaps the infinite divisibility condition in the complete first-order theory of intervals (Needham 1981) should be abandoned in favour of a condition analogous to (4) restricting the domain to intervals during which a temperature can be sustained by something. Or rather, bearing in mind that some intervals may be so long that nothing maintains a warmth relation that long, any interval must contain a subinterval of which this is true:

$$(4) \qquad \forall t\exists t' \subseteq t\exists\pi\exists\rho W(\pi, \rho, t').$$

Once more, a modal formulation might be thought more appropriate:

$$(4') \qquad \forall t\exists t' \subseteq t\exists\pi\exists\rho\Diamond W(\pi, \rho, t').$$

4. THE RELATION OF QUANTITIES TO SPACE

Formally, by laying down general conditions ensuring that all quantities are macroscopic quantities, it has not been necessary to introduce a spatial restriction on the distributivity condition. In fact, there is nothing in the assumptions introduced so far requiring that mereological parts be specifically interpreted as *spatial* parts. They might be. But an apparatus is needed to express that this is the case, since proper parts of quantities are by no means always proper spatial parts. Equilibrium thermodynamics assigns a quantity an internal energy, U, which is in general a function $U(S, V, N_1, \ldots, N_r)$ of the entropy, S, the volume, V, and the amount of each component substance, N_i, $1 \leq i \leq r$. A quantity of matter may, in other words, be a mixture of several substances, and the theory has something to say about how changes in some of the features are related to changes in the amounts of the various substances. But although there may be several substances in the quantity, only one volume is specified, the volume of the whole. And when determining the concentrations of the various substances, which are the variables used to follow the course of a reaction, it is the mass of the substance concerned in the volume of the whole that is at issue. There is no justification in the formalism for interpreting the different parts of the original quantity each comprising all and only one of the various substances mixed in the original quantity as having a volume smaller than that of the entire original quantity. In other words, a mixture may be partitioned into a collection of non-overlapping parts each of one particular substance kind in the mixture. Non-overlapping quantities have no part in common. But the separate (i.e. non-overlapping) quantities in such a partition all occupy the same place. They have their location in common.

In order to talk about spatial parts we need a notion of a quantity occupying a place, and spatial regions, $p, q, r, \ldots$, must be considered to have a mereological structure of parts. Classical mereology is again assumed. Since what occupies one place at one time might occupy another at another time, occupying is a time-dependent notion expressed by a triadic relation, for which purpose $Occ(\pi, p, t)$ is introduced for "π occupies region p for t". With this predicate to hand, it is laid down as an axiom that every quantity is always somewhere:

(5) $\exists p\, Occ(\pi, p, t)$.

This expression serves as a kind of temporal existence predicate applicable to quantities. For a quantity to exist at a time it suffices that it occupies somewhere. There is no need to pervert the notion of an occurrent by referring to the occupying of somewhere during a time as a process. In fact, there is no necessity for "vital" processes of any kind involving a quantity in order for it to occupy anywhere.

The "occupies" predicate is specifically interpreted to mean "occupies exactly the region in question", and so satisfies the principle

(6) $Occ(\pi, p, t) \wedge q \neq p \,.\supset\, \sim Occ(\pi, q, t)$.

Quantities may become dispersed, when their parts occupy non-abutting spatial regions. Accordingly, (5) requires that arbitrary sums of regions be considered as regions. They satisfy the usual mereological criterion of identity, so that $q \neq p$ in (6) means that q and p don't have all their parts in common. One may be a proper part of the other,[3] or more generally they may *overlap* (have some part in common), or perhaps they don't even overlap (and are entirely *separate*). In view of the existence and uniqueness postulates (5) and (6), the occupies relation $Occ(\pi, p, t)$ can be expressed in functional form, $Occ(\pi, t) = p$, where $Occ(\pi, t)$ is *the* region occupied by π at t.

Given what was said at the beginning of this section, a quantity ρ might be a *proper part* of a quantity π and yet occupy the same place. The salt in a brine solution,[4] for example, occupies the same place as the brine. Conversely, we can't in general infer from occupying a proper part to being a proper part, i.e. the possibility of cooccupancy also blocks the following as a general principle:

$$Occ(\rho, t) \subset Occ(\pi, t) \supset \rho \subset \pi.$$

For suppose ρ is a proper part of the salt in a brine solution and π the water. Then ρ occupies a proper part of the region occupied by π, yet they are completely separate and ρ is certainly not a proper part of π. A further restriction is needed to infer a proper part of a quantity from a proper part of a region occupied, namely that the quantities concerned are of the same kind of substance. In fact, a proper part of the same kind must be a proper spatial part:

(7) $$SameKind(\rho, \pi, t) \supset .Occ(\rho, t) \subset Occ(\pi, t) \equiv \rho \subset \pi.$$

("$\equiv$" stands for biimplication, "if and only if".) In the absence of information about sameness of kind, there is in general no implication either from $\rho \subset \pi$ to the region occupied by ρ being a proper part of that occupied by ρ or conversely. All we can say is

(8) $$\rho \subset \pi \supset Occ(\rho, t) \subseteq Occ(\pi, t)$$

since proper parts certainly do not extend beyond the spatial extent of the whole. Note that this is equivalent with $\rho \subset \pi \supset \forall t(Occ(\rho, t) \subseteq Occ(\pi, t))$ – proper parts of a quantity never extend beyond the spatial extent of the whole.

The fact that times are thought of as intervals with a mereological structure rather than point entities or instants must be taken into account. Should a quantity move during a time, the region it occupies during this time will be the region swept out, being the sum of all regions swept out by the quantity at subintervals of the time in question. Thus, a quantity that for a proper part of t occupies a proper part of the region occupied at t is in motion. But in general, a quantity may occupy the same place at different times.

Spatial parthood is defined in terms of this time-dependent notion of occupying a place at a time. It is therefore itself a time-dependent notion, and ρ being a *proper spatial part* of π at t is defined as

$$\rho \subset \pi \wedge Occ(\rho, t) \subset Occ(\pi, t),$$

where the time-dependency of spatial parthood reflects the free time variable in the right-hand side. No part of the salt dissolved in a quantity of water at t was a spatial part of the water before t (though it was, of course, a spatial part of its sum with the quantity of water in which it is dissolved at t). So it is not in general true that

$$Occ(\rho, t) \subset Occ(\pi, t) \supset \forall t(Occ(\rho, t) \subset Occ(\pi, t)),$$

and the temporal dependency introduced in the definition of spatial parthood is not redundant.

Summarising, given the mereological criterion of identity – same parts, same quantity – quantities don't gain or lose parts with the passage of time, though they may well become dispersed or assembled, and become spatial parts of other quantities when entering into solution without necessarily sustaining this relation eternally.

5. BODIES

Given the possibility of several quantities occupying the same place at the same time, it will be useful to distinguish a special kind of quantity called a *body* (cf. Denbigh 1981, p. 6) as a quantity which

(i) is the mereological sum of all quantities occupying a *cohesive region* at t. This cohesive region is the region occupied by the body at t.
(ii) is closed during t.

A *cohesive* region is one which can't be exhaustively divided into two non-abutting regions. The definition can be made precise in a mereologically-based axiomatisation of space.[5] Intuitively, it must be possible to get from any part of a cohesive region to another without leaving the region. A quantity of matter π is said to be *closed* at time t iff no quantity of matter either enters or leaves the region occupied by π during t. A quantity which moves into, or out of, a cohesive region p during some part of t does not occupy p during t given the uniqueness condition (6) on occupying, but rather sweeps out some region overlapping, but distinct from, p during t. Accordingly, such quantities are not included in the mereological sum of quantities occupying p during t, and the additional clause (ii) excludes any such overlapping of a body. We may therefore speak of *the* body occupying p, since it is a mereological sum of *all* quantities occupying p. This body may move during t, and still it contains no parts occupying any region outside p for any part of t. The term "open" might be thought of as meaning "not closed". But it is perhaps more natural to call a region rather than a quantity of matter open, even if the region is one confined within the walls of a containing body in the sense of being the region swept out by these walls.

Bodies are said to be *simple* in the sense that

(i) all parts are of the same phase, and
(ii) there are no internal barriers constraining parts to be warmer than, or exert higher pressure than or sustain higher concentrations of particular substances, than other parts.

The notion of an internal barrier was introduced into classical thermodynamics to enable a theory whose fundamental properties are not defined in non-equilibrium conditions to articulate the criteria of equilibrium – i.e. specify how equilibrium situations differ from non-equilibrium situations. Removal of partitions by quasi-static processes proceeding at all times through equilibrium states allows the body to come to equilibrium under new conditions which the theory can describe. A complex body is one partitioned into spatial parts characterised either by being separated by barriers maintaining differing degrees of intensive properties, or by virtue of constituting different *phases*. A phase is the mereological sum of all those parts of a body having a given phase property, and different phases occupy separate regions because phase properties are spatially exclusive. Note, finally, that the closure condition applies only to the whole complex body, and not to each of its single phases.

Ordinary objects like chairs, rivers and people are not bodies in the present sense. These things remain the same despite addition and loss of quantities of matter. Quantities of matter are not parts of such things in the sense in which the part relation is used here. I call such things individuals and say that they are *composed* of a quantity of matter at a time. I won't enter into a lengthy discussion here about the detailed interpretation of this relation. Suffice it to say that variable composition can be accommodated by analogy with the interpretation of the region occupied by a moving quantity: x is composed at t of the sum of the quantities of which it is composed at any subinterval of t. Again by analogy with "occupies", x is composed at t of exactly one quantity, and not of any other not identical with it. This precludes the distributive condition: proper parts of quantities composing an individual don't compose that individual at the same time.

Further, if "composed" is a relation between an individual and a quantity of matter at a time, then one individual cannot be composed of another. The organs of a human body, or the arms of a chair, are individuals which stand in a distinct relation of *belonging to* another individual at a time. This accommodates some of the features Seibt (2000, p. 257; 2001, §2.3) addresses by making her part relation non-transitive, so that some of our differences are merely terminological. Whether she remains closer to the everyday usage of "part" is a moot point; but my principal concern is with ordinary usage in chemistry.

6. INTRODUCING PROCESSES

Quantities of matter participate in processes – mechanical, thermal and electrical, for example, as well as chemical transformations of substances – all of which are treated by thermodynamics. I have argued elsewhere (Needham 1999, 2000) that, in terms of the interpretation given in ordinary textbooks, heating is a process – an ontological category of things which can't be reduced to continuants and their properties. This is not simply because there are no such things as quantities of caloric which are transferred from one body to another at some rate or other. As Duhem (1895) points out, it is equally directed against the competing idea of heat as a measure of the amount of motion of microscopic particles composing a body. Both ideas required that heat be a reproducible feature of a body – something a body can absorb or relinquish in an amount directly related to the state of the body before and after its change in heat. But the amount of heating involved in the heating of one body by another is not in general a function of the states of the bodies involved before and after the process; the states also depend on the amount of mechanical working of the one body on the other taking place simultaneously. The standard formalism of thermodynamics, which aims to satisfy whatever conditions are required for the application of standard tools of mathematical analysis, has no processes but introduces an ontology of states and overcomes the problem that the amount of heating is not a finite difference between some feature of initial and final states (of the bodies) by following a continuous path over (instantaneous) states from the initial to the final state. Moreover, bodies themselves only figure as sets of such states. But as indicated above, the ordinary interpretation doesn't even recognise instants and predications of instants, let alone accept that ordinary objects are reducible to instantaneous states of affairs. Recognising continuants as autonomous objects of the macroscopic ontology obliges us to take processes just as seriously.

Heating and working are processes which usually result in changes in the states of bodies as a whole, paradigmatically their degree of warmth (temperature) and the amount of space they occupy (volume), respectively. There are also processes involving the various substances in a body. Substances may be redistributed over the body by diffusion, or move across a semi-permeable membrane which divides the body and allows some but not all the substances to pass. They may also participate in chemical transformations which result in changes in the total amounts of the various substances present in a body. Heating air in a closed vessel to 1873 °K leads to the formation nitric oxide, for example, in a process which reaches equilibrium after sufficient time. Moreover, material involved in one of these processes might be involved in several others at the same time. The conversion of air to nitric oxide with, as it is often put, the absorption of heat is a case in point. Literally speaking, of course, there is no caloric to be absorbed; rather, the quantity of matter within the vessel originally comprised entirely of air is heated by some other body. Again, a body formed by adding a quantity of concentrated sulphuric acid to a quantity of water forms a homogeneous mixture or solution, generating, as it is often put, a good deal of heat in the process. But no such thing is literally generated; rather, the mixing matter heats surrounding bodies. There can also be cross-effects between several simultaneous processes; the rate of diffusion of one substance, for example, is governed not only by the distribution of concentration of that substance, but also on the rates of diffusion of other substances in the same body. Complex effects of this sort are not the place to begin a discussion of processes, however. A more appropriate starting point is the general form taken by descriptions of the elementary processes which serve as the basis of the mereology of processes explored here. This will involve challenging some ideas about occurrents and causation widely held in philosophical circles and calls for some discussion.

Two general observations get the discussion off the ground. First, processes take time. Using lower case letters from the beginning of the Greek alphabet as variables over processes, the fundamental principle is

$$(9) \qquad \forall\alpha\exists t\, TP(\alpha, t),$$

where the "takes place" predicate TP functions in effect as a sort of time-dependent existence predicate. Just as the "occupies" predicate means "occupies exactly the region in question", taking place at a time involves exactly that time:

$$(10) \qquad TP(\alpha, t) \wedge t \neq t'. \supset \ \sim TP(\alpha, t').$$

The takes place predicate is therefore a function, and $TP(\alpha, t)$ will sometimes be written in functional form, $TP(\alpha) = t$.[6]

The second observation is that processes involve the interaction of quantities. In simple cases they are interactions between two quantities, and descriptions of processes take the form $\varphi(\alpha, \pi, \rho)$, for "$\alpha$ is a φ-ing by π of ρ". For example, α is a heating by π of ρ (cf. Needham 1999, 2000), or α is a diffusing of π through ρ. Heating radiation in a vacuum by a body would be an exception to the rule, taking the form $H(\alpha, \pi, p)$, where p is a region of space, if we follow the thermodynamic theory which treats this region as not occupied by any substance. Where more than two quantities seem to be involved, sums might sometimes be used to reduce the quantities to two, much as Leonard and Goodman (1940) suggested coping with multigrade relations. But there is no general restriction in principle of processes to interactions between just two quantities.[7]

Now the word "interaction" suggests causation, and the examples of heating, diffusing and chemical reactions are naturally thought of as causal processes. The two initial moves here would seem, however, to be at odds with a long tradition within philosophy which treats causation as a dyadic relation between two occurrents. Further, a particular philosophical theory of causation, the so-called regularity theory, needs some way of pairing off a cause with its effect in such a way as to enable a relation of constant conjunction to be defined between events of one kind (those resembling the cause) and another (those resembling the effect) without circularity – i.e. without appealing to their causal connection. Hume's classical solution to this problem was to define the appropriate sense of occurring together in terms of spatio-temporal contiguity: abutment in both space and time. And many philosophers accept this contiguity condition, or at least the prerequisite that cause and effect are occurrents with definite spatial locations, even if they don't accept the regularity theory of causation. But the present approach makes provision neither for a dyadic relation of causation nor for a spatial location of a process.

In what has traditionally been taken as a paradigm case of colliding billiard balls, an event taken to comprise the approach of the white cue ball is called the cause and the subsequently moving red ball is taken to be an event and called the effect. The location of the balls at the appropriate times is then taken to define the location of the events in which they are involved. But if all that is meant by the location of an event is simply the region occupied by the single continuant presumed to be involved, we might with less obfuscation speak directly of the region occupied by the continuant instead. Now this information is not lost on the present approach. An apparatus for saying what region is occupied by the quantities involved in a process is already available, without necessitating any change in what was said above about taking place. Just as the time at which a process occurs can be indicated by a conjunction – that α is a diffusion of π through medium ρ during t, for example, is indicated by a conjunction $D(\alpha, \pi, \rho) \wedge TP(\alpha, t)$ – so the regions occupied by any quantities involved in a process can also be specified by adding further conjunctions. There is nothing here that requires making the "takes place" predicate triadic. In fact, information about how the regions concerned are related can also be incorporated. For example, a diffusion would normally satisfy

$$D(\alpha, \pi, \rho) \wedge TP(\alpha, t) \wedge Occ(\pi, t) \subseteq Occ(\rho, t),$$

whereas a heating would normally satisfy

$$H(\alpha, \pi, \rho) \wedge TP(\alpha, t) \wedge Occ(\pi, t) | Occ(\rho, t).$$

("|" is the mereological relation of separation.) Here we see that the regions involved in these two examples stand in incompatible relations, and only the latter is consistent with abutment (contiguity). Further examples relating to this point are taken up in connection with chemical reactions in the section 8. Unimaginatively restricting examples to cases involving billiard balls provides no argument for the universality generally assumed for the contiguity condition unless backed by some further argument for the reduction of all processes to simple mechanical collision processes. But even in the heyday of the regularity theory, it was reckless to challenge Newton's theory of gravitation in this way. Now the claim is totally implausible. Classical mechanics has been superseded by quantum mechanics, which provides no support for such naive ideas about contiguity. In any case, the assumption of the present treatment is that macroscopic ontology be treated autonomously.

As for the analysis of causation as a dyadic relation, it seems once more to be a case of generalising from the kind of example paradigmatically illustrated by billiard ball collisions without any argument demonstrating its universal applicability. But even this paradigm case of the billiard ball collision can be viewed along the lines of the present approach. The mechanical process is a colliding, involving the mutual compression of two bodies in contact followed by their pressing one another away from each other. A three-body collision would be described on the present view along the same general lines, but with a four-place predicate applying to a process and three continuants. On the other hand, expressions like "the rolling of the white billiard ball", or "the rotating of the earth", as long as they are thought of as motions maintained by inertia, refer at best to states, certainly not processes. Only frictional interactions with the table or the atmosphere, involving several bodies, are processes.

Be this as it may, understanding thermodynamic examples of primary interest here doesn't depend on how mechanical collisions are viewed. Consider a heating process caused by a temperature difference between two proximate bodies. The temperatures of each body might be maintained at fixed values by suitable ancillary heating and cooling arrangements. It is the states of the bodies which, in the circumstances, give rise to the heating at any stage in the continuous process. Even if the temperatures of the bodies weren't held fixed, but allowed to approach one another, it certainly wouldn't be the change in the temperature of the one body which causes the change in the other. It is the states of the interacting bodies, whether changing or not, which, in the circumstances, give rise to the heating. Where the situation is allowed to go to equilibrium, what is brought about at the end of the process is a changed state in each body, from an initial to a different final state. The change might not be a change in temperature. As just noted, there need be no change in temperature if the two bodies are suitably thermostatically controlled. Alternatively, the heated body might be a gas which expands isothermally, or a piece of ice at 0 °C at normal pressure which melts at the same temperature. Yet it is still true that the fact that one body is warmer than the other brings about a heating process. Moreover, the temperature of a gas might be changed adiabatically, for example raised by compression (i.e. mechanical working) when thermally insulated from the surroundings, and thus not raised by heating.[8] So it is a mistake to identify heating with a change in temperature. In fact it is a mistake to identify anything with a change. Grammatical appearance notwithstanding, expressions such as "the change in temperature of a body", and "the change in quality φ", for any quality φ, are not proper referring expressions.[9]

A heating process is not necessarily an interaction between bodies at different temperature. The thermoelectric phenomenon known as the Peltier effect involves a heating ("heat flow") along a copper wire inserted in an aluminium wire circuit through which an electric current passes where the aluminium/copper junctions are maintained at the same temperature. This is an example of a so-called cross effect, where the potential gradient giving rise to its "own" flux (the electrical current in this case) also affects other fluxes. (In this sense of "own", a temperature gradient would have a heating as its "own" flux, and a concentration gradient of a particular substance a diffusion of that substance.) In general, fluxes depend on all the "affinities" or thermodynamic forces in the system, although most strongly on their own affinities. The Seebeck effect involves a temperature gradient between the junctions of dissimilar metal wires forming a circuit giving rise to a potential difference.

Although processes arise, according to the present view, as a result of the states of continuants and their dispositions to interact with one another, examples such as those of the last paragraph illustrate that interactions do occur between processes too. There is no joy in this for the view of causation as a dyadic relation between occurrents, however. Quantities each of different substances may be simultaneously diffusing in the same medium, and these diffusions affect one another. Heating processes within a single body arising from a temperature gradient are similarly interconnected with diffusion processes. Irreversible thermodynamics studies such interactions, guided by the Onsager theory of the reciprocal relations between the constants of proportionality in the rate equations for the individual processes. Given the present approach to the individual processes, this might suggest that there are "second order" processes Γ amounting to an interaction of two "first order" processes, which should be described in the form $\varphi(\Gamma, \alpha, \beta)$. But this doesn't seem to be in the spirit of the usual treatment of such interactions in science. A simple diffusion process is driven by the chemical potential[10] of a substance unevenly distributed in another (the medium), forcing the evening out of the concentration gradient. Also in a multidiffusion case, the chemical potential of each substance drives the distribution of concentration of each substance. But according to the Gibbs-Duhem theorem of classical thermodynamics, the chemical potentials of a multi-component system are not independent. Consequently, nor are the thermodynamic forces driving the movement of each substance. Irreversible thermodynamics describes how the flows of each substance are determined not only by the nature of the substance, as reflected in its chemical potential, but also by this mutual influence of the flows of each substance on the others arising from the non-independence of the chemical potentials. Rather than involving a second order process, it seems that two processes interact in such a way that they are somewhat modified compared to what they would be like were there no interaction. The rates of diffusion of two substances in the same medium influence one another, for example. But each process is still a diffusion of a quantity of a particular substance which would have occurred were the other substance not present, although at a somewhat different rate.

7. THE MEREOLOGY OF PROCESSES

Processes are assumed to have a mereological structure. With the exception of the sum principle, discussed shortly, classical mereology is again assumed. Now it was mentioned at the beginning that processes have temporal parts (without implying that processes only have temporal parts).

The first part of the diffusion process was fast, and later parts were slower. Although the reaction was initially very slow (i.e. the first stages were very slow), it gradually accelerated and became much faster (i.e. the later stages were fast). In general, it would seem to be a reasonable principle that

$$(11) \quad t \subset TP(\alpha) \supset \exists\beta(\beta \subset \alpha \wedge TP(\beta, t)).$$

If α is a diffusing of π through ρ, for example, then there will in general be temporal parts, β, of α which are diffusings of π through ρ. *Proper temporal parts* of processes can be defined by

$$\text{Def.} \quad \alpha \subset_{TP} \beta \equiv . \alpha \subset \beta \wedge TP(\alpha) \subset TP(\beta).$$

And where $t' \subset_{ITP} t$ means "t' is an initial part of t",[11] an initial stage or part of a process can be defined by

$$\text{Def.} \quad \alpha \subset_{ITP} \beta \equiv . \alpha \subset \beta \wedge TP(\alpha) \subset_{ITP} TP(\beta),$$

and similarly for *final temporal part*:

$$\text{Def.} \quad \alpha \subset_{FTP} \beta \equiv . \alpha \subset \beta \wedge TP(\alpha) \subset_{FTP} TP(\beta).$$

But just as quantities may have parts which are not spatial, so there may well be other parts of processes too, in addition to the temporal parts. This is a simple consequence of allowing mereological sums of processes. Summation is not quite arbitrary, at least if the restriction on sums of times to binary sums of abutting or overlapping times in Needham (1981) is followed. (The restriction excludes times with gaps, and doesn't allow the entire expanse of time to count as a time.) Given the general principle (9) requiring that every process take place at some time, the summation of processes must be restricted in order to ensure the existence of a time when the sum takes place. Accordingly, saying that times which abut or overlap are connected, abbreviated "*Con*", a restriction on the existence axiom for sums of processes is required along the lines of:

$$(12) \quad Con(TP(\alpha), TP(\beta)) \; .\supset \; \exists\delta(\forall\gamma(\gamma|\delta \equiv . \gamma|\alpha \wedge \gamma|\beta) \wedge TP(\delta, TP(\alpha) \cup TP(\beta))).$$

Although this doesn't allow processes with temporal gaps – "temporarily suspended processes" – it is still a very liberal summation principle.[12] There is no spatial restriction on the continuants involved, and sums of any two processes taking place at same time, for example, exist by principle

(12). The two processes forming such sums would be examples of non-temporal parts of a process, namely their sum. Such a liberal principle might be thought unreasonable. But it is not the only source of non-temporal parts. If α is a diffusing of π through ρ, then even during the time α takes place there will be diffusings of parts of π through parts of ρ. Whether the same can be said for heating I leave unsaid. But for diffusings, at any rate, we have

$$(13) \quad D(\alpha, \pi, \rho) \wedge \pi' \subset \pi \,.\!\supset \exists \alpha' \exists \rho' (\alpha' \subset \alpha \wedge \rho' \subseteq \rho \wedge D(\alpha', \pi', \rho') \wedge TP(\alpha) = TP(\alpha')).$$

($\rho' \subseteq \rho$ is written here rather than $\rho' \subset \rho$ because π' might be one of several kinds of substance in π and occupy the same place as π.) Since some simple processes have non-temporal parts in addition to the temporal parts, even if not all do, the mere possibility of non-temporal parts can't reasonably be made the basis of an objection against the summation principle (12). Complex processes like the multidiffusions mentioned above are sums of simultaneous diffusions of single substances, even if these parts interact. But interaction can't be the general principle sanctioning summation of processes, since a process is the sum of its temporal parts, although temporal parts are not the cause of succeeding temporal parts. An initial part of a heating is not the cause of any succeeding part, for example. Heating is a continuous process generated by the states of the bodies involved in the heating process. There would have to be some point, some reason for finding the arbitrary sums allowed by (12) objectionable, motivating a search for some other principle restricting sums. But I don't see any such objection, even if it is more intimately related parts that are summed in the more interesting complex processes, such as chemical reactions taking place in a body where a number of elementary reactions are interlinked, the one producing intermediate substances consumed in others, whose total effect is the overall change of initial reactants to final products.

Given that processes have a mereological structure, are process kinds analogous to substance kinds? They don't have the corresponding mass predicate characteristics of substance predicates such as satisfying the appropriate analogue of the distributive condition (2). If α is a diffusing of π through ρ, for example, then *it* is not a diffusing of anything more or less than π through anything more or less than ρ:

$$(14) \quad D(\alpha, \pi, \rho) \wedge (\pi' \neq \pi \vee \rho' \neq \rho) \,.\!\supset \sim D(\alpha, \pi', \rho'),$$

and similarly for heating. What is true is that heatings and diffusings are *simple* in the sense that all their parts are of the same *kind*, satisfying the schema

$$(15) \quad \varphi(\alpha, \pi, \rho) \wedge \beta \subset \alpha \,.\supset \exists \pi' \subseteq \pi \exists \rho' \subseteq \rho \varphi(\beta, \pi', \rho').$$

Heatings, for example, don't have parts which are not heatings of something by something. As we will see, this is not always the same as saying none of its parts are of a different kind. *Complex processes* are processes with parts not of the same kind as the whole. As mentioned above, they may be the result of arbitrary summation; but processes involving quantities within the same body, or a sequence of temporal parts of different kinds, may also be complex. Chemical reactions provide examples of processes involving both these types of complexity.

8. CHEMICAL REACTIONS

Considered as an interaction, a chemical reaction involves a quantity of matter comprising one or more substances brought together within a region being transformed into a quantity of matter comprising one or more substances under the condition that, as Lavoisier put it, "an equal quantity of matter exists both before and after the experiment" (1789, p. 130). Lavoisier's principle is understood here to imply that the sum of all quantities involved in the transformation at all stages of the process is the same.

A chemical reaction is represented in the general form

$$(16) \quad a_1 A_1 + \cdots + a_n A_n \rightarrow b_1 B_1 + \cdots + b_m B_m,$$

where a_i, b_j are the stoichiometric coefficients representing the combining proportions and A_i, B_j denote the distinct kinds of substances. The transformation of air into nitric oxide, for example, is expressed by

$$N_2 + O_2 \rightarrow 2NO.$$

The idea that a heating involves an affecting and an affected body might suggest that a reaction be described as a process standing in a three-place relation to reactants and products. But although these notions of reactant and product should be appropriately captured, such an approach loses sight of the idea that a process is an interaction between quantities of matter. With a view to retaining this idea, the format of a three-place predicate is in general replaced by a description $R(\alpha_1, \pi_1, \ldots, \pi_n)$ of a reaction of kind (16) involving an $n+1$-place relation for a reaction with n interacting substances.

What, exactly, are the quantities of reactants, $\pi_1, \ldots, \pi_n$, entering into such an interaction? Each π_i comprises not all the stuff in the universe of kind A_i, but that occurring in some particular region r_i. A further consideration is that quantities involved in a reaction change from one kind of substance to another during the course of the reaction. Yet the distributive condition (6) implies that if $A_i(\pi, TP(\alpha))$, then every part of π is A_i for every stage in the process α. The problem is analogous to that of allowing the notion of the region occupied by a quantity to accommodate the possibility of change – of the quantity moving – by understanding it to be the region swept out by the quantity during the time in question. Here the quantity of reactant A_i is identified with the sum of those quantities which, for some part of the duration of the reaction, are one of the reactants in the region r_i. Accordingly, writing $\Sigma\sigma\varphi(\sigma)$ for the sum of everything with the property φ, the quantity of reactant, π_i, of kind A_i is the sum

$$(17) \quad \Sigma\sigma\exists t \subseteq TP(\alpha)(A_i(\sigma, t) \wedge Occ(\sigma, t) \subseteq r_i).$$

The reactants as a whole – i.e., the *quantity of reactants* – is the sum of the quantities of each particular reactant. Chemists speak of "the reactants"; but note that a single thing is at issue.

The region r_i occupied by the quantity of reactant of kind A_i is assumed given at the outset. But it might be understood as the sum of those regions occupied by some of the quantity of reactant participating in the reaction at any time during the course of the reaction. Different quantities of reactants are not necessarily located at the same place during the course of the reaction. Consider, for example, an open fire, which draws oxygen from the atmosphere to combust with wood piled on the ground. A first thought might be that the scene of the reaction can be understood as the mereological product of the regions swept out by the quantities of the various reactants during the course of the reaction. This assumes, however, that we are not dealing with a situation like a forest fire which covers ground while it proceeds, or a laboratory situation involving a moving reaction vessel. A second thought is that, in general, the scene of a reaction is a sum of products of regions where some parts of the quantities of reactants spatially overlap. The contents of a rotating reaction vessel, for example, when sweeping over a region the size and shape of that occupied by the stationary vessel, spatially overlap one another, and a mereological product of the regions occupied by each component during this time exists. The sum of such products during the course of a reaction would be the scene of the reaction. But whether this does justice to the intricacies of heterogeneous reactions will not be pursued further here.

As defined here, the quantity of reactants isn't depleted as the reaction proceeds. It delimits the total quantity of matter which, in accordance with Lavoisier's principle, is conserved throughout the reaction. The *quantity of products* (of the chemical reaction (16)), defined analogously as the sum of each quantity, ρ_j, of products of kind B_j given by

$$\Sigma\sigma\exists t \subseteq TP(\alpha)(B_j(\sigma,t) \wedge Occ(\sigma,t) \subseteq r_j),$$

is similarly never augmented. If the entire quantity of reactants were converted in the course of the reaction, it would be identical with the quantity of products. Otherwise, the quantity of products is a proper part of the quantity of reactants for a reaction which doesn't go to completion. Now consider how the reaction progresses through successive reaction stages $\alpha_1, \alpha_2, \ldots$, such that $\alpha_1 \subset_{ITP} \alpha_2 \subset_{ITP} \alpha_3 \ldots$, where $R(\alpha_1, \pi_1^1, \ldots, \pi_n^1)$, $R(\alpha_2, \pi_1^2, \ldots, \pi_n^2)$, and so forth. Each stage is a reaction of the same kind (16) described here by "R", with the same quantity of reactants: $\pi^1 = \pi^2 = \cdots = \pi^n$ (given that the stages are all initial parts of the succeeding stages). But the quantity of each product kind is continually augmented with new quantities of that kind, as, therefore, is the quantity, ρ^k, of all product kinds, and we have $\rho^1 \subset \rho^2 \subset \cdots \subset \rho^{n-1} \subset \rho^n$. Any reconversion due to the reverse reaction,

$$b_1B_1 + \cdots + b_mB_m \rightarrow a_1A_1 + \cdots + a_nA_n,$$

would not, however, lead to a depletion of the quantity of products on the suggested definition. To rectify this failing, the definition of the quantity of products of kind B_j can be elaborated with an added clause to the effect that the time concerned in the sum is not just any subinterval of the total reaction time, but one which extends to the end of the reaction time:

$$(18) \quad \Sigma\sigma\exists t \subset_{FTP} TP(\alpha)(B_j(\sigma,t) \wedge Occ(\sigma,t) \subseteq r_j).$$

By the distributive condition, which is satisfied by substance predicates like B_j, only what is B_j for an entire final subinterval is included in the sum. With this additional restriction, the amount of matter in ρ_j at any stage of the reaction (measured in moles and divided by the stoichiometric coefficient b_j) would be a measure of the extent of the reaction. A reaction which proceeds to completion is one in which the quantities of reactants and products are identical. One which reaches equilibrium or is otherwise drawn to a halt before completion leaves the mereological difference, $\pi - \rho$, of the quantity of reactants less the quantity of products unconverted back to reactants by the reverse reaction. In either case, Lavoisier's condition holds; quantities of matter are neither created nor destroyed, but merely converted from one or more kinds to others.

A chemical reaction conducted within the confines of a body can come to equilibrium. If there is a change in volume – for example, if the reaction proceeds entirely in the gas phase, in an arrangement which allows expansion or contraction in order to maintain the external pressure – then the quantity of reactants will also participate in a mechanical working on or by the environment. Similarly, if conducted within the confines of diathermal walls, an exothermic or endothermic reaction will lead the quantity of reactants to participate in a heating. But if the body is constrained to maintain a fixed volume, and adiabatically insulated from the surroundings, then the quantity of reactants will not participate in additional processes of mechanical working or heating. The constraints on the body will usually have repercussions on the extent of reaction. Thus, Denbigh (1981, p. 73) describes the idealistic device called the van 't Hoff equilibrium box in connection with the combustion of carbon in oxygen yielding carbon dioxide. The carbon is contained in a reaction vessel equipped with two windows, one permeable only to oxygen and the other permeable only to carbon dioxide. A piston and cylinder arrangement feeds oxygen slowly into the vessel through the first window, and a similar arrangement removes carbon dioxide via the other window. Controlling the partial pressure ratio of oxygen and carbon dioxide infinitesimally less than the equilibrium ratio allows the reaction to proceed reversibly while a net amount of mechanical work is performed by the movement of the pistons. The regions involved in the above definitions would not be confined to that occupied by the reaction vessel in this case, but include the piston and cylinder arrangements. In practice, power is raised commercially from this reaction under thermodynamically less efficient conditions, allowing fuel to burn freely and irreversibly, obtaining energy in the form of heating which is subsequently harnessed in the performance of mechanical working, rather than directly in the form of working under the reversible conditions of the van 't Hoff box. That part of the free energy change due to entropy change which is lost is not great; but the efficiency of the subsequent conversion of the heating to mechanical working is limited by the thermodynamic conversion factor, and by less than perfect engines even more so. The regions occupied by the reactants and products would not, in this case, be confined to the reaction vessel, but include that region containing whatever air is pumped into the vessel during the course of the reaction.

Energy from chemical reactions can sometimes be directly channelled into electrical working, avoiding heating processes and the conversion factor, when the reaction involves the production of ionic substances and can be set up as a galvanic cell. The Daniel cell, for example, requires separation of zinc and copper sulphate solutions by a porous barrier which slows the mixing so that ionic copper cannot be deposited on the zinc electrode immersed in the zinc sulphate solution but is deposited on the copper electrode immersed in the copper sulphate solution. In this arrangement, the transformation of metallic zinc into zinc sulphate and copper sulphate into metallic copper can drive a current round an external circuit.

Where the reaction proceeds in a body which is not isolated, but in diathermal contact with another body, σ, then a heating may take place. Since it is the reaction that is endothermic or exothermic, the heating is the same process as the chemical reaction. But then different bodies are involved, adding to the difficulties already mentioned in the last section of identifying a place where a process takes place. As a chemical reaction, a process α is an interaction between reactants $\pi_1, \ldots, \pi_n$, whereas it is a heating of something else, σ, by the quantity of reactants $\pi = \pi_1 \cup \cdots \cup \pi_n$. Conceivably, the diathermal contact might be so arranged that if the reacting body expands, it does work against a different body, τ – say a piston – which is not the body being heated. Then we have

$$(19) \qquad R(\alpha, \pi_1, \ldots, \pi_n) \wedge H(\alpha, \pi, \sigma) \wedge W(\alpha, \pi, \tau).$$

The combustion of carbon in the van ’t Hoff box described above is a mechanical working but not a heating, whereas the combustion of carbon in commercial power plants is a heating but not a mechanical working. The reaction

$$Zn + CuSO_4 \rightarrow ZnSO_4 + Cu$$

can be conducted in a complex body so constrained that the reaction is also an electrical working. If, on the other hand, the reacting body is isolated, then the process itself won’t also be a heating or a working.

There is no limit, of course, to the number of properties an entity might have. But one and the same process may thus be nontrivially *multifarious* by being of several significant kinds without necessarily involving any complexity in the sense outlined above. Multifarious processes may well be simple, as characterised by the satisfaction of a condition like (15) to the effect that all of their parts are of the kinds instanced by the whole process. But they may be complex.

Chemical reactions which proceed by way of a number of elementary steps[13] which can be treated as independent reactions are ordinarily understood as complex, and this agrees with the present sense of complexity. Oxidation of solid carbon to produce carbon dioxide comprises two reactions

$$E_1 \quad O_2 \text{ (gas)} + 2C \text{ (solid)} \rightarrow 2CO \text{ (gas)}$$

$$E_2 \quad O_2 \text{ (gas)} + 2CO \text{ (gas)} \rightarrow 2CO_2 \text{ (gas)},$$

the joint net effect of which is the conversion of oxygen and carbon to carbon dioxide. A process of the heterogeneous reaction kind

$$E \quad O_2 \text{ (gas)} + C \text{ (solid)} \rightarrow CO_2 \text{ (gas)}$$

may thus comprise proper parts of the kinds E_1 and E_2 taking place simultaneously, with the presence of some additional intermediary substance continually being produced and consumed.

Complex reactions sometimes evolve in such a way that their temporal parts are of different kinds. Clock reactions are so called because they evolve at definite rates, depending on initial concentrations and other conditions, with an induction time during which the intensity of some feature builds up to some value sufficient to instigate another process. The classic reaction discovered by Landolt in 1886 involves iodate and bisulphite ions in aqueous solution with starch indicator. The reactants are colourless. After a long induction period during which iodate ion is slowly reduced and the iodide concentration slowly builds up by the reaction

$$IO_3^- + 3HSO_3^- \rightarrow I^- + 3HSO_4^-,$$

there is a rapid acceleration in the rate of production of I^-, leading to a spectacularly sharp change to blue in the starch indicator. Sudden ignition of haystacks months after building, or of a cloth drenched in linseed oil, can be understood as involving an exothermic reaction of a quantity of reactants in a particular region proceeding initially very slowly. The material in the region in question becomes warmer as the reaction proceeds. But this material is not perfectly adiabatically insulated and there is a cooling effect as the surroundings are heated, the magnitude of which depends on features of the situation such as the surface-to-volume ratio of the region and the temperature excess over the immediate surroundings. If the cooling doesn't keep pace with the warming of the region, as is more likely if the cloth soaked in linseed oil is crumpled rather than flattened out, the region will slowly become warmer, eventually reaching the ignition temperature when the material bursts into flame.

These are examples of chemical processes taking initial reactants through to final products, or settling down in an equilibrium state in which the concentration of all participating substances, including intermediaries, is so adjusted that there is no net rate of production or destruction – forward and reverse rates for all steps in the overall process become equal. There are other cases where the chemical process is repetitive, as in the now famous Belousov-Zhabotinsky reaction. Once the reactants specified in a standard "recipe" for the reaction are mixed, oscillatory behaviour begins after an initial induction period in which colourless ceric ions, Ce^{4+}, are consumed and replaced by yellow cerous ions, Ce^{3+}, or ferrous ions in the complex ferroin are converted into ferric ions in ferriin and back when the solution oscillates between red and blue. According to the Field-Körös-Noyes mechanism, a process A removes bromide ions:

$$\text{A} \quad BrO_3^- + 2Br^- + 3H^+ \rightarrow 3HOBr,$$

comprising two steps involving production and consumption of $HBrO_2$:

$$\text{A}_1 \quad BrO_3^- + Br^- + 2H^+ \rightarrow HBrO_2 + HOBr,$$

$$\text{A}_2 \quad HBrO_2 + Br^- + H^+ \rightarrow 2HOBr.$$

When the bromide concentration is reduced sufficiently, process B, in which $HBrO_2$ reacts instead with bromate ion, comes into play:

$$\text{B} \quad BrO_3^- + HBrO_2 + 2Ce^{3+} + 3H^+ \rightarrow 2HBrO_2 + 2Ce^{4+} + H_2O.$$

This process comprises two steps involving production and consumption of the radical $BrO_2^{\cdot}$:

$$\text{B}_1 \quad BrO_3^- + HBrO_2 + H^+ \leftrightharpoons 2BrO_2^{\cdot} + H_2,$$

$$\text{B}_2 \quad BrO_2^{\cdot} + Ce^{3+} + H^+ \rightarrow HBrO_2 + 2Ce^{4+}.$$

Process B is an autocatalytic process, causing the concentration of $HBrO_2$ to increase and accelerating the rate of oxidation of the Ce^{3+} ion. It is inhibited by the presence of bromide ion, which competes for the $HBrO_2$ and prevents oxidation of Ce^{3+}. But when bromide ion concentration falls below a critical level due to reaction A, reaction B takes over and would be the end of the story were there not a mechanism restoring the bromide ion concentration and reducing the metal ion back to its lower oxidation state. Process C enters the picture at this point, involving bromination of malonic acid by HOBr produced in process A, perhaps via the production of bromine, to produce bromomalonic acid which, as well malonic acid itself, reduces the catalyst ion Ce^{4+} back to Ce^{3+}:

$$\text{C} \quad 2Ce^{4+} + CH_2\,(COOH)_2 + BrCH(COOH)_2 \rightarrow f\,Br^- + 2Ce^{3+} + \text{other products},$$

Bromide ion builds up in concentration, and the process returns to the point where reaction A can take place, inhibiting the autocatalysis reaction B by consuming $HBrO_2$.

9. CONCLUDING REMARKS

The exploration of the nature of macroscopic processes in the second part of this paper has relied upon prior notions connected with continuants introduced in the first, particularly notions relevant to articulating the concept of a chemical substance. But the dependence is not so one-sided as this would suggest. The importance of processes of chemical transformation to the general notion of substances made its mark early on in the discussion with the requirement of the time-dependency of substance predicates. Moreover, the ability to participate in chemical transformations and form compounds is the basis of the distinctly chemical measure of the amount, according to which there is as much oxygen in water as there is sulphur in hydrogen sulphide despite the differing gravimetric proportions of 8 : 1 and 16 : 1, respectively. If π and ρ are each single substances, then π and ρ are said to be of the same amount, or equivalent, iff either π and ρ are both of the same kind and have same volume, or π and ρ separately combine with or replace the same volume of hydrogen, all comparisons of volume being made at the same temperature and pressure. This classical notion of *equivalent* amounts needs to be extended to cover all cases of single substances along well understood lines. But this suffices to indicate the interdependence of continuants and occurrents, or at least quantities and processes of the kinds considered here, as was suggested in the Introduction.

NOTES

1 "[W]e need not appeal to 'ontological emergence', the 'supernatural', or other questionable notions in order to get our epistemological autonomy," say Scerri and McIntyre, who maintain that chemistry "is a science whose ontological dependence upon physics is not in doubt" (1997, pp. 224–5, 227). More recently, Vemulapalli and Byerly have endorsed "[t]he idea of supervenience[, which] is to enjoy ontological reduction while allowing

epistemological emergence," since "[m]ultiple realizability of concepts undercuts epistemological reduction but does not preclude the ontological reductions posited by physicalism" (1999, p. 35).

[2] For a discussion, see Needham (2003).

[3] The *proper part* relation, written "$\subset$", excludes the possibility of identity, and is defined in terms of the relation "$\subseteq$ by the schema:

$$\text{Def.} \quad x \subset y \equiv . \, x \subseteq y \wedge x \neq y,$$

where x and y are the appropriate kinds of variables.

[4] or whatever it is that the salt becomes when in solution.

[5] In Tarski's (1983) system, for example, any two separate regions can be defined as abutting if there is a sphere overlapping both but none other separate from both.

[6] Since the "takes place" predicate doesn't locate processes in space, processes are not "concrete particulars" in the full sense of being "filled space-time regions or the discrete particular fillings of space-time regions" (Seibt 2001, §2.1).

[7] Note that "interaction" is not introduced as a formal predicate, but rather as an informal characterisation of predicates which are introduced to describe kinds of process.

[8] Ladyman is wrong to suggest that, for a gas heated in a sealed container, "the temperature increase caused the pressure rise and not the other way around" (2002, p. 205). The causal process involved the heat source and the gas, which when constrained to constant volume increases both in pressure and temperature. If instead it was constrained to constant temperature (by diathermal walls) and pressure (by a movable piston under constant external force such as a weight), it would increase in volume. As in the case of the relation of the period of a pendulum and its length, laws of covariation like the Charles-Boyle gas law cannot serve as the basis of symmetry counterexamples to the DN-model of explanation by appeal to the cause unless the the underlying causal process (gravitation, heating, etc.) is properly identified.

[9] In Russell's famous formula, a change is a circumstance expressed by a *conjunction* of two *sentences* and not a referring term: "Change is the difference, in respect of truth or falsehood, between a proposition concerning an entity and a time T, and a proposition concerning the same entity and another time T', provided that the two propositions differ only by the fact that T occurs in the one where T' occurs in the other" (Russell 1937, p. 469).

[10] The chemical potential of a substance expresses how the energy would vary with the variation in the amount of the substance from one point in the medium to another, and has different values at different points in a medium in which the concentration of the substance is uneven.

[11] I.e., t is a part of t' and there is an interval abutting both t and t' on the earlier end.

[12] If processes with temporal gaps are allowed, either the restriction on times must be dropped or the "takes place" relation adjusted, say to a multigrade relation read "α takes place at t, and then at t', and then at …".

[13] It is only for elementary reactions that the rates of reaction are proportional to the concentrations of the reactants and products raised to the power of their stoichiometric coefficients.

Cartwright, Helen Moris: 1970. 'Quantities', *Philosophical Review* **79**, 25–42.

Cartwright, Helen Moris: 1975. 'Amounts and Measures of Amounts', *Noûs* **9**, 143–164.

Denbigh, Kenneth: 1981. *The Principles of Chemical Equilibrium*, 4th. edn. Cambridge: Cambridge University Press.

Dretske, Fred: 1967. 'Can Events Move?', *Mind* **76**, 479-492.

Duhem, Pierre: 1892. 'Notation atomique et hypothèses atomistiques', *Revue des questions scientifiques* **31**, 391–457. Translated by Paul Needham as 'Atomic Notation and Atomistic Hypotheses', *Foundations of Chemistry* **2** (2000), 127–180.

Duhem, Pierre: 1895. 'Les Théories de la Chaleur: I. Les Précurseurs de la Thermodynamique', *Revue des deux mondes* **129**, 869–901. Translated in *Mixture and Chemical Combination, and Related Essays*, trans. and ed. by Paul Needham, Kluwer, Dordrecht, 2002; pp. 121-147.

Johnson, W.E.: 1921. *Logic*, Vol. I. Cambridge: Cambridge University Press.

Kitcher, Philip: 1984. '1953 and All That. A Tale of Two Sciences', *Philosophical Review* **93**, 335-373.

Ladyman, James: 2002. *Understanding Philosophy of Science*. London: Routledge.

Leonard, H. and Nelson Goodman: 1940. 'The Calculus of Individuals and Its Uses', *Journal of Symbolic Logic* **5**, 45–55.

Nagel, Ernest: 1961. *The Structure of Science*. London: Routledge and Kegan Paul.

Needham, Paul: 1981. 'Temporal Intervals and Temporal Order', *Logique et Analyse* **93**, 49–64.

Needham, Paul: 1999. 'Macroscopic Processes', *Philosophy of Science* **66**, 310–331.

Needham, Paul: 2000. 'Hot Stuff', in Jan Faye, Uwe Scheffler and Max Urchs (eds.), *Facts, Things, Events*, Poznan Studies in the Philosophy of the Sciences and the Humanities, Vol. 76. Amsterdam: Rodopi, pp. 421–46.

Needham, Paul: 2003. 'Chemical Substances and Intensive Properties', *Annals of the New York Academy of Sciences* **988**, 99–113.

Roeper, Peter: 1983. 'Semantics for Mass Terms With Quantifiers', *Noûs* **17**, 251–265.

Russell, Bertrand: 1937. *The Principles of Mathematics*, 2nd. edn. London: George Allen and Unwin.

Scerri, Eric and Lee McIntyre: 1997. 'The Case for the Philosophy of Chemistry', *Synthese* **111**, 213–232.

Seibt, Johanna: 2000. 'The Dynamic Constitution of Things', in Jan Faye, Uwe Scheffler and Max Urchs (eds.), *Facts, Things, Events*, Poznan Studies in the Philosophy of the Sciences and the Humanities, Vol. 76. Amsterdam: Rodopi, pp. 241–278.

Seibt, Johanna: 2001. 'Formal Process Ontology', in C. Welty and B. Smith (eds.), *Formal Ontology in Information Systems: Collected Papers From the Second International Conference*. Ogunquit: ACM Press, pp. 333–345.

Simons, Peter: 2000. 'Continuants and Occurrents', *Aristotelian Society, Supplementary Volume* **74**, 59–75.

Tarski, Alfred: 1983. 'Foundations of the Geometry of Solids', trans. of original 1926 article in *Logic, Semantics and Metamathematics*. Indianapolis: Hackett Publishing company.

van Brakel, Jaap: 2000. *Philosophy of Chemistry: Between the Manifest and the Scientific Image*. Leuven: Leuven University Press.

Vemulapalli, G.K. and Henry Byerly: 1999. 'Remnants of Reductionism', *Foundations of Chemistry* **1**, 17–41.

WIM CHRISTIAENS

THE EPR-EXPERIMENT AND FREE PROCESS THEORY

ABSTRACT. As part of the 'creation-discovery' interpretation of quantum mechanics Diederik Aerts presented a setting with macroscopical coincidence experiments designed to exhibit significant conceptual analogies between portions of stuff and quantum compound entities in a singlet state in Einstein–Podolsky–Rosen/Bell-experiments (EPR-experiments). One important claim of the creation-discovery view is that the singlet state describes an entity that does not have a definite position in space and thus 'does not exist in space'. 'Free Process Theory' is a recent proposal by Johanna Seibt of an integrated ontology, i.e., of an ontology suitable for the interpretation of theories of the macrophysical and microphysical domain (quantum field theory). The framework of free process theory allows us to show systematically the relevant analogies and disanalogies between Aerts' experiment and EPR-experiments. From free process ontology it also follows quite naturally that the quantum compound entity described by the singlet state 'does not exist in space.'

1. INTRODUCTION

Louis de Broglie reports Einstein as saying: "It should be possible, aside from all calculations, to illustrate physical theories, with images of such simplicity, that a child should be able to understand them".[1] One can distinguish two kinds of such simple examples of a theory T: ones that bring out what is distinctly different about T, and examples that make T more familiar and transparent. In the case of quantum theories, the first kind of illustrative purpose is easily fulfilled, while the second is not. It is not that hard to find illustrations that bring out what is distinctly different about quantum theories.[2] In contrast, it is notoriously difficult to come up with examples that would bring quantum theory close to our common understanding of physical reality. It is well known that quantum mechanics (QM) violates Bell-inequalities and that the experiments support the quantum-mechanical predictions. If a theory confirms a Bell-inequality it is local, otherwise the theory is non-local. There have been a number of experiments confirming the quantum mechanical predictions, the most famous one being the experiments by Alain Aspect and his collaborators in Paris in 1982.[3] In such experiments ensembles of pairs of quantum entities are considered. These compound entities are prepared in a uniform way

J. Seibt (ed.), Process theories: Crossdisciplinary studies in dynamic categories, 267–283.

(they are produced by a source in a singlet state), which leads us to expect statistical correlations between the measurements results performed on each of the entities in a pair. But the measurements are performed in apparatuses that are widely separated in space. (Henceforth we will refer to these experiments as EPR-experiments.) In this paper we will proceed from the commonly accepted view that the cause of the violation of Bell-inequalities is the violation of *outcome independence*.[4] Usually we tend to believe that when there is a correlation between kinds of event, if the events have a common cause, the common cause makes the events statistically independent (they are screened off from each other by the common cause). Outcome independence identifies the uniformly prepared states coming from the source as the common (stochastic) causes of the measurement results performed on each of the components in of one of the compound entities (Redhead (1987, p. 102)). A violation would mean, first, that the state coming form the source is not the common cause of the measurement results in both wings, secondly, that there is a direct stochastic causal link between the two particles. But then we are not taking into account the following fact: "Definite correlations are there if I do not poke or disturb the system in any way, but they are too delicate ...to reflect a causal connection" (Redhead (1989, p. 440)). So basically the violation of outcome independence only tells us that probability distributions of properties possessed by the one particle would depend essentially on the property possessed by the other particle. QM is non-local in the sense of violating outcome independence. In the eighties Diederik Aerts presented a simple macroscopical thought experiment with coincidence measurements on communicating vessels of water, which he claims to be strongly analogous ('identical') to EPR-experiments (Aerts 1981; Aerts 1982; Aerts 1983b).[5] It is clear that Aerts' experiment (hereafter: the Aerts-experiment) belongs to the second kind of illustration: it describes a macrophysical scenario with familiar entities that aims to give us a simple image of what is actually happening in an EPR-experiment. The Aerts-experiment is part of the creation-discovery view on quantum mechanics, developed by Aerts and his co-workers. According to the creation-discovery view on quantum mechanics, the entity described by the singlet state in EPR-experiments acts as the common cause of the measurement results in both wings, however without making the measurement results statistically independent. Part and parcel of the creation-discovery view is the following assumption:[6]

CD: The entity described by the singlet state has no definite position in space, i.e. it 'does not exist in space', which explains why it is able to act in two spatially separated regions of space and why it does not make its separate effects statistically independent.

Our focus here will be on what might be considered the crucial *lacuna* in Aerts' attempt to provide a macrophysical illustration of the EPR-paradox. Aerts does not state this explicitly but he operates on the assumption that water and quantum entities are sufficiently analogous in ontological respects. The main aim of this paper is to show that this assumption is justified and this leads to a confirmation of CD.

As philosophers of science like to point out, theories are underdetermined by empirical data, and as QM has taught us more than any other physical theory, theories underdetermine their interpretation (an interpretation is here understood as a description of what the theory says there is in the world). Let us very briefly look at this last aspect. There are physicists who have adopted the language of philosophy here, speaking of 'ontologies' for QM (see Herbert (1985) for example). Ontology is traditionally understood as the (or a) theory about what there is. The ontology of a theory T makes explicit and systematizes what there is according to T. But is this not what T already does? In other words, should we not be able to 'read off' the ontology of T from T? Unfortunately that is not always possible, for at least two reasons. First, in order to determine the ontological commitments of a theory T, there should not exist competing formulations of T. Second, within the representative formulation of T it should be clear which parts are to be considered pure 'machinery' or, to use Redhead's term, "theoretical surplus structure." The ontological interpretation of QM confronts both of these obstacles. That QM lacks conceptual transparency is not due solely to the fact that the quantum world is not directly observable and thus cannot be exemplified with an intuitive mental picture close to common sense experience (a characteristic not unique for QM); despite its impressive experimental confirmation it is not clear which of the competing mathematical formalisms should be chosen for interpretation and moreover which parts of the chosen mathematical formalism are ontologically significant. In recent years particular efforts have been spent in interdisciplinary discussion to make some headway on the conceptual clarification and ontological interpretation of theories of the quantum domain.[7] There are a number of descriptive frameworks that are currently discussed as underlying ontologies of quantum theories, based on events, tropes, Tellerian 'quanta,' Whiteheadian occasions, or structures (James Ladyman and Steven French). The theory of free processes is one of the candidate ontologies for the quantum domain; but unlike others its primary application and heuristic origin is the interpretation of macrophysical stuffs and subjectless activities. Thus it suits the purposes of the Aerts-experiments particularly well.[8]

We will show two things: (A) where and how the Aerts-experiment clarifies the EPR-experiments (i.e. specify the relevant analogies and disanalogies) using free process ontology; (B) free process ontology supports CD. In a sense free process ontology completes Aerts' attempts to understand the quantum world on the model of entities familiar from everyday experience. Water and entangled quantum states are ontologically similar and can be considered to be specifications of the same ontological category. With the theoretical embedding in free process theory the Aerts-experiment thus becomes more than just a nice illustrative *metaphor*. Rather, in combination with free process theory the Aerts-experiment can be taken to provide a *literal* illustration: a macrophysical case of 'quantum behavior'. Since free process theory aspires to be an integrated ontology accommodating macrophysical and microphysical entities as cases of the same ontological category, the Aerts-experiments supports, vice versa, free process theory with an additional argument. (Seibt so far only has explored to what extent free process theory could be used for the interpretation of *free* quantum fields, cf. Seibt 2002 and forthcoming.)

2. FREE PROCESS ONTOLOGY

In this section some basic notions of free process ontology are looked at, which will then be applied to the analogy between the Aerts-experiment and EPR-experiments in the next section.

2.1. *Free processes*

Suppose we start from the idea that the world consists of individuals. Although we have a tendency to understand this statement as saying that the world consists of countable things (like billiard balls and stars) or countable events (like collisions and supernovae), we can and should resist this tendency. As Seibt has argued, the features 'individuality,' 'countability,' and 'particularity' are "not an ontological package deal." There are concrete individuals that are neither countable nor particular, i.e., uniquely locatable. While Seibt introduces the notion of a free process via a criticism of substance ontology, for present purposes it suffices to note the differences between the following properties of the axiomatic version of free process theory, the mereological system FPT (cf. Seibt 2002, p. 25):

FPT(1): x can be *continuously amassed* iff x is of kind K and for any y of kind K, the mereological sum of x and y is again of kind K.

FPT(2): x can be *discretely amassed* or *measured* iff x is of kind K and x is part of S which is coextensive with a sum of non-overlapping P_i of kind K, and for any y and z part of S, the sum of y and z is also part of S.

FPT(3): x can be *aggregated* iff x can be discretely amassed and for any sum S_i of entities of kind K there is exactly one S_k coextensive with S_i and the former is a sum of non-overlapping K-'atoms' (i.e. parts of kind K which do not have parts of kind K).

FPT(4): x can be '*cardinally counted*' iff x is aggregable and there is a function from the K-atoms in x into the natural numbers.

FPT(5): x can be '*ordinally counted*' iff x can be cardinally counted and any two K-atoms in x are distinct from each other.

Let us get clearer on the contrast between FPT(1)–FPT(2) and FPT(5) by looking at some simple examples. Take the referent of a mass term like 'water'. It can be continuously amassed: if one adds water to water, one obtains water. As such it cannot be counted, i.e. it is a non-countable entity. Usually spatio-temporality serves as ultimate 'principium individuationis'; for indistinguishable things for example. In contradistinction to things, a stuff or mass is the same individual even when it appears in two different locations: for example we can say this bottled water is the same source water they sell in the States. In free process ontology a necessary condition for individuation is having descriptive properties *and* being *reidentifiable* on the basis of these descriptive properties. The referent of 'water' does not imply or contain any unique spatio-temporal location. Yet it is arguably (a) an individual and (b) a concrete individual, since (a) we can refer to it, reidentify it, and predicate of it, and (b) ascribe to it concrete properties: 'water is wet', 'water was added to water.' Note the difference between 'water' and an expression like 'the water in the glass on the table', 'the swirling water in the river', 'a mole of H_2O' – the referents of the latter expressions are *amounts* of water, which *can* be counted, as we will see presently. A stuff is not countable because is it not individuated via the occupation of a determinate spatial (or spatiotemporal) region. Stuffs are thus individuals that may be multiply-occurrent in space because they do not have as such, before amounts are measured out, a determinate spatio-temporal location. The generality and indeterminateness of pure free processes is intrinsic to the individuals and in some instances irreducible. Seibt has argued that activities, in particular subjectless activities (e.g., the referent of 'it is snowing'), share the spatial indeterminateness and functional individuatedness of stuffs. Based on the symmetric features of stuffs and activities, she introduces a unified category for concrete, general or indeterminate, individuals. Such 'four-dimensional stuffs' are called 'free processes.' Continuous amassable entities like stuffs and activities contrast with things and events which are always at least cardinally countable and have unique and determinate spatio-temporal locations (FPT(5)).[9]

The property FPT(1), so fundamental for free processes, is closely related to the Property of *homomerity* or *like-partedness*. Both can serve to (a) distinguish free processes from countables and (b) define countables (e.g., things and classical particles) as limiting cases of non-countables (Seibt 2002:32).

FPT(1-a): x is normally homomerous iff x is of kind K and has $n \leq 4$ dimensions and there is at least one m-dimensional part of x, $m \leq n$, which is again of kind K.
FPT(1-b): x is maximally homomerous iff x is of kind K and has $n \leq 4$ dimensions and for all m-dimensional parts of x, $m \leq n$, such that, it is again of kind K.
FPT(1-c): x is minimally homomerous iff x is of kind K and has $n \leq 4$ dimensions and there is no m-dimensional part of x, such that $m \leq n$, and the part is again of kind K.

A part of water is again water, but a part of the billiard ball is not again a billiard ball. The 'anhomomerity' of the billiard ball can be taken as a case of 'minimal homomerity.' We can distinguish two kinds of countables: things or particles and events or structured developments. An event is for example the shower we were in just yet: it has a beginning and an end, a specific development and it happened, i.e., it is minimally homomerous in time. All countables are simply minimally homomerous free processes and so the ontological dichotomy between non-countables (e.g., stuffs and activities) and countables (e.g., things and events) is avoided. Normally homomerous entities can involve, or be constituted by, or even spatiotemporally overlap with minimally homomerous entities. Water is again exemplary: it is normally homomerous, but spatiotemporally overlaps with a set of H_2O-molecules each of which is minimally homomerous. Stuffs like water, wood, etc. not only are continuously amassable, they also can be measured: they have both the property FPT (1) and the property FPT(2). There are examples of stuffs or activities with merely property FPT(1), i.e., merely continuously amassable but not measurable stuffs and activities, such as anger, friendship, enjoyment, or the weather.

When we say 'the billiard ball is black', 'the billiard has momentum **p**' ... then the individual 'billiard ball' carries the properties 'black', 'momentum **p**' ... Not so in free process ontology: a free process is unsupported by a particular entity. The entity *is* the process. This is sometimes evident in the way we talk. We say: 'It is raining', not: 'x is raining'. The basic entities of the world – so runs the basic tenet of free process ontology – are such 'subjectless' or 'carrier-less' four-dimensional stuff-activities, which are more or less homomerous, and more or less determinate with respect to their descriptive properties. Classical particles, assumed to be (a) fully determinate with respect to their descriptive properties including location, and (b) anhomomerous, are considered to be the limiting case of the homomerity and determinateness of a free process.

2.2. *Amounts of free processes and quantities of free processes*

With respect to a typical stuff like water we can look at the referents of 'the water in the glass on the table', 'the swirling water in the river' etc. instead of the referent of 'water'; and with respect to a typical activity like raining, we can look at the referents of 'the rain shower we were just in', 'the storm above Ghent', etc., instead of the denotation of 'raining'. 'The water in the glass on the table', 'the swirling water in the river', 'the shower we were just in', 'the storm above Ghent', etc. are all expressions which refer to amounts of free processes. An *amount* of x is situated in a determinate and unique location in space-time and is spatio-temporally bounded, i.e. it is ordinably countable in the sense of FPT(3) so that one can say *this* amount of water as distinct from *that* amount. In free process theory amounts are distinguished from *quantities*. Amounts have a determinate location, quantities do not – they are just specified with respect to measure (volume, temporal extent etc.) but still multiply occurrent. While *this cup of coffee* refers to an amount of coffee, the expression *a cup of coffee* refers to a quantity; *today's hour of music* talks about an amount of music, *an hour of music* is about a quantity of music; etc. Unlike amounts, quantities are only cardinally countable in the sense of FPT(4) – since they do not have a determinate location we cannot say 'which is which.' An amount of a free process x can be written as a collection of measurable quantities plus a location. For example, consider a certain piece of wood, a stuff with normal homomerity. This piece of wood, which is an amount of the free process wood, denoted [wood], can be understood as a collection of measurable properties:[10]

> *Amount*: [wood] = (location l, volume a, weight b, ..., inflammable (yes/no), buoyant (yes/no))

In the spatiotemporal region occupied by this piece of wood there is a certain quantity of wood:

> *Quantity*: {wood} = (volume a, weight b, ..., inflammable (yes/no), buoyant (yes/no))

We can distinguish two kinds of properties: (a) occurrent properties, i.e., those that state a determinate value of a certain measure for measuring dimensions such as volume, weight, time etc.; (b) mutually dependent dispositional properties that are, first, *qua* dispositional properties, indeterminate with respect to their value and, second, depend on what other measurements have been performed on at least one other constituent quantity of the amount in question. For example the *inflammability* and *buoyancy* of a certain piece of wood are related in this way. If one sets the piece of wood on fire, one has destroyed that amount of wood, thereby making it impossible to render the property of buoyancy determinate. If one lets the piece of wood float on water then it becomes humid and does not burn any more. Both on a theoretical level and in the practice of the laboratory one supposes that for the classical world, all these kinds of incompatibility and all indeterminateness can be alleviated by giving a more fine-grained description or describing the phenomenon at a deeper level. Notoriously, this is quite different in the quantum domain.

One peculiarity of free process theory consists in the fact that *any* specification of a free process identifies another free process (mereologically related to the first). *Water* refers to one free process, *hot water* refers to another. Similarly for specifications of measurable units or locations the expressions *a cup of water* and *the water in my cup here* are about two different free processes which are each different from the free process referred to by *water*. In general, an amount of x or a quantity of x are themselves free processes different from (yet mereologically related to) x.

To conclude these cursory remarks about FPT and to move to our task, consider what I think would be the FPT-description of an electron. One electron is an amount of a minimally homomerous free process, 'electroning' as one might allow oneself to say for didactic purposes. Let us take the case of a spin 1/2-entity. We leave out QFT-considerations and stay within QM.

$$[\text{electron}] = (\text{location l, negative charge, } \pm\tfrac{1}{2}(\mathbf{S.n}), \pm\tfrac{1}{2}(\mathbf{S.n'}), \ldots)$$

The numbers **S.n, S.n'**, ... are spin-components along directions **n, n'** ..., each having two-fold degeneracy which gives rise to a two-dimensional orthonormal base in two-dimensional Hilbert space. With $\frac{1}{2}(\mathbf{S.n})$ we mean spin $\frac{1}{2}$ along **n**, $\frac{1}{2}(\mathbf{S.n'})$ means spin $\frac{1}{2}$ along **n'**, $-\frac{1}{2}(\mathbf{S.n})$ means spin $-\frac{1}{2}$ along **n**, etc. (It is well known that for quantum entities one cannot describe the spin state by giving the direction of angular momentum, but only by giving the component of angular momentum along some direction.) Some measurable quantities are already determinate by taking the amount of the free process, for example the charge. Others, like spin, are not, because not all spin observables can be determinate at the same time.

3. AERTS' THOUGHT-EXPERIMENT AND THE ONTOLOGY OF QUANTUM MECHANICS

Let us begin with a reminder about some essential components of the EPR-experiment. First, two spin 1/2-entities are produced by the decay of a single spin 0-entity S at a central point (called the source); second, spatially well-separated coincidence measurements of spin are performed, when performed along the same direction in space on S1 and S2 giving the results $s_m = -1/2, 1/2, \ldots$ for S1, in perfect anti-correlation with the results $t_n = \frac{1}{2}, -1/2, \ldots$ for S2; third, before the measurements the entity S can only be described by an "entangled" state k (a non-factorizable state in the tensor product of the state spaces of S1 and S2).

The EPR-experiment is mysterious since the correlations of simultaneous and spatially well-separated measurements calls for an explanation. Is there a direct (stochastic) causal link between the two wings of the experiment, is it the state coming from the source that causes the measurement results, or is there some other ingenious explanation? We already touched on this in the introduction. However, quite independently of the question of how to explain the correlations violating outcome independence, one might pursue the modest task of trying to understand the type of entity the experiment talks about. Even though in descriptions of the EPR-setting we speak of two spin 1/2-entities, the compound entity is in a non-factorizable state. The experiment highlights a general feature of quantum entities that one can formulate as follows: if two distinguishable entities S1 and S2 are joined into a whole S, then S can very easily exhibit states that do not allow to distinguish S1 and S2 within S.[11]In these cases one could say: when S exists, S1 and S2 do not exist anymore. Classical particles do not exhibit this feature: when two parts are joined into a whole, they can be distinguished within the whole also afterwards, or vice versa, the parts of a whole can be distinguished even before the whole is *de facto* divided into its parts.

On the other hand, classical stuffs behave in this respect rather similar to quantum entities. Suppose volume is the only measurable property. When we join two amounts of water each containing a quantity of 10 liters of water, the combined amount – call it W – contains a quantity of 20 liters. Even though the measurable property determining the quantity of water contained in W is determinate in this case, the amount has indeterminate parts. Let us assume that the quantity of water contained in W is obtained by adding two (uniquely located) amounts of water, W1 and W2, containing each a quantity of 10 liters. If at a later time one extracts 10 liters from W the question: "Are these the first or the second 10 liters that were combined to make up W?" makes no sense. There are two (uniquely located) amounts W3 and W4 containing each a quantity of 10 liters. As long as W1 and W2 existed, W did not exist, and as long as W exists, W3 and W4 do not exist. Within W there exist two quantities of 10 liters which cannot be distinguished. They become distinguishable only by 'dishing out' twice from W a quantity of 10 liters, i.e., by creating two amounts combining a quantity of 10 liters with two different locations.

The idea of the Aerts-experiment should by now appear quite straightforward: exploit the analogy between on the one hand the confluence and separation of amounts of water and on the other hand the entanglement and separation of quantum entities, in order to construct a macroscopic experiment with water exhibiting outcome dependence comparable to the outcome dependence in EPR-experiments. There are two vessels, V1 on the left and V2 on the right, connected to each other by a tube Tu. We will refer to V1, V2, Tu together as V. The tube is attached to the bottom of V1 and V2. If it stays hidden, then we will not be aware that what appears to be two amounts of water can be considered as one connected amount of water.[12] We fill V1 with 10 liters of water and we fill V2 with 10 liters of water. The state k of the compound entity is 20 liters. For the experimentalist who fills V1 with 10 liters of water and V2 with 10 liters of water, it seems as if V1 contains 10 liters of water and V2 contains 10 liters of water. At V1 there is a vessel R1 and a siphon to extract water from V1 and at V2 there is a vessel R2 and a siphon to extract water from V2. Suppose we perform the following coincidence-measurement: we extract once more than 10 liters from V1 and once more than 10 liters from V2. If we extract more than 10 liters from V1 we state the outcome as $s_m = 1/2$ otherwise we state the outcome as $s_m = -1/2$; if we extract more than 10 liters from V2 we state as outcome $t_n = 1/2$, otherwise we state as outcome $t_n = -1/2$. If carried out the experiment will give the following measurement results stated in liters of water measured at R1 and R2: (12,8), (7,13), ... In terms of the special convention for stating these measurement results the combined outcome will be: (1/2, −1/2), (−1/2, 1/2), ... We will never see the values (1/2, 1/2). The single outcomes for R1 and R2 depend on each other.[13]

Without the specifications of free process ontology all of this stays more or less on the level of an intuitive analogy illustrating the possibility of a common cause that has effects which it does not make statistically independent (i.e. which it does not screen of). With free process ontology, however, the analogy becomes more precise and more profound. Let us re-describe the experiment in terms of the FPT-notions introduced above and try to make good on the two goals mentioned in section 1, namely, (A) to show where and how the Aerts-experiment clarifies the EPR-experiments (i.e. specify the relevant analogies and disanalogies) using free process ontology; (B) to show how free process ontology supports CD.

(A) The water in the Aerts-experiment is a particular amount of water and can consequently be written as a collection of measurable quantities (weight, volume, ...) with a location. We are only interested in the quantity of volume contained in the amount W:

$$\{W\} = (20 \text{ liters})$$

Compare this with the FPT-representation of the singlet state in the EPR-experiment. We will write spin along a particular direction as before. For the singlet state we have zero-amount of spin for all directions, so the quantity of spin contained in the amount E of 'electroning' that is the compound of two electrons is this:

$$\{E\} =(0\ (\mathbf{S.n}), 0\ (\mathbf{S.n'}), \dots)$$

The individual E has zero spin for all directions *and* these measurable quantities are indeterminate, since they are mutually incompatible.

There are three disanalogies between the entities involved in the two experiments, which we need to identify:

(a) Aerts' convention hides the fact that there is an infinite number of ways we can divide the water W in two quantities of water:

$$\{W\} = (5 \text{ liters}, 15 \text{ liters}) = (4 \text{ liters}, 16 \text{ liters}) = (12 \text{ liters}, 8 \text{ liters}) = .$$

In contrast there are only two possible ways of dividing 0(**S.n**): in 1/2(**S.n**) − 1/2(**S.n**) and −1/2(**S.n**) + 1/2(**S.n**). That is:

$$\{E\} = (0(\mathbf{S.n}), 0(\mathbf{S.n'}), \dots) = (1/2(\mathbf{S.n}) - 1/2(\mathbf{S.n}), 0(\mathbf{S.n'}), \dots)$$
$$=(-1/2(\mathbf{S.n}) + 1/2(\mathbf{S.n}), 0(\mathbf{S.n'}), \dots) = \dots$$

To restate, while in the Aerts experiment we have an infinite number of ways of dividing 20 liters, there are only two ways of dividing the zero amount of spin.

(b) There is only a finite number of measurable quantities in our idealized example of water W: volume, weight, ... which are all compatible and thus determinate. In contrast, there are an infinite number of measurable spin-quantities: **S.n, S.n'**, ...

(c) The disanalogies mentioned under (a) and (b) do not strike me as significant. A more problematic difference might be, however, that in the Aerts-experiment there is nothing that corresponds to the incompatibility of spin-properties in the EPR-experiment. But it is certainly admissible to add something to the Aerts-experiment that establishes an incompatibility of properties in the macroscopic setting. What about if we were to exchange the occurrent property 'has volume of value v' for the dispositional (operational) property 'yields value v when measured for volume', which we denote as A(yes/no), and then added another dispositional property, for example 'vaporizes when heated up'. We would then have:

$$\{W\} = (\text{A(yes/no), vaporizes(yes/no), } \ldots)$$

The amount of water can be indeterminate, because of the incompatibility with the other dispositional property.

(B) We promised to show that CD(2) follows from free process ontology. First, note that in a way the stuff water is a stranger entity than a compound quantum entity described with $\otimes_i H_i$. The reason is that the Hilbert-spaces are labeled $i = 1, 2 \ldots$, which suggest that, despite the non-factorizable states, it is in principle still possible to actually count the entities in the entangled state, while water has no parts that can be counted in the way we presented it above (water has the property FPT(2)).[14] In QM we are considering cardinally countable minimally homomerous entities, i.e. entities with property FPT(4). Such entities are in the same class as *quantities* of water (not amounts of water, which have the property of ordinal countability FPT(5)). The existence of non-occurrent states in Bose–Einstein statistics is a strong argument in favour of the Fock-space formalism of QFT (Teller (1995)). QFT, which is probably the more fundamental theory, makes ordinal countability of quanta impossible. So both QM and QFT suggest that quantum entities are *quantities* of free processes in the sense of FPT. To be sure, for any *amount* of a classical stuff like water, there is always the physical *possibility* (but see endnote 13) of complete particularization at the level of molecules, i.e., to consider the stuff or quantity as an assembly of ordinally countable, uniquely located (particular) entities. But recall that this does not hold for *quantities* of classical stuffs. A quantity of stuff (10 liters of water) does not, as such, have a determinate unique spatiotemporal location. It can be *assigned* a location, thereby turning the quantity into an amount containing a quantity. Such an assignment does not, however, consist in 'determining which quantity goes where' or 'which amount has which quantity.' Both of these sentences make no sense since quantities, being only cardinally countable and not uniquely located in spacetime, are not traceable in spacetime. When dividing the 20 liters of water into two vessels (i.e., two amounts: uniquely located or particular, ordinally countable entities) we cannot meaningfully ask where the first quantity of 10 liters went. So even while it is possible to determine *how many* quantities a certain amount of water (e.g., the left vessel) contains (e.g., two quantities of 5 liters), it is not possible to determine *which* quantities of 5 liters it contains, because quantities are not individuated in this sense of 'which'. This is strongly parallel to the situation in QFT. In QFT it is a measurement interaction that 'induces' the superposition state to be projected on one of the eigenvectors, and even then we cannot say which quantum has which eigenvalue, but only how many.

We already drew attention to the difference between concrete entities that are uniquely located in spacetime, i.e., have determinate spacetime locations (= particular, = ordinally countable entities, = amounts) and those concrete that are in an indeterminate sense located in spacetime, that are simply somewhere in spacetime (= cardinally countable, = non-particular, = quantities). Due to the conceptual separation between particular individuals, which are uniquely and determinately located, and non-particular individuals, which are indeterminately located (just *somewhere*), free process theory is capable of accommodating quantum entities into an integrated ontology.

Remember that CD is the claim that the entity described by the singlet state has no definite position in space, which is also formulated as the claim that it 'does not exist in space'. If we interpret this along the lines of the previous paragraph, we have carried out what we promised in the introduction. But CD can also be taken as a stronger claim: the EPR-entity is not located *anywhere in space*, not just located indeterminately. Could the process-ontological interpretation of EPR-phenomena also be put to use in order to articulate such a strong reading of CD? This is certainly possible. In this case we need to construe the zero quantity of spin in the usual QM-description as *no* quantity of spin, i.e. the compound entity in the singlet state has no amount (because of the possible indeterminateness of the position variable) and no quantity of angular properties. Given that in FPT it holds that any free process that is spatially located is either an amount (if determinately located) or a quantity (if indeterminately located), free processes that are neither amounts nor quantities are not spatially located. If we take the compound quantum entity E in a non-separable state of the EPR-experiment to be a free process (rather than a quantity of amount of a free process) we characterize it in effect – just as the strong reading of CD says – as a real individual (we can refer to E and reidentify it for further reference) which does not exist in space. Measurements produce amounts of such EPR-processes: individuals that are uniquely and determinately located in space.

ACKNOWLEDGMENTS

The author is post-doctoral researcher at Flanders Fund for Scientific Research. He would like to thank Bob Coecke, Maarten Van Dijk, Jean-Paul Van Bendegem and especially Johanna Seibt for helpful comments.

NOTES

[1] My translation of a remark by Einstein in a personal conversation with Louis De Broglie reported in De Broglie (1956, p. 236): «Il me dit encore qu'il avait peu de confiance dans l'interpretation indéterministe et qu'il blâmait l'orientation trop formelle que commençait a prendre la Physique quantique; forgant peut-être quelque peu sa pensée, il me disait que toute théorie physique devrait pouvoir, en dehors de tout calcul, être illustrée par des images si simples «qu'un enfant même devrait pouvoir les comprendre.»»

[2] Compare for example the ingenious Mermin-contraption (for example) in Mermin (1985).

[3] See chapter four of Redhead (1987).

[4] The name outcome independence was coined by Abner Shimony. The inventor named the condition *completeness*. See Redhead (1987) for references and the precise mathematical formulation.

[5] Both philosophers and physicists have done the same thing. See Teller's analogy between quantum entities and amounts of money in the bank in Teller (1995, pp. 29–30) and the references in the footnote on p. 29 of Teller (1995). Note that the money analogy is actually due to Schrödinger (as Teller acknowledges).

[6] This specific way of presenting the creation-discovery view was given in more detail in Christiaens (2002). I do not elaborate here on other elements of the creation-discovery view, such as the idea of hidden measurements and hidden correlations. The Aerts-experiment was proposed before the hidden-measurements view on quantum measurement and can therefore be discussed independently of the hidden measurement/hidden-variables component of the creation-discovery view.

[7] See for instance Cao 1999 and Kuhlmann et al. 2002.

[8] FPT is the formal scheme in which Seibt's Free Process Theory is couched. Core ideas of Free Process Theory and sketches of FPT are to be found in Seibt (1990), Seibt (1995), Seibt (1996a), Seibt (1996b), Seibt (2000), Seibt (2001a), Seibt (2002). I am drawing here in particular on Seibt (2001a) and Seibt 2002.

[9] The reach of the concept of free process goes beyond the particularities of common parlance. QFT may provide one of the scientific theories to which free process ontology can be applied to. The state of a simple physical system typical for QFT – a free quantum field – is described by a state vector in a Fock space. A typical state is superposition of ket vectors

$|n_1, n_2, \ldots, n_i, \ldots>$ spanning the Fock space. The number n_i is the occupation number, telling you how many quanta posses the eigenvalue q_i. There are two remarkable facts here. The first is that the ket vector allows only to say that n_i quanta have eigenvalue q_i, but not which ones. Secondly, a linear combination of these ket vectors makes it impossible to tell how many quanta have a particular eigenvalue; we only know that if we would measure and we would obtain a particular ket-vector, then we would have the occupation numbers specified by that ket-vector. But superposition is ubiquitous and frequent in the quantum world, which means that a lot of the time FPT(4) in not an option in the quantum world. QFT is a fundamental theory of matter, arguably the fundamental theory of matter. In other words, there is no 'classical' reality underlying these facts. For an accessible exposition of quantum field theory, see Teller (1995).

[10] I simplify here Seibt's presentation of amounts, applying FPT to Aerts' analysis of 'a piece of wood' in Aerts (1981), Aerts (1983a) and Aerts (1983b).

[11] The state space of a compound quantum entity is the tensor product of the state spaces of the compounds. For a specific observable of the compound entity there is an orthonormal collection of eigenvectors in the state space of the compound entity. These states are factorizable. Note that in his 1981 and 1982 Aerts proved with quantum logical tools that QM cannot describe separated entities.

[12] But note that even if we see the tube we could still say that there are two amounts, the one on the right hand side up to the middle of the tube, and the one on the left hand side, up to the middle of the tube. Suppose one would calibrate the sides of the vessels V1 and V2, then it is easy to locate the amounts in which the 20 l are separated. If you get (5 l, 15 l), then these two amounts were already there before the measurements.

[13] The Aerts-experiment is 1) a nice illustration of a common cause that does not screen off its effects – i.e. the measurements results obtained in both "wings" of the Aerts-experiment – because they are carried out on an entity that is one whole, but appears to be two separate entities; 2) an illustration of a common cause, which is not easily recognized as such, since the effects appear to be unconnectable as long as the two measurements are taken to happen on two spatially separate entities. When we identify the common cause, we see that it does not make its separate effects statistically independent, which usually leads to k being discarded as possible common cause. In both experiments we have a violation of factorization: $\Pr(s_m \;\&\; t_n) \neq \Pr(s_m).\Pr(t_n)$. In both experiments it holds that the probabilities for the measurement results are also not rendered probabilistically independent if the common cause k is taken into account: $\Pr(s_m \;\&\; t_n) \neq \Pr(s_m/k).\Pr(t_n/k)$.

[14] Unless one would immediately turn to the molecular level in the case of water. This is not obvious, since it involves physicalist reductionism (the idea that water can be completely described in the vocabulary of physics, i.e., that all properties of water can be reduced to the most fundamental physical description). See the section on reductionism in Boyd et al. 1991.

REFERENCES

Aerts, D.: 1981, *The One and the Many. Towards the Unification of the Quantum and the Classical Description of One and Many Physical Entities*, Brussels: Free University of Brussels (VUB) doctoral dissertation.

Aerts, D.: 1982, 'Example of a Macroscopical Situation that Violates Bell Inequalities', *Lett. Nuovo Cimento* **34**, 107–111.

Aerts, D.: 1983a, 'Classical Theories and Non Classical Theories as a Special Case of a More General Theory', *Journal of Mathematical Physics* **24**, 2441–2453.

Aerts, D.: 1983b, 'The Description of One and Many Physical Systems', in C. Gruber (ed.), *Foundations of Quantum Mechanics*, Lausanne: A.V.C.P., Geneva.

Aerts, D.: 1990, 'An Attempt to Imagine Parts of the Micro-World', in J. Mizerski, A. Posievnik, J. Pykacz, and M. Zukowski (eds.), *Problems in Quantum Physics II*, World Scientific, Singapore, pp. 3–25.

Boyd, Richard, Philip Gasper and J.D. Trout (eds.): 1991, *The Philosophy of Science, A Bradford Book*, Cambridge, MA: The MIT Press.

Cao, T. Y. (ed.): 1999, *Conceptual Foundations of Quantum Field Theory*, Cambridge University Press.

Christiaens, W.: 2002, 'Some Notes on Aerts' Interpretation of the EPR-paradox and the Violation of Bell-inequalities', in D. Aerts, M. Czachor and T. Durt, *Probing the Structure of Quantum Mechanics: Nonlinerity, Computation and Axiomatics*, Singapore: World Scientific.

De Broglie, L.: 1956, *Nouvelles Perspectives en Microphysique*, Paris : Editions Albin Michel.

Herbert, F: 1985, *Quantum Reality. Beyond the New Physics*, New York: Doubleday, Anchor Books.

Kuhlmann, M., H. Lyre and A. Wayne (eds.): 2002, *Ontological Aspects of Quantum Field Theory*, Singapore: World Scientific.

Mermin, N. D.: 1985, 'Is the Moon There When Nobody Looks? Reality and the Quantum Theory', *Physics Today*, pp. 38–47. Reprinted in: Richard Boyd, Philip Gasper and J. D. Trout (eds.), 1991, *The Philosophy of Science, A Bradford Book*, Cambridge, MA: The MIT Press.

Redhead, M.: 1987, *Incompleteness, Nonlocality and Realism. A Prolegomenon to the Philosophy of Quantum Mechanics*, Oxford, UK: Oxford University Press.

Redhead, M.: 1989, 'The Lakatos Award Lecture: The Nature of Reality', *British Journal for the Philosophy of Science* **40**, 429–441.

Seibt, J.: 1990, *Towards Process Ontology: A Critical Study in Substance-Ontological Premises*, Pittsburgh: University of Pittsburgh Ph.D. Dissertation, Michigan: UMI-Dissertation Publication.

Seibt, J.: 1995, 'Individuen als Prozesse: Zur prozess-ontologischen Revision des Substanzparadigmas', *Logos***5**, 303–343.

Seibt, J.: 1996a, 'Non-countable Individuals: Why One and the Same Is Not One and the Same', *Southwest Philosophy Review* **12**, 225–237.

Seibt, J.: 1996b, 'Existence in Time: From Substance to Process', in J. Faye, U. Scheffler, M. Urs, *Perspectives on Time, Boston Studies in Philosophy of Science*, Dordrecht: Kluwer, pp. 143–182.

Seibt, J.: 2000, 'The Dynamic Constitution of Things', in J. Faye et al. (eds.): *Facts and Events. Poznañ Studies in Philosophy of Science* **72**, 241–278.

Seibt, J.: 2001a, 'Formal Process Ontology', in C. Welty. and B. Smith (eds.), Formal Ontology in Information Systems: Collected papers from the second international conference, ACM Press, Ogunquit, pp. 333–345. (http://www.hum.au.dk/filosofi/filseibt).

Seibt, J.: 2001b, 'Processes in the Manifest and Scientific Image', in U. Meixner (ed.), *Metaphysic im postmetaphysischen Zeitalter/Metaphysics in the Post-Metaphysical Age*, Wien: "bv&hpt, pp. 218–230.

Seibt, J.: 2002, 'Quanta, Tropes, or Processes: On Ontologies for QFT beyond the Myth of Substance', in Kuhlmann et al. 2002, pp. 53–93.

Seibt, J.: forthcoming, *Free Process Theory. A Study in Revisionary Ontology*, Habilitationsthesis to be submitted at the University of Konstanz.
Teller, P.: 1995, *An Interpretative Introduction to Quantum Field Theory*, Princeton: Princeton University Press.

RICCARDO MANZOTTI

A PROCESS-BASED ARCHITECTURE FOR AN ARTIFICIAL CONSCIOUS BEING

ABSTRACT. A conscious being is a system that experiences (feels) something. In order to build an artificial conscious being we need to give an account of what it is to experience or feel something. Any project that aims to design an artificial conscious being thus needs to be concerned with the notion of experience or feeling. As I argue in the following, for the purposes of robotics this task can be profitably approached if we leave behind the dualist framework of traditional Cartesian substance metaphysics and adopt a process-metaphysical stance. I begin by sketching the outline of a process-ontological scheme whose basic entities are called 'onphenes'. From within this scheme I formulate a series of constraints on an architecture for consciousness. An architecture abiding by these constraints is capable of ontogenesis driven by onphenes. Since an onphene is a process in which the occurrence of an event creates the conditions for the occurrence of another event of the same kind, an onphene-based architecture allows for external events to provoke the repetition of other events of the same kind. In an artificial conscious being, this propensity to repeat events can be considered as a functional reconstruction of motivation. In sum, if we base the architecture for an artificial conscious being on onphenes, we receive a system that experiences (feels) and is capable of developing new motivations. In conclusion I present some experimental results in support of this claim.

1. IS CONSCIOUSNESS INDEED AN IRRESOLVABLE MYSTERY?

What is consciousness? How is it possible that a part of reality (the conscious subject) has an experience of some other part of reality? Why is the phenomenon of consciousness so elusive within a physicalist frame of reference? One of the main aims of this paper is to suggest that consciousness loses some of its theoretical mysteriousness if one abandons two traditional assumptions: (i) that reality and its representation are two different kinds of entities, and (ii) that reality is constituted by static things (which are occasionally engaged in dynamic interactions). In short, I want to argue here for the claim that the problematic character of consciousness diminishes to the extent to which we succeed in relinquishing (i) a dualistic stance in (ii) the traditional object-ontological (substance-ontological) framework. As I shall try to show in the following, it is possible to challenge both of these assumptions and present an alternative ontological framework in which basic elements of a theory of the mind can be profitably formulated, and

J. Seibt (ed.), Process theories: Crossdisciplinary studies in dynamic categories, 285–312.

from which some guidelines for experimental work in the field of artificial agents can be derived.

The traditional disciplines which examine the problem of consciousness are philosophy, cognitive science and neuroscience (Chalmers 1996; Editor 2000; Crick and Koch 2003; Zeki 2003). However, more recently there is growing awareness about the central role of robotics for the study of consciousness: "to understand the mental we may have to invent further ways of looking at brains [and we] may even have to synthesize artifacts resembling brains connected to bodily functions in order fully to understand those processes" (Edelman and Tononi 2000). Since the construction of a conscious artifact should help us to understand the processes of thought itself, the engineering approach to the problem of consciousness – i.e., the attempt to design and build an artificial conscious being – is receiving increasing attention (Steels 1995; Aleksander 1996; O'Brien and Opie 1997; Manzotti et al. 1998; Schlagel 1999; Aleksander 2000; Martinoli et al. 2000; Togawa and Otsuka 2000; Aleksander 2001; Buttazzo 2001; Manzotti and Tagliasco 2002; Perruchet and Vinter 2002).

Most of the past engineering approaches to the construction of cognitive agents did not address the issue of phenomenal experiences as such, but rather focused exclusively on the analysis of behavior (Brooks 1991; McFarland and Bosser 1993; Arkin 1999). However, phenomenal experience is one aspect of agents that cannot be reduced merely to their behavior. The philosopher David Chalmers argued that there are two orders of problems in the study of the mind: one is the cognitive-behavioral-functional problem which he labels the 'easy problem', the other is the phenomenal problem labeled the 'hard problem' (Chalmers 1996)). The engineering approach to consciousness cannot avoid tackling the 'hard problem' for a conscious being is an entity that *feels*, not an entity that *does* something. The 'hard problem' is even more of an obstacle in this context since engineers have traditionally constructed 'objects' and consequently lack even the conceptual tools required to deal with the design of 'subjects'. Until now, the world of engineering simply had no place for phenomenal experiences.

That consciousness and phenomenal experience exist we know by introspection. We are 'open to the world', we experience the smell of flowers, the color of the sky, the meaning of a sentence. Furthermore we do not have any evidence that our consciousness could not be realized in anything else but a human body. There are no theoretical arguments tying phenomenal consciousness exclusively to DNA-based organisms. This lack of negative evidence opens up the new research area of artificial consciousness.

Instead of adopting the traditional idioms and tools used in the discussion of the 'hard problem', researchers in artificial consciousness do best to begin by putting the standard conceptual framework under close critical scrutiny. This is the thrust of my overall argument in this paper, which I present in five steps. In a first step I outline the limitations of the current conceptual framework for the understanding of consciousness. Then I sketch an alternative framework based on a kind of process named 'onphene'. In a third step I present core elements of an account of consciousness based on onphenes. In step four I describe the architecture of an artificial device implementing an onphene-based account of consciousness, i.e., an architecture based on motivations. In the last and final step I explain the design of the experimental setup. To anticipate the overall conclusion of this argument, I will suggest that consciousness as something different from reality is an invention which stems from a impoverished conception of reality itself. The problem of consciousness has been created by the hypothesis of an abstract physical world of purely quantitative objects or substances.

2. REPRESENTATION AND PHYSICAL WORLD: AN UNNECESSARY DICHOTOMY

In order to model conscious experience, we need to concentrate on three characteristics that in combination furnish a working hypothesis about the nature of conscious experience. As conscious beings we know that experiences (i) are always an experience of something, i.e., every experience has a given content and (ii) can be phenomenally distinguished on the basis of their content. (For instance, there is a difference in content between the experience of drinking lemonade and looking at a painting by Mondrian.) Further, setting aside extreme forms of skepticism, it is an introspective part of our natural epistemic stance that (iii) the content experienced represents external events (Dennett 1969; Block 1988; Dennett 1988; Fodor 1990; Chalmers 1996; Bickhard 2001), at least in ordinary perception. In a sufficiently open sense of the term, introspection commits us to the 'representationalist' standpoint. The claim that conscious experience occurs, that it has a phenomenal content, and that phenomenal content is tied to what happens, can be summed up as the thesis that to be conscious of something is to have a representation of that something.

On the basis of these considerations, we can put forth two hypotheses: (1) there are occurrences of events that correspond to contentful phenomenal experiences; (2) these occurrences represent something. According to this point of view, an event has a phenomenal content only if it has a representational role (Fodor 1981; Millikan 1984; Dretske 1993; Dretske 1995; Clark and Tornton 1997; Bickhard 1999). Hence conscious experience is the occurrence of events with phenomenal content and, by implication, with a representational role.

The notion of a representation is commonly taken to be more than an abbreviation for the claims (i) through (iii) above. Rather, it is used as an explanatory notion with a meaning of its own – a representation is something that presents (or re-presents) something else. If we adopt an object ontology and assume that the world is composed only of objects, then a representation is an object which presents another object. Yet how could we make sense of the kind of 'presentation' involved here, how could we think of an object 'referring' to another? One way to approach this question is by taking representation to be somehow effected by similarity. This approach – termed the 'copy theory of representation' by Nelson Goodman (Goodman 1974) – does not get us anywhere, however, since the identity principle establishes that an object is just itself, and that no object can be another object at the same time.

From within an object-ontological setting the only other way to understand representation is to view it as a relation. The representing entity and the represented one are linked by some kind of abstract relation (semantics, aboutness, or intentionality). However, this is in conflict with a commitment to physicalism which admits one relation only, namely, causality. Between events there are only causal relations, there are no intentional, teleological, formal or semantic relations. It is not a coincidence that most of the attempts to naturalize semantics, perception and representation are based on some kind of causation (Grice 1961; Armstrong and Malcom 1984; Haybron 2000). However, if we accept that representation is a species of causation then representation becomes a ubiquitous phenomenon.

If we want to hold on to the insight that phenomenal experiences correspond to the occurrence of representations, while at once holding on to a commitment to physicalism, a physical interpretation of representation must be found. Obviously the classical dualist Cartesian model of representation as a relation between physical and mental items is not helpful here. However, the dualistic model of representation (the representing item is something different from what it represents) is unavoidable given an underlying ontology that divides the world in separate objects.

Two entities are separate if their existences, in a given instant t, are mutually independent. (If you destroy your computer, nothing would therefore happen to mine in the same instant). All objects which do not stand in parts/whole relations are separate. If an object ontology is accepted, the dualist model is the only way coherently to make sense of the introspective data of conscious experience. As long as we describe ourselves as entities that are separate from what they are conscious of, we have taken on board the supposition that experience is some kind of duplication of the external world inside the internal domain of the subject.

In fact, the dualism of the external domain and the internal domain is independent of the additional qualification of the external as physical and the internal as mental. It is a contingent detail that in the XVIIth century the internal domain could only be assessed as mental while nowadays we also have a neurophysiological description of this domain. Given an object ontology, the function of representation will always imply the existence of a dualist counterpart (either a copy or a relatum) of the world, no matter how that counterpart is characterized. In this respect current neuroscientists assign to the brain the same role Descartes attributed to the *res cogitans*. However, it is quite clear that the brain taken as a physical object cannot contain copies or isomorphic relational counterparts of the external world (i.e., of something with completely different physical properties). Thus it remains a mystery how the brain can represent the external world.

For an object-based account of representation the only viable strategy is to leave physicalism behind. One might suppose, then, that in the brain there are qualia or pure phenomenal qualities, the modern version of the secondary qualities, which can only be identified in terms what they represent and thus are a label for a problem rather than a solution. Alternatively, one might embrace a functionalist stance according to which brain states are representation of something in the world due the fact that they a certain functional role for the organism. The functional domain is an abstract domain of input/output relations built on top of, and always extraneous to, physical events. However, functional roles do not account for the introspective difference between conscious and subconscious perception.

In short, traditional object or substance ontologies are committed to explaining representations (and hence the mind) in a dualist fashion. Conversely a Cartesian dualist standpoint is naturally compatible with a substance ontology (even though it does not entail the latter). Consider the following three claims:

(a) the world consists of separate substances or objects
(b) the mind represents the world (or the mind is equivalent to a set of representations of the world)
(c) representations are different from what they represent (dualism)

As worked out consistently in XVII century metaphysics, a commitment to (a) and (b) entails a commitment to (c). If the mind represents the world, and the world and the mind are made of separate entities, the mind must be a separate entity from the world; thus representations must be separate (and hence different) from what they represent.

However, is claim (a) indeed an *a priori* truth? We can reject the assumption that the world is made of separate substances for at least the following three reasons. First, there is enough evidence from microphysics to militate against classical objects or substances as a type of fundamental entities (Cramer 1988; Zohar 1990; Stapp 1998; Auletta 2000). Second, the claim is not logically necessary – as we shall see presently, a different ontology can be formulated. Third, besides the problem of representation there are a number of fundamental *ontological* difficulties arising for object ontologies (Seibt 1990).

If (a) is rejected, it is important to settle for the right type of alternative ontology. An event ontology, for example, gets us from the frying pan into the fire. From a scientific point of view, the idea of a single event is a nomological absurdity. In contemporary science we cannot speak of anything which is not the object of an experiment, the result of a measurement, the target of an observation, or a postulated interaction whose results we observe. If something is not directly or indirectly observed, it is not part of what is empirically known. However, in order to be observed an event must be in relation with other events. It must in some fashion 'present' itself to other events. But then we must admit that in science there are no singular events – events are derived entities, namely, interactions of processes. Singular events or autonomous static objects are abstractions: like the Euclidean point or line, which are not part of the real world. Unfortunately these abstractions have been misunderstood as the real world, in the sense of Whitehead's 'fallacy of misplaced concreteness' (Whitehead 1925).

In the following I will argue that once we replace the traditional object-ontological framework with a suitable process-based alternative we can deal with the relation between mind and the world without falling in the dualistic trap. I advocate the following alternative set of assumptions:

(a) the world is an assembly of processes which are not necessarily separate
(b) the mind represents the world (or the mind is equivalent to a set of representations of the world)
(c) representations are not different from what they represent (monism)

Assumption (b) has remained unchanged, while (a) has been changed and, as a result, (c) also. In the following I will try to show how a process ontology permits us to: (i) account for representation without dualism, (ii) treat the mind as a set of representations; (iii) ensure true knowledge of the world without solipsism or dualism.

3. A PROCESS ONTOLOGY FOR REPRESENTATION: RECIPROCAL CAUSATION OR 'ONPHENE'

In order to introduce the basic notions of our process ontology let us start out with some simple observations about everyday physical processes in which the absence of any form of dualism, or even of duality of domains, is intuitively clear. Subsequently we will use these illustrations to introduce a new categorization for conscious experience.

The canonical illustration for a physical phenomenon in which the physical continuity between the represented object/event and the representing object/event is quite evident is the rainbow (Figure 1). When the sun is sufficiently low on the horizon and sheds its rays at a right angle on a sufficiently big volume of water droplets suspended in the air, an observer (either a human being or a camera) can see (record) a colored arch. All drops of water reflect the sunlight in the same manner, yet only those that are in a particular geometrical relation between the observer position and the direction of the sun rays are seen as part of the rainbow. The rainbow cannot be defined in any meaningful sense other than from the point of view from which it is seen. In this sense the rainbow, although constituted by a set of physical entities (drops of water in space reflecting light in a certain way), cannot be defined without knowing where and how it will be seen. For instance, it is not possible to see oneself as flying under or around a rainbow. Furthermore, the rainbow is a private physical phenomenon because two different observers always see two (however slightly) physically different rainbows. Since two separate observers occupy two different positions in space, they select different rays of light and accordingly different drops. Rainbows thus cannot be said to exist independently of the act of observation. In fact, if there were no eyes (or camera) looking at the rainbow, the relevant set of droplets of water would not produce any effect and the phenomenon called 'rainbow' would not exist at all. Even if it were possible to argue that an expert physicist who knew the drops' position, the sun's position and the observer's position might be able to calculate the projection of the rainbow on the observer's retina, such a calculation would require the knowledge of the observer's position as an essential prerequisite. The rainbow occurs only when it is seen. The cause (the arch of drops) is not there as a distinctive whole until it produces an effect (the projection in the observer's retina). Here we cannot separate the cause from the event, in the same way in which we cannot separate the events from their relation. The effect is responsible for the existence of the cause.

Figure 1. A rainbow. Does it exist without an observer? Does its observer exists without the drops of water? What is the cause and what is the effect?

This is quite familiar from conscious representations, of course. For instance, if there are six points on a wall, in a hexagonal arrangement, these do not exist as an arrangement or a whole as long as do not produce any effect as a whole. If there were no human observer to see them, it would be extremely improbable that they would produce any effect as hexagonal arrangement in purely causal interaction with physical entities. Thus we can say that the hexagonal arrangement comes into existence only when it is *seen* for the first time. Generally speaking, it is impossible to define the physical existence of a structural arrangement without referring to a cognitive system which processes it as that arrangement. (Of course, the cognitive system in question does not need to be a human observer. The arrangement of six dots can be processed by much simpler systems as well. More complex arrangements, however, like a face or a word, require interaction with more complex systems, and in general there is a continuum of representational unities created by interacting agents.)

The unity of the arrangement, its 'objecthood', is a consequence of its capability of producing an effect as a whole. In fact, many objects can produce a joint effect only by means of the interaction with a cognitive agent. The whole, which is the content of the observer's perception, is at the same time 'inside' and 'outside' of the observer. The whole 'arises' or comes about because of the act of observation and its occurrence is the very act of observation in itself. There is no inner event and outer event: the physical process that is responsible both for the six dots and their recognition as hexagon. The six dots as a whole arise when they produce a joint effect.

Even though most obvious in conscious representations, the same situation obtains in all perceptual events. Whenever we have a representation, there is no real distinction between the represented event and the representing one. Both occur conjointly as different aspects of one physical process. This process fits the requirements of representation since it contains what it has to represent: there is no need to assume a duplication of reality as in the Cartesian framework. On the other hand, all the problems surrounding traditional versions of identity theories (i.e., theories endorsing the identity between brain processes and mind processes) are undercut since the mind is no longer a neural activity located exclusively in the brain (a property of a substance) and separated from what it should refer to: the physical substratum of representations. Rather, the mind is a process starting in the external world and ending in the brain. Any process is extended in time and space.

The puzzle of representation does not arise within the new framework since there is no longer any separation between the outside world and the internal world. In this way a mind no longer corresponds to an emergent property of a system duplicating external reality by means of some internal code. The mind '*enlarges*' to cover that part of reality which it represents. In fact, representation is no longer a *representation* but is just a *presentation* (and thus tantamount to occurrence) of reality inside a system of events.

If we weigh the empirical (introspective) evidence concerning experience as well the degree of internal coherence of traditional theories, we might well say that the hypotheses of a purely quantitative physical object which is outside of the domain of our experience and (at the same time) the efficient cause of our perception is not well supported. This holds at least in the sense that there is no good empirical and theoretical reason that would militate against the radical approach of introducing a new account of representation and conscious experience based on processes – spatio-temporally extended dynamic entities. The envisaged new kind of entity is nothing more than the occurrence of reality without the division between "real reality" and "experienced reality".

In order to avoid all unnecessary and misleading connotations I will use a new term: 'onphene' to refer to an entity of this type. 'Onphene' is the contraction of *ontos* (existence), *phenomenon* (representation), and *episteme* (being in relation with) (Figure 2). The choice of words is motivated by the fact that the onphene is (i) a physical process (something that exists or is ontic), (ii) corresponds to a phenomenal content (a representa-

Figure 2. A visual analogy to explain the role of the onphene. It can be seen under three main perspectives: as a *phenomenon* (what appears?), as *ontos* (what is?) and as *epistemé* (what is in relation with?). The three standpoints, traditionally separated, can be seen as three manifestation of the same underlying principle.

tion) and (iii) is in relation with other entities (nonseparate). Similar types of processes have been proposed by a number of authors: Whitehead's prehension (Whitehead 1933), reciprocal causation (Newman 1988; Hausman 1998), intentional relation (Manzotti 2000; Manzotti 2001). There are analogies also with Maturana and Varela's autopoiesis and Merlau-Ponty's circular causation (Merleau-Ponty 1945/2002; Maturana and Varela 1980; Maturana and Varela 1987/1998).

The notion of the onphene allows us to formulate a unified theory of mind, body and environment. Following an onphene-based approach, the distinction between a representing brain and a represented body plus environment is arbitrary and unnecessary. No pure disembodied mind has ever been experienced. On the other hand, instead of postulating that the brain has a dual state, an invisible property corresponding to a phenomenal experience even more mysteriously linked to the external events, the mind is an activity that reaches beyond the physical area occupied by the brain at a certain time t. The mind is 'enlarged' to cover all those events that constitute the content of the conscious mind – these are physically part of the mind. There is no more dualism: there is just one reality. There are no more representations: there are just events constituted by the interaction of processes. We feel something because the process that we are is extended to comprehend those events we experience. Events that previously had no part in our developmental history become entangled in our internal process. The traditional picture of a boundary between an internal domain and an external domain is replaced by the image of the occurrence of an immensely complex fabric of processes continuously merging and dividing.

In an onphene-based ontology there are also events, but these figure as a derived category. Above I argued that events must be relational entities since they are essentially the result of interactions. The existence of such interactions is guaranteed by the fact that onphenes are by nature entangled entities in the sense that they cannot happen without being in interaction with others. Since onphenes are characterized as transferences with reciprocal causation (the existence of the effect effecting the existence of something that is cause for this very effect) an onphene not interactively entangled with other onphenes would be a contradiction in terms. An onphene-based ontology is naturally projected onto a relational structure where all parts of reality contribute to each other's identity and occurrence to different degrees. Events are abstractions of interactions, but they do not always carry this relational nature on their conceptual sleeves: onphenes and events are related like the trajectory of a bullet in physical space and the Euclidean points on the metrical counterpart of that trajectory, or like the south pole of a magnet to the magnetic field.

Returning to the example of the rainbow, we can abstract two events from the rainbow-onphene: the cause, i.e., the reflection of sunlight from an arch of drops in the cloud, and the effect, the perception of the arch in the observer's retina. These two events belong to the same process (as well as the many other events along the path from the arch-shaped reflection to the perception in the retina which we could abstract). Strictly speaking, an event is a second-order interaction: an interaction of an interaction and a measuring process, since in order to ascertain the identity and existence of an event (the drops in the cloud or the activation in the retina), we need other onphenes, other processes (for instance a probe that measures the density of chemical compounds in the retina). As we shall see presently, even though the names of events denote ontologically derived or secondary entities, terms for events are useful tools in the formulation of an onphene-based framework.

4. THE ENLARGED MIND: ONPHENES AND MOTIVATIONS

If the world is an assembly of onphenes, why and how do 'subjects' come about? A subject is a 'knot' of onphenes, but why do such 'knots' form and what ties them together? The onphene-based framework allows us to define a particular causal structure in terms of which we can demarcate those onphenes that constitute a conscious agent. The causal structure in question is at the core of what we mean by 'motivation'.

Any onphene has an abstract projection as a causal chain of events. This projection does not exhaust the nature of the onphene but provides a useful simplification. From the causal point of view an onphene is a process that links the occurrence of the cause with the occurrence of the effect in the form of a reciprocal causation. It corresponds to a situation in which the occurrence of an event E1 produces (a) an effect E2 and (b) the condition for the occurrence of the causal relation between that kind of cause (E1) and that kind of effect (E2) (compare again as examples the rainbow or a pattern). Talk about onphenes in terms of their causal abstractions, i.e., in terms of causal structures on events, provides a convenient tool to formulate a criterion for consciousness, as I shall explain now.

In an onphene-based ontology a subject S is a complex bundle of onphenes which consists of all those onphenes that are S's experiences at every instant of S's conscious life. To explain the emergence of a conscious mind is to explain how onphenes interact together to engender the occurrence of more and more complex onphenes. The model for this process is 'ontogenesis', the formation of complex units by linking together different causal chains into a unified causal process (or, to use the non-causal idiom, the progressive entanglement of more and more onphenes). Ontogenesis is familiar from the origination of planets – just as planets form without there being an a priori center of gravity, conscious beings may form without there being an a priori subject or transcendental Ego. Planets form due to conditions under which a large number of particles mutually attract. Similarly, conscious beings form due to conditions under which more and more phylogenetically induced motivations and goals converge into that giant unified causal process which is the subject. The conscious mind is the product of such a development, but a processual product: a kind of process or a way of events taking place.

A conscious being, then, is a bundle of onphenes linked in the way in which onphenes entangle during ontogenesis. This linkage may be considered as a goal of natural selection – a system capable of incorporating its past causal relations and their relata. If we view ontogenesis as an evolutionarily 'necessitated' developmental procedure, the consciousness, i.e., the ontogenesis of onphenes, also receives an evolutionary explanation. Onphenes entangle in the way of ontogenesis since this proved an evolutionary advantage and in doing so conscious beings were formed.

Motivations play the key role in the ontogenesis of conscious beings. A motivation is an internally produced criterion for the control of developments. In fact, a subject can be viewed as the process resulting from the incremental aggregation of onphenes elicited by motivations. We distinguish between fixed or hard-wired motivations and acquired or ontogenetic motivations. Fixed motivations are a priori coded and hence do not depend on the ontogenetic history of a system. By contrast, acquired or ontogenetic motivations must be the result of the interactions with the environment. In the following we will refer exclusively to the latter. *A (ontogenetic) motivation is here defined as a process whose probability of occurrence is increased* due to its own occurrence, in combination with the existence of certain embedding conditions.

To illustrate, let us suppose you see a face and hear that this is Sigourney Weaver. As a result you will be able to recognize her again and again. A process that has happened just once (the perception of a face *as* Sigourney Weaver's face) has produced itself as a future possibility. You are conscious of her because her face has become a part of the processes that are your ontogenetic make-up. Your mind has enlarged itself by incorporating a new process that remains intertwined into what you are. The occurrence of such processes of recognition may be a matter of chance – they happen just in case you happen to see that face. But as soon as the occurrence of your recognition of Weaver's face increases the probability (frequency) of the occurrence of such recognitions, motivation has set in. The process of recognizing Sigourney could become a target for your future action – you might buy cinema tickets to repeat the recognition.

Motivations and phenomenal experiences are very similar in their causal structure. However they play a different role with respect to other onphenes. The crucial difference is, however, that motivations are processes that 'call' for their repetition. A phenomenal experience does not produce a modification in the subject's causal structure, while a motivation propagates its effects through time in the subject's history due to a stable modification in the subject's causal structure. A phenomenal experience (onphene) is an event C whose effect E is the cause of a causal relatedness of a cause of kind of C and an effect of the kind of E. Simplifying we might say, a phenomenal experience is an event Q which generates a causal relation of the kind Q instantiates. Motivations are just like onphenes with an additional element of causal structure. A motivation is an onphene which generates *future* occurrence of onphenes of this kind. More precisely, switching to the causal idiom, a motivation is an event C whose effect E is the cause of the causal relatedness of a cause of kind of C and an effect of the kind of E, *and of future occurrences of a causal relatedness of this kind.*

To have a motivation for getting something means to be in relation with an event: the target event (what we want to achieve) becomes the cause of all or most of our action. How we act is not caused by a 'final cause' but by our past, i.e., by the processes which entered a past processual constitution of the subject that we are. However, once the spectre of final causation is properly replaced it is admissible to speak of a system's goals. For instance, assume that the system has been exposed to the presence of Susan, and as a result the system aims at having Susan in its field of view. The system will behave with the goal of repeating as much as possible the process of seeing Susan. It was the process of seeing Susan that modified the structure of the system in such a way that among its goals there is 'having Susan in the field of view'. The occurrence of the process 'seeing Susan' has increased the probability of its own repetition. If Susan would not have come into the system's field of vision, the system would not have modelled its criteria for 'preferred visual object' around her visual appearance. The Susan process became entangled into the ontogenetic history of the observing subject by adding a new motivation.

A caveat: what increases its possibility to happen again is not the appearance of Susan (which depends only on Susan), but the process of Susan's being seen and recognized by the system. Motivation is not a causal reflection of conditions in environment – rather, it is a self-reinforcing activity. Playing tennis is a way to become fond of playing tennis.

The view of the motivation I have set out here acquires its full meaning only within the broader process-based approach I have sketched here. Without this background, the suggested account of motivation suggested can still be read in purely causal terms, but in doing so its explanatory power with respect to the constitution of a self will be lost. Let us review the main conceptual elements of this process-based approach encountered in this and the previous section:

5. A PROCESS BASED ARCHITECTURE

An architecture is the description of the essential features of an embedded system in order to produce a certain phenomenon. We can now present a general architecture for an artificial device designed to engender the occurrence of the above described processes. To this effect the architecture must work with a physical body (sensory and motor systems). The goal is to build something that will become part of the 'external' flow of processes. Thus the body and its sensory and motor equipment are necessary – a computer without the capacity of interacting directly with the physical aspects of its environment would not suffice.

The general idea is the following. The system must be capable of self-organizing the flow of incoming stimuli and at the beginning it must do so driven by pre-defined criteria. This is consistent with what happens during

TABLE I

A list of used terms and their definitions

Term	Definition
Process or onphene	The basic unit of reality
Event	An abstraction from the interaction of processes
Cause	What takes part in a process
Effect	What can be part of other processes after the occurrence of a process
Content of an onphene	The cause of an onphene
Reciprocal causation	When the occurrence of the effect is responsible for the occurrence of the cause or rather when the occurrence of an event C whose effect E effects a causal relatedness of events of the kind of C and E, respectively (an onphene viewed in objectivistic terms.)
Represented event or object	An onphene from the point of view of the cause
Representing event or representation	An onphene from the point of view of the effect
Phenomenal experience	A representation, i.e., an onphene
Motivation	A process whose occurrence creates the condition of its own repetition
Action	An event provoked by a subject as a result of a motivation

the ontogenesis of a biological being in a natural environment. The fixed goals of phylogenesis put constraints on the complete freedom of ontogenesis (which is conditioned only by the environment and the experience). There is obviously a trade-off between flexibility and adaptivity (to new situations and potentially unpredictable events) and the control that phylogenesis can exert upon individuals. The primary objective is to design and implement an architecture into which processes 'get trapped' and find a way to repeat themselves over and over.

At the beginning the system merely contains some bootstrapping criteria aimed at orienting its ontogenesis towards specific and useful classes of stimuli. These criteria are the equivalent of instincts in a biological being. Otherwise the system is literally a *tabula rasa*. When something happens it is processed by the system; however the system has still to build its own internal perceptual categories. If the phylogenetic criteria give their approval, the system begins to build its ontogenetic categories. At first they will be just perceptual categories, subsequently some of them can be selected as newly generated criteria for controlling further ontogenetic development. Generally speaking the architecture is composed of three parts, in the following referred to as modules (Figure 3): a phylogenetic module to bootstrap the system, a module to store new categories, and an ontogenetic module to determine which events have to become new ontogenetic criteria (motivations) for the system.

Whenever something happens and it is part of the sensory experience of the system, it can produce an effect. Further this effect is the condition for the future repetition of the same kind of causal processes. For instance, the system will become capable of recognizing faces because faces have been part of its past. Faces will become objects of perception because the system will have developed structures to recognize them. The process structure of the ontogenetic development of the system is further reinforced by the development of new ontogenetic criteria which will control the future choices of new classes of objects.

To offer a more comprehensive illustration, consider an experimental set up where at the beginning a series of stimuli with different colors and shapes are presented. They are perceived as sets of features taken as a whole (including shape). The stimuli are categorized on the basis of all their properties (shape, size, texture, orientation, spatial frequency) according to (a) criteria contained in the Phylogenetic Module and (b) criteria created by the system. The Phylogenetic Module provides only a color criterion. Yet the system might create a category for stimuli of similar shapes or similar colors or similar texture. By a category we mean a class of different stimuli that are perceived by the system as the same stimulus. In order to build category the complexity of sensory stimulation must reduced. In the first phase of development only those categories are created which have a ready-made criterion in the Phylogenetic Module If a colorless stimulus were introduced, albeit equipped with other properties including shape, the Phylogenetic Module will pass it over. In the second phase the Ontogenetic Module comes into play. The Ontogenetic Module can transform a number of categories into new criteria. Subsequently such criteria supplement the criteria of the Phylogenetic Module. If brightly colored triangles are shown to the system, they will become a new category (triangle). After a while, this category will become a suitable candidate for transformation into a new criterion. From then on the Ontogenetic Module will accept even a colorless triangle in the category triangle (colored and not). In turn, the colorless triangle category shall be transformed by the Ontogenetic Module so as to become a further category in itself. In this way new categories can be formed which are connected only to the new ontogenetic properties (shape) regardless of the phylogenetic properties (color).

The criteria play a fundamental role in the Ontogenetic and Phylogenetic Modules. The criteria are implemented by a 'Relevant Signal' which manages the creation and allocation of incoming stimuli into the relevant categories. For instance, if the incoming stimulus corresponds to a brightly colored object, the system will produce a strong Relevant Signal. If the incoming stimulus corresponds to a dull grey object, the Relevant Signal will be weaker. A criterion depends on the value the system gives to the incoming stimulus with respect to the whole past ontogenetic history. The content of an ontogenetic criterion is given by a category. An ontogenetic criterion is a motivation. Technically a motivation is implemented by selecting a given category. All categories developed during the phylogenetic phase potentially provide the content for the same number of criteria. Only certain categories shall become criteria (motivations).

The described architecture must be implemented by a physical structure that is activated by, and develops motivations, on the basis of incoming events. The architecture makes use of elementary associative processes, simple Hebbian learning and case-based reasoning. The events occurring nearby the motivation-based architecture become the seeds for motivations. Due to the existence of the architecture the events in its environment become entangled in a growing network of onphenes and so we can say in turn that the physical implementation of the architecture organizes the environment.

In the following I will introduce the architecture in somewhat greater detail to describe how this architecture at once engenders the self-organization of incoming stimuli and uses them both to categorize reality and to develop criteria on the basis of which new categories can be introduced. We will see how the system gradually modifies its structures and overcomes its initial limitations by developing new criteria. The architecture is aimed at mimicking the development of motivations in human beings. For instance, a human might develop an interest in cars even if nothing in his/her phylogenetic code is explicitly directed towards cars. By contrast, an insect cannot develop new motivations but must follow its genetic blueprint: it has no ontogenetic development. One of the issues of this architecture is to divide explicitly the ontogenetic part from the phylogenetic one.

In a nutshell, the architecture's three main modules are: the Category Module, which is basically a pattern classifier; the Phylogenetic Module which contains the a priori criteria; and the Ontogenetic Module which applies Hebbian learning and develops new criteria by using the patterns stored in the Category Module. The incoming stimuli are categorized in the Category Module on the basis of the Relevant Signal coming from

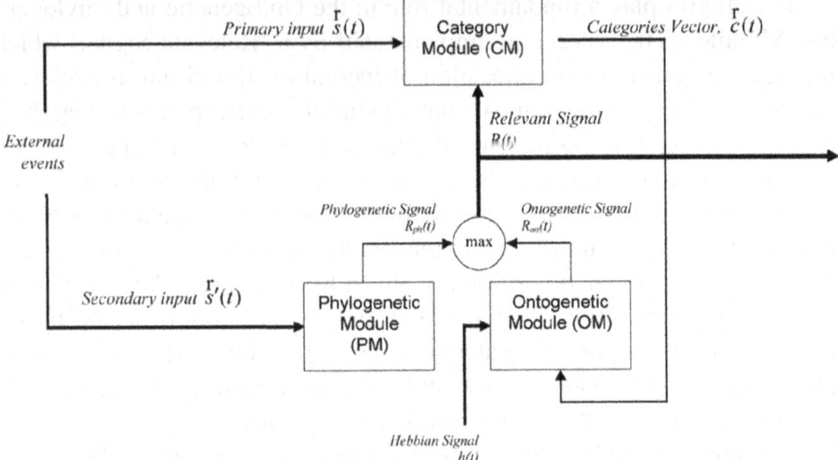

Figure 3. Motivation-based architecture scheme.

the Phylogenetic Module and the Ontogenetic Module. At the beginning, the Relevant Signal depends on those properties of the incoming stimuli that are selected by the Phylogenetic Module; later the Relevant Signal is flanked by the new signals coming from the Ontogenetic Module.

5.1. *Category Module*

The Category Module has the role of grouping in clusters the stimuli coming from the external events. The process of cluster definition is based on an internally built-in criterion for clustering and on the presence of a Relevant Signal (Figure 3). Whenever an incoming stimulus is received, a Categories Vector, which is the output of the CM, is computed; the elements of this vector provide an indication of which cluster best represents the current stimulus. The Categories Vector is empty at the beginning and eventually becomes larger and larger adding new categories. Each of its components measures how much the incoming stimulus matches the corresponding category. The CM tunes its activity to the Relevant Signal (the sum of the Relevant Ontogenetic Signal and the Relevant Phylogenetic Signal).

If and only if the Relevant Signal is active, every time a signal is received, the CM performs the following actions:

(i) if the stimulus is too similar to the already stored stimuli, do nothing;
(ii) if the stimulus is sufficiently similar to one of the previously created clusters, the stimulus is added to that cluster;
(iii) if the stimulus is not sufficiently similar to any of the stimuli already stored, a new cluster is created.

By storing a stimulus only if the Relevant Signal is active, the system does not assign new resources to every incoming signal (the first rule is useful to avoid storing equivalent stimuli).

5.2. *Phylogenetic Module*

This module is the only one that has some built-in criteria concerning the relevant properties of the incoming signal. Functionally, it has the same role as the genetic instincts in biological systems. A Phylogenetic Module autonomously produces a signal on the basis of some external events (the presence of soft or brightly colored objects). For instance, a baby of 2 months looks with more curiosity at brightly colored objects than at dull colorless objects, independently of any past experience. This behavior requires the existence of a hardwired function looking for a relevant property of images (saturated colors). The Phylogenetic module provides criteria that can be used to select correct actions (for instance those actions that maximize the presence of the interesting stimuli). If the system were composed of just the PM and the CM, the system would be a reinforcement learning system.

5.3. *Ontogenetic Module*

Whereas the Phylogenetic Module has built-in criteria about the nature and the relevant properties of the incoming signal, the Ontogenetic Module selects new criteria on the basis of experience. Functionally it has the same role as the acquired ontogenetic criteria in biological systems. The main goal of the Ontogenetic Module is to transform a subset of categories into criteria. Not all the categories built by the CM will become criteria. For instance, if an infant is exposed to colored stimuli of a given shape, she will develop a particular perceptual sensibility for that kind of colored shapes. After a while, the shapes alone (not colored shapes) will become a category. Under certain conditions, the category (colorless shape) will be transformed into a criterion: the Ontogenetic Module will produce an active Relevant Signal even in absence of colors when the specific shape will be present. If she has spent a lot of time looking at colored triangles, it is possible that she will become interested in triangles, independently of their color. She could eventually be interested in grey triangles. The Relevant Signal gives a measure of how much the incoming stimuli is part of the ontogenetic history. By means of Hebbian learning (roughly: what happens together is reinforced), the Ontogenetic Module communicates to the system to what extent each cluster of the Category Vector has been correlated in the past with the signal produced by the Phylogenetic Module.

The main goal of the architecture is to create a structure that can be changed completely by the architecture's own 'experiences'. In the architecture there is a clear-cut division between the phylogenetic part (the *a priori* section) and the ontogenetic part produced by the interaction with the environment. Whenever an event is capable of being recognized by the CM and then selected by the OM, it becomes part of the ontogenetic structure of the developing agent. The event is responsible for the occurrence of a process, whose occurrence will increase the probability for such a process to occur again. The events that become the content of the system motivation are those events that have been able to modify the agent structure. They are abstractions from processes that have become entangled in the system history and perpetuate themselves by means of the system itself.

The processes made possible by the existence of the system can be considered logically and physically continuous with the environment. Furthermore, they shape and modify what the environment is – in fact they create new kinds of objects by creating the conditions in which the new objects can exert their effects. Since they are the result of the environment itself, they can be considered as the result of the self-organization of the environment.

The entire architecture operates under one single general directive: if a process passes through the architecture, the probability for the occurrence of that process has to increase.

5.4. *A comparison with Pavlov's classic conditioning*

It might be instructive to compare the motivation-based architecture I have sketched in the previous paragraphs with Pavlov's classic experiment of conditioning. There are two good reasons for such a comparison: (i) there are strong similarities; ii) there is evidence that many cognitive learning processes could be reduced to Pavlov's associationism (Pavlov 1955/2001; Perruchet and Vinter 2002).

Pavlov focused on modifications of the relation between a given stimulus and a given response. Although Pavlov's test animal was able to select a different stimulus (the ring of the bell), the focus was more on the fact that the animal is capable of linking it to a behavior (the salivary response) than on its capability of selecting a given stimulus from the continuum of the environment. In Pavlov's experiment, there are two hardwired receptors for two different kinds of stimuli (sound of a bell and meat powder): one is a neural structure capable of recognizing the presence of food and another is a neural structure capable of recognizing the ring of a bell. Before the conditioning process, the behavioral response (the salivation) was

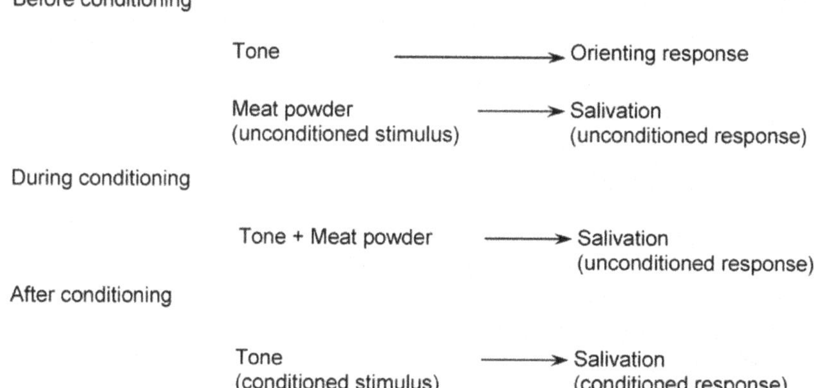

Figure 4. The three stages of conditioning in the classical Pavlov experiment.

only connected with the presence of food. During the training the conditioned response became stronger: more drops of saliva were secreted. The learning consisted in the creation of a connection between the conditioned stimulus and the response.

In our case, the conditioned stimulus does not exist before the conditioning process. The machine is not capable of recognizing the unconditioned stimulus (the shape of an object). It only recognizes colored objects. At first sight our experiment might resemble Pavlov's experiment. The Phylogenetic Stimulus and the Ontogenetic Stimulus could be taken to correspond to the Unconditioned Stimulus and the Conditioned Stimulus, respectively. The Developmental Signal could be counted as the Response (first Unconditioned and then Conditioned). However, the analogy is not sufficiently smooth. In our case, since the color was presented conjointly with the shape of an object, a new ontogenetic stimulus (the shape) is added to the machine's repertoire of stimuli. In other words, the *Umwelt* of the machine is increased and enlarged by a new kind of event. In the case of the motivation-based architecture as described two things happen: (i) the machine learns to recognize something which was previously unknown to it; (ii) the machine links this new stimulus to a given motor behavior (Figures 4 and 5).

Briefly, Pavlov's experiment highlighted the fact that the test animal is capable of establishing a new association between an already familiar stimulus to a motor response. The goal of our experiment is to model the development of the capability of recognizing new stimuli.

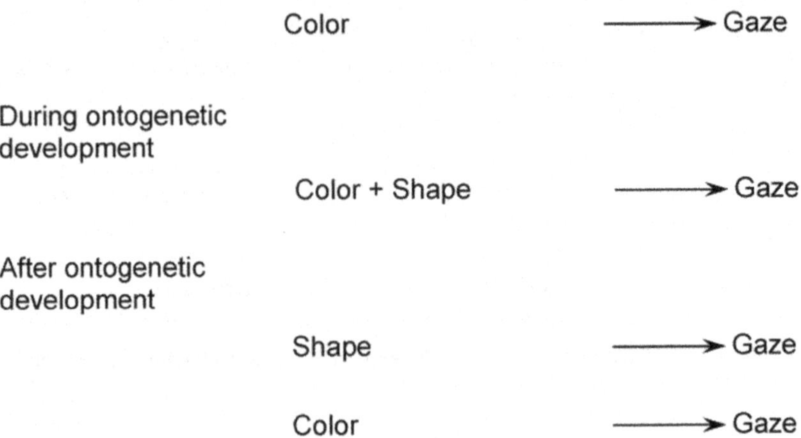

Figure 5. The three stages of ontogenetic development from a process based standpoint (our experiment.

6. EXPERIMENTAL RESULTS: THE EMERGENCE OF MOTIVATIONS

To test the architecture, we conducted an experiment in which a robot implementing the described motivation-based architecture evidently develops a new motivation on the basis of its own experiences. In the experiment, an incoming class of visual stimuli (not coded inside the architecture) produces a modification in the system's behavior that changes not only *what* the system is doing (behavior) but also *why* it does what it does (motivation at the basis of behavior). Something happening in the environment (the appearance of a class of shapes) becomes part of the agent's behavior.

The system has, in this preliminary experiment, a single behavioral choice: to direct or not its gaze towards objects. A series of different shapes associated with colors were presented to the robot. The system was equipped with a phylogenetic motivation that is aimed at brightly colored objects; a colorless stimulus, independently of the shape, did not elicit any response. Since the system has an Ontogenetic Module it develops further motivations directed towards classes of stimuli different from those relevant for its Phylogenetic Module. After a period of interaction with the visual environment (i.e., the presentation of a series of elementary colored shapes), the robot was motivated also by colorless shapes. The category of shape alone had been accepted by the Ontogenetic Module. The system showed the ability to develop a motivation (by directing its gaze towards

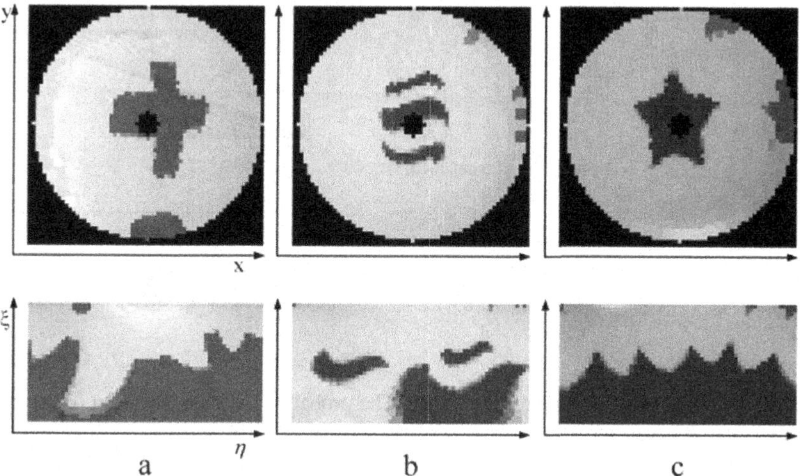

Figure 6. The Cartesian (upper row) and log-polar (lower row) images for a cross (a), a wave (b), and a star (c).

the stimulus) that was not envisaged at design time and that is the result of the ontogenetic development.

For the robotic set-up a robotic head with two degrees of freedom had been adopted which is equipped with a videocamera capable of acquiring logpolar images (Sandini and Tagliasco 1980; Sandini et al. 2000), see Figure 6, i.e., images like those perceived in human beings (with a fovea and a periphery) (Figure 6). The robotic head has two degrees of freedom: the camera is capable of a tilt and pan independent motion (Figure 7a). The robotic head is programmed to make random saccades; a Motor Module generates saccades on the basis of an input signal that controls the probability density of the amplitude r. If λ is low (near to 0), the probability density is almost constant, otherwise, if λ is higher, a small amplitude is more probable (Figure 7b). This probability schema is to ensure that the motor unit mimics an exploratory strategy. When a visual system explores a field of view, it makes large random saccades. When it fixates an interesting object, it makes small random saccades.

We presented different sets of visual stimuli to the system. A first set consisted in a series of colorless geometrical figures as shown in Figure 8a on the left (Figure 8) The frequency with which the system looked at different points was measured. The system spent more time on stimuli corresponding to its motivations by reducing the amplitude of its saccades. At the beginning the system gazed around completely randomly with large saccades since its Ontogenetic Module was unable to catch anything relevant and the Phylogenetic Module was programmed to look

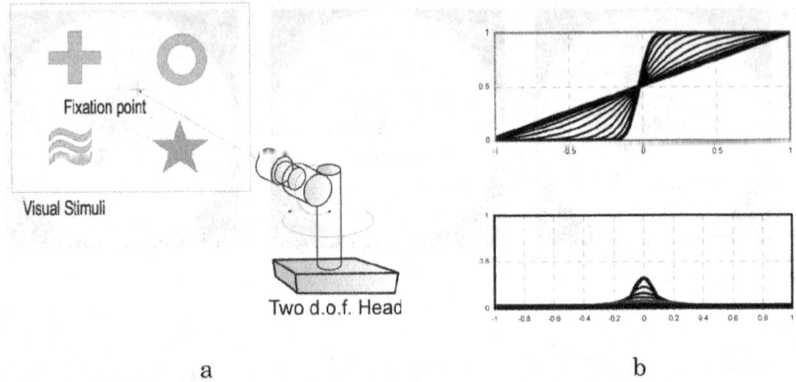

Figure 7. (a Sensory and motor set-up. (b) The probability density function on the basis of the control parameter λ.

for very saturated colored objects, which were absent in the first set. To get a qualitative visual description of how much time was spent by the system on each point of its field of view, we assigned to each point of the visual field an intensity value proportional to the normalized time the system gazed at it. The images in the centre of Figure 6a–b, and c were generated after 103 saccades (equivalent to about 500 sec). The brighter a point of the image, the more frequently the system gazed at it. With the first set of visual stimuli, the resulting image is reproduced in Figure 6a. The system does not show any polarization towards a specific part of the field of view. Subsequently we presented a different stimulus: a series of colored figures. The difference is shown in Figure 8b. The head spent more time on the colored shapes instead of on the white background because of the phylogenetically implanted rule. Finally we repeated the initial stimulus (the set of colorless shapes). The system spent more time on the colorless shapes than on the background (Figure 8c). The behavior of the system changed since the system added a new motivation (shapes) to the previous ones (saturated colors).

7. CONCLUSION

The outlined theory of consciousness is a theory of phenomenal experience – consciousness as phenomenal experience is here defined in terms of the nature of processes involved in development of a conscious agent. Unlike many theories that address the problem of consciousness and phenomenal experience, the account I have sketched here can be tested in the laboratory. Even though my hypotheses gained by philosophical reflection rather than by induction, they have counterparts in the design and implementation of

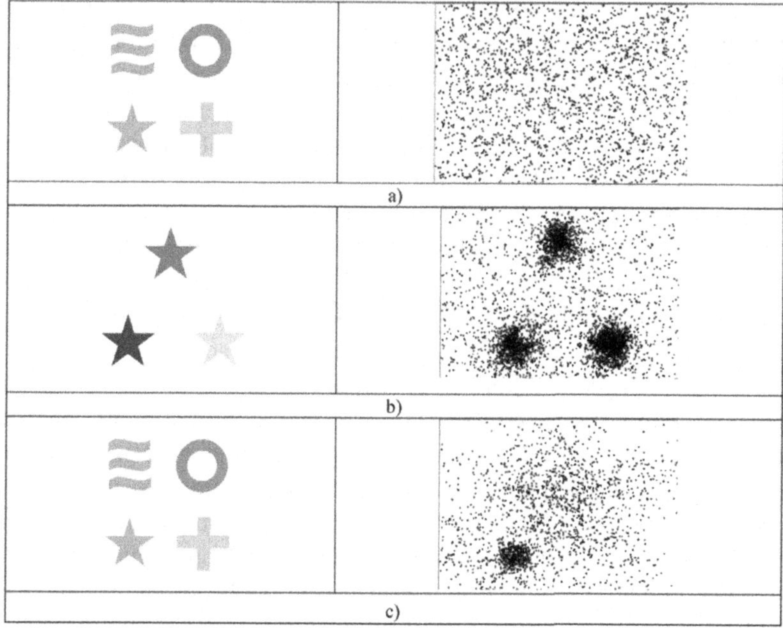

Figure 8. Experimental results.

an architecture which put them to the test in various experiments with machines.

I have suggested here that the subject is the result of the self organization of the environment of which the developing subject is a part. If the world were devoid of subjects, it would be a very different place in causal terms. If shapes or drawings were to be carved in stone by a capricious whirl of wind, their appearance would not increase the probability of their repetition. On the other hand, in a world populated by subjects, every process that becomes part of the structure of a subject increases the probability of the recurrence of that kind of entanglement. Thus the presence of a subject is tantamount with the presence of a certain type of process-the subject is nothing more than a collection of processes of a certain kind.

Due to the existence of subjects, processes propagate themselves in a new and interesting way. The phenomenal experience of shapes, colors, behaviors, and actions ensures self-repetition countless times wherever subjects are located. According to the account outlined here, the subject is as a set of processes (onphenes) that, by means of motivations, becomes progressively integrated during the subject's development. The content of the conscious mind, i.e., the content of the phenomenal experiences of a subject, is not 'inside the head'; rather consciousness is how the world is organized due to the existence of the subject.

To restate the main steps of my argument, I pointed out that classic physicalistic object ontology is saddled with presuppositions that inevitably lead to dualist accounts of representation. Instead of maintaining such ontology and then trying to justify the existence of the mind by adding new hypotheses (like qualia or dual aspect of the world) in a dualistic fashion, I suggested a wholesale rejection of the traditional object ontological framework. I introduced a new type of ontological entity – the onphene – that contains both 'ontic' and 'phenomenal' aspects and is best understood as a 'presenting' of the most general sort. The notion of the onphene undercuts the traditional dualism of a conscious mind separated from the physical world. I then offered an account of motivation, often considered the hallmark of conscious agents, in terms of onphenes. Motivation can be ascribed to a system if that system supports onphenes with a certain causal role, namely, self-propagating onphenes. In terms of self-propagating onphenes I defined criteria for the existence of representing subjects-subjects exist if self-propagating onphenes (motivation) exist. On the basis of the suggested definition of motivation I sketched a general architecture in which self-propagating onphenes can be 'induced'. Finally, I reported on experimental findings that confirm the presence of such processes (motivation-based learning), in a simple learning situation.

Given the line of argument presented, the architecture described can be considered as the general recipe for the design of a conscious machine: it must be able to 'catch onphenes', promote and control a large number of them, and to integrate them. The first step is implemented by means of the cooperation between a Category Module and an Ontogenetic Module inside a Motivation-based Architecture Module. Here I described only the implementation of the first step for a simple test case where the input capability is limited to shapes and colors.

Future work will concern the building of a network of Motivation-based Architecture Modules, the implementation of other different motor and sensor capabilities, and the improvement of the general process-based framework. The implementation of a mechanism of progressive unification and integration of a large number of processes should eventually lead to the development of an artificial subject.

ACKNOWLEDGEMENTS

I would like to thank Johanna Seibt for many helpful comments on a first draft of this paper. The work reported here has been supported by the ADAPT Project, IST-2001-37173.

Aleksander, I.: 1996, *Impossible Minds: My Neurons, My Consciousness*, London: Imperial College Press.

Aleksander, I.: 2000, *How to Build a Mind*, London: Weidenfeld & Nicolson.

Aleksander, I.: 2001, 'The Self "out there" ', *Nature* **413**, 23.

Arkin, R.C.: 1999, *Behavior-Based Robotics*, Cambridge, MA: MIT Press.

Armstrong, D.M. and N. Malcom: 1984, *Consciousness and Causality: A Debate on the Nature of Mind*, Oxford: Blackwell.

Auletta, G.: 2000, *Foundations and Interpretation of Quantum Mechanics*, Singapore: World Scientific.

Bickhard, M.: 1999, 'Representation in Natural and Artificial Agents', in E. Taborsky (ed.), *Semiosis, Evolution, Energy: Towards a Reconceptualization of the Sign*, Aachen: Shaker Verlag.

Bickhard, M.: 2001, 'The Emergence of Contentful Experience', in T. Kitamura (ed.), *What Should be Computed to Understand and Model Brain Function?*, Singapore: World Scientific.

Block, N.: 1988, 'What Narrow Content is Not', in B. Loewer and G. Rey (eds.), *Meaning in Mind: Fodor and his Critics*, Oxford: Blackwell.

Brooks, R.A.: 1991, 'New Approaches to Robotics', *Science* **253**(September), 1227–1232.

Buttazzo, G.: 2001, 'Artificial Consciousness: Utopia or Real Possibility', *Spectrum IEEE Computer* **18**, 24–30.

Chalmers, D.: 1996, 'The Components of Content', in D. Chalmers (ed.), *Philosophy of Mind: Classical and Contemporary Readings*, Oxford: Oxford University Press, pp. 608–633.

Chalmers, D.J.: 1996, *The Conscious Mind: In Search of a Fundamental Theory*, New York: Oxford University Press.

Clark, A. and C. Tornton: 1997, 'Trading Spaces: Computation, Representation and the Limits of Uninformed Learning', *Behavioral and Brain Sciences* **20** 57–90.

Cramer, J.G.: 1988, 'An Overview of the Transactional Interpretation of Quantum Mechanics', *International Journal of Theoretical Physics* **27**(227).

Crick, F. and C. Koch: 2003, 'A Framework for Consciousness', *Nature Neuroscience* **6**(2), 119–126.

Dennett, D.C.: 1969, *Content and Consciousness*, London: Routledge & Kegan Paul.

Dennett, D.C.: 1988, 'Quining Qualia', in A. Marcel and E. Bisiach (eds.), *Consciousness in Contemporary Science*, Oxford: Oxford University Press.

Dretske, F.: 1993, 'Conscious Experience', *Mind* **102**(406), 263–283.

Dretske, F.: 1995, *Naturalizing the Mind*, Cambridge, MA: MIT Press.

Edelman, G.M. and G. Tononi: 2000, *A Universe of Consciousness. How Matter Becomes Imagination*, London: Allen Lane.

Jennings, C.: 2000, 'In Search of Consciousness', *Nature Neuroscience* **3**(8), 1.

Fodor, J.A.: 1981, *Representations: Philosophical Essays on the Foundations of Cognitive Science*, Cambridge, MA: MIT Press.

Fodor, J.A.: 1990, *A Theory of Content and Other Essays*, Cambridge, MA: MIT Press.

Goodman, N.: 1974, *Language of Art.*

Grice, P.: 1961, *The Causal Theory of Perception.*

Hausman, D.M.: 1998, *Causal Asymmetries*, Cambridge: Cambridge University Press.

Haybron, D.M.: 2000, 'The Causal and Explanatory Role of Information Stored in Connectionist Networks', *Minds and Machines* **10**, 361–380.

Manzotti, R.: 2000, *Intentionalizing Nature*, Tucson 2000, Tucson: Imprint Academic.
Manzotti, R.: 2001, 'Intentional Robots. The Design of a Goal Seeking, Environment Driven, Agent', *DIST*, Genova: University of Genoa.
Manzotti, R.: 1998, *Emotions and Learning in a Developing Robot. Emotions, Qualia and Consciousness*, Napoli, Italy: World Scientific.
Manzotti, R. and V. Tagliasco: 2002, 'Si può parlare di coscienza artificiale?', *Sistemi Intelligenti* **XIV**(1), 89–108.
Martinoli, A., O. Holland et al.: 2000, *Internal Representations and Artificial Conscious Architectures*, California Institute of Technology.
Maturana, H.R. and F.J. Varela: 1980, *Autopoiesis and Cognition: The Realization of the Living*, Dordrecht: D. Reidel Publishing Company.
Maturana, H.R. and F.J. Varela: 1987/1998, *The Tree of Knowledge: The Biological Roots of Human Understanding*, Boston: Shambhala.
McFarland, D. and T. Bosser: 1993, *Intelligent Behavior in Animals and Robots*, Cambridge, MA: MIT Press.
Merleau-Ponty, M.: 1945/2002, *Phenomology of Perception*, London: Routledge.
Millikan, R.G.: 1984, *Language, Thought, and other Biological Categories: New Foundations for Realism*, Cambridge, MA: MIT Press.
Newman, A.: 1988, 'The Causal Relation and its Terms', *Mind* **xcvii**(388), 529–550.
O'Brien, G. and J. Opie: 1997, 'Cognitive Science and Phenomenal Consciousness', *Philosophical Psychology* **10**, 269–286.
Pavlov, I.P.: 1955/2001, *Selected Works*, Honolulu, Hawaii: University Press of the Pacific.
Perruchet, P. and A. Vinter: 2002, 'The Self-Organizing Consciousness', *Behavioral and Brain Sciences*.
Sandini, G., P. Questa et al.: 2000, *A Retina-Like CMOS Sensor and its Applications*, SAM-2000, Cambridge, USA: IEEE.
Sandini, G. and V. Tagliasco: 1980, 'An Anthropomorphic Retina-Like Structure for Scene Analysis', *Computer Vision Graphics and Image Processing* **14**, 365–372.
Schlagel, R.H.: 1999, 'Why not Artificial Consciousness or Thought?', *Minds and Machines* **9**, 3–28.
Seibt, J.: 1990) *Towards Process Ontology: A Critical Study on the Premises of Substance Ontology*, Ph.D. dissertation, University of Pittsburgh, Michigan: UMI-Publications.
Stapp, H.P.: 1998, *Whiteheadian Process and Quantum Theory of Mind*, Claremont, CA: Silver Anniversary International Conference.
Steels, L.: 1995, in G. Trautteur (ed.), *Is Artificial Consciousness Possible? Consciousness: Distinction and Reflection*, Napoli: Bibliopolis.
Togawa, T. and K. Otsuka: 2000, 'A Model for Cortical Neural Network Structure', *Biocybernetics and Biomedical Engineering* **20**(3), 5–20.
Whitehead, A.N.: 1925, *Science and the Modern World*, New York: Free Press.
Whitehead, A.N.: 1933, *Adventures of Ideas*, New York: Free Press.
Zeki, S.: 2003, 'The Disunity of Consciousness', *Trends in Cognitive Sciences* **7**(5), 214–218.
Zohar, D.: 1990, *The Quantum Self*, New York: Quill.

CLAUS EMMECHE

CAUSAL PROCESSES, SEMIOSIS, AND CONSCIOUSNESS

ABSTRACT. The evolutionary emergence of biological processes in organisms with inner, qualitative aspects has not been explained in any sufficient way by neurobiology, nor by the traditional neo-Darwinian paradigm – natural selection would appear to work just as well on insentient zombies (with the right behavioral input-output relations) as on real sentient animals. In consciousness studies one talks about the 'hard' problem of *qualia*. In this paper I sketch a set of principles about sign action, causality and emergent evolution. On the basis of these principles I characterize a concept of cause that would allow for a naturalistic explanation of the origin of consciousness. The suggested account of causation also turns the 'hard problem' of *qualia* into the easier problem of relating experimental biology to experiential biology.

1. NEW APPROACHES TO LIFE AND CONSCIOUSNESS

The past 15 years have witnessed a considerable increase in scientific and philosophical consciousness studies, including research into the material processes related to phenomena of consciousness. This is well reflected in the recent development of cognitive science. Cognitive science studies information processing in the mind in a cross-disciplinary fashion, drawing on research in neuroscience, psychology, logic, and artificial intelligence (especially conceptual modelling based on neural networks). Even though researchers in cognitive science originally did not focus on the study of consciousness, they found they were unable to escape philosophical questions concerning conceptualization, the functioning of symbols, intentionality, reference, and knowledge. In brief, cognitive science found itself saddled with the problem of how to account for the *aboutness* aspect of consciousness-conscious processes (like the processing of symbols and similar intrinsically intentional phenomena) are *about* something, and usually refer to something other than itself. Semioticians have not hesitated to point out that these concepts pertain to *significance*, and thus are located within the sphere of interest of any theory of sign processes.

Later on in the 1990s, 'consciousness studies' established itself as a field of research with separate journals and large conferences. Consciousness studies tries to overcome the traditional sceptical position of the 'hard'

J. Seibt (ed.), Process theories: Crossdisciplinary studies in dynamic categories, 313–336.

sciences which supposed that one could not deal in any serious theoretical fashion with subjective phenomena – i.e., with phenomena which hitherto were studied only phenomenologically 'from within' (or even by very naive forms of 'introspection'), or were investigated by relating data 'from without' of human brain activity (gained by various scanning methods) with the verbal reports of experimental subjects communicating their simultaneous experiences 'from within' of doing different tasks. In the same period, traditional philosophy of mind seemed to 'rediscover' its proper object (Searle 1992) and again became a flourishing area of research. Indeed, philosophy of mind was inspired by cognitive science to state (or reformulate) the so-called 'hard' problem of consciousness (Chalmers 1996). Similarly, cognitive semantics (Lakoff and Johnson 1999) and 'new AI' or new robotics (Ziemke and Sharkey 2001) increased the interest in new conceptions of knowledge and language as phenomena that are always strongly tied to the condition of being realized through a body ('embodied knowledge') – as 'enacted' phenomena in interaction with a surrounding environment in specific situations ('situated cognition') and expressed in sign systems where meaning is grounded in basic metaphors relating to the body and the specific context in which that local agent is embedded.

Furthermore, within the philosophy of biology interest shifted from a narrow focus on problems within a neo-Darwinian conception of evolution towards a more semiotic perspective. Within neo-Darwinism, the evolution of species is taken to be the result of natural selection of the 'fittest' variants of the set of phenotypes (or 'interactors'), themselves being an ontogenetic and molecular product of inherited genotypes (or 'replicators'). The neo-Darwinian paradigm operates with an account of evolution as an algorithmic and mechanist process and due to this fact the emergence of physical systems capable of processing experience and signification remains a deep explanatory problem. From the neo-Darwinian point of view natural selection works on insentient zombies just as well as on sentient animals, provided they have the same behavioral input-output relations and the same functional architecture as sentient animals. The natural history of signification remains unexplained, and it is this lack of explanation, or at least the inconceivability of such a process within a paradigm constrained by a mechanist metaphysics, that biosemiotics seeks to remedy. Biosemiotics[1] does not contest the concrete findings and explanations of neo-Darwinism as a limited scientific research programme, but questions any assertion as to the completeness of that framework vis-à-vis all aspects of evolutionary processes. As a corrective theoretical enterprise, biosemiotics contributes to an investigation of those questions that have been dismissed due to the materialist and reductionist assumptions of neo-

Darwinism, such as the question about the emergence of consciousness. Other research areas of theoretical biology – e.g., the new form of interactionism in the evolution-development debate called 'developmental systems theory' (see Oyama et al. 2001) and the applications of complexity research to modern biophysics by Stuart Kauffman and others – contribute in tandem other 'missing links' for a more coherent theory of evolution, and serve as additional inspirations for the biosemiotic project. These theoretical developments open up new perspectives on the processes linking consciousness, body, organism, and environment, including the qualitative aspects of consciousness that have been neglected due to traditional metaphysical and methodological presumptions of natural science.

The purpose of this paper is to offer some suggestions about the concept of cause that is needed for a biosemiotic understanding of the origin of consciousness in evolution. Thus is a much more limited project than trying to sketch any specific theory about the evolutionary emergence of consciousness. Within the current biosemiotic literature there are various vaguely formulated ideas about an alternative concept of cause which could be used to overcome the problem of dualism, and to integrate physical, behavioural, and phenomenological descriptions of the phenomenon of consciousness. Given that semiotics takes its departure from the work of C.S. Peirce, the latter's own-very general-concept of final causation in Nature is an obvious point of departure (Santaella-Braga 1999; Hoffmeyer 2002). However, the Peircean notion of final causation needs to be reassessed in view of the results gained in nearly hundred years of subsequent research in physics, biology, and process philosophy. In particular, I shall here consider the possibility of integrating (a) the understanding of sign action and interpretation within biosemiotics with (b) a special elaboration of the concept of cause from the point of view of non-linear dynamical systems theory and complexity research (see Emmeche 1997 for a brief introduction to this field). Thus, I will investigate whether and how a reformulation of the notion of causation, as biosemiotics and complexity research suggest, can contribute to an understanding of the origin of consciousness in evolutionary history.

2. GENERAL PRINCIPLES FOR A NATURAL SEMIOTICS OF CAUSES

Consider the following six principles.

1. There are several types of causes.
2. Causes are real on several levels.
3. Signs act in nature and enter into networks of causes.

4. By emergent evolution new types of causes are generated.
5. Causes are associated with levels of signs.
6. The causes within a complex include causes within the components.

I hold that these principles (which are not logically independent – e.g., (5) can be derived from 3 and 4) highlight central aspects of the notion of cause and causation needed in a comprehensive theory of evolutionary history. My main task in this section is to elaborate on these six principles and to try to render them plausible.

1. *The principle of causal pluralism.* Complex things are the outcome of complex processes and thus they have many kinds of causes: effective, organizational, material, semiotic. It is a physicalist presumption that only elementary particles harbour the causal powers of the universe in which we are situated. On the contrary, we must allow that *causes* – by which we mean real powers in nature, in mind, and in society which change things, process information, and develop the richness of phenomena in the world – can have a plurality of characteristics, and that we can achieve an understanding of these by different forms of inquiry. This idea is not new but goes back (although in a substance-metaphysical framework) to Aristotle who distinguished between different types of causes.

2. *The principle of causal realism on several levels.* The causes are located in nature, not merely in our description of nature, and nature has several levels. Physicists talk about the quantum ladder spanning from quanta to atoms, molecules, and so on. Biologists talk about cells, organs, organisms, populations and species. The properties of the phenomena at higher levels cannot normally be reduced to the properties of the entities or processes at the lower levels. A 'thing' or entity at level n may have its own causal powers interacting with other entities at that same level. Such entities may be found to be an organized processual product of interacting components belonging to level $n - 1$ (e.g., an organism, being composed of cells, causes changes in other organisms). The items in what we vaguely call the physical world have causal powers, but the same applies to phenomena in what we just as vaguely call the psychic[2] world. Also social phenomena like institutions, and abstract entities like numbers and rules of inference, are governed by constraints with causal consequences. In an institution the individuals are constrained by social rules, and in her thoughts the mathematician is constrained by abstract rules, and indeed, even within a cell the molecules are constrained by their functional relations to other molecules defined by the whole causal network of metabolism.[3] A thought can cause the following thought, and in their scientifically accessible modes of being thoughts are associated with (i) biological states in the parts of an organism (especially those parts in the

nervous system that process signs, though not exclusively in the brain), and (ii) the environment of the organism. There is nothing mysterious about conscious and physical phenomena both having causal powers. Apart from the mentioned physicialist bias there is nothing in the concept of a cause that would prohibit assigning causal powers to non-physical items.

3. *The sign principle.* Causes exist in nature, including that very large part of nature that brings about (generates, processes, and interprets) signs. This view of causation is the semiotic account of signs as very general processes, active in nature as well as in mind, which constitute the very precondition for human beings – and living beings in general – to know their worlds, their Umwelten. This approach to signs – versions of which are known since antiquity – received its most comprehensive elaboration in Peirce, for whom signs are triadic relations developing in time[4] through an interplay between lawlike tendencies and spontaneous random perturbations, and mediated by an interrelationship to other signs; or, as one may say, through the interplay with historically determined coding systems. Signs are not simply mental or psychic constructions of individual persons' brains, they are relationally extended within the physical space (so that something physical has to instantiate or realize them); even 'virtual' signs in information technology systems have this 'material' aspect. Yet this 'materialist' aspect of the sign principle does not commit us to physicalist reductionism, as higher order 'spaces' are embedded within, yet irreducible to, the physical space. (The physics cannot be ignored, of course, since physical laws set limits on the amount of information that can be contained in a limited space defined by some number of atoms, and on how fast signs can be transmitted; physics cannot tell much, however, about the meaning of signs on higher levels of signification.) Furthermore, the reality of signs does not imply any thesis about correspondence or simple isomorphism between the signs of nature and our theoretical knowledge. The semiotic realism of the sign principle is not a claim about simple correspondence between language and reality, or about the truth of a sentence being reducible to its truth-conditions. It is a way to express the reality of the existence of a plenitude of signs in a human universe; the fact that the human universe is filled with signs connecting nature and culture in hybrid ways. Implied in the claim of the reality of signs is that signs have potential or actual causal roles, understood here as a capacity for determination, which is a causal notion broader than efficient causality (cf. Santaella-Braga 1999). This is evident Peirce's definition of a sign. "A Sign, or Representamen, is a First which stands in such a genuine triadic relation to a Second, called its Object, as to be capable of determining a Third, called its Interpretant, to assume the same triadic relation to its Object in which

it stands itself to the same Object" (CP. 2.274). This doctrine of the causal reality of signs has a *principle of symmetry* built into it. When we claim that parts of nature create, process, and interpret signs, we can just as well read such a statement as the claim that the signs themselves (in ways that are open to scientific inquiry) generate, manipulate, and interpret natural processes. It is an anthropocentric presupposition to restrict all agency to human subjects. The signs deal with us, just as much as we deal with the signs.

4. *The principle of emergent evolution.* Throughout the natural history of the universe a continuous 'creative' evolution has taken place; an evolution of new types of systems and process types has appeared on higher levels based upon already existing conditions and simpler components and processes. These levels are described today by the empirical natural and human sciences in a mosaic of stories about, among other things, the splitting of the four physical forces in the early universe and the separation of matter and radiation; the generation of new stars and galaxies and clusters of galaxies; the generation of solar systems with planets with individual geophysical and geomorphologic characteristics; the generation of life in a few 'lucky' places (such as here); the creation of the first multicellular organisms; the rise of animals with mental representations (a new kind of 'inner life'); the arrival of social systems; the emergence of human beings with their language and culture; the generation of states and higher forms of civilization. For each coming of a new type of system there appear some new properties, processes, patterns and forms of movement, which in comparison with the former types are *emergent* in the following sense. They are (a) radically new, that is, with new properties characterizing the macro-level system rather than its component parts; (b) they are non-predictable from knowledge about the initial conditions and the guiding laws or tendencies; (c) they function as real causal constraints for the component processes that partake in this new whole structure. In (a), 'radically new' typically means 'irreducibility' in (i) principle (*de iure*), and (ii) praxis (*de facto*). The generation of new system types and the generation of a causal dynamics that characterizes them are simply two sides of the same coin.

5. *The principle of emergent sign levels.* There are different levels of nature's handling of information, that is, generation of signs, translation, coding, re-coding and interpretation of signs, all operative within the organism as well as between the organism and its emergent environment. Sign processes at a certain level (the 'focal' level) can have specific characteristics that cannot by any simple method be deduced from (or reduced to) lower levels of sign processing. Such processes would then be emer-

gent compared to their parts, which means they have (at least seemingly[5]) irreducible properties. As emergent relational entities, these signs have a real existence, *sui generis*, and partake in a causal network together with other signs on the same focal level. In this world of signs, which is simply the signifying aspect of what normally happens in the material world, there are both continuous transitions and gradations of intensity[6] of the various meanings, as well as more sudden 'jumps' between levels of signification; for instance, in the contrasts between different coding systems (as in any 'semiosic architecture', cf. Taborsky 2002). This dialectic between continuity and borders between levels is not something unintelligible and is not true only for signs. It can also be seen in simple self-organizing systems. For instance, in oil heated in a frying pan one can see the formation of heat convection cells (drop in some thyme powder or pepper, then it's easy to observe): a single molecule can be constrained to the middle of a convection cell, or circulate around in the periphery of the cell, and eventually be transported from one cell to its neighbor as the cells are continuously connected. Yet there are two levels: a level of the continuous liquid of high viscosity constituted by an enormous aggregation of individual molecules, and a higher level forming the pattern of convection cells, the level that introduces distinctions in the continuum, distinctions of cells, bordering zones, centers, peripheries, ordered directions of movement. [7] The oil that is organized into these 'cells' (which are far from being alive in any biological sense) can be understood as a form of 'proto-semiotics', or physiosemiosis (as sign activity occurs in the non-living chemical and physical realm, "in the background" as it were, "throughout the material realm" (Deely 1990, p. 30)). Suddenly the different regions of the liquid are ascribed ('objectively' as it were, not due to the ascriptions of an external observer) a new significance, namely, to be center or periphery of this or that cell. Obviously, we get more complicated relationships on higher (bio- as well as glottosemiotic) levels between the emergent meaning of a whole and its component parts, where the parts combine to determines the meaning of the whole, and the whole conversely determines the meaning of the parts. The meaning of a DNA sequence depends, among other things, upon its neighbouring sequences, as the meaning of the individual words in a sentence depends on the meaning of the whole sentence, and in fact on a wider pragmatic context. However, the meaning of the whole is also determined by its parts. The individual DNA sequences co-define an organism (together with an abundance of extra-genetic factors), as the meaning of individual words co-determines the meaning of a sentence. This interplay between wholes and parts is a general organicist principle (cf. Gilbert and Sarkar 2000).

6. *The principle of inclusion.* The higher levels presuppose and include the lower. Yet knowledge and understanding of the lowest levels presuppose as a rule the higher.[8] The principle of inclusion is important both (A) generally, regarding the emergent levels, and (B) particularly, regarding signs.

(A) With respect to the emergent levels the implication is that the biological includes the physical, even though physics does not fully explain all biological phenomena. A bacterial cell is an organized system of physical processes and doesn't stop being so while unfolding its biotic and semiotic activities; the biological 'laws', habits or regularities governing its metabolism do not in any way break the laws of physics (a vitalist belief in some non-physical life force governing metabolism or the embryologic form-generation is rejected). But this does not imply that physics could specify concepts like "genome", "flagella", "cell wall" or "signal transduction", needed to describe the bacterial way of life. The bacterium is one peculiar way to organize a system of physical processes, and physics cannot fully account for the peculiarities of this organization. There are biological principles (like the cell's regulatory memory encoded in DNA and the overall structure of the cell) governing the physics of a bacterium. We have a parallel situation when we look at the psychic level. The psyche of a human being is an organization of experiential, conscious, and subconscious processes being realized in a biological (and physical) system, but this organization is emergent relative to the biological and the physical. It is in this sense that we should consider psychic processes as included or embedded within biologic processes.

(B) Another important form of inclusion is the semiotic. Sign processes come in various types, and the higher forms include the lower. This insight can be drawn from the classifications (and tri-partitionings) of signs in Peirce's writings. The following talk about sign classification easily could give the impression that a token of a sign type is a particular entity, which is false. However, to re-emphasize, according to the nature of the sign, a particular sign is a triadic relational process (and Peirce was indeed a process thinker). A basic partitioning applies to what the very sign is 'in itself' (i.e., a highly virtual being, apart from its functions as signifying the object and as generating another sign, the interpretant): The sign in itself can be (1) *qualisign*, (2) *sinsign*, or (3) *legisign.* A qualisign is a sign of a mere quality like 'redness'. A sinsign is a singular sign of such a quality, like a token I may experience of particular redness here and now. A legisign is a sign that is a type, like the word 'redness', of which the present text has several tokens (i.e., sinsigns as individual replicas of the legisign 'redness'). In this semiotic context, the principle of inclusion (cf. Liszka

1996) implies that the higher categories of signs include the lower: The sinsign incudes the qualisign, and the legisign includes the sinsign in the sense that it has to be realized through particular existing sinsigns. Without going into details it should be mentioned that inclusion also applies for the higher trichotomies of signs, that is, the tri-partitioning according to the (similarity-, or referential, or lawlike) relation between the sign and its object, i.e., (a) *icon*, (b) *index* and (c) *symbol* (so that an index includes an aspect of iconicity; a symbol involves an index of some sort); as well as the tri-partitioning according to the sign's relation to the interpretant, i.e., being (I) a *rheme* (a sign which for its interpretant is a word-like sign of qualitative possibility), (II) a *dicisign* (a sign which for its interpretant is a proposition-like sign of actual existence), and (III) an *argument* (a sign which, for its interpretant is an inference-like sign of a general regularity, habit or law). Thus the activity of complex signs includes the activity of less complex signs; for instance, if an argument is put forth, this involves the processing of singular propositions and individual words. We return later on to the connection between this principle and its application to emergent levels and to signs.

3. ELEMENTS FOR A THEORY OF THE NATURAL HISTORY OF EXPERIENCE

From the principles set out in the previous section we can begin to catch a glimpse of the contours of an evolutionary theory of the emergence of experience in nature's history. The normative idea of *experimental biology* has for long been one of an objective science based upon the conduct of well-controlled experiments on observable properties of organisms; properties any researcher could access from without as being part of a public sphere of observation. The fact that organisms, including the researcher as a person, always have an 'inner' experiential sphere – we experience phenomena in a way that has an intrinsic qualitative value – was not thought to have any role in the idea of biology after Darwin. Yet this aspect of life, the subject of what we could call *experiential biology*, cannot (or at least not without great difficulty) be accounted for by means of the 'objective' methods of science. Some philosophers talk about 'qualia' to denote the special subjective character of experiences: the fact that roses (or the molecules they emit?) have this particular attractive scent; and that a wet dog has a distinctive other scent; or that light of a wave length of 600 nanometer is experienced as the colour quality of orange.

Questions such as "how do particular properties of the physical acquire particular irreducible experiential qualities?", or "what is the causal connection between the physical universe and our subjective experience?", are often perceived as old and unsolvable philosophical conundrums. The idea that a handful of principles like the above mentioned should enable us to solve such questions may sound rather far-fetched. Yet let us inspect for a moment an early proposal to this effect: that of Uexkull's. A biosemiotic pioneer, Baron Jakob von Uexküll already tried something similar a long time ago by founding what he called "Umwelt-research", i.e., research into the subjective Umwelten of animals, the Umwelt being the subjective aspect of the world experienced by the organism (Emmeche 1990; Kull 2001). However, the conservative Uexküll would not have embraced the 4th principle as he did not fancy the Darwinian theory of evolution. One of the advantages of biosemiotics in its contemporary form, apart from its basic evolutionism, is that it does not force upon us a dualist metaphysics that separates the phenomena into two distinct worlds or realms which are afterwards difficult to reconnect again. Peirce, to booth, was a monist – he did not believe in the existence of radically different ontological domains or 'types of substances' and developed his own form of process thinking (Rescher 1996). His monism was semiotic, and conjoined with an ontological category theory, based on the categories of firstness (possibility), secondness (existence), and thirdness (reality). The third category includes phenomena of a lawlike nature; processes; the generation of habits, and semiotic phenomena.[9] Let us see how we can use some of the conceptual building blocks of the Peircean system in contemporary thinking about life and consciousness.

The leading idea is to construe the appearance of consciousness in evolution as a process that in many ways resembles the emergence of other complex systems on higher levels – and here we can draw on insights from biology, complexity research and cognitive science – but without tying the description to a physicalist and objectivist ontology (implied by many contributions from the mentioned areas) that allows us to account only for the world's 'outer' or behavioral properties. From the outset we will acknowledge that some complex systems (as for instance animals) may have emergent phenomenal properties which are only directly accessible from within those very systems themselves, yet such phenomenal properties being no less real than behavioural properties accessible to public observers. This assumption needs to be backed up by Peircean semiotics (seeing signs in organisms as including qualitative experiential aspects) as embedding theoretical framework. As John Searle (1992, p. 97ff) pointed out, the standard model of observation for natural science – presupposing a clear distinction drawn between a subjective observer and the object observed-breaks down in the case of consciousness. This distinction, and the model of scientific observation that rests on it, does not hold for consciousness because of its specific mode of being real, i.e., because of it being at once observer and observed. The traditional model of observation is a basic obstacle to a biological understanding of consciousness (a conclusion Searle was not ready to draw). Even an understanding of complex processes of living systems other than consciousness demands multiple, inner as well as outer, perspectives (Van de Vijver et al. 2003). Thus there are good reasons to include the semiotic approach as a fully valid method on a par with traditional 'objective' methods, in particular – but not exclusively – in order to account for direct conscious experiences. The idea about signs, or sign action, as genuine and real processes with qualitative aspects is an important key to an alternative research framework in consciousness studies – and a key to an alternative philosophy of science with a broader view of what the set of acceptable methods may comprise.

Let me begin with a brief explanation of how the generation of a complex system affects its parts. In a sense, we are dealing here with the "mind over matter" formula, though purged of any spiritistic mysticism and dualism. There is nothing mysterious about a pattern having the power to govern its parts. The cell-like or beehive-like patterns that emerge in a process of self-organizing convection cells on a frying pan will to a large extent determine the trajectories of the individual molecules. After the initiation of the self-organization of the pattern of convection cells, the movements of the molecules are constrained by the pattern (cf. Swenson 1999). A lipid molecule in the oil cannot any longer move around by random diffusion (as in a liquid where the molecules realize Brownian random motion), but is now forced into an emergent pattern of movement. This is what the notion of 'downward causation' implies, a modern form of the Aristotelian idea of a final cause (Emmeche et al. 2000). The form (pattern, or mode) of movement constitutes a higher level constraining or 'governing' the movements of the entities at the lower level. Complex systems studies within physics is rich in examples of this kind, and once one is familiar with this way of conceiving causality, form, and interactions between levels, one finds few cases in biology where this principle is not at work, as biosystems intrinsically involve several levels of causality. (A further example of downward causation might even be the influence of such a paradigmatic idea upon our capacity to identify individual empirical cases as a further example of downward causation). A certain species in an ecosystem has to adapt continuously to fluctuations in climate, nutrients, competitive interactions with other species, and so forth, but if a species is decimated or even driven to extinction by a competitor, this does not necessarily have to change the overall dynamics of an ecosystem; a new species may simply take-over and occupy almost the same niche as the former (Ulanowicz 1997). Thus, there are structures in the ecosystem, e.g., in food webs, that allow particular elements to be substituted and define criteria for their remaining in the system, and in that way constrain the possible trajectories of evolution for the elements of the system. In a similar way, an enzyme within a cell will not chemically react with anything in any way, but is functionally bound to realize very specific catalytic processes; thus one may conceive of function as the active ascription of significance by the whole to the individual biological parts (Emmeche 2002).

Downward causation is a form category of cause that must not be confused with the time-sequential 'effective' cause. The renaissance critique of the Aristotelian variant of the principle of causal pluralism left one single type of cause as the only legitimate candidate for use in causal explanations: the effective cause, interpreted as sequential in time, "if A (cause) now, then B (effect) thereafter". In contrast, downward causation 'from' the emergent level 'to' the individual parts of the system is not something to be understood as being extended in sequential time; it is rather the form of movement (through 'phase space') which the whole system forces upon the individual elements. The analysis of the physics of complex systems offer some good analogies for an understanding of this kind of structural causation. From the physical point of view, phase space is structured; it contains regions of specific types of attractors (e.g., point attractors, cyclical attractors, 'strange' or chaotic attractors) who delimit the possibilities that a system in its movement through phase space may choose, even when 'choices' are real (as in bifurcations). Even though such analogies most often can be drawn only in a loose and metaphorical fashion, it is still possible to achieve a schematic understanding of conscious processes – e.g., the visual consciousness of a prey animal and its pragmatic decisions inferred from the percepts of a predator – by considering such processes as a system's movement through an abstract space of possible brain states. This abstract space is a phase space with certain attractors that govern the activity of local clusters of neurons (within a certain range of variation determined by random processes). In such a 'weak' version of downward causation (Emmeche et al. 2000) consciousness manifests itself as a pattern governing the behaviour of the individual neurons or neuron cluster. In fact, consciousness can be perceived as a form of movement at a high level that is co-determining the behaviour at the micro-level of the individual cells and molecules in an organism and its surroundings. In other words, the idea is to use a dynamical approach to cognition (cf. Skarda and Freeman 1987; Port and van Gelder 1995; see also Newman 2001) but without eliminating or reducing the qualitative aspects of the phenomena of consciousness.

4. CONSCIOUSNESS AND EXPERIENCE

It may sound bizarre to treat self-organizing heat convection cells on a frying pan as something that should have any bearing upon our understanding of consciousness, but the basic idea is not some form of panpsychism, according to which consciousness should be found – though of very low intensity – even in such a simple and purely physical system. What is expected to be found in lower intensities are specific sign processes, that is, signs producing and mediating other signs. Consciousness is, in contrast, an emergent phenomenon, associated with particular forms of sign action in particular kinds of systems: self-moving autonomous organisms-animals. The concept of experience describes the continuous scale between very simple and very complex forms of sign activity. The concept of consciousness describes a jump in this continuum. *Experiences* – understood as traces of particular significant interactions between a system and its surroundings that for some period is represented within the system – exist as coded even in pure physical systems (like moon craters who indexically 'encode' earlier meteor impacts), although it is only with animals as a system type that the full implications of this concept is unfolded. In this way we may be able to settle the controversy between, on the one hand, a pansemiotics that somewhat unprecisely claims every phenomenon in the universe to be semiotic (which is hardly true since both pure chance and the brute facticity of here-and-now – in Peirce's terms, the phenomena of firstness and secondness – are not yet semiotic), and on the other hand, a restrictive variety of biosemiotics, claiming any semiotic activity to be co-extensive with biological activity, that is, that one should not be able to conceive of sign processes before the advent of life on Earth. By contrast I suggest that we neither consider everything to be signs nor believe only biological life to realize semiosis. One must distinguish between semiotic processes in both the physical, the biological and the psychological systems, and thus conceive of the formation of experience as a fundamental requirement for the particular type of semiosis that during the course of evolution is intensified as consciousness.

In other words the thesis is that there are comparable semiotic processes at the physical and biological level. Our contemporary physical universe with it characteristic chemical elements is a particular way of 'coding' the energy of the universe (Christiansen 2002; Taborsky 2002) – i.e., this energy is not dispersed in a big undifferentiated porridge, but is exactly differentiated into matter and radiation (a difference that really makes a difference), and the matter again is differentiated into the well-known elements (the periodic table being our rational symbol of the universe's own coding of the elementary particles into different kinds of atoms). Similarly the biological phenomena are a particular way of 'coding' organic chemistry. The organisms are not dispersed into a big undifferentiated porridge of macromolecules (lipids, proteins, nucleic acids, carbohydrates etc.); these substances are codified into the particular autocatalytic system of the cell with its network of biomolecular signs, which continuously maintain both the network and its boundary, the cell membrane. The evolutionary experiences acquired by the cell in its natural history are coded, partly in the form of sequence information in DNA, and partly in the whole complex structure that a 'modern' (pro- or eukaryotic) cell embodies. If we move further up some levels of organization, to animals with a nervous system, we can observe in a similar manner that new, psychic, processes (processes like proprioception, movement, motor coordination, action, perception, attention, consciousness) do not simply constitute an undifferentiated continuum of signs, information processes, or 'computations', but are organized into emergent structures, in which the animal's experiences with its surrounding milieu are continuously transformed, re-built, confirmed, encountered, felt, challenged, re-interpreted, and which forms a rich (emotive, volitive and cognitive) structure of feelings, desires and thoughts; something genuinely semiotic and crucial for the continuation of life of the organism.

Let us take a closer look upon the generation of the specific form of signs we called experiences, within a macro-evolutionary perspective.[10] We can do so by formulating a new principle, closely connected to the earlier ones:

In macro-evolution experiences are intensified from movement to consciousness. The idea is (a) to relate consciousness to certain especially salient 'jumps' (or instances of emergence) seen in physical and biological evolution, primarily the transition from 'plants' (loosely conceived) to animals, i.e., from multicellular organisms with no nervous system to those with such a system; and (b) to relate this jump to a leap within the semiotic aspect of the same macro process – from experiences as simply past-directed and fossil-like signs, to active, sensing, feeling, and future-directed signs (intentions, as intentionality is often a directedness towards possible future state of affairs not yet realized). Let us call this:

7. *The principle of semiotic intensification.* Signs are found at the physical, the biological as well as the psychic level, and the same applies to experiences (as we have broadly construed them here). Experiences are

fossilized signs (cf. Hansen 2000) or quasi-stable forms of movement that organize the system's past forms of movement in such a way as to have significant consequences for the system's future movement. Since they are, they necessarily involve relationshis between the following three elements: the physical carrier of 'the fossil' (the representamen); its reference to its significance (the object); and its potential or actual future-directed effects (the interpretant(s)). The intensification of the sign process takes place at several levels, it is at once physical, biological and psychic (see below). With the emergence of coded autocatalytic life on cell form, the semiotic freedom[11] is intensified at the biological level. Here, semiotic intensification manifests itself both by the appearance of qualitative irritability[12] (in cells who selectively can respond to stimuli) and by the emergence of code-duality in the form of cell-lines (with a digital as well as an analog aspect, cf. Hoffmeyer 1996) incorporating past experiences into the future. This semiotic freedom is greatly expanded later on with the neurally based forms of sign action we observe with the arrival of animals. Intensification is to be understood both as qualitative and quantitative. One might conceive of measuring the informational band width of an instance of semiotic processing such as cognition in chimpanzees and attempt to operationalize this as information transmission per time unit in the brain modules where the processing is going on. But apart from the theoretical and methodological problems that would be generated by this endeavour, such a quantitative measure does not catch the qualitative and content-related aspects of this kind of sign action. Yet we seldom doubt that a chimpanzee experiences a content of its sensing or perceptual 'measuring' of the environment. The idea of semiotic intensification is an attempt to make explicit the intuition that animals experience their world with greater depth and diversification of content than plants, and that something similar applies when we compare elephants to flatworms, or grown-up animals to embryos (fully acknowledging the fundamental difficulties involved in these kinds of comparisons). In a very general sense of the word *experience* one can say that all these systems, even the purely physical, *experience* something, get 'irritated' or affected by their surroundings, and store this influence, even when such stimuli are quite evanescent or produced by chance. In that generalized sense, process and experience are interrelated in all situations where the *process* of interaction between one subsystem (corresponding to an agent) and another subsystem (corresponding to the environment of the agent) leaves traces in one of the subsystems. But only in complex living systems (showing history, multiple levels, and built-in genetic, neural, or psychic mechanisms of selection as an element of the coding processes or memory) the formation of experience has been able to achieve an intensified form that makes it reach forward in time. This renders such a system anticipatory, i.e., endows it with the capacity of operating with models of possible future states, including what has been called 'mental models'. Such models have both a formal outer aspect (as when a neural code within the visual cortex is described by scientists in terms of algorithms for edge detection, object recognition, etc.), and an aspect of being amenable to sensory experience from within the system (for example, the model can be sensed as colours, smells, sounds, touches and so on). Thus experiences can be described both objectively in terms of (grammatical) third person predicates, as when we investigate Peter's or the chimpanzee's mental model of a banana, and subjectively in the form of descriptions of the grammatical first person, giving others indirect access to one's own phenomenological experience of a banana tree.

In a physical system like a tornado (an open, metastable, dissipative, self-organized system) there is no marked distance between outer and inner, nor between past and future. Talking about a tornado's spatial differentiation of 'eye' and 'body' is not meant to imply any truths about a rich experiential life of tornados. For the tornado, there is little separation between a reference of the experience to the conditions that makes the same experience possible, the processes it realizes, and its immediate occurrence. The movement is hardly evolutionary, there is no difference between the units of selection and the unit of evolution; everything in the system is being 'selected' for continuing self-organizing movement given that the boundary conditions for such a type of movement of system are satisfied. The movement is identical to a simple time evolution of the system.[13] (This is why moon craters as traces of experience have a different status relative to us and relative to the moon, which is an important distinction for us 'moderns' who have no empathy with the scarred man in the moon).

In biological systems like the cell, experiences are, among other things, the genetic 'fossils' in DNA witnessing the specific proteins that were functionally participating in earlier ancestor cell lines to maintain the metabolic form of movement. Here, a high degree of temporal separation of past movement and present structure is achieved. This is due to the fact that the digital code provides stable representations of, for instance, early active but now passive genes (so-called 'pseudogenes' which have had immediate significance, but now only have potential significance for the cell life or the species as a ressource of variation and mutation). In addition, the system's boundary to its environment is sharper and functionally effective. The cell membrane represents the organismic information about a primary difference between 'inner' and 'outer' which intensifies the significance of molecular systems for measuring changes in what is 'outer' in relation to the states of the 'inner'. The constant threat of the draining of energy reserves (and thus by death) constitutes a *telos* within the system, that is, the goal of survival, a need, an overall interpretant corresponding to the future dual possibility of the death or continued life of the system. Each of these two possibilities, life or death, organize the developmental trajectories of the elements of the system around a particular attractor, of which only one of those, namely life, has a biological description in addition to the purely physical one. (The physical description of the phase space, allowing for statistical measures, has to be supplied with a 'an additional' biological or quasi-semiotic description of (minimally) a historically contingent 'sequence space' of digital codes, cf. Küppers 1992). This goal or need of survival, which already appears to emerge for free-living single cell organisms (simply constituting a continuous line of cell divisions) receives more complicated elaborations, both by the exchange of fragments of experience between the cells (e.g., by bacterial conjugation DNA plasmids can be transferred from one cell to another, a kind of sex), and by the generation of multicellular organisms with life cycle, alternation of generations and sexual reproduction, i.e., species in a 'modern' sense.

In a psychic (and thereby biological) system like a multicellular animal, experiences are sign processes that temporarily 'fossilize' as quasi-stable representations of outer forms and their relations to the organism and its inner, and thereby create traces in the form of neurally stored patterns of memory. Through the evolution of multicellular organisms, especially of animals with a nervous system, an additional intensification is achieved – partly by irritability[14] (which gets differentiated into neurally based systems of representation with outer as well as inner aspects), and partly by sign based strategies for reproduction and ecological 'competencies' (such as food search patterns). The semiotic intensification transforms merely vegetative organisms into animals, that is, it endows them with dynamic forms – corresponding to what Aristotle[15] called a soul of movement – i.e., a semiosic active systems, that through self-movement acquire experiences, cognitively process these, and have an emergent phenomenal inner world. Movement includes autonomously governed changes of form and position of parts of the organism (like muscles) based upon sign processes like proprioception and sensori-motor coordination. Movement must be distinguished from merely physical change of position over time; rather, the course of movement in animals is always governed by semiotic codes based within the animal body (an idea elaborated in detail by Hoffmeyer (2000), and Sheets-Johnstone (1998)). Movement is an externally observable process that is also well internally sensed. In simple animals the movement-governing models are identical to the immediate coordination of sensory signs from the environment and proprioceptive signs of the body, signifying states and movements of the muscles. Thereby an *Umwelt* is formed as functional circles which dynamically represent flexible interactions between the animal and its environment, i.e., a species-specific 'cut' – mediated by the sensory organs – of relevant features of the organism's physical environment is formed. The simple kinds of experiences generated in this process can later on (in animals with more elaborated systems for neural representation) be incorporated as a source for higher-order anticipatory models, not only including here-and-now coordination of movement, but also longer sequences of movement, based upon 'choice' among (or inference to the best consequences of) several possible routes of escape, or other kinds of action. Such models are symbolic in form to the extent that the experiences govern the relevant inferences in a law-like manner. (We shall not discuss the concept of symbol here, but see Stjernfelt (2001)). As a general tendency in the course of macro-evolution one can observe the intensification of experiences from movement to consciousness.

Consciousness appears as the present moment's qualitative feature of a moving animal which experiences a process of complex relations between sensing the movements of its own body and sensing the corresponding changes of the environment. Consciousness is an emergent higher order patters which (a) has genuine causal power in its own right (just like the movement patterns that are based upon experiences and govern the behavior of the organism) and (b) has a qualitative, phenomenal aspect (just like irritability). I cannot go here into details on the neurobiology of memory, proprioception and perception, or the electrochemically based processing of information in the nervous system. But it is possible to conceive of consciousness as a specific property of the dynamical interpretation of experiences, and of experiences of experiences, including proprioception; this interpretation is an ongoing affair, continuously modulated against the habit-like traces that earlier experiences have deposited in the neural codes of the body. Like any sign, consciousness is a dynamic, relational, and intentional phenomenon; consciousness connects signs outside to signs inside the organism – consciousness is nothing in itself, it emerges only in connection with the general semiosic make-up of the experiences of the body.[16] The subjective nature of experience is rooted in a semiotic intensification of those qualisigns that are parts of the subjective aspects of irritability (and 'appetite', see Brogaard (1999)) that are found in even simple organisms. The signs themselves have both formal outer aspects allowing us to re-represent them as an algorithm or a logic model, but their formalization does not exhaust their qualia character. Normally this character is shadowed by their dynamical and formal aspects, but it can be observed in direct immediate experience. This means that semiotics as a set of methods must include not only the construction of empirically testable models of dynamic and formal aspects of consciousness (as it is already done to a large extent within cognitive science) but also phenomenological approaches to the qualitative experiential processes connected to sign action. The situation calls for a *qualitative organicism* (Emmeche 2001), according to which complex systems are both emergent and capably of supporting phenomenal experience. They are considered as emergent patterns of movement with downward causality (in the sense mentioned above in which the emergent pattern of movement organizes the dynamics of the parts through new boundary conditions for their unfolding). In addition to this, complex systems as exemplified by animal bodies are also taken to realize phenomena such as telos, semiotic intentionality (cf. Peirce's notion of 'final causation') and experience formation within an Umwelt. The phenomenal aspect of sign interpretation is simply the experienced quality, the inner side of those transformations in the neural state space that are generated with this higher form of semiosis.

Thus a new concept of *qualitative complexity* is abducted: A system of processes is qualitative complex if (i) the system is self-organizing and has emergent properties and downward causality, (ii) the system has an Umwelt – the subjective aspect of the world experienced by the organism – giving rise to experience-based qualia (such qualia having the character of qualisigns as defined above), (iii) these qualia have causal efficacy (not in the sense of efficient causation but as a form of final causation). Merely epiphenomenal interpretations of qualia are accordingly excluded – for certain aspects of the system the fulfillment of criterion (ii) is a necessary condition for fulfilling criterion (i).

In other words, an externalist description of the motion (changes of spatial positions) of an animal within an environment is not sufficient for understanding the specific animate, flexible and graceful form of movement governed by an interactive experiential Umwelt. Within the Umwelt's embodied process of experiential becoming (incorporating a phenomenal dimension grounded in the qualisigns included in the system's semiosis), consciousness emerges as a causally consequential form of orchestrating (by downward causation) the correlative 'self' of the system, making it cohere and giving the movement its 'animate form'.

This is of course an insufficient sketch of the causal principles that must be taken into account in a future processual and biosemiotic theory of consciousness, yet I hope the vague approximation to such principles

I have offered here suffices to outline the project. To restate, we have the following principles that may be used to formulate a more detailed theory about the emergence of conscious processes in evolution: (1) There are several types of causes, and one can see consciousness and mental signs as quasi-autonomous formal and final causes that are active within the complete system of an animal in its surroundings. (2) The causes are active at several levels and one can consider the psychic as the level for conscious sign action (this level also include non-conscious signs). (3) Signs act in nature and enter into networks of causes; within the body conscious signs enter together with other signs in causal networks. (4) By emergent evolution new types of causes arise, and experience-based movement in animals is such a type. (5) Causes are related to sign levels, and have, like the signs themselves, an outer as well as an inner side. (6) Causes in complex phenomena include causes in simple phenomena; and causes that regulate consciousness include causes regulating physical and biological processes in the animal as well as sign processes related to sensing and irritability. (7) In the course of macro-evolution experiences get intensified from movement to consciousness, and one can conceive of consciousness as the experiential aspects of sign processing (production, coding and interpretation of signs) in self-moving systems. (8) Consciousness is the present qualitative moment of a continuously running future-directed experiential process.

ACKNOWLEDGEMENTS

I would like to thank Frederik Stjernfelt, Stefanie Jenssen, Jesper Hoffmeyer and Johanna Seibt for critical comments on an earlier draft of this piece.

NOTES

1 It is beyond the scope of this paper to introduce biosemiotics in any detail. The basic idea of biosemiotics is to consider living systems not so much as organized molecular systems, but rather as semiotic systems (sign processing systems) where the molecular structure functions so as to mediate semiosis, or sign action. During the last decade more and more theoretical biologists have been influenced by biosemiotic ideas. For brief comprehensive introductions, see Hoffmeyer (1996), Emmeche et al. (2002), Kull ed. (2001).

2 The adjective 'psychic' (and the noun 'psyche') is used here for what we in English just as vaguely call "psychological phenomena" in order to emphasize that it pertains to the psyche as that emergent property of some organisms having an animate, experiential or 'inner' world.

[3] A detailed exposition of the inter-level relations in a living cell can be found in Bruggeman et al. (2002), see also Boogerd et al. (2002).

[4] One might think that temporally developing relations are not necessarily processes; they may be sequences of states. However, in Peirce's philosophy, process is of the nature of Thirdness, i.e., the metaphysical category of mediation, and Peirce considered signs or semiotic processes of interpretation as a temporally continuous developmental phenomenon (in accordance with his synechism, i.e., philosophy of continuity).

[5] The restriction 'seemingly' is due to the fact that even though emergent properties are defined in terms of genuine irreducibility, there may be cases where our claims concerning genuine reducibility may be changed by developments within science.

[6] It may be possible to define *semiosic intensity* precisely, but in the context of the present exposition this concept is only implicitly defined (it be clearer from what I say on the 7th principle below). It connotes (but does not equal) the semantic distinction between intensional and extensional, and the notion of intentionality in philosophy of mind. However, semiosic intensity (or intensity of meaning) is more like a measure of the number of possible experiential qualities of a sign process, and thus of the richness of its interpretation. Thus, it is related to the notion of semiotic freedom in Hoffmeyer (1996).

[7] For a more fine grained analysis of convections cells, see Swenson (1999).

[8] This second, 'upward' direction of inclusion demands a separate treatment, but is not crucial for the argument of the present paper. "Downward" inclusion as, e.g., in (A), a biologic process includes physical processes, and, in (B), arguments including propositions, is primarily an ontological property, while "upward" inclusion is a mixture of epistemic and ontological characteristics; e.g., the laws of physics (not in their ontological sense but in their mode of existence as objects of knowledge) presupposing a knowing inquirer.

[9] "By the third, I mean the medium or connecting bond between the absolute first and last. The beginning is first, the end second, the middle third" (. . .) "Continuity represents Thirdness almost to perfection. Every process comes under that head", Peirce (CP. 1.337).

[10] *Micro-evolution* designates evolutionary processes within a species while *macro-evolution* comprises processes like speciation, generation of longer trends and overall patterns of form relationships in evolution.

[11] Semiotic freedom can be thought upon as a generative combinatorics of significations, see Hoffmeyer (1996).

[12] The term irritability, denoting the capacity of certain parts of the body to contract when stimulated, was introduced by the English physician Francis Glisson (*c* 1597–1677) who saw it as a property of all the body's fibres independent of consciousness and the nervous system (cf. Lawrence 1981). It has played an important role in debate between mechanicists and vitalists over the basic definition of life.

[13] This is compatible with a physical (and trivial) sense of 'time evolution' in which no reference to specific biological phenomena like natural selection based upon variation and inheritance is implied.

[14] Cf. footnote 12 on irritability. It is useful to remember that Peirce (in his 1890 manuscript "A Guess at the Riddle") viewed the irritability of the "protoplasm" as an example of Firstness: "The properties of protoplasm are enumerated as follows: contractility, irritability, automatism, nutrition, metabolism, respiration, and reproduction; but these can all be summed up under the heads of sensibility, motion, and growth. These three properties are respectively first, second, and third" (CP.1.393). Here, the phenomenal aspect of irritability (or sensibility), as a first, can be seen to correspond to the phenomenal aspect of very simple forms of signs, the qualisigns (see below). Thus, we can conceive of simple irritability as

already having outer, behavioural aspects (like a capacity to contract responsively upon a stimulus) as well as inner aspects (feeling, pain, itching).

[15] Compare Aristotle's biologically based psychology in *De Anima* (see for instance Everson 1995). No need to say that the purpose here is not to give an exposition of the hierarchic system of 'souls' in Aristotle, but to let his system inspire the interpretation of what has been called here the semiotic intensification as an evolutionary process. On the combination of Aristotle and biosemiotics, see also Brogaard (1999).

[16] Describing consciousness as being nothing in itself and only understandable as a situational, relational and embodied phenomenon, may seem to contradict my claim above that consciousness is "an emergent higher order pattern with a genuine causal power in its own right", yet, this discussion can only be clarified within a general treatment of the nature of emergence and downward causation (see, e.g., Emmeche et al. 2000). The fact that an item is a higher order phenomenon, such as a mental image, and has a causal power *sui generis*, does allow for the possibility that the item is also essentially dependent upon its constituent parts and processes and its (semiotic and material) relations to its environment.

REFERENCES

Baas, Nils A. and Claus Emmeche: 1997, 'On Emergence and Explanation', *Intellectica* 1997/2, no. 25, pp. 67–83.

Boogerd, Fred, Frank Buggeman, Cathjolin Jonker, Huib Looren de Jong, Allard Tamminga, Jan Treuer, Hans Westerhoff and Wouter Wijngaards: 2002, 'Inter-level Relations in Computer Science, Biology, and Psychology', *Philosophical Psychology* **15**(4), 463–471.

Brogaard, Berit O.: 1999, 'An Aristotelian Approach to Animal Behavior', *Semiotica* **127**(1/4), 199–213.

Bruggeman, F.J.; H.V. Westerhoff and F.C. Boogerd: 2002, 'BioComplexity: A Pluralist Research Strategy is Necessary for a Mechanistic Explanation of the "Living" State', *Philosophical Psychology* **15**(4), 411–440.

Chalmers, David: 1996, *The Conscious Mind. In Search of a Fundamental Theory*, Oxford: Oxford University Press.

Christiansen, Peder Voetmann: 2002, 'Habit Formation as Symmetry Breaking in the Early Universe', *Sign Systems Studies* **30**(1), 347–360.

Deely, John: 1990, *Basics of Semiotics*, Bloomington: Indiana University Press.

Emmeche, Claus: 1990, 'Kognition og omverden – om Jakob von Uexküll og hans bidrag til kognitionsforskningen', *Almen Semiotik* **2**, 53-67.

Emmeche, C.: 1997, 'Aspects of Complexity in Life and Science', *Philosophica* **59**(1), 41–68. (www.nbi.dk/~emmeche/cePubl/97g.complisci.html)

Emmeche, C.: 2001, 'Does a Robot have an Umwelt? Reflections on the Qualitative Biosemiotics of Jakob von Uexküll', *Semiotica* **134**(1/4), 653–693.

Emmeche, C.: 2002, 'The Chicken and the Orphean Egg: On the Function of Meaning and the Meaning of Function', *Sign Systems Studies* **30**(1), 15–32.

Emmeche, C., F. Stjernfelt and S. Koppe: 2000, 'Levels, Emergence, and Three Versions of Downward Causation', in: P.B. Andersen, C. Emmeche, N.O. Finnemann and P.V. Christiansen (eds.), *Downward Causation. Minds, Bodies and Matter*, Aarhus: Aarhus University Press, pp. 13–34.

Emmeche, Claus; Kalevi Kull and Frederik Stjernfelt (2002): *Reading Hoffmeyer, Rethinking Biology*. (Tartu Semiotics Library 3). Tartu: Tartu University Press.

Everson, Stephen: 1995, ‘Psychology’, in Jonathan Barnes (ed.), *The Cambridge Companion to Aristotle*, Cambridge: Cambridge University Press, pp. 168–194.

Gilbert, Scott F. and Sahotra Sarkar: 2000, ‘Embracing Complexity: Organicism for the 21st Century’, *Developmental Dynamics* **219**, 1–9.

Hansen, Jens Morten: 2000, *Stregen i sandet, bolgen på vandet. Stenos teori om naturens sprog og erkendelsens grænser*, København: Fremad.

Hoffmeyer, Jesper: 1996, *Signs of Meaning in the Universe*, Bloomington: Indiana University Press.

Hoffmeyer, Jesper: 2000, ‘The Biology of Signification’, *Perspectives in Biology and Medicine* **43**(2), 252–268.

Hoffmeyer, Jesper: 2002, Biosemiosis som årsagsbegreb. *Kritik* nr. 155/156, s.101–119.

Kauffman, Stuart: 1993, *The Origins of Order. Self-Organization and Selection in Evolution*, Oxford: Oxford University Press.

Kull, Kalevi (ed.): 2001, ‘Jakob von Uexküll: A Paradigm for Biology and Semiotics’ (special issue), *Semiotica* **134**(1/4), 1–828.

Küppers, B.-O.: 1992, ‘Understanding Complexity’, in A. Beckermann, H. Flohr and J. Kim (eds.), *Emergence or Reduction? Essays on the Prospects of Nonreductive Physicalism*, Berlin and New York: Walter de Gruyter, pp. 241–256.

Lakoff, George and Mark Johnson: 1998, *Philosophy in the Flesh: The Embodied Mind and Its Challenge to Western Thought*, New York: Basic Books.

Lawrence, Christopher J.: 1981, ‘Irritability/Sensibility’, in W.F. Bynum, E.J.Browne and Roy Porter (eds.), *Macmillan Dictionary of the History of Science*, London: Macmillan Press, p. 214.

Liszka, James Jakób: 1996, *A General Introduction to the Semeiotic of Charles Sanders Peirce*, Bloomington: Indiana University Press.

Newman, David V.: 2001, ‘Chaos, Emergence, and the Mind-Body Problem’, *The Australasian Journal of Philosophy* **79**(2), 180–196.

Oyama, Susan, Paul E. Griffiths and Russell D. Gray: 2001, ‘Introduction: What is Developmental Systems Theory’, in Susan Oyama, Paul E. Griffiths and Russell D. Gray (eds.), *Cycles of Contingency. Developmental Systems and Evolution*, Cambridge, MA: The MIT Press, pp. 1–11.

Peirce, Charles Sanders: 1931–1958, *Collected Papers of Charles Sanders Peirce*, Vol. 1–6, Charles Hartshorne and Paul Weiss (eds.); vols. 7–8, Arthur W. Burks (ed.). Cambridge, MA: Harvard University Press. [References: CP, followed by vol., and paragraph number].

Port, Robert F. and Timothy van Gelder (eds.): 1995, *Mind as Motion. Explorations in the Dynamics of Cognition*, Cambridge, MA: MIT Press.

Rescher, N.: 1996, *Process Metaphysics. An Introduction to Process Philosophy*, New York: State University of New York Press.

Skarda, Christine A. and Walter J. Freeman: 1987, ‘How Brains Make Chaos to Make Sense of the World’, *Behavioural and Brain Sciences* **10**, 161–195.

Santaella-Braga, Lucia: 1999, ‘A New Causality for Understanding the Living’, *Semiotica* **127**(1/4), 497–519.

Sheets-Johnstone, Maxine: 1998, ‘Consciousness: A Natural History’, *Journal of Consciousness Studies* **5**(3), 260–294.

Searle, John: 1992, *The Rediscovery of the Mind*, Cambridge, MA: The MIT Press.

Stjernfelt, Frederik: 2001, ‘Biology, Abstraction, Schemata’, in Berit Brogaard and Barry Smith (eds.), *Rationality and Irrationality*, Vienna: Hölder-Pichler, pp. 341–361.

Swenson, Rod: 1999, 'Epistemic Ordering and the Development of Space-Time: Intentionality as a Universal Entailment', *Semiotica* **127**(1/4), 567–597.

Taborsky, Edwina: 2002, 'Energy and Evolutionary Semiosis', *Sign Systems Studies* **30**(1), 361–381.

Ulanowicz, Robert E.: 1997, *Ecology, the Ascendent Perspective* (Complexity in Ecological Systems Series). New York: Columbia University Press.

Van de Vijver, Gertrudis; Linda Van Speybroeck and Windy Vandevyvere: 2003, 'Reflecting on Complexity of Biological Systems: Kant and Beyond?', *Acta Biotheoretica* **51**, 101–140

Ziemke, Tom and Noel E. Sharkey: 2001, 'A Stroll Through the Worlds of Robots and Animals: Applying Jakob von Uexküll's Theory of Meaning to Adaptive Robots and Artificial Life', *Semiotica* **134**(1/4), 701–746.

Springer Theses

Recognizing Outstanding Ph.D. Research

For further volumes:
http://www.springer.com/series/8790

Weronika Walkosz

Atomic Scale Characterization and First-Principles Studies of Si_3N_4 Interfaces

Doctoral Thesis accepted by
the University of Illinois – Chicago,
Chicago, USA

Author
Dr. Weronika Walkosz
Argonne National Laboratory
Argonne, IL
60439
USA
e-mail: walkosz@anl.gov

Supervisors
Dr. Robert Klie
Department of Physics
University of Illinois–Chicago
845 W. Taylor Street, M/C 273
Chicago, IL 60607
USA
e-mail: rfklie@uic.edu

Dr. Serdar Ogut
Department of Physics
University of Illinois–Chicago
845 W. Taylor Street, M/C 273
Chicago, IL 60607
USA
e-mail: ogut@uic.edu

Dr. Juan C. Idrobo
Materials Science and Technology
Division
Oak Ridge National Laboratory
1 Bethel Valley Rd.
Oak Ridge, TN 37831
USA
e-mail: jidrobo@gmail.com

ISSN 2190-5053

e-ISSN 2190-5061

ISBN 978-1-4614-2857-2

ISBN 978-1-4419-7817-2 (eBook)

DOI 10.1007/978-1-4419-7817-2

Springer New York Dordrecht Heidelberg London

Softcover reprint of the hardcover 1st edition 2011

Cover design: eStudio Calamar, Berlin/Figueres

Printed on acid-free paper

Springer is part of Springer Science+Business Media (www.springer.com)

Supervisor's Foreword

This Ph.D. thesis is the result of more than 4 years (mid 2005 through 2009) of dedicated and meticulous research by Weronika Walkosz at the Physics Department of the University of Illinois at Chicago (UIC). During this time, Weronika worked in two complementary areas of Condensed Matter Physics, experimental and computational, with two advisors (Klie and Ogut at UIC) and a third supervisor (Juan Idrobo at Oak Ridge National Laboratory, who was the actual PI on the grant) at different institutions in different states of the United States. That she was able to do this and finish her thesis successfully is a tribute to her intellectual capacity, persistence, ambition, and dedication.

Weronika initially started her Ph.D. research in computational materials modeling with one of us (Ogut) and worked, in collaboration with Juan, on first principles modeling of a variety of systems including $SrTiO_3$/GaAs interfaces, elastic properties of InN, and optical absorption spectra of Au clusters. With the arrival of one us (Klie) at UIC in the Fall of 2006, Weronika became more and more interested in learning and using electron microscopy techniques, in addition to her computational studies. As such, one unique aspect of Weronika's Ph.D. research, a significant portion (but not all) of which made it into this thesis, is related to her ability to strive in both computational and experimental platforms. While many Ph.D. students typically focus on either experimental or theoretical condensed matter physics, Weronika realized early on that the complex problem of the ordering transition in crystal/amorphous interfaces, such as those observed in Si_3N_4/rare-earth oxide and SiO_2 interfaces, can only be understood by using both theoretical first principles calculations, as well as experimental transmission electron microscopy. As a result, her thesis describes both theoretical and experimental study of the disorder/ordering transition in structural ceramic materials.

For many years, scientists have studied crystalline amorphous interfaces, since they play an important role in many technological materials, from semiconductors to structural ceramics and vitrified nuclear waste. In spite of the intense focus, to date an atomic-scale description of the disorder/ordering transition at ceramic interfaces has not been well understood and the models developed in Weronika's thesis are challenging the common notion of atomically abrupt crystal/amorphous interfaces.

In her computational studies, Weronika modeled bare Si_3N_4 surfaces as well as the attachment of oxygen and different rare-earth atoms to these surfaces by utilizing first principles calculations within the framework of density functional theory (DFT). She performed an exhaustive search (testing more than 350 different interfacial structures), and constructed a phase diagram for oxygenated β-Si_3N_4 surfaces as a function of relevant chemical potentials of oxygen and nitrogen. Her calculations revealed that the structures commonly observed in Si_3N_4/oxide interfaces are not energetically favorable, not even in the presence of sub-surface oxygen or rare-earth elements attached to the surface, challenging both her experimental and theoretical skills to resolve this puzzle. In the case of α-Si_3N_4, Weronika's calculations revealed that oxygen impurities could diffuse into the bulk at interstitial positions, and potentially stabilize the α-phase at higher temperature and pressures.

Her experimental work was focused on utilizing state-of-the-art aberration-corrected scanning transmission electron microscopy (STEM) to characterize Si_3N_4 hetero-interfaces and bulk defects. In particular, Weronika focused on characterizing, using high-angle annular dark-field imaging (HAADF), annular dark-field imaging (ABF) and electron energy-loss spectroscopy (EELS), the structures she predicted using DFT modeling. Some of her experimental results are the first of its kind showing that the crystalline/amorphous transition at the Si_3N_4/SiO_2 interfaces is not atomically abrupt. Instead, the first few monolayers of atoms in the amorphous SiO_2 exhibit some semi-crystalline ordering, in accordance with the structures predicted by first-principles modeling. This discovery was made possible only by the recent development of aberration-corrected STEM at low acceleration voltages to enable the visualization of individual atoms even in beam sensitive materials, such as the Si_3N_4/SiO_2 interface.

The research described in this thesis is challenging, both experimentally and theoretically. For modeling hetero-interfaces and determining the lowest energy configuration and electronic structure of Si_3N_4/rare-earth oxide interfaces, large super-cells are required, consisting of many atoms. Experimentally, these interfacial structures are very sensitive to the electron beam and only a few seconds of exposure can dramatically alter the interfacial atomic arrangement. Therefore, the fact that in the end both first-principles modeling and experimental STEM imaging and spectroscopy agree on the interfacial and bulk structures of Si_3N_4 is an amazing tribute to Weronika's ability as a condensed matter physics graduate student, who has mastered both the experimental and theoretical aspects of her project.

We believe that Weronika's unique ability of combining atomic-level structural characterization with realistic first-principles modeling of interfacial structures will have wide-ranging consequences on our understanding of many other systems. For example, in dielectric interfaces in semiconductor devices, which are approaching the quantum limit of thickness, a partial crystalline ordering at the Si/amorphous oxide interface will have significant effects of the device performance and influence the end of the roadmap scenario dramatically. The techniques and methods developed by Weronika in this thesis will aid scientists in the future in assessing the nature of crystalline/amorphous interfaces and help develop devices taking advantage of the semi-crystalline ordering discovered in this thesis.

On behalf of Juan and ourselves, we would like to thank Weronika for being a great Ph.D. student to work with and for the opportunity to see her grow into an exceptional condensed matter physicist who bridges the gap between theoretical and experimental materials research. We also would like to thank the National Science Foundation (Grant No. 0605964), in particular Dr. Lynnette D. Madsen, for supporting this research for the last 4 years, and believing in our ideas of combining first-principles modeling, aberration-corrected scanning transmission electron microscopy and electron energy-loss spectroscopy to study the fundamental structure-property relationships in ceramic materials. Finally, we are grateful to Dr. Stephen J. Pennycook for hosting Weronika at Oak Ridge National Laboratory and providing her with the opportunity of using the STEM group's outstanding facilities.

Chicago, August 31, 2010 Serdar Ogut and Robert Klie

Acknowledgments

I would like to thank my advisors Dr. Serdar Ogut and Dr. Robert Klie from UIC, and Dr. Juan C. Idrobo from Vanderbuilt University and Oak Ridge National Laboratory for their outstanding mentoring during my Ph.D. work. Also, I would like to thank the RRC group, in particular Dr. Alan Nicholls, Dr. Ke-bin Low, and John Roth who have taught me various microscopy techniques. I am very grateful to Dr. Stephen Pennycook and his group at Oak Ridge National Laboratory for allowing me to use their microscopy resources. On a personal note, I would like to thank my friends from the UIC Physics Department for making my graduate experience very enjoyable, and my whole family for their love and constant support. I especially thank my mom.

Contents

Chapter 1
Silicon Nitride Ceramics

1.1 Introduction

The widespread interest in silicon nitride ceramics stems from their desirable physical and mechanical properties in many high temperature and pressure applications [1–5]. Good resistance to oxidation and corrosive environments, low coefficient of friction and thermal expansion, negligible creep, and high decomposition temperature are some of these important properties. Because of these, silicon nitride (Si_3N_4) (especially its β polymorph) is widely used in gas turbines, engine parts, bearings, dental drills and gauges, and cutting tools. In addition, thin Si_3N_4 films and coatings have been studied in relation to high-speed memory devices [6–10] and optical waveguide applications [11].

Being a highly covalent material with a low self-diffusion coefficient for the nitrogen atoms (6.3×10^{-20} cm^2/s at $1,400°$ C), Si_3N_4 is very difficult to densify by conventional techniques as it decomposes to a Si melt and N_2 gas at high temperatures [12]. Therefore, the process of liquid-phase sintering with various additives such as SiO_2, SiC, MgO, Al_2O_3, Y_2O_3, HfO_2, ZrO_2, serving to enhance the mass transport of Si_3N_4, is commonly used in the fabrication of the ceramic [5, 13–23]. The process involves making Si_3N_4 from powder under conditions where the solid particles coexist with a wetting liquid at high temperatures until the particle adhesion is achieved. The liquid enhances the rate of interparticle bonding, while reducing the material's porosity [24].

An intrinsic characteristic of Si_3N_4 ceramics, on the other hand, is their brittleness, which compromises their use and reliability as structural components. To overcome this problem, microstructural and compositional designs with rare-earth oxides (REO) sintering additives are used. The resulting microstructure, shown in Fig. 1.1, consists of elongated matrix grains that are randomly oriented, interlocked, and interspersed with secondary REO intergranular phases (or glassy networks), forming either thin intergranular films (IGFs) or large triple pockets [23, 25–31]. The IGFs are the regions of about 1–2 nm thickness separating the

W. Walkosz, *Atomic Scale Characterization and First-Principles Studies of Si_3N_4 Interfaces,* Springer Theses, DOI: 10.1007/978-1-4419-7817-2_1,

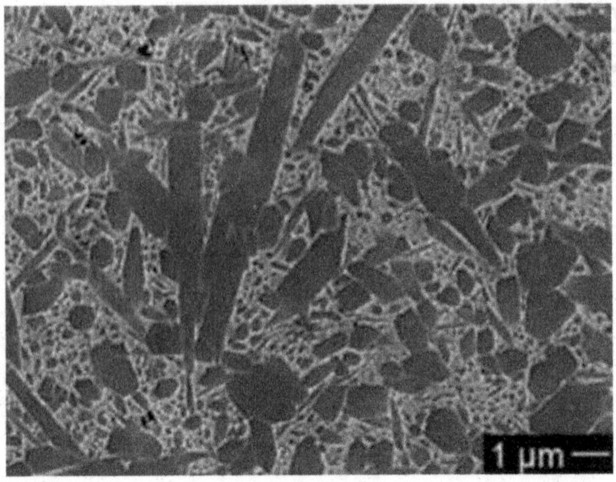

Fig. 1.1 Microstructure of polycrystalline Si_3N_4 containing Lu_2O_3. Si_3N_4 grains are the *dark* regions in the image, while the intergranular film formed by Lu_2O_3 is *bright*. Adapted with permission from [23]

ceramic grains, while the triple pockets are juctions of more than just two grains [31]. The intergranular phases in Si_3N_4 are the key microstructural element influencing the materials densification process, phase transformations, crystal growth rate, and most of the mechanical properties, in particular, its strength (or the resistance to deformation) and toughness (the resistance to fracture or crack propagation when stressed) [15, 16, 21, 22, 32–47]. The presence of REO promotes the formation of a toughened microstructure and assist in debonding processes along the interface to form a crack bridging reinforcement [23, 33, 48]. Studies aimed at achieving a microscopic understanding of the atomic composition and local bonding at the Si_3N_4/REO interfaces are of both fundamental and technological interests.

1.2 Overview of Previous Studies on Si_3N_4

It is well known that the major contribution to toughness in Si_3N_4 is the energy consumed in creating a more torturous crack path between the elongated grains, acting as reinforcing agents, and a weaker interface [15, 17, 49–51]. Highly anisotropic coarse grained materials present the highest toughness and very short crack extensions, which is advantageous in avoiding catastrophic fractures [50, 52]. However, they have an adverse effect on the strength of the material [53]. Therefore, in optimizing the performance of Si_3N_4 both the size and number of the reinforcing grains, in addition to the matrix grain size, must be controlled. Furthermore, the toughening mechanisms (crack bridging, deflection, and pull-out) depend upon retention of these reinforcing grains during crack extension by an interfacial debonding process [23, 48, 54, 55]. Indeed, both the microstructure and the debonding along the interface with the glass network are critical for enhancing the fracture toughness of Si_3N_4. To obtain conditions suitable for high-temperature applications, the crystallization of the amorphous phases at the grain boundary is also essential [56–58]. Early studies have shown that the crystallization can be achieved with heat treatment procedures after sintering, but it can alter the interface chemistry and the intergranular film thickness [59, 60]. All of these changes, both in the morphology and properties of the ceramic, can be controlled with the sintering additives used during its densification.

In the particular case of REO additives (formed with: RE= La, Ce, Nd, Sm, Gd, Er, Yb, Lu), the microstructure morphology of Si_3N_4 and the viscosity of the intergranular phase, or the resistance to deformation, correlate with the atomic size of the rare-earth element. Si_3N_4 grains with larger aspect ratios and intergranular phases with lower viscosities are formed for rare-earth elements with smaller atomic size [18, 34]. As a consequence of their weaker grain boundaries, the resulting Si_3N_4 ceramic presents enhanced toughness [23]. Because of the critical role of the intergranular films in Si_3N_4, in particular those composed of the rare-earth oxides, there has been a considerable interest to characterize these films structurally, at the atomic level.

Early computational studies mainly focused on the bulk properties of Si_3N_4, the bonding sites of additive atoms at the interface, and the resulting electronic structure [61–76]. Calculations describing an atomic model encompassing the entire width of the IGF, on the other hand, were less prevalent, as essential experimental data about atom positions within the intergranual phases were not available. The breakthroughs in Scanning Transmission Electron Microscopy (STEM) [77, 78] in the last decade, permitting the probing of atomic structures with a sub-Ångstrøm resolution, however, have provided great insight into how these IGFs are affected by the presence of the rare-earth atoms. In particular, these studies have demonstrated that the structure of the intergranular film (IGF) formed by REOs has a partial ordering, with the RE elements arranged periodically at the Si_3N_4/IGF interface [61, 79–85]. Furthermore, the observed attachments of the RE elements at the interface were correlated with their atomic radii and the presence of oxygen at the interface [80]. However, the precise information on the bonding characteristics of the RE elements, their valence states, and the electronic configurations at the interfaces were not revealed in those studies. In addition, the roles of lighter atoms (Si, N, O, and other cations used for sintering purposes) in controlling the attachment of RE elements at the interface were not assessed. Furthermore, recent computational attempts aimed at describing the interface between Si_3N_4 and rare-earth oxide IGF have always assumed N terminated Si_3N_4 surface in contact with the RE elements, with no consideration for oxygen or silicon. The lack of detailed work on the interface had left these experimental and theoretical claims unchallenged, providing a motivation for new studies to be performed.

1.3 Overview of the Present Study

This Ph.D. research study integrates quantum modeling using first principles techniques based on Density Functional Theory (DFT) [86–88] and atomic scale characterization methods in the aberration-corrected STEM [77, 89–92] to develop a fundamental understanding of the structure-property relationships in Si_3N_4/REO (RE=Ce, Lu) interfaces and to address some of the outstanding problems listed above. The experimental techniques chosen in this research study, specifically Z-contrast imaging and Electron Energy Loss Spectroscopy (EELS) in the STEM, have been shown to offer unique information on the structure, composition, and bonding characteristics of many systems [61, 80, 84, 89, 93]. Their strength lies in the ability to provide microstructural and chemical information at a sub-Ångstrøm scale, while combining both the real- and the reciprocal-space information. Similarly, DFT has been one of the most widely used theoretical tools over the last 20 years for investigating physical properties of many materials of varying structural and chemical complexity [86, 87, 94]. Because of its accurate predictions when compared to experiments, it has established a very good track record, offering, at the same time, detailed microscopic information that can aid in the physical interpretation of the results. In this thesis, the experimental techniques are used to reveal atomic and electronic configurations at the Si_3N_4/REO interfaces, while the theoretical calculations serve to model these interfaces focusing on the question of how the light atoms and the RE elements bond to the grains of Si_3N_4. A detailed overview of these techniques is given in Chap. 2. In this thesis, I focus mostly on the β phase of Si_3N_4, which is the most stable and commonly used phase of the ceramic, although the α phase is also considered.

Figure 1.2 shows the crystal structures of Si_3N_4 in the α-, β-, and the recently discovered γ- phase [95–100]. Both the α- and β- phases have a hexagonal lattice with Si and N atoms having tetrahedral (fourfold) and trigonal (threefold) coordination, respectively. β-Si_3N_4 has a periodic stacking structure of ABAB... in the direction of the c-axis, while in the case of α-Si_3N_4 the periodicity is ABCDABCD... Both phases can be synthesized under ordinary pressure. At high

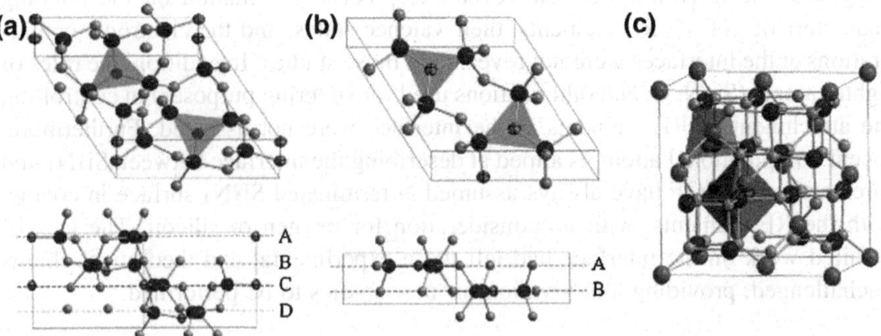

Fig. 1.2 Crystal structures of **(a)** α - **(b)** β - and **(c)** γ - Si_3N_4. In case of **(a)** and **(b)**, the projections in [1120] direction are shown below each structure. Si and N atoms are illustrated by *blue* and *green* spheres, respectively. The *triangles* show the trigonal coordination of $N - Si_3$, parallel to the basal plane. In case of γ - Si_3N_4 **(c)**, the Si atoms have tetrahedral and octahedral coordinations denoted by the small and big polyhedra, respectively. Adapted with permission from [99]

temperatures ($\sim 1,900°C$) α to β transformation takes place and the process is irreversible expect for highly concentrated alloys that prefer the formation of the α- phase. On the other hand, the γ- phase is a high-pressure product with a cubic spinel structure. All of N atoms have tetrahedral coordination, while the Si atoms can have either tetrahedral or octahedral (sixfold) coordination. This phase is characterized by a large bulk modulus, defined as the pressure change with respect to a volume compression, and is investigated in relation to superhard materials [97, 99]. A comparison of the Helmholtz free energies calculated from first principles methods shows that β-Si_3N_4 is more stable than the α polymorph in the temperature range from 0 to 2,000 K [99]. This suggests that α-Si_3N_4 is not the low-temperature phase, but it is a metastable phase under ordinary pressure. Possible factors stabilizing this phase will be discussed in Chap. 7.

The surface of β-Si_3N_4 considered in the theoretical part of this study is the $(10\bar{1}0)$ surface. Although it is not the lowest energy surface predicted theoretically using the Wulff construction [101], this surface was shown to exhibit an abrupt interface with the rare-earth oxide additives in all recent STEM experiments [61, 79, 80–85]. In order to understand the observed preferential attachment of the REOs to the interface between this surface and the IGF and to elucidate the role of lighter elements at the interface, a Si_3N_4 sample prepared with $CeO_{2-\delta}$ was used in this thesis. The choice of cerium as the rare-earth element was the key in this study because cerium has the ability to easily change its oxidation state (and so its radius) in different local oxygen environments [102, 103]. This allowed for testing both the role of the rare-earth element as well as oxygen in controlling the Si_3N_4/REO interface. For comparison, the interface formed between Si_3N_4 and the SiO_2 sintering additive was also studied to examine the morphology of Si_3N_4 surfaces in the absence of RE elements and to assess the importance of RE elements in stabilizing various Si_3N_4 surface terminations.

Ther rest of this thesis is organized as follows: In the following two chapters, detailed descriptions of the theoretical and experimental methods employed in this thesis are given. Next in Chap. 4, I present my theoretical results on the structural energetics of β-$Si_3N_4(10\bar{1}0)$ surfaces modeled in the presence of oxygen and the RE elements (RE=Ce, Lu). In Chaps. 5 and 6, I present my atomic resolution studies on the interfaces between $Si_3N_4(10\bar{1}0)$ surface in the presence of $CeO_{2-\delta}$ and SiO_2, respectively. In Chap. 7, I report the results on α-Si_3N_4 and its stability. Finally, in Chap. 8, I summarize and discuss of the presented results with directions for future studies.

References

1. Chen, I.W., Becher, P.F., Mitomo, M., Petzow, G., Yen, T.S.: Silicon nitride ceramics scientific and technological advances. Mater. Res. Soc. (MRS Proc.) **287**, 147–158 (1993)
2. Hoffmann, M.J.: Analysis of microstructural development and mechanical properties of Si_3N_4 ceramics. In: Hoffmann, M.J., Petzow G. (eds.) Tailoring of Mechanical Properties of Si_3N_4 Ceramics, pp. 59–72. Kluwer Academic Publishers, Dordrecht (1994)

3. Cahn, R.W., Hassen, P., Kramer, J.: Materials Science and Technology, Structure and Properties of Ceramics. Wiley, Weinheim (1994)
4. Richerson, D.W.: The Magic of Ceramics. The American Ceramics Society, Westerville (2000)
5. Riley, F.L.: Silicon nitride and related materials. J. Am. Ceram. Soc. **83**(2), 245–265 (2000)
6. Liu, L., Xu, J.P., Chen, L.L., Lai, P.: A study on the improved programming characteristics of flash memory with Si_3N_4/SiO_2 stacked tunneling dielectric. Microelectron. Reliab. **49**, 912–915 (2009)
7. Saraf, M., Akhvlediani, R., Edrei, R., Shima, R., Roizin, Y., Hoffman, A.: Low thermal budget $SiO_2/Si_3N_4/SiO_2$ stacks for advanced SONOS memories. J. Appl. Phys. **102**, 054512 (2007)
8. Berberich, S., Godignon, P., Morvan, E., Fonseca, L., Millan, J., Hartnagel, H.L.: Electrical characterisation of Si_3N_4/SiO_2 double layers on p-type 6H-SiC. Microelectron. Reliab. **40**, 833–836 (2000)
9. Wang, Y.Q., Hwang, W.S., Zhang, G., Yeo, Y.C.: Electrical characteristics of memory devices with a high-$k$$HfO_2$ trapping layer and dual SiO_2/Si_3N_4 tunneling layer. IEEE Trans. Electron Devices **54**(10), 2699–2705 (2007)
10. Santussi, S., Lozzi, L., Passacantando, M., Phani, A.R., Palumbo, E., Bracchitta, G., De Tommasis, R., Alfonsetti, R., Moccia, G.: Properties of stacked dielectric films composed of $SiO_2/Si_3N_4/SiO_2$ tunneling layer. J. Non-Cryst. Solids **245**, 224–231 (1999)
11. Kazmierczak, A., Dortu, F., Schrevens, O., Giannone, D., Vivien, L., Marris-Morini, D., Bouville, D., Cassan, E., Gylfason, K.B., Sohlstrom, H., Sanchez, B., Griol, A., Hill, D.: Light coupling and distribution for Si_3N_4/SiO_2 integrated mutichannel single-mode sensing system. Opt. Eng. **48**(1), 014401 (2009)
12. Kijima, K., Shirasaki, S.I.: Nitrogen self-diffusion in silicon nitride. J. Chem. Phys. **65**(7), 2668–2671 (1976)
13. Lange, F.: The sophistication of ceramic science through silicon nitride studies. J. Ceram. Soc. Jpn. **114**(11), 873–879 (2006)
14. Dutta, S., Buzek, B.: Microstructure, strength, and oxidation of a 10 wt% zyttrite-Si_3N_4 ceramic. J. Am. Ceram. Soc. **67**(2), 89–92 (1984)
15. Pezzotti, G.: Si_3N_4/SiC-platelet composite without sintering aids: a candidate for gas turbine engines. J. Am. Ceram. Soc. **76**, 1313–1320 (1993)
16. Li, H., Komeya, K., Tatami, J., Meguro, T., Chiba, Y., Komatsu, M.: Effect of HfO_2 addition on sintering of Si_3N_4. J. Am. Ceram. Soc. Jpn. **109**, 342–346 (2001)
17. Campbell, G.H., Rühle, M., Dalgleish, B.J., Evans, A.G.: Whisker toughening: a comparison between aluminum oxide and silicon nitride toughened with silicon carbide. J. Am. Ceram. Soc. Jpn. **73**, 521–530 (1990)
18. Becher, P.F., Ferber, M.K.: Temperature-dependent viscosity of SiREAl-based classes as a function of N:O and RE:Al ratios (RE = La, Gd, Y, and Lu). J. Am. Ceram. Soc. **87**, 1274–1279 (2004)
19. Park, H., Kim, H.E., Niihara, K.: Microstructural evolution and mechanical properties of Si_3N_4 with Yb_2O_3 as a sintering additive. J. Am. Ceram. Soc. **80**, 750–756 (1995)
20. Hong, Z.L., Yoshida, H., Ikahara, Y., Sakuma, T., Mishimura, T., Mitomo, M.: The effect of additives on sintering behavior and strength retention in silicon nitride with RE-disilicate. J. Eur. Ceram. Soc. **22**, 527–534 (1997)
21. Guo, S.Q., Hirosaki, N., Yamamoto, Y., Nishimura, T., Mitomo, M.: Strength retention in hot-pressed Si_3N_4 ceramics with Lu_2O_3 additives after oxidation exposure in air at 1500 degrees C. J. Am. Ceram. Soc. **85**, 1607–1609 (2002)
22. Guo, S.Q., Hirosaki, N., Yamamoto, Y., Nishimura, T., Mitomo, M.: Hot-pressed silicon nitride ceramics with Lu_2O_3 additives: elastic moduli and fracture toughness. J. Eur. Ceram. Soc. **23**, 537–545 (2003)
23. Satet, R.L., Hoffmann, M.J.: Influence of the rare-earth element on the mechanical properties of RE-Mg-bearing silicon nitride. J. Am. Ceram. Soc. **88**(9), 2485–2490 (2005)
24. German, R.M.: Liquid Phase Sintering. Plenum Press, New York (1985)

25. Clarke, D.R.: On the equilibrium thickness of intergranular glass phases in ceramic materials. J. Am. Ceram. Soc. **70**(1), 15–22 (1987)
26. Kleebe, H.J., Hoffmann, M.J., Rühle, M.: Influence of secondary phase chemistry on grain boundary film thickness in silicon nitride. Zeitschrift fur Metallkunde **83**(8), 610–617 (1992)
27. Kleebe, H.J., Cinibulk, M.K., Cannon, R.M., Rühle, M.: Statistical analysis of the intergranular film thickness in silicon nitride ceramics. J. Am. Ceram. Soc. **76**, 1969 (1993)
28. Clarke, D.R., Shaw, T.M., Philipse, A.P., Horn, R.G.: Possible electrical double-layer contribution to the equilibrium thickeness of intergranular glass films in polycrystalline ceramics. J. Am. Ceram. Soc. **76**, 1201 (1993)
29. Tanaka, I., Kleebe, H.J., Cinibulk, M.K., Bruley, J., Clarke, D.R., Rühle, M.: Calcium concentration dependence of the intergranular film thickness in silicon nitride. J. Am. Ceram. Soc. **77**, 911 (1994)
30. Wang, C., Pan, X., Hoffmann, M.J., Rühle, M.: Grain boundary films in rare-earth-glass-based silicon nitride. J. Am. Ceram. Soc. **79**, 788 (1996)
31. Subramaniam, A., Koch, C.T., Cannon, R.M., Rühle, M.: Intergranular glassy films: an overview. Mater. Sci. Eng. A **422**(1–2), 3–18 (2006)
32. Sanders, W.A., Miekowski, D.M.: Strength and microstructure of sintered Si_3N_4 with rare-earth-oxide additions. J. Am. Ceram. Soc. **64**, 304–309 (1985)
33. Sun, E.Y., Becher, P.F., Plucknett, K.P., Hsueh, C.H., Alexander, K.B., Waters, S.B., Hirao, K., Brito, M.E.: Microstructural design of silicon nitride with improved fracture toughness: II, effects of yttria and alumina additives. J. Am. Ceram. Soc. **81**, 2831–2840 (1998)
34. Satet, R.L., Hoffmann, M.J.: Grain growth anisotropy of β-silicon nitride in rare-earth doped -oxynitride glasses. J. Eur. Ceram. Soc. **24**, 3437–3445 (2004)
35. Choi, D.J., Scott, W.D.: Devitrification and delayed crazing of SiO_2 on single-crystal silicon and chemically vapor-deposited silicon nitride. J. Am. Ceram. Soc. **70**, 269–272 (1987)
36. Part, J.Y., Kim, J.R., Kim, C.H.: Effects of free silicon on the α to β phase transformation in silicon nitride. J. Am. Ceram. Soc. **70**, 240–242 (1987)
37. Burns, G.T., Chandra, G.: Pyrolysis of preceramic polymers in ammonia: preparation of silicon nitride powders. J. Am. Ceram. Soc. **72**, 333–337 (1989)
38. Choi, D.J., Fishbach, D.B., Scott, W.D.: Oxidation of chemically-vapor-deposited silicon nitride and single-crystal silicon. J. Am. Ceram. Soc. **72**, 1118–1123 (1989)
39. Kleebe, H.J., Ziegler, G.: Influence of crystalline secondary phases on the densification behavior of reaction-bonded silicon nitride during postsintering under increased nitrogen pressures. J. Eur. Ceram. Soc. **72**, 2314–2317 (1989)
40. Tanaka, I., Pezzotti, G., Okamoto, T., Miyamoto, Y., Koizumim, M.: Hot isostatic press sintering and properties of silicon nitride without additives. J. Am. Ceram. Soc. **72**, 1656–1660 (1989)
41. Mitomo, M., Tsutsumi, M., Tanaka, H., Uenosono, S., Saito, F.: Grain growth during gas-pressure sintering of β-silicon nitride. J. Eur. Ceram. Soc. **73**, 2441–2445 (1990)
42. Mitomo, M., Uenosono, S.: Microstructural development during gas-pressure sintering of α-silicon nitride. J. Eur. Ceram. Soc. **75**, 103–108 (1992)
43. Watari, K., Hirao, K., Toriyama, M., Ishizaki, K.: Effect of grain size on the thermal conductivity of Si_3N_4. J. Am. Ceram. Soc. **82**, 777–779 (1999)
44. Kitayama, M., Hirao, K., Tsuge, A., Watari, K., Toriyama, M., Kanzaki, S.: Thermal conductivity of beta-Si_3N_4: II, effect of lattice oxygen. J. Am. Ceram. Soc. **83**, 1985–1992 (2000)
45. Shen, J.Z., Zhao, Z., Peng, H., Nygren, M.: Formation of tough interlocking microstructures in silicon nitride ceramics by dynamic ripening. Nature (London) **417**, 266–269 (2002)
46. Cinibulk, M.K., Thomas, G., Johnson, S.M.: Strength and creep behavior of rare-earth disilicate-silicon nitride ceramics. J. Am. Ceram. Soc. **75**, 2050–2055 (1992)
47. Hoffmann, M.J., Gu, H., Cannon, R.M.: Interfacial engineering for optimized properties II. In: Hall, E.L., Carter, C.B., Briant, C.L. (eds.) MRS Proceedings, p. 65. Pittsburgh, Pennsylvania, Mater. Res. Soc. (2000)

48. Satet, R.L., Hoffmann, M.J., Cannon, R.M.: Experimental evidence of the impact of rare-earth elements on particle growth and mechanical behaviour of silicon nitride. Mater. Sci. Eng. A **422**, 66–76 (2006)
49. Tanaka, I., Pezzotti, G., Matsushuta, K.I., Miyamoto, Y., Okamoto, T.: Impurity-enhanced intergranular cavity formation in silicon nitride at high temperatures. J. Am. Ceram. Soc. **73**, 752–759 (1990)
50. Ohji, T., Hirao, K., Kanzaki, S.: Fracture resistance behavior of highly anisotropic silicon nitride. J. Am. Ceram. Soc. **78**, 3125–3128 (1995)
51. Becher, P.F., Sun, E.Y., Plucknett, K.P., Alexander, K.B., Hsueh, C.H., Lin, H.T., Waters, S.B., Westmoreland, C.G., Kang, E.S., Hirao, K., Brito, M.E.: Microstructural design of silicon nitride with improved fracture toughness: I, effects of grain shape and size. J. Am. Ceram. Soc. **81**, 2821–2830 (1998)
52. Tajima, Y.: Development of high-performance silicon nitride ceramics and their applications. In: Chen, I.W. (ed.) Silicon Nitride Scientific and Technological Advances, p. 189. Pittsburgh, USA, journal = Mater. Res. Soc. (MRS Proc.) (1993)
53. Hoffmann, M.J.: Relationship between microstructure and mechanical properties of silicon nitride ceramics. Pure Appl. Chem. **67**(6), 939–946 (1995)
54. Becher, P.F., Painter, G.S., Shibata, N., Satet, R.L., Hoffmann, M.J., Pennycook, S.J.: Influence of additives on anisotropic grain growth in silicon nitride ceramics. Mater. Sci. Eng. A **422**, 85–91 (2006)
55. Becher, P.F., Painter, G.S., Shibata, N., Water, S.B., Lin, H.T.: Effect of rare-earth (RE) intergranular adsopriton on the phase transformation, microstructure evolution, and mechanical properties in silicon nitride with RE_2O_3 + MgO additives: RE = La, Gd, and Lu. J. Am. Ceram. Soc. **91**(7), 2328–2336 (2008)
56. Bonnell, D.A., Tien, T.Y., Rühle, M.: Controlled crystallization of the amorphous phase in silicon nitride ceramics. J. Am. Ceram. Soc. **70**, 460–465 (1987)
57. Lee, W.W., Hilmas, G.E.: Microstructural changes in β-silicon nitride grains upon crystallizing the grain-boundary glass. J. Am. Ceram. Soc. **72**, 1931–1937 (1989)
58. Greil, P.: Analysis of Microstructural Development and Mechanical Properties of Si_3N_4 Ceramics. In: Taylor, D. (ed.) High-Temperature Strengthening of Silicon Nitride Ceramics, p. 645. Stoke-on-Trent, Canterbury (1987)
59. Bergström, L., Pugh, R.J.: Interfacial characterization of silicon nitride powders. J. Am. Ceram. Soc. **72**, 103–109 (1989)
60. Keeble, H.J.: Structure and chemistry of interfaces in Si_3N_4 ceramics studied by transmission electron microscopy. J. Ceram. Soc. Jpn. **105**, 453–475 (1997)
61. Shibata, N., Pennycook, S.J., Gosnell, T.R., Painter, G.S., Shelton, W.A., Becher, P.F.: Observation of rare-earth segregation in silicon nitride ceramics at subnanometre dimensions. Nature **428**(6984), 730–733 (2004)
62. Benco, L.: Chemical bonding at grain boundaries: MgO on β-Si_3N_4. Surf. Sci. **327**, 274–284 (1995)
63. Liu, A.Y., Cohen, M.L.: Structural properties and electronic structure of low-compressibility materials: $\beta - Si_3N_4$ and hypothetical $\beta - C_3N_4$. Phys. Rev. B **41**, 10727 (1990)
64. Nakayasu, T., Yamada, T., Tanaka, I., Adachi, H.: Local chemical bonding around rare-earth ions in $\alpha-$ and $\beta - Si_3N_4$. J. Am. Ceram. Soc. **80**, 2525–2532 (1997)
65. Nakayasu, T., Yamada, T., Tanaka, I., Adachi, H.: Calculation of grain-boundary bonding in rare-earth-doped $\beta - Si_3N_4$. J. Am. Ceram. Soc. **81**, 565–570 (1998)
66. Dudesek, P., Benco L.: Cation-aided joining of surfaces of β-silicon nitride: structural and electronic aspects. J. Am. Ceram. Soc. **81**, 1248–1254 (1998)
67. Bermudez, V.M.: Theoretical study of the electronic structure of the Si_3N_4(0001) surface. Surf. Sci. **579**(1), 11–20 (2005)
68. Wang, L., Wang, X., Tan, Y., Wang, H., Zhang, C.: Study of oxygen adsorption on beta-Si_3N_4(0001) by the density functional theory. Chem. Phys. **331**(1), 92–95 (2006)
69. Belkada, R., Shibayanagi, T., Naka, M.: Ab initio calculations of the atomic and electronic structure of β-silicon nitride. J. Am. Ceram. Soc. **83**, 2449 (2000)

70. Matsugana, K., Iwamoto, Y.: Ab initio molecular dynamics study of atomic structure and diffusion behavior in amorphous silicon nitride containing boron. J. Ceram. Soc. Jpn. **84**, 2213–2219 (2001)
71. Pezzzotti, G., Painter, G.S.: Mechanisms of dopant-induced changes in intergranular SiO_2 viscosity in polycrystalline silicon nitride. J. Am. Ceram. Soc. **85**, 91–96 (2002)
72. Painter, G.S., Averill, F.W., Becher, P.F., Shibata, N., Van Benthem, K., Pennycook, S.J.: First-principles study of rare earth adsorption at beta-Si_3N_4 interfaces. Phy. Rev. B **78**, 214206 (2008)
73. Painter, G.S., Becher, P.F.: Bond energetics at intergranular interfaces in alumina-doped silicon nitride. J. Am. Ceram. Soc. **85**, 65–67 (2002)
74. Yoshiya, M., Tatsumi, K., Tanaka, I., Adachi, H.: Theoretical study on the chemistry of intergranular glassy film in Si_3N_4-SiO_2 ceramics. J. Am. Ceram. Soc. **85**, 109–112 (2002)
75. Painter, G.S., Becher, P.F., Shelton, W.A., Satet, R.L., Hoffmann, M.J.: Theoretical study on the chemistry of intergranular glassy film in Si_3N_4-SiO_2 ceramics. Phys. Rev. B **70**, 144108 (2004)
76. Rulis, P., Chen, J., Ouyang, L., Ching, W.Y., Su, X., Garofalini, S.H.: Electronic structure and bonding of intergranular glassy films in polycrystalline Si_3N_4: ab initio studies and classical molecular dynamics simulations. Phys. Rev. B **71**, 235317 (2005)
77. Pennycook, S.J., Jesson, D.E.: High-resolution incoherent imaging of crystals. Phys. Rev. Lett. **64**, 938–941 (1990)
78. Nellist, P.D., Pennycook, S.J.: The principles and interpretation of annular dark-field Z-contrast imaging. Adv. Imag. Elect. Phys. **113**, 147–203 (2000)
79. Shibata, N., Painter, G.S., Becher, P.F., Pennycook, S.J.: Atomic ordering at an amorphous/crystal interface. Appl. Phys. Lett. **89**(5), 051908 (2006)
80. Ziegler, A., Idrobo, J.C., Cinibulk, M.K., Kisielowski, C., Browning, N.D., Ritchie, R.O.: Interface structure and atomic bonding characteristics in silicon nitride ceramics. Science **306**, 1768–1770 (2004)
81. Ziegler, A., Idrobo, J.C., Cinibulk, M.K., Kisielowski, C., Browning, N.D., Ritchie, R.O.: Atomic-resolution observations of semicrystalline intergranular thin films in silicon nitride. Appl. Phys. Lett. **88**(4), 041919 (2006)
82. Van Benthem, K., Painter, G.S., Averill, F.W., Pennycook, S., Becher, P.F.: Experimental probe of adsorbate binding energies at internal crystalline/amorphous interfaces in Gd-doped Si_3N_4. Appl. Phys. Lett. **92**, 163110 (2008)
83. Winkelman, G.B., Dwyer, C., Marsh, C., Hudson, T.S., Nguyen-Manh, D., Döblinger, M., Cockayne, J.H.: The crystal/glass interface in doped Si_3N_4. Mater. Sci. Eng. A **422**, 77–84 (2006)
84. Winkelman, G.B., Dwyer, C., Hudson, T.S., Nguyen-Manh, D., Döblinger, M., Satet, R.L., Hoffmann, M.J., Cockayne, J.H.: Arrangement of rare-earth elements at prismatic grain boundaries in silicon nitride. Philos. Mag. Lett. **84**, 755–762 (2004)
85. Walkosz, W., Klie, R.F., Öğüt, S., Borisevish, A., Becher, P.F., Pennycook, S.J., Idrobo, J.C.: Atomic resolution study of the interfacial bonding at $Si_3N_4/CeO_{2-\delta}$ grain boundaries. Appl. Phys. Lett. **93**, 053104 (2008)
86. Hohenberg, P., Kohn, W.: Inhomogeneous electron gas. Phys. Rev. **136**(3B), B864–B871 (1964)
87. Payne, M.C., Teter, M.P., Allan, D.C., Arias, T., Joannopoulos, J.D.: Iterative minimization techniques for ab initio total energy calculations: molecular dynamics and conjugate gradients. Rev. Mod. Phys. **64**(9), 30–35 (1992)
88. Kohn, W., Sham, L.J.: Self-consistent equations including exchange and correlation effects. Phys. Rev. **140**(A4), A1133–A1138 (1965)
89. Muller, D.A.: Structure and bonding at the atomic scale by scanning transmission electron microscopy. Nat. Mater. **8**, 263–270 (2009)
90. Pennycook, S.J.: Structure determination through Z-contrast microscopy. In: Merli, P.G., Vittor-Antisari, M. (eds.) Advances in Imaging and Electron Physics, vol. 123, p. 173. Academic Press, New York (2002)

91. Pennycook, S.J.: Z-contrast imaging in the scanning transmission electron microscope. In: Zhang, Z.F., Zhang, Z. (eds.) Progress in Transmission Electron Microscope 1: Concepts and Techniques, pp. 81–111. Springer, Tsinghyua (2001)
92. Egerton, R.F.: Applications of energy-loss spectroscopy. In: Electron Energy-Loss Spectrosocy in the Electron Microscopy 2nd edn., pp. 59–72. Plenum Press, New York (1996)
93. Idrobo, J.C., Öğüt, S., Yildirim, T., Klie, R.F., Browning, N.D.: Electronic and superconducting properties of oxygen-ordered MgB_2 compounds of the form $Mg_2B_3O_x$. Phys. Rev. B **70**, 172503 (2004)
94. Bernholc, J.: Computational materials science: the era of applied quantum mechanics. Phys. Today **52**(9), 30–35 (1999)
95. He, H., Sekine, T., Kobayashi, T., Hirosaki, H., Suzuki, I.: Shock-induced phase transition of β-Si_3N_4 to c-Si_3N_4. Phys. Rev. B **62**(17), 11412–11417 (2000)
96. Ching, W.Y., Ouyang, L., Gale, J.D.: Full ab initio geometry optimization of all known crystalline phases of Si_3N_4. Phys. Rev. B **61**, 13 (2000)
97. Hao, S., Delley, B., Veprek, S., Stampfl, C.: Superhard nitride-based nanocomposites: role of interfaces and effect of impurities. Phys. Rev. Lett. **97**, 086102 (2006)
98. Mo, S.D., Ouyang, L., Ching, W.Y., Tanaka, I., Koyama, Y., Riedel, R.: Interesting physical properties of the new spinel phase of Si_3N_4 and C_3N_4. Phys. Rev. Lett. **83**(24), 5046–5049 (1999)
99. Kuwabara, A., Matsunaga, K., Tanaka, I.: Lattice dynamics and thermodynamical properties of silicon nitride polymorphs. Phys. Rev. B **78**, 064104 (2008)
100. Paszkowicz, W., Minikayev, R., Piszora, P., Knapp, M., Bähtz, C., Recio, J.M., Marques, M., Mori-Sanchez, P., Gerward, L., Jiang, J.Z.: Thermal expansion of spinel-type Si_3N_4. Phys. Rev. B **69**, 052103 (2004)
101. Idrobo, J., Iddir, H., Öğüt, S., Ziegler, A., Browning, N.D., Ritchie, R.O.: Ab initio structural energetics of $\beta - Si_3N_4$ surfaces. Phys. Rev. B (Rapid Communications) **72**, 241301 (2005)
102. Skorodumova, N.V., Ahuja, R., Simak, S.I., Abrikosov, I.A., Johansson, B., Lundqvist, B.I.: Electronic, bonding, and optical properties of CeO_2 and Ce_2O_3 from first principles. Phys. Rev. B **64**, 115108 (2001)
103. Andersson, D.A., Simak, S.I., Johansson, B., Abrikosov, I.A., Skorodumova, N.V.: Modeling of CeO_2, Ce_2O_3, and CeO_{2-x} in the LDA+U formalism. Phys. Rev. B **75**, 035109 (2007)

Chapter 2
Theoretical Methods and Approximations

This chapter describes the theoretical approaches and approximations used in standard first principles calculations. First, the Born Oppenheimer approximation is stated. Next, an overview of the basic formulation of density functional theory (DFT) is given, underlying its merits in describing various materials and their properties, but also pointing out the aspects which need further improvements. The Kohn–Sham ansatz, which provides a means of calculating properties of many-body system using independent-particle methods, is presented along with practical schemes for solving the resulting Kohn–Sham equations. Finally, the DFT+U method and its implementation within a pseudopotential-plane wave framework are introduced.

2.1 Born Oppenheimer Approximation

The adiabatic approximation of Born and Oppenheimer allows one to treat the electrons and nuclei of a real system separately as a consequence of the large mass difference between them [1]. Being much lighter than the ions, the electrons move in a solid much faster than the nuclei, which makes it possible to consider the electronic configuration as completely relaxed in its ground state at each position of the ions during their motion. Thus the nuclei can be treated adiabatically, leading to a separation of the electronic and nuclear degrees of freedom in the many-body wavefunction. The only exception to this are the lightest elements (especially H), where the ions require quantum mechanical treatments. Although the Born Oppenheimer approximation provides a major simplification to the complicated many-body problem by transforming it into the solution of the electronic dynamics in some frozen configuration of nuclei, further approximations are required to perform efficient and accurate total energy calculations. These include *density functional theory* (DFT) to model the electron–electron

W. Walkosz, *Atomic Scale Characterization and First-Principles Studies of Si_3N_4 Interfaces*, Springer Theses, DOI: 10.1007/978-1-4419-7817-2_2,

interactions, *pseudopotential theory* to describe the electron–ion interactions, and *supercell approach* to model systems with aperiodic geometries.

2.2 Density Functional Theory

Since its theoretical foundation in the mid-1960s [2], DFT of Hohenberg and Kohn has demonstrated a large predictive power in first principles studies of the study of the ground state properties of real materials. Its conceptual advantage lies in reformulating the problem of an interacting electron gas under an external (ionic) potential in terms of the ground state electronic charge density, thereby bypassing the necessity to calculate the many-body electron wavefunctions. The N particle system of interacting particles with $3N$ degrees of freedom is reduced to a significantly more tractable problem, which deals with a function (density) of only three variables. The many-body effects incorporated in the exchange-correlation potential are typically approximated within either the local density approximation or the generalized gradient approximation. The formulation applies to any system of interacting particles in an external potential $V_{\text{ext}}(\mathbf{r})$, including any problem of electrons and fixed nuclei, where the Hamiltonian can be written as

$$H = -\sum_i \frac{\hbar^2}{2m_e}\nabla^2_{\mathbf{r}_i} + \sum_i V_{\text{ext}}(r_i) + \frac{e^2}{2}\sum_{i\neq j}\frac{1}{|\mathbf{r}_i - \mathbf{r}_j|} + \sum_{i\neq j}\frac{Z_i Z_j e^2}{|\mathbf{R}_i - \mathbf{R}_j|} \tag{2.1}$$

The first term in this equation corresponds to the kinetic energy of the interacting electrons, the second term is the external potential acting on the electrons due to the ions, the third term is the electron Coulomb interaction, and the last term is the interaction energy of the nuclei. The latter is called Madelung energy and, as far as the electron degrees of freedom are concerned, is simply a constant. For such a system, Hohenberg and Kohn demonstrated a one-to-one correspondence between the external potential and the ground state density $n(\mathbf{r})$, which allows to express the former as a functional of the latter. Since the Hamiltonian is thus fully determined (except for a constant shift of the energy), it follows that all properties of the system can be found given only the ground state density $n(\mathbf{r})$. This result allowed Hohenberg and Kohn to further prove the existence of an energy functional of the density $E[n(\mathbf{r})]$, which assumes its minimum value for the correct ground state density. For practical purposes, it is convenient to write this energy functional $E[n(\mathbf{r})]$ in terms of a contribution due to the external potential acting on the particles and a universal functional $F[n(\mathbf{r})]$ representing both the kinetic energy and the Coulomb interaction,

$$E[n(\mathbf{r})] = \int V_{\text{ext}}(\mathbf{r})n(\mathbf{r})d\mathbf{r} + F[n(\mathbf{r})] \tag{2.2}$$

The minimization of energy functional $E[n(\mathbf{r})]$ with respect to the charge density with the constraint of fixed number of electrons gives the ground state energy and the ground state charge density from which all other physical properties can be extracted. It should be pointed out, however, that in spite of the universality of $F[n(\mathbf{r})]$ no explicit expressions for this functional are known to date.

In 1965, Kohn and Sham readdressed the problem of minimizing the Hohenberg-Kohn density functional (Eq. 2.2) directly with an improved strategy that maps the original interacting problem into an auxiliary non-interacting one [3]. This is achieved by expressing the charge density $n(\mathbf{r})$ as

$$n(\mathbf{r}) = \sum_i |\psi_i(\mathbf{r})| \tag{2.3}$$

where ψ_i's are the single-particle wavefunctions for the non-interacting electron gas with ground state charge density $n(\mathbf{r})$, and the sum is over all occupied single-particle states.

The $F[n(\mathbf{r})]$ functional is now expressed as

$$F[n(\mathbf{r})] = T_s[n(\mathbf{r})] + \frac{e^2}{2}\int \frac{n(\mathbf{r})n(\mathbf{r}')}{|\mathbf{r}-\mathbf{r}'|}\mathrm{d}\mathbf{r}\mathrm{d}\mathbf{r}' + E_{xc}[n(\mathbf{r})] \tag{2.4}$$

where the first term corresponds to the kinetic energy of a non-interacting electron gas at the same density $n(\mathbf{r})$, the second term is the classical Coulomb interaction energy (the Hartree term), and the last term $E_{xc}[n(\mathbf{r})]$ represents the quantum mechanical exchange-correlation energy. This term accounts for the differences between the non-interacting fictitious system and the real interacting one, collecting the contributions from the non-classic electrostatic interaction and the differences in their corresponding kinetic energies. The success of the Kohn–Sham approach ultimately lies in the fact that $E_{xc}[n(\mathbf{r})]$, which contains the many-body contributions, is a small fraction of the total energy, and although not known exactly, it can be *approximated* surprisingly well. These approximations, discussed in Sect. 2.3, are at present the strength and the limitation of DFT, providing efficient yet not exact reformulation of the quantum mechanical problem, respectively.

If the energy functional defined in Eq. 2.2 and Eq. 2.4 is now varied with respect to the wavefunctions ψ_i's subject to the orthonormality constraint, the following set of Schrodinger equations is obtained

$$\left[-\frac{\hbar^2}{2m_e}\nabla_{\mathbf{r}}^2 + v_{\mathrm{eff}}(\mathbf{r}, n(\mathbf{r}))\right]\psi_i(\mathbf{r}) = \varepsilon_i\psi_i(\mathbf{r}) \tag{2.5}$$

where the effective potential $v_{\mathrm{eff}}(\mathbf{r}, n(\mathbf{r}))$ is given as

$$v_{\mathrm{eff}}(\mathbf{r}, n(\mathbf{r})) = V_{\mathrm{ext}}(\mathbf{r}) + e^2\int \frac{n(\mathbf{r}')}{|\mathbf{r}-\mathbf{r}'|}\mathrm{d}\mathbf{r}' + \frac{\delta E_{xc}[n(\mathbf{r})]}{\delta n(\mathbf{r})} \tag{2.6}$$

Equations 2.5 and 2.6 are called the Kohn–Sham equations and have to be solved self-consistently because of the dependence of $v_{\rm eff}(\mathbf{r})$ on $n(\mathbf{r})$. It should be emphasized here that the Kohn–Sham procedure introduces a one-body Hamiltonian representing a single-particle electron in the mean field created by the nuclei and by all other electrons. However, it assigns no formal interpretation to the calculated orbitals and the eigenvalues. Indeed, the sum of the single-particle Kohn–Sham eigenvalues does not give the total electronic energy because this would overestimate the effect of the electron–electron interactions in the Hartree energy and in the exchange-correlation energy. More specifically, the Kohn–Sham eigenvalues are not the energies of the single-particle electron states, but rather the derivatives of the total energy with respect to the occupation numbers of these states [4]. The only exception is the highest eigenvalue in a finite system (atomic or molecular), which represents the unrelaxed ionization [5].

2.3 Approximations for the Exchange-Correlation Energy Functional

In principle, the solution of the Kohn–Sham equations would yield the exact ground state energy of the interacting electron gas problem. However, as mentioned earlier, the exact exchange-correlation functional $E_{\rm xc}[n(\mathbf{r})]$ for an inhomogeneous interacting electron gas is not known for general $n(\mathbf{r})$. To proceed further, approximations to this functional are required. The most common and extensively tested approximation is the local density approximation (LDA), in which $E_{\rm xc}[n(\mathbf{r})]$ for the inhomogeneous system is constructed from a parametrized form of the exchange-correlation energy density of the homogeneous electron gas $\varepsilon_{\rm xc}^{\rm hom}$ [3],

$$E_{\rm xc}[n(\mathbf{r})] = \int \varepsilon_{\rm xc}(\mathbf{r})n(\mathbf{r})d\mathbf{r} \tag{2.7}$$

and

$$\frac{\delta E_{\rm xc}[n(\mathbf{r})]}{\delta n(\mathbf{r})} = \frac{\partial[n(\mathbf{r})\varepsilon_{\rm xc}(\mathbf{r})]}{\partial n(\mathbf{r})} \tag{2.8}$$

with $\varepsilon_{\rm xc}(\mathbf{r}) = \varepsilon_{\rm xc}^{\rm hom}[n(\mathbf{r})]$. Several parametrizations for the exchange-correlation energy of a homogeneous electron gas exist in the literature. They are based on Monte Carlo simulations and many-body perturbation theory, and use interpolation formulas to link exact results for the exchange-correlation energy of high-density electron gases and those corresponding to the intermediate and low-density electron systems [3, 6, 7, 8, 9]. Because of the local nature of the exchange-correlation energy functional within the LDA, this approach is expected to be valid for systems with slowly varying electron density such as simple metals or intrinsic semiconductors. In all other cases, the LDA is indeed unpredictable and its justification relies mainly on its ability to reproduce experimental ground state properties of many solids, especially covalent or metallic. Although recent work has shown that the success of LDA can be partially attributed to the fact that the LDA gives the correct sum rule for the exchange-correlation hole (that is, there is a total electronic charge of one electron excluded from the neighborhood of a given electron), there is no general argument explaining why it works remarkably well [10, 11, 12].

In spite of the success of the LDA in capturing the exchange-correlation effects in many systems, the approximation generally overestimates the cohesive energies of solids and predicts shorter equilibrium bond lengths than found experimentally. Errors in the structural properties calculated within the LDA are usually small for crystals with covalent or metallic bonds, but it is well known that the hydrogen bond cannot be described accurately within this approximation. These shortcomings have led to the development of various generalized gradient approximations (GGAs) with marked improvement over the LDA in many cases [13, 14]. In the GGA, the $E_{\text{xc}}[n(\mathbf{r})]$ is a functional density as well as its local spatial variations:

$$E_{\text{xc}}^{\text{GGA}}[n(\mathbf{r})] = \int \varepsilon_{\text{xc}}^{\text{GGA}}(n(\mathbf{r}), |\nabla n(\mathbf{r})|)n(\mathbf{r})d\mathbf{r} \tag{2.9}$$

The approximation usually gives a better description of the structural properties of real materials (in particular, the binding energies) as compared to the LDA. It is expected to work better in the case of inhomogeneous systems, like transition metals, where the LDA typically fails. Several expressions of the exchange energy density in the GGA formulation are available in the literature; two of most popular are the Perdew–Wang-91 (PW91) and the Perdew–Burke–Ernzerhof (PBE) functionals [5, 11]. Although both functionals produce similar results in the calculations of lattice constants, bulk moduli, and atomization energies, they can lead to numerical differences when surface effects are present. In this thesis, both the PW91 and the PBE formulations of the GGA will be used.

2.4 Periodic Systems: Bloch's Theorem

In the previous section it was established that certain observables of the many-body problem can be mapped into equivalent observables in an effective single-particle system. In spite of the great simplification of the original problem, two difficulties still remain for practical and efficient computations: a wavefunction must be calculated for each of the $\sim 10^{23}$ number of electrons in the system, and the correct description of the electronic wavefunctions, extending over the entire solid, requires an infinite basis set. To overcome both problems, calculations involving periodic systems are performed using Bloch's theorem to describe the electronic wavefunctions.

The theorem states that in a solid characterized by the periodic potential $V(\mathbf{r}+\mathbf{R}) = V(\mathbf{r})$, where $\mathbf{R}$ is a direct lattice vector, the electronic wavefunction can be written as the product of a cell-periodic part and a wave-like part

$$\psi_{n\mathbf{k}}(\mathbf{r}) = \exp[\mathrm{i}\mathbf{k}\cdot\mathbf{r}]u_{n\mathbf{k}}(\mathbf{r}) \tag{2.10}$$

where $\mathbf{k}$ is a vector in the first Brillouin zone (BZ) of the reciprocal lattice, n is the band index classifying states corresponding to the same $\mathbf{k}$-vector, and $u_{n\mathbf{k}}(\mathbf{r})$ is a function with the periodicity of the unit cell, i.e., $u_{n,\mathbf{k}}(\mathbf{r}+\mathbf{R}) = u_{n,\mathbf{k}}(\mathbf{r})$. This cell-periodic part can be expanded using a discrete set of plane waves whose wave vectors are reciprocal lattice vectors of the crystal,

$$u_{\mathbf{k},n}(\mathbf{r}) = \sum_{\mathbf{G}} c_{\mathbf{k},n}(\mathbf{G})\exp[\mathrm{i}\mathbf{G}\cdot\mathbf{r}] \tag{2.11}$$

Effectively, the single-particle wavefunction can be expressed as a sum of plane waves,

$$\psi_{\mathbf{k}}(\mathbf{r}) = \sum_{\mathbf{G}} c_{\mathbf{k},n}(\mathbf{G})\exp[\mathrm{i}(\mathbf{k}+\mathbf{G})\cdot\mathbf{r}] \tag{2.12}$$

Electronic states are allowed only at the set of $\mathbf{k}$ points imposed by the boundary conditions. The density of these $\mathbf{k}$ points is proportional to volume of the solid, but only a finite number of electronic states are occupied at each $\mathbf{k}$ point. By virtue of Bloch's theorem, the infinite number of wavefunctions of an extended system is now represented by a finite number of electronic wavefunctions at an infinite set of $\mathbf{k}$ points.

2.5 k-Point Sampling

The calculation of many physical quantities such as the total energy of a solid requires integration of functions with Bloch vector $\mathbf{k}$ performed over the BZ. Using the symmetry of the crystal, the integration can be conveniently confined in a smaller region of the BZ, the irreducible wedge of the Brillioun zone (IBZ). The approach can be further improved by the use of the special k-point integration techniques, which allow to perform reciprocal space integrations with a finite number of $\mathbf{k}$-vectors in the IBZ. For example, in a total energy calculation, after obtaining the single-particle wavefunctions, $\psi_{\mathbf{k}n}(\mathbf{r})$, one has to construct the corresponding charge density $n(\mathbf{r})$ first summing over bands and then integrating over the BZ. The typical practice for this integration is to replace it by discrete sum over a few carefully chosen special $\mathbf{k}$ points. That is,

$$n_{\mathbf{k}}(\mathbf{r}) = 2\sum_{n} |\psi_{\mathbf{k}n}(\mathbf{r})|^2 \tag{2.13}$$

$$n(\mathbf{r}) = \frac{\Omega}{(2\pi)^3}\int n_{\mathbf{k}}(\mathbf{r})d\mathbf{k} \approx \sum_{i=1}^{N} \omega(k_i)n_{\mathbf{k}_i}(\mathbf{r}) \tag{2.14}$$

where $\mathbf{k}_i \in \text{IBZ}$ and $\omega(\mathbf{k}_i)$ is the relative weight associated with $\mathbf{k}_i$. Several schemes to construct these points have been proposed in the literature with the sampling accuracy easily checked by the convergence tests involving denser sets of **k** points [15, 16, 17].

2.6 Plane Waves, Pseudopotentials, and the Projector Augumented Wave Method

As shown in Eq. 2.12, the single-particle wavefunctions at each **k** point in the BZ of a given crystal can be expanded in a discrete set of plane waves. In principle, an infinite plane wave basis set is required to expand these wavefunctions. However, in practical calculations this set is truncated by an energy cutoff E_{cut} such that only plane waves with kinetic energy $\frac{\hbar^2}{2m}|\mathbf{k}+\mathbf{G}|^2$ less than E_{cut} are included. The choice of E_{cut} affects the accuracy of the calculations and should be carefully checked with simple convergence tests, i.e. the physical results obtained from the calculations should be converged with respect to this cutoff.

When plane waves are used as a basis set, the Kohn–Sham equations take on a particularly simple form

$$\sum_{\mathbf{G}'}\left[\frac{\hbar^2}{2m_e}|\mathbf{k}+\mathbf{G}|^2\delta_{\mathbf{G}\mathbf{G}'}+v_{\text{eff}}(\mathbf{G}-\mathbf{G}')\right]c_n(\mathbf{k}+\mathbf{G}')=\varepsilon_{\mathbf{k}n}c_n(\mathbf{k}+\mathbf{G}) \qquad (2.15)$$

Solution of the Hamiltonian matrix in Eq. 2.15 proceeds by conventional diagonalization techniques, taking advantage of the Fast Fourier Transformations. Since the size of the matrix is determined by the choice of E_{cut}, its diagonalization can be formidable for systems that contain both valence and core electrons. This is because a very large number of plane waves are required to expand the tightly bound core orbitals and to follow the rapid oscillations of the wavefunctions of the valence electrons in the core region. To circumvent this problem the pseudopotential approximation [18, 19], which allows the electronic wavefunctions to be expanded using a much smaller number of plane wave basis states, is widely used. The approximation is based on the assumption that most physical and chemical properties of a system depend, to a very good approximation, only on the distribution of the valence electrons, while the ionic cores (the nuclei and the most internal electronic cloud) can be considered as frozen in their atomic configurations. The strong ionic core potential is thus replaced by a weaker pseudopotential acting on a set of nodeless pseudowavefunctions rather than the true valence wavefunctions.

In addition to reducing the number of plane wave basis sets needed to expand the electronic wavefunctions, the pseudopotential approximation has many other advantages. The removal of the core electrons, which contribute most to the total energy of the system (typically a thousand times more than the valence electrons), means that the total energy differences between various ionic configurations will be taken between much smaller numbers. This relaxes the required accuracy for the total energy calculations as compared to the all-electron calculations. In addition, the relativistic effects can be easily incorporated into the potential while treating the valence electrons non-relativistically.

In this work, the projector augmented wave (PAW) method of Blöchl [20], combining the advantages of full potential augmented plane wave approach (by keeping the nodal structure of the wavefunctions inside the core) and the simplicity and numerical efficiency of pseudpotentials, will be used. The computational times for such calculations are slightly higher than for pseudopotentials, while the accuracy is at the level of most accurate full potential linear augmented plane wave methods since PAW provides access to the full charge and spin density. The method is based on a mapping between the true valence wavefunctions $\psi_n(\mathbf{r})$ with their complete nodal structure and the smooth auxiliary wavefunctions $\tilde{\psi}_n(\mathbf{r})$ having a rapidly convergent plane wave expansion via a linear transformation T,

$$|\psi_n\rangle = \hat{T}|\tilde{\psi}_n\rangle \tag{2.16}$$

The transformation is assumed to be unity except for a sphere centered on the nucleus (so-called augmentation region), whose cutoff radius r_c^a should be chosen such that there is no overlap between neighboring augmentation spheres. The transformation can be written as

$$\hat{T} = 1 + \sum_a \hat{T}^a \tag{2.17}$$

where a is an atom index. All physical quantities like $\langle\psi_n|A|\psi_n\rangle$ can be easily calculated in the pseudo Hilbert space representation $\langle\tilde{\psi}_n|\tilde{A}|\tilde{\psi}_n\rangle$ (with $\tilde{A} = \hat{T}^\dagger A\hat{T}$) rather than directly from the true wavefunctions. Similarly, the variational principle for the total energy,

$$\frac{\partial E[\hat{T}|\tilde{\psi}\rangle]}{\partial\langle\tilde{\psi}|} = \varepsilon\hat{T}^\dagger\hat{T}|\tilde{\psi}\rangle \tag{2.18}$$

gives an equivalent of the KS equations for the pseudowavefunctions, and so the calculations for the ground state can now be carried out in the pseudo Hilbert space, which can be efficiently spanned by the plane wave expansion. For more details, please refer to Appendix C.

2.7 Aperiodic Structures

Most of the standard electronic structure algorithms rely on the periodic nature of the studied systems for actual calculations. Semi-periodic structures such as surfaces are typically modeled using a slab geometry, in which a number of bulk layers separated by a vacuum region are repeated periodically. This approach allows the use of standard methods described in the previous sections to model surfaces, provided that a large vacuum region and a sufficient number of bulk layers are used in the calculations to ensure that the introduced artificial periodicity does not affect the physical results.

Relative structural stabilities of various atomic arrangements on a given surface can be analyzed using well-established formalisms in terms of chemical potentials of the relevant species [21]. For a bare surface, for example, the surface formation energy E_{surf} is written as

$$E_{\text{surf}} = E_{\text{surf}}^{\text{tot}} - \sum_i \mu_i N_i \tag{2.19}$$

where $E_{\text{surf}}^{\text{tot}}$ is the total energy of the slab, N_i is the number of atoms of type i in the slab, and μ_i is the corresponding chemical potential. Assuming that the surface is in equilibrium with the bulk, which sets the upper limits on the μ_i's, a phase diagram can be calculated to determine stabilities of different surfaces as a function of stoichiometry.

2.8 DFT+U

In spite of the great success of DFT in predicting the fundamental properties of molecules and materials, there still exist classes of systems and phenomena for which current forms of the theory fail or are inadequate. These include the strongly correlated electron materials, such as actinides and middle-to-late transition metal oxides, which contain electrons in d or f shells. The d and f electrons are inherently localized on each metal atom resulting in large intra-atomic exchange and Coulomb energies [22]. The conventional DFT with approximate exchange and correlation does not properly remove the ficticious electron self-interaction; as a consequence, the electron–electron repulsion is predicted to be too large. To reduce this repulsion energy, the electrons incorrectly delocalize, causing invalid predictions such as giving a metallic ground state for FeO and a small band gap for MnO, whereas these materials are believed to be wide gap insulators.

Since its introduction more than two decades ago, the DFT+U method by Anisimov and coworkers has remained to be the most efficient method to describe strong correlations of electrons confined in an atomic region [23, 24, 25]. In this method, the interactions between electrons localized on the same atomic centers (the on-site interactions) are described with a Hartree Fock (HF)-like formalism, while the remaining interactions are treated within DFT [22]. In real calculations, the on-site interaction energy is obtained from a parameterized Hamiltonian (coming from the Hubbard Hamiltonian), rather than from an explicit HF calculation. This Hamiltonian includes the average Coulomb and exchange interactions between electrons of the same angular momentum that are localized on the same atom, labeled as U_{Il} and J_{Il}, respectively. The index I denotes the atomic center on which the electrons are localized, and l represents their angular momentum. The accuracy of the DFT+U calculations lies in selecting appropriate values for the U and J parameters, which are system-specific. These values can be treated as tunable parameters chosen to reproduce known properties of the system of interest (empirical approach) or can be extracted from constrained DFT calculations.

The energy functional within the DFT+U framework can be written as

$$E^{\text{DFT+U}}[n(r), n_{Ilm\sigma}] = E^{\text{DFT}}(n) + E^{\text{on-site}}[n_{Ilm\sigma}] - E^{dc}[n_{Il\sigma}] \tag{2.20}$$

where $E^{\text{DFT+U}}$ is the total energy of the system, E^{DFT} is the DFT energy of the system based on the total electron density $n(\mathbf{r})$, $E^{\text{on-site}}$ is the HF energy arising from on-site interactions between localized electrons, and E^{dc} is the double counting term which corrects for contributions to the total energy that are included in both E^{DFT} and $E^{\text{on-site}}$. The $E^{\text{on-site}}$ term depends on the number of electrons that occupy localized orbitals centered on atom I and is described by angular momentum l, magnetic number m, and spin σ. These atom-centered, localized orbitals consist of spherical harmonics multiplied by a radial function selected to accurately represent the electronic states of the non-spin-polarized atom. The occupation numbers $n_{Ilm\sigma}$ can be obtained by projecting the Kohn–Sham DFT orbitals for the total system onto a set of these localized orbitals, while the n_{Ilm} term corresponds to the total number of electrons for a given spin and angular momentum that are localized on I, i.e. $n_{Il\sigma} = \sum_m n_{Ilm\sigma}$.

To evaluate Eq. 2.20, the form of $E^{\text{on-site}}$ and E^{dc} must be decided. One approach is to use the rotationally invariant formulation proposed by Dudarev *et al.* [26], which yields the following expression for the total energy functional

$$E^{\text{DFT+U}}[n(r), n_{Ilm\sigma}] = E^{\text{DFT}}(n) + \sum_{Ilm\sigma} \frac{U_{Il} - J_{Il}}{2}(n_{Ilm\sigma} - n^2_{Ilm\sigma}) \tag{2.21}$$

The first term on the right-hand side of this equation corresponds to the DFT energy obtained using the total electron density and includes the on-site interactions. The second term corrects for the first term by driving the system toward electron densities in which the localized states have occupation numbers of either 0 or 1 ($n_{Ilm\sigma}$ lies between 0 and 1) [22]. This approach has been shown to lead to an improved description of the electronic structure of many strongly correlated electron systems and will be used in this thesis in the calculations involving rare-earth elements.

References

1. Hohenberg, P., Kohn, W.: Zur Quantentheorie der Molekeln. Ann. Phys. (Leipzing) **84**, 456 (1927)
2. Hohenberg, P., Kohn, W.: Inhomogeneous electron gas. Phys. Rev. **136**(3B), B864–B871 (1964)

3. Kohn, W., Sham, L. J.: Self-consistent equations including exchange and correlation effects. Phys. Rev. **140**(4A), A1133–A1138 (1965)
4. Janak, J.F.: Proof that $\frac{\partial E}{\partial n_i} = \varepsilon$ in density-functional theory. Phys. Rev. B. **18**(12), 7165–7168 (1978)
5. Perdew, J.P., Parr, R.G., Levy, M., Balduz, J.L.: Density-functional theory for fractional particle number: derivative discontinuities of the energy. Phys. Rev. Lett. **49**(23), 1691–1694 (1982)
6. Wigner, E.: On the interaction of electrons in metals.: Phys. Rev. **46**(11), 1002–1011 (1934)
7. Vosko, S.H., Wilk, L., Nusair, M.: Accurate spin-dependent electron liquid correlation energies for local spin-density calculations—a critical analysis. Can. J. Phys. **58**, 1200 (1980)
8. Ceperley, D.M., Alder, B.J.: Ground state of the electron gas by a stochastic method. Phys. Rev. Lett. **45**(7), 566–569 (1980)
9. Perdew, J.P., Zunger, A.: Self-interaction correction to density-functional approximations for many-electron systems. Phys. Rev. B. **23**(10), 5048–5079 (1981)
10. Harris, J., Jones, R.O.: The surface energy of a bounded electron gas. J. Phys. F Metal Phys. **4**, 1170–1186 (1974)
11. Gunnarsson, O., Lundqvist, B.I.: Exchange and correlation in atoms, molecules, and solids by the spin-density-functional formalism. Phys. Rev. B **13**(10), 4274–4298 (1976)
12. Langreth, D.C., Perdew, J.P.: Exchange-correlation energy of a metallic surface: wave-vector analysis. Phys. Rev. B. **15**(6), 2884–2901 (1977)
13. Perdew, J.P., Burke, K., Ernzerhof, M.: Generalized gradient approximation made simple. Phys. Rev. Lett. **77**(18), 3865–3868 (1996)
14. Perdew, J.P., Wang, Y.: Accurate and simple analytic representation of the electron-gas correlation energy. Phys. Rev. B **45**(23), 13244–13249 (1992)
15. Chadi, D.J., Cohen, M.L.: Special points in the Brillouin zone. Phys. Rev. B **8**(12), 5747–5753 (1973)
16. Monkhorst, H.J., Pack, J.D.: Special points for Brillouin-zone integrations. Phys. Rev. B **13**(12), 5188–5192 (1976)
17. Evarestov, R.A., Smirnov, V.P.: Special points of the Brillouin-zone and thier use in the solid-State theory. Phys. Status Solidi B **119**, 9–40 (1983)
18. Phillips, J.C.: Energy-band interpolation scheme based on a pseudopotential. Phys. Rev. **112**(3), 685–695 (1958)
19. Yin, M.T., Cohen, M.L.: Theory of ab initio pseudopotential calculations. Phys. Rev. B **25**(12), 7403–7412 (1982)
20. Blöchl, P.E.: Projector augmented-wave method. Phys. Rev. B **50**(24), 17953–17979 (1994)
21. Qian, G.X., Martin, R.M., Chadi, D.J.: First-principles study of the atomic reconstructions and energies of Ga- and As-stabilized GaAs(100) surfaces. Phys. Rev. B **38**(11), 7649–7663 (1988)
22. Mosey, N.J., Carter, E.: Ab initio evaluation of Coulomb and exchange parameters for DFT+U calculations. Phys. Rev. B **57**, 1505–1509 (1998)
23. Anisimov, V.I., Zaanen, J., Andersen, O.K.: Band theory and Mott insulators: Hubbard U instead of Stoner I. Phys. Rev. B **44**(3), 943–954 (1991)
24. Anisimov, V.I., Solovyev, I.V., Korotin, M.A., Czyzyk, M.T., Sawatzky, G.A.: Density-functional theory and NiO photoemission spectra. Phys. Rev. B **48**(23), 16929–16934 (1993)
25. Solovyev, I.V., Dederichs, P.H., Anisimov, V.I.: Corrected atomic limit in the local-density approximation and the electronic structure of impurities in Rb. Phys. Rev. B **50**(23), 16861–16871 (1994)
26. Dudarev, S.L., Botton, G.A., Savrasov, S.Y., Humphreys, C.J., Sutton, A.P.: Electron-energy loss spectra and the structural stability of nickel oxide: An LSDA+U study. Phys. Rev. B **57**, 1505–1509 (1998)

Chapter 3
Overview of Experimental Tools

This chapter reviews a range transmission electron microscopy (TEM) techniques, including conventional electron microscopy, High-Resolution Electron Microscopy (HREM), Z-contrast imaging, Electron Energy-Loss Spectroscopy (EELS), and spectrum imaging (SI). The latter three will be discussed in relation to Scanning TEM (STEM). The key aspects of these techniques will be described in detail, emphasizing their strengths and drawbacks. Moreover, various simulation techniques will be discussed in the chapter, and a short summary of advances in electron optics will be provided.

3.1 Conventional Transmission Electron Microscope

In its basic form, the conventional TEM consists of an electron source, a series of condenser lenses for parallel illumination of a sample with high-energy electrons, the objective lens, and stage system where all the beam-specimen interactions take place, several post-specimen lenses that magnify the resulting signal from these interactions, adjustable apertures, and a detector [1, 2]. Of these, the most important is the objective lens itself, which takes the electrons emerging from the exit surface of the specimen, disperses them to create a diffraction pattern in its back focal plane, and recombines them to form specimen's first image in the image plane. The quality of this image controls the resolution of the final image recorded in the detector. Therefore, optimizing operating conditions of the objective lens is an essential step in the TEM, especially for high-resolution imaging discussed in detail in Sect. 3.2

At low and medium magnifications, imaging in the electron microscope is quite similar to that of an optical microscope in which image contrast is caused by the variation in optical absorption within different parts of the specimen [3]. The imaging process of conventional TEM, referred to as amplitude or diffraction contrast imaging, can be interpreted in a similar way, with "absorption" referring to

W. Walkosz, *Atomic Scale Characterization and First-Principles Studies of Si_3N_4 Interfaces*, Springer Theses, DOI: 10.1007/978-1-4419-7817-2_3,

the interception of the electron scattering by an aperture placed below the objective lens. In thin specimens, highly energetic electrons are actually never absorbed; they can only be absent from the image by large angle scattering (diffraction) exceeding the semi-angle of the selected objective aperture or as a result of energy loss and wavelength change in the specimen [3]. The latter process, comprising the inelastic scattering, causes the electrons to be focused on a plane far distant from the one corresponding to the elastic scattering characterized by no energy loss, and contributes only a uniform background to the elastic image. The two processes by which an incident electron interacts with a specimen, the elastic and inelastic scattering, have different angular distributions and carry unique information about the specimen. The elastic scattering (scattering angle $\theta_{el} > 1°$) corresponds to the interaction of the electrons with the atomic nucleus and is the primary basis for imaging and electron diffraction (ED) in the TEM. On the other hand, the inelastic scattering refers to the Coulomb repulsion between the incident and the atomic electrons and is characterized by low-angle distribution ($\theta_{inel} < 1°$). It manifests itself in many processes including X-ray production, secondary electron emissions, collective oscillations (plasmons and phonons), and inner-shell ionizations, and it is favored in light elements. In crystalline samples, the ED pattern is used to ensure that the orientation of the specimen relative to the direction of the incident electron beam will satisfy strongly diffracting conditions defined by Bragg's law: $n\lambda = 2d \sin \theta_B$, where λ is the electron wavelength, d is the spacing between atomic planes, and n is an integer that represents the order of reflection. Using the small-angle approximation, the equation can be written as $\lambda \approx 2d\theta_B = \theta d$, where θ is the scattering angle of the electron. When θ exceeds the semi-angle of the objective aperture, the diffracted electrons will be absorbed by it and the crystalline region will appear dark in the TEM (bright-field) image. In the case of amorphous materials, the amplitude contrast comes from mass-thickness considerations only. Thicker regions of the sample scatter a higher fraction of the incident electrons. Many of them become absorbed by the aperture, making the corresponding regions in the image appear dark. Moreover, areas of higher atomic number also appear dark because of an increase in the amount of elastic scattering [4].

In practice, image formation in the conventional TEM is more complicated than in the optical microscope. Strong fields of the magnetic lenses used for focusing the electron beam cause the electrons to take a spiral trajectory through the lens field. In addition, the unavoidable aberrations of round electron lenses as well as mechanical and environmental instabilities compromise the microscope performance and affect image resolution. A survey of different aberrations of the electron microscope lenses is given in Sect. 3.4

3.2 High-Resolution TEM

In High-Resolution TEM (HRTEM), a much larger objective aperture (or sometimes none at all) is used to transfer high spatial frequencies [3]. The directly transmitted beam can then interfere with one or more diffracted beams giving rise to the contrast across the image (called the bright-field image), which depends on the relative phases of the various beams. The technique is often referred to as phase-contrast imaging and can be used to resolve crystalline lattices, columns of atoms or, in the case of the most modern aberration-corrected microscope, sub-Ångstrøm imaging of the lattice and single atoms [5–7].

In an ideal electron microscope the phase shifts come exclusively from the interactions between the specimen and the incident electron beam. However, in a real microscope, phase shifts also arise from lens aberrations and changes in the objective lens defocus. The net phase error due to these aberrations and defocus is represented by the aberration function $\chi(k)$ given by [8],

$$\chi(k) = \frac{\pi}{2}C_s\lambda^3k^4 - \pi\Delta f\lambda k^2 + \pi f_{a2}\lambda k^2\sin[2(\phi - \phi_{a2})] + \frac{2\pi}{3}f_{a3}\lambda^2k^3\sin[3(\phi - \phi_{a3})] + \cdots \tag{3.1}$$

where C_s is the spherical aberration of the objective lens, λ is the electron wavelength, k is the spatial frequency, f_{a2} is the twofold astigmatism, f_{a3} is the threefold astigmatism, and the ϕ's are the azimuthal orientations of the two astigmatisms aberrations. The effect of this function can be conveniently understood by the Phase-Contrast Transfer Function (CTF) defined as

$$H(k) = \sin\chi(k). \tag{3.2}$$

The basic form of CTF is independent of both specimen and microscope, so that a single set of universal curves can be used to describe the transfer characteristics of the objective lenses. Figure 3.1 shows CTFs for the optimum defocus of the objective lens, calculated by minimizing $\chi(k)$, of a typical 200 kV HRTEM. The two curves represent coherent and partially coherent incident electron illumination. It is important to note that the CTF has an oscillatory nature, which means that electrons scattered at different angles undergo reversals in phase. These oscillations cause spurious details in the final image that may easily be misinterpreted for real atomic structures. When $H(k)$ is negative, positive phase contrast results, meaning that atoms appear dark against a bright background in the bright-field images; when $H(k)$ is positive, atoms are bright and the background is dark; when $H(k)$ is zero, there is no detail in the image for this value of k. The CTF is also focus dependent, implying that further phase changes occur when the focus is changed. Moreover, loss of coherence resulting from focal spread (temporal coherence) and finite beam divergence (spatial coherence) will dampen the CTF at larger scattering angles as shown in Fig. 3.1 [9, 10].

Although the performance of high-resolution electron microscopes can be optimized by working at Scherzer defocus, where the CTF has the largest possible band of spatial frequencies without any phase reversal, employing aberration correctors (see Sect. 3.5), and using highly coherent field-emission electron gun, the interpretation of HRTEM coherent images remains difficult. Multiple electron

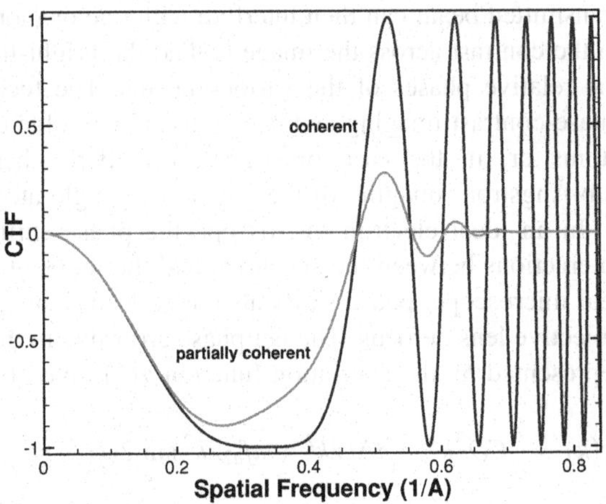

Fig. 3.1 The TEM BF phase-contrast transfer function with and without partial coherence. The astigmatism was assumed to be 0

scattering with large phase changes is rather typical, complicating the quantitative interpretation of image features. Therefore, image simulations are considered as essential in HRTEM.

3.3 Image Simulations

The purpose of performing image simulations is to understand experimental images better by comparing them with the simulated images, where specimens 'features and microscopes' properties and settings can be easily changed and controlled. Such comparisons can provide detailed information about the observed atomic structures in real images by identifying their source, i.e., unraveling what is due to the specimen itself and what is due to the microscope. Different methods have been developed over the past several years to calculate HRTEM images, diffraction patterns, and Z-contrast images in STEM. Among them, the Bloch wave and the multislice methods are the most popular. In the Bloch wave approach [11–14], electron wavefunctions are expanded in Bloch waves inside a crystalline specimen. These wavefunctions have the periodicity of the specimen and satisfy the Schrödinger equation, which is then solved for the eigenvectors and eigenvalues of the incident electron inside the crystal. In spite of the success of this method in reproducing many experimental images [15, 16], the approach can be only applied to periodic systems. Also, it works best for small, defect-free crystals since the computer time for performing the calculation scales is N^3, where N is the number of Bloch waves or Fourier components (http://www.people.ccmr.cornell.edu/davidm/weels/summer06/mtutor.pdf). Smaller N, in turn, makes the calculations inaccurate.

The multislice method ([8, 17, 18]; http://www.people.ccmr.cornell.edu/davidm/weels/summer06/mtutor.pdf), on the other hand, is generally much more efficient and easier to implement numerically. It is also flexible enough to simulate defects and interfaces. The storage requirements for the multislice method scale as N or $N \ln N$ when the fast Fourier transforms are used. In this method the specimen is divided into thin two-dimensional slices along the beam direction. At each slice the electron wavefunction experiences a phase shift due to the projected atomic potential of all atoms in the slice expressed as [8]

$$\upsilon_z(\mathbf{r}_{xy}) = \lambda \frac{m}{m_o} \frac{1}{ab} \int \sum_j f_{ej}(k_x, k_y) \exp[2\pi i \mathbf{k}_{xy} \cdot (\mathbf{r}_j - \mathbf{r}_{xy})] \mathrm{d}\mathbf{k}_{xy} \tag{3.3}$$

where m/m_o is the relativistic mass ratio of the electron, ab is the total area of the specimen being simulated (with supercell dimensions $a \times b$), and f_{ej} is the electron scattering factor in the first Born approximation for atom j, which can be obtained from relativistic Hartree–Fock calculations. The object function $O(\mathbf{r}_{xy})$, which modulates the wavefunction, is calculated as

$$O(\mathbf{r}_{xy}) = \exp[i\sigma \upsilon_z(\mathbf{r}_{xy})] \tag{3.4}$$

where σ is defined as $\sigma = 2\pi$ $me\lambda/h^2$ and m is the relativistic electron mass. The electron probe is then recursively transmitted through each slice and finally projected onto the detector to form an image. Similarly to Bloch method, the multislice algorithms have proven to correlate very well with experimental images [19–21]. In this thesis, the Bloch wave simulation method will be applied to bulk α-Si_3N_4.

3.4 Z-Contrast Imaging in STEM

As discussed in the previous sections, a major drawback of the HRTEM coherent imaging method is that the resulting images are not directly interpretable, i.e., the observed intensity at a given position may not relate to actual atomic arrangement. On the contrary, the technique of Z-contrast imaging (also called High-Angle Annular Dark-Field imaging or HAADF imaging) in STEM has the advantage that the intensity I in the image is related to atomic arrangements in a simple fashion, namely, $I \sim Z^n$, where Z is the atomic number and n is a constant ranging from 1.6 to 2 for different collection conditions [16, 22]. Owing to the incoherent nature of this technique, Z-contrast images have no phase problem for structure determination, the resolution is a factor of two higher than in coherent images, and there are no contrast reversals with crystal thickness or defocus. Furthermore, the HAADF detector, collecting the high-angle scattered electrons used for Z-contrast images, does not interfere with the low-angle scattering. This allows simultaneous acquisition of the bright-field images, as well as various inelastic signals that can be used for spectroscopic work.

The basic idea behind imaging and spectroscopy in a STEM (shown schematically in Fig. 3.2) is as follows: a high-energy electron beam is focused down to a small spot and passed through a thin sample, while scanning it across. Signals from

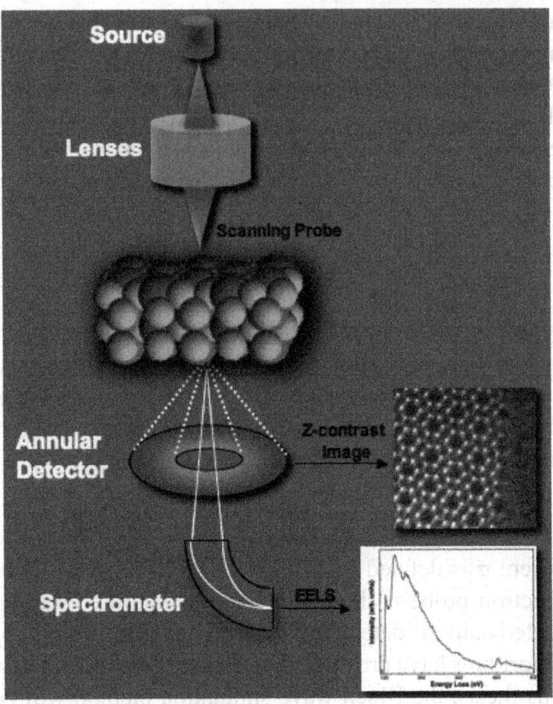

Fig. 3.2 Schematic showing major elements of a scanning transmission electron microscope. A high-brightness electron source produces the electron beam, which passes through round magnetic lenses and is focused to a spot size of between 0.5 and 2.0 Å. As the beam is scanned across the sample, both elastic and inelastic signals can be collected simultaneously for Z-contrast imaging or atomic resolution electron-energy-loss spectroscopy

scattered electrons and ionized atoms are recorded by various detectors and used to build up a two- dimensional map. Elastically scattered electrons with large angle deflections (40–100 mrad at 200 kV) are collected by the annulus-shaped HAADF detector placed below the sample to form Z-contrast images. Detecting the high-angle scattering and integrating over a large angular range effectively average coherence effects between neighboring atomic column in the specimen. Thermal vibrations, on the other hand, reduce the coherence between atoms in the same column to residual corrections between near neighbors. This allows each atom to be considered as an independent scatterer with a cross-section approaching a Z^n dependence. Mathematically, the image intensity $I(\mathbf{R})$ in the incoherent Z-contrast image can be described as a convolution between the object function $O(\mathbf{R})$, which is sharply peaked at the atomic columns, and the probe intensity profile $P^2(\mathbf{R})$,

$$I(\mathbf{R}) = O(\mathbf{R}) * P^2(\mathbf{R}) \tag{3.5}$$

The small width of the object function ($\sim$0.1 Å) implies that the spatial resolution is limited only by the size of the electron probe. This result is also valid for thicker specimens [16, 23]. The large angular spread of the electron probe and the specific detector setup (i.e. large inner angle and large detector area) lead to the highly bound 1*s* Blochn states to contribute mostly to the measured image intensity, because only these states have transverse momentum components that extend to the high-angle detector [24]. With only one dominant Bloch state, dynamical diffraction effects are removed and manifest as a columnar channeling effect, maintaining the thin specimen description of the image given in Eq. 3.5. Since no contrast reversals with changing sample thickness or defocus occur, atomic sites can be easily and unambiguously identified during the experiment, making the Z-contrast technique suitable for the examination of various structures like surfaces, interfaces, defects, and bulk samples [19, 25–29]. It should be emphasized that the incoherent nature of atomic resolution imaging in the STEM is a consequence of the detector geometry rather than the electron probe, which must remain coherent, i.e. it is, strictly, the detector size that determines whether an image is best described as coherent or incoherent (see Appendix B for more details). A small detector gives a more coherent image, while a large detector renders more incoherent one. Intermediate angles display an initial coherent dependence on thickness, changing to an incoherent one as an atomic column becomes significantly longer than the correlation length [29]. The condition on the specific detector size limits the application of the incoherent Z-contrast technique in the conventional TEM. Although technically possible in the form hollow-cone illumination, the method requires significant modifications in the illumination system, making it rather impractical [30].

Lastly, it should be mentioned that a significant fraction of the thermal displacements of the atoms, breaking the intracolumn coherence, are due to zero point fluctuations which will not disappear on cooling the sample. Thus, the scattering will not become coherent on cooling, but to maintain its incoherent nature, the detection angles will need to be increased [29]. In this thesis, the technique of Z-contrast imaging will be used to image β-$Si_3N_4/CeO_{2-\delta}$ and β-Si_3N_4/SiO_2 interfaces, as well as bulk α-Si_3N_4.

3.5 Annular Bright-Field Imaging in STEM

In addition to the Z-contrast imaging, the STEM geometry allows for a simultaneous acquisition of the signal for the Annular Bright-Field (ABF) imaging. Electrons scattered to intermediate angles ($\sim$11–22 mrad) are collected by the annulus-shaped detector placed below the sample and are used to form the ABF image. Although the principles behind this new method are in the course of detailed investigations, the resulting ABF images do not show contrast reversals with thickness or defocus values, implying that they are incoherent in nature (S.D. Findlay 2009, private communication). This technique will be applied to study β-Si_3N_4/SiO_2 interfaces.

3.6 Probe Formation

The most important aspect of imaging in STEM is the formation of the smallest possible electron probe with a sufficient probe-current to allow dwell times of

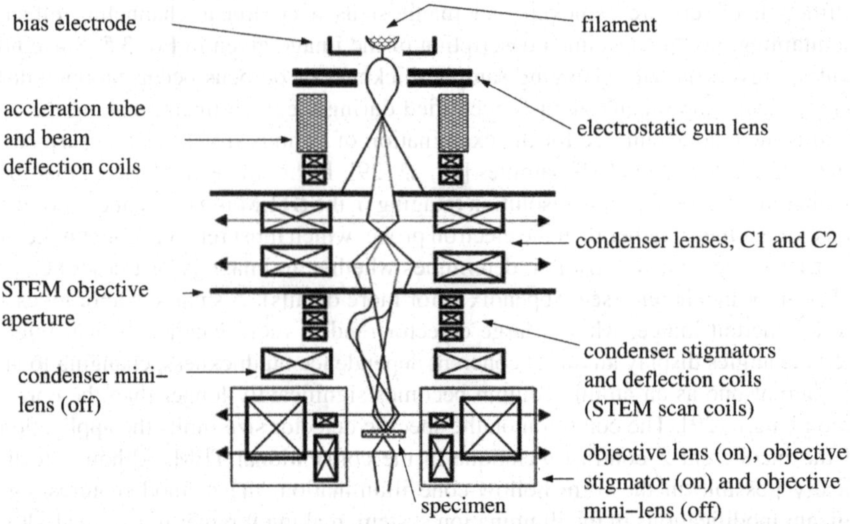

Fig. 3.3 Simplified drawing of the electron-optical arrangement for the formation of the STEM probe. The probe is rastered across the specimen by the deflection coils. The *dashed lines* indicate a deflection path. Adapted with permission from [28]

$\sim 100\,\mu$sec per pixel or less. This is accomplished through demagnification of the electron source with probe-forming lenses. Figure 3.3 shows a simplified diagram of the electron-optical arrangement for the formation of the electron probe in JEOL-2010F. The electrons are emitted by the Schottky field emission source and are focused by the gun lens to a cross-over, reducing its size. The source is further demagnified by a set of condenser lenses comprising of C1 and C2 lenses, with the highest demagnification achieved when C1 is near the maximum excitation. The C2 lens is then used to project the cross-over from C1 to a fixed cross-over for the objective lens, which projects the image of the electron gun tip onto the specimen. The total demagnification of the probe-forming lens can thus be calculated from $M_{\mathrm{tot}} = M_{\mathrm{gun}} M_{\mathrm{cond}} M_{\mathrm{obj}}$, where M_{gun}, M_{cond}, and M_{obj} correspond to the demagnification of the gun lens, the condenser lens, and the objective lens, respectively. In spite of improving the resolution of STEM techniques by forming a small probe, a large demagnification reduces the amount of electron current at low angles in the electron probe, thus compromising the quality of images and EELS spectra. Moreover, since more electrons pass the objective lens at high angles, larger aberrations are introduced into the electron beam, necessitating the use of an objective aperture, which limits the ultimate probe size.

The tuning and optimizing the electron probe can be most easily done with the electron Ronchigram or shadow image formed at the microscope Fraunhofer diffraction plane [31, 32]. This image, representing the undiffracted disk of electrons at the center of diffraction pattern, varies considerably with convergence angle and is sensitive to lens aberrations and defocus. Therefore, any misalignment in the electron-optical system can be readily observed in the Ronchigram. If the spherical aberration is the limiting aberration, the Ronchigram will depart from its ideal circular symmetry as the excitation of each illumination optical component changes. Furthermore, the presence or absence of interference fringes in the pattern indicates the amount of incoherent probe broadening due to instabilities and the effect of finite source size.

Figure 3.4 shows a series of Ronchigrams at the amorphous edge of a specimen taken on the JEOL 2010F (uncorrected) microscope [33]. At large defocus, the electron cross-over is at relatively large distance from the specimen plane, and a projection image is observed. As Gaussian focus is approached, an angular dependence on the magnification emerges due to lens aberrations and the extent to which they change the phase of the electron beam. At slight underfocus (Fig. 3.4b), the azimuthal and radial circles of infinite magnification can be

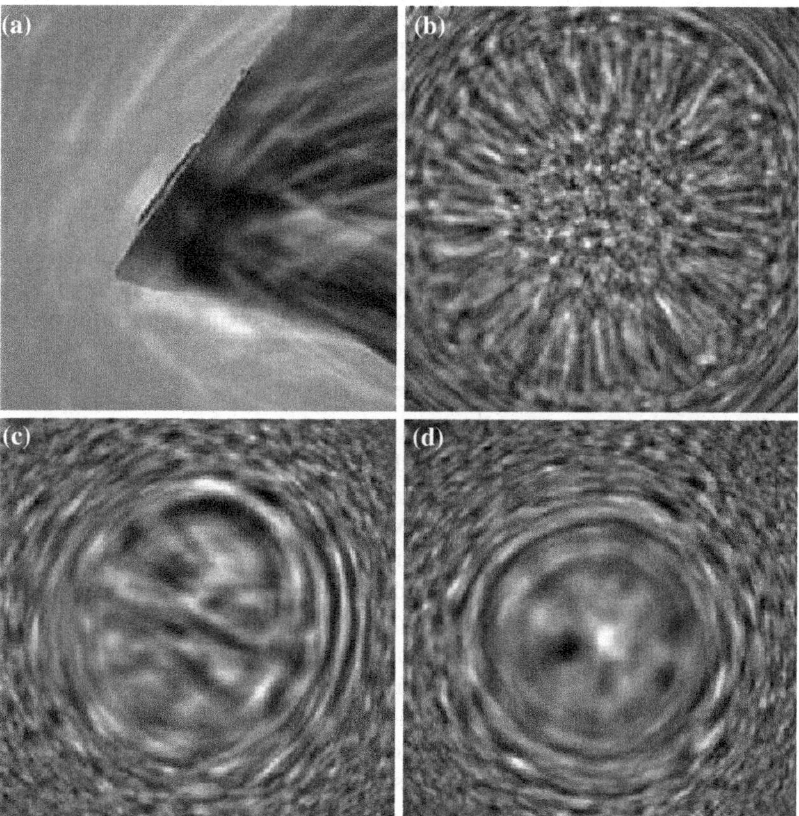

Fig. 3.4 Experimental electron Ronchigram from crystalline sample with amorphous edge formed with a stationary probe showing: **a** large underfocus with Kikuchi lines of crystalline bulk; **b** small underfocus; **c** departure from circular symmetry due to astigmatism; **d** ideal Gaussian focus. Adapted with permission from [33]

observed [31]. At this setting, axial astigmatism can be easily corrected by exciting the stigmator coils and ensuring that all the features in the pattern are circularly symmetric as shown in Fig. 3.4c. As the beam is focused (Fig. 3.4d), the center of the Ronchigram displays the highest magnification. The comma-free axis is clearly defined by this position and all alignment and positioning of detectors and apertures can be performed with respect to this spot. Next, the illumination beam alignment can be checked by wobbling first the condenser lens excitation and then the microscope high tension. Any misalignment of the beam between the condenser and objective lenses, showing up as a periodic translation of Ronchigram features when the wobbling takes place, can be corrected with the condenser alignment coils. If a crystalline specimen is used and the probe is coherent enough, diffraction effects in the form of the lattice fringes will be present in the Ronchigram [34]. It is the movement of these fringes across the STEM detectors as the probe is scanned that gives rise to image contrast. At large defocus, the Kikuchi lines resulting from inelastic scattering are visible and can be used to align the specimens crystal orientation with the coma-free axis. The intensity distribution Full Width at Half Maximum (FWHM) of the aligned and stigmated probe depends on the focus setting, the illumination convergence angle (set by an aperture), the spherical aberration of the objective lens, and the level of any incoherent broadening of the probe by instabilities in the microscope. In the approximation of the spherical aberration being the limiting aberration, the optimum convergence angle and the defocus are given by $\alpha_{\text{opt}} = (4\lambda/C_s)^{1/4}$ and $\Delta f = -(\lambda C_s)^{1/2}$, respectively, where C_s is the spherical aberration of the objective lens and λ is the electron wavelength [33]. The probe size and shape can be calculated for a given convergence angle and defocus by the following equation [8]:

$$\phi(r) = \sqrt{A} \int A(k) \exp[i\chi(k) - 2kr] \mathrm{d}^2 k \tag{3.6}$$

where the value of A is chosen such that the total intensity of the probe, $|\phi(r)|^2$, is normalized to one and $\chi(k)$ is the spherical aberration function defined in Eq. 3.1. It should be noted here that the FWHM of the probe profile can be reduced significantly by changing the defocus, but at the expense of creating undesirable second maxima, which contribute to the delocalization of the electron signal. The ideal defocus, therefore, represents a compromise between the smallest probe diameter and negligible secondary maxima. Figure 3.5 shows the calculated probe intensity profile for a 200 keV electron beam, with a convergence angle of 15 mrad and different defocus values. The appearance and increase in the second maxima can be clearly seen as the focus deviates from its optimum value. Once the probe is properly tuned, it should be possible to achieve a STEM image with a spatial resolution mostly limited by the probe diameter.

It should be noted that the optimized Ronchigrams of the aberration-corrected microscopes are no longer circular, but rather hexagonal or octagonal depending on the corrector-lens design. Several approaches have been developed to diagnose and correct the wave aberrations in the electron microscopes with a high degree of

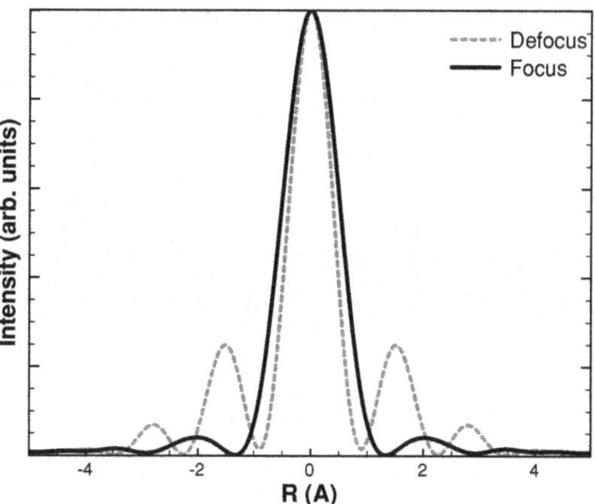

Fig. 3.5 Calculated probe profile for conditions close to focus and small defocus. A spherical aberration of 0.5 mm and a defocus of 361 nm were assumed

precision [35–37]. They depend on either the numerical analysis of several Ronchigrams taken at different positions of the electron probe on an amorphous area of the specimen [38, 39], or on the observations of the intensity along so-called achromatic rings formed at points equidistant from the centers of two overlapping convergent beam ED disks [40].

3.7 Electron Energy-Loss Spectroscopy/Energy-Filtered STEM

As mentioned before, in their interaction with a specimen, electrons undergo both elastic and inelastic scattering. Elastically scattered electrons are the basis for imaging, whereas inelastically scattered electrons are used in various spectroscopic methods, such as EELS and X-ray absorption spectroscopy, to extract chemical information on the electronic structure of the specimen. Figure 3.6 summarizes the energy-level diagram of the primary and secondary inelastic processes probing various energy levels of a solid [41]. The interaction of a fast electron with an inner-shell electron results in discrete transition of the core electron to an unoccupied state above the Fermi level. The transitions are governed by the dipole selection rule requiring that the change in the angular momentum quantum number, Δl, between the initial and the final state be equal to ± 1 [2, 41]. In the de-excitation process, the atoms' outer-shell electrons (or the inner-shell electrons of lower binding energy) undergo a downward transition to the vacant core holes emitting either X-rays or Auger electrons. Outer-shell electrons, on the other hand, can experience both single-electron and collective excitations induced by the fast electrons. In the case of an insulator or semiconductor, single-electron excitations involve intraband transitions of the valence electrons across the energy gap, whereas in metals it is the conduction electrons that make transitions to higher states, possibly within the same energy band. If the final state of these transitions

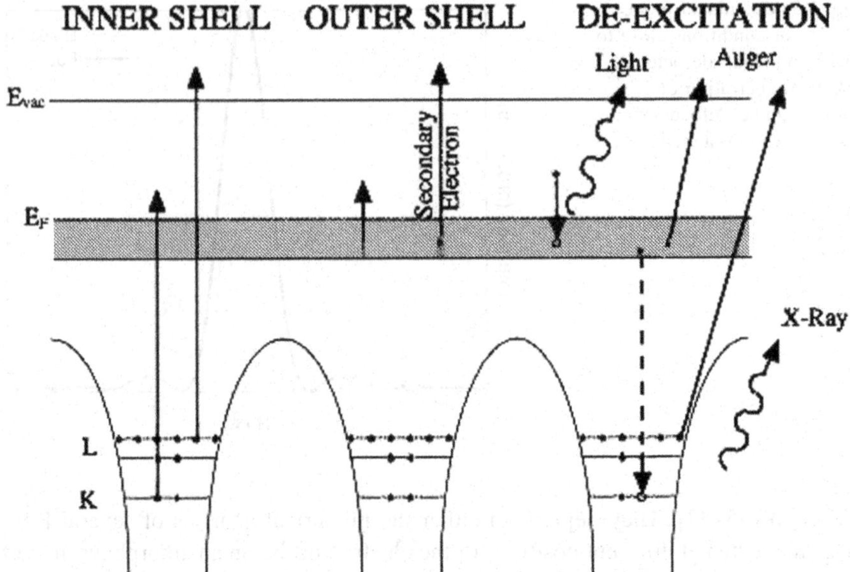

Fig. 3.6 Interaction between the incoming electron beam and the specimen, including K- and shell core levels and a valence band of delocalized states (*shaded*). Inner-shell excitation promotes highly localized electrons into unoccupied states above the Fermi level. Outer-shell excitation advances electron from the conduction band into unoccupied states above the Fermi level. X-rays and Auger electrons are produced as by-products of the de-excitation processes. Adapted with permission from [41]

lies above the vacuum level of the solid and if the excited atomic electron has enough energy to reach the surface, it may be released as a secondary electron. In the de-excitation process, the atoms might emit electromagnetic radiation in the visible region (cathodoluminescence) or the infrared spectrum as heat. Collective excitations of the outer-shell electrons involve plasmons and take the form of longitudinal oscillations of weakly bound electrons. The oscillations are observed predominantly in metals, particularly the ones with large Fermi surfaces, although they occur to a greater or lesser extent in all materials.

Electron Energy-Loss Spectroscopy, used in this thesis, analyzes the primary processes of electron excitations. A typical EEL spectrum, showing the scattered intensity as a function of the decrease in kinetic energy of the fast electron interacting with the sample, is illustrated in Fig. 3.7. The spectrum can be divided into three regions (the zero-loss peak, the low-loss, and the high-loss regions), each carrying unique information about the sample [2, 41]. The zero-loss peak consists of electrons transmitted without any measurable energy loss, including those which are scattered elastically in the forward direction, as well as those which have excited phonon modes for which the energy loss is less than the experimental energy resolution. It is the most intense feature of a spectrum and can be used to

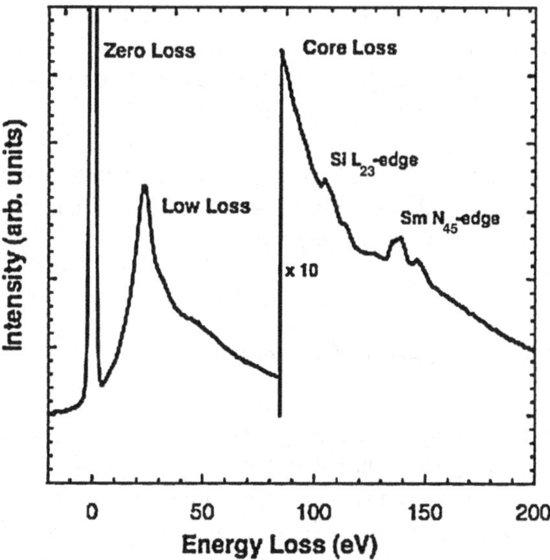

Fig. 3.7 A schematic EEL spectrum illustrating the zero-loss peak as well as the low-loss and the high (core)-loss regions. Adapted with permission from [42]

define its energy resolution. The low-loss region (typically < 50 eV), corresponding to the interaction of an incident electron with the outer-shell atomic electrons, on the other hand, can be used to reveal specifics such as the thickness of the specimen, band gap, valence-electron density, and surface and interface states through analysis of the plasmon losses. Lastly, the high-loss region of the EELS spectrum (typically 50–1,000 eV) can be used to analyze the atom type through analysis of characteristic ionization edge energies representing transitions of the core electrons into unoccupied levels; elemental concentration and distribution from the integrated intensity of the edge; the chemical state, local structure, coordination, and bonding through analysis of the shape of the near-edge structure; and bond distances through the analysis of the extended energy-loss fine structures [43].

Electrons used for the EELS analysis are collected by a spectrometer consisting of a magnetic prism that deflects, disperses, and focuses them according to their kinetic energy in the plane of an energy-selecting slit. Alternatively, an image-forming spectrometer, consisting of an energy-selecting slit and a two-dimensional CCD array detector, can be used to form an energy-filtered image representing a two-dimensional map of a given element within the sample. Variations of this general technique can be used to form thickness maps, chemical state maps, and three-dimensional spectrum images, in which each pixel in an image is associated with a full energy spectrum.

As mentioned before, due to the low angular distribution of the inelastically scattered electrons, the HAADF detector does not interfere with the post-column spectrometer used in EELS, allowing to use Z-contrast images to position the electron probe over a particular structural feature for acquisition of a spectrum. To be able to correlate the EEL spectra with the structural features seen in the image, it is essential for the spectrum to have the same atomic resolution as the

Z-contrast image. In order to achieve this resolution, the spatial range over which a fast electron can cause an excitation must be less than the interatomic spacing [44]. Hydrogenic and more elaborate models show that for the major edges accessible by conventional spectrometers ($\Delta E < 2$ keV), the object function is localized within ~1 Å of the atom cores [45, 46]. This enables the electronic structure information obtained from the EELS spectra to correspond to individual atomic columns, making the combined Z-contrast imaging and EELS approach a unique tool to study various materials.

3.8 Aberration Correctors

As mentioned in the previous sections, the resolving power of modern electron microscopes is largely limited by various lens aberrations. Among them, spherical and chromatic aberrations, shown in Fig. 3.8, as well as astigmatism are most common. Spherical aberration is caused by the lens field acting inhomogeneously on the off-axis electrons. As a consequence, electrons traveling at higher angles to the optical axis are focused too strongly and a point object is thus imaged as a disk of finite size. Chromatic aberration, on the other hand, causes the focusing to depend on energy, i.e. electrons with different kinetic energies will be focused at a different distance from a lens. The energy variations depend on the electron gun, fluctuations in the potential applied to accelerate the electrons, and energy losses due to inelastic scattering. Lastly, astigmatism occurs when the electrons sense a nonuniform magnetic field as they spiral round the optic axis, contributing to image distortions. Although the last defect can be easily corrected using stigmators (octupoles that introduce a compensating field to balance inhomogeneities), spherical and chromatic aberrations require more sophisticated solutions. Both defects are unavoidable for rotationally symmetric lenses as a consequence of Laplace's equation, where the electric and magnetic potentials used to focus the electron beam cannot assume arbitrary shapes in free space. As a result, the achievable spatial resolution is limited to about 50 times the wavelength of the imaging electrons. This fundamental limit was first identified in 1936 by Otto Scherzer, who later proposed several methods to deal with the aberration problem [48–50]. In particular, Scherzer suggested an arrangement of non-static, non-round lenses, a lens-mirror combination, or placing free charges in the path of the beam. Of these, the most practical approach has proven to be departing from rotational symmetry and adding a series of multipole optical elements that correct the leading terms in an increasing power-series expansion of the wavefront distortion. At present, there are two main types of aberration correctors, aptly named after the type of multipole they employ: the quadrupole–octupole (QO) correctors and the round lens-hexapole correctors [35, 36, 39, 51–53]. Both classes of correctors consist of correcting parts (octupoles or hexapoles), and first-order elements (quadrupoles or round lenses), whose primary function is to steer the electron

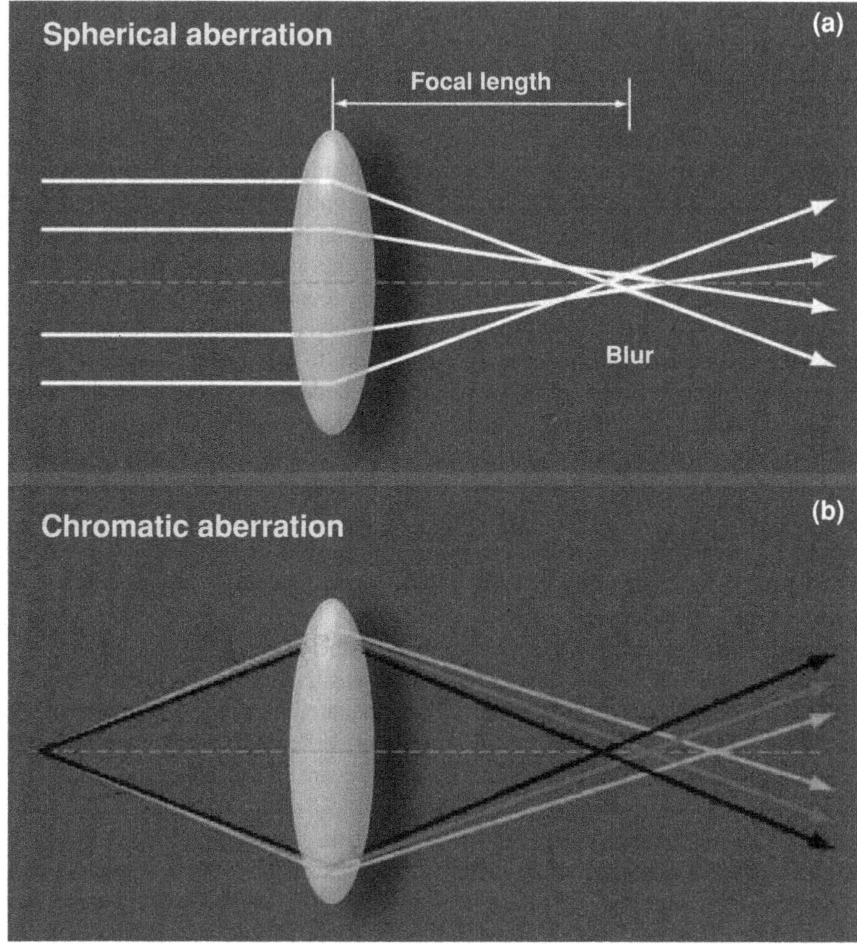

Fig. 3.8 The effect of geometrical aberrations on the ray paths: **a** spherical aberration; **b** chromatic aberration. Adapted with permission from [47]

trajectories. In a QO corrector, the octupoles have a field that varies with distance from the optic axis in the same way as does spherical aberration, but have an undesirable fourfold rotational symmetry with an alternating sign on rotation by 45° [54]. A series of quadrupoles are therefore used to deform the beam in order to pick up the desired components of the phase change from three suitably excited octupoles, but leave a round beam after correction. One advantage of this type of corrector is that the addition of electrostatic elements allows the focusing to depend on the energy, and thus provide the correction for the chromatic aberration [55]. Hexapole correctors were shown to correct spherical aberration, but at the cost of introducing an undesirable threefold symmetry. The correction by hexapoles relies on extended multipoles where the field changes the trajectories, which means that the electrons go through different parts of the field and are further deflected. Pairs of round lenses are used to project the beam from one hexapole to the next and lastly into the objective lens, while retaining the correction and ensuring that the final beam is round.

It took nearly 50 years after Scherzer's original work for the wavelength limited electron be realized due to the required mechanical stability of the multipole systems (better than 1 ppm). Additionally, the requirement for the multipoles to match the original lenses in position, orientation, and tilt has only recently become possible, enabled by better electron optics and the availability of highly stable electronics and inexpensive computer platforms for control of the optical elements. The benefits of the aberration correctors are, however, tremendous ranging from the smaller probe size and better image resolution, to a larger convergence semi-angle for the electron probe in STEM without the undesirable increase in the probe diameter by spherical aberration. In turn, this implies higher probe current and larger signals for the Annular Dark-Field (ADF) and X-ray Energy Dispersive Spectrometry (XEDS) detectors, as well as the EEL spectrometer, allowing for the elemental analysis with higher sensitivity. At present, aberration-corrected electron microscopes obtain a resolution limit of about 0.5 Å [56–58] and 1.0 Å in STEM and TEM modes, respectively [5–7]. Moreover, single-atom detection is possible [59].

Higher current densities ($\sim 100\,\mathrm{A/cm^2}$) in the aberration-corrected microscopes can, however, cause damage to beam-sensitive materials in the form of breaking chemical bonds between their constituent atoms (*radiolysis*) or displacing the atoms from their sites (*knock-on damage*) [2]. Radiolysis involves inelastic scattering, mainly ionization, and can be very severe in inorganic materials and oxides. The knock-on damage, on the other hand, results from the momentum transfer in elastic scattering between the fast electron and the atomic nucleus in the sample, and it occurs primarily in metals. At voltages about 50% less than knock-on thresholds, sputtering of surface atoms is possible [60]. This effect can be, however, cured by coating samples with a carbon layer. Working with lower beam voltages, on the other hand, can greatly reduce knock-on damage, but it does little to reduce the ionization damage. Because the latter cannot be completely avoided, it sets the ultimate practical limit to TEM/STEM spatial resolution and elemental mapping sensitivity.

3.9 Microscopes Used

Figure 3.9 shows the electron microscopes used in this thesis. They include JEOL-2010F located at the University of Illinois at Chicago, FEI Titan S and VG HB603U located at Oak Ridge National Laboratory, and JEOL-ARM200F located in Tokyo, Japan (Fig. 3.9d). The JEOL-2010F shown in Fig. 3.9a is equipped with a Schottky field-emission gun operated at 200 kV, an ultra-high-resolution

Fig. 3.9 Electron microscopes used in this research: **a** JEOL-2010F located at UIC, **b** FEI Titan S located at Oak Ridge National Laboratory (ORNL), **c** VG HB603 located at ORNL, **d** JEOL-ARM200F in Tokyo, Japan

objective lens-pole piece ($C_s = 0.55$ mm), and a post-column Gatan Imaging Filter (GIF). The Titan S (Fig. 3.9b) has a Schottky field-emission gun operated at 300 kV, CEOS aberration corrector, and a post-column GIF. The VG HB603U (Fig. 3.9c) is a cold-field-emission microscope operated at 300 kV and is equipped with Nion aberration corrector and Gatan Enfina electron energy loss-spectrometer. Lastly, the JEOL-ARM200F has a Schottky field-emission gun with variable accelerating voltage (80–200 kV), an aberration corrector, and an Enfina EELS spectrometer.

3.10 Multivariate Statistical Analysis

As discussed in the previous section, elemental and compositional mapping in EELS can be done with the SI technique, representing a three-dimensional map of the selected area within the sample. In this technique, full energy spectra are acquired at each pixel as the probe moves across the specimen. While providing more detailed information about the sample as compared to the conventional point or line-scan approaches, the technique suffers from weak (noisy) signal collection due to the relatively shorter acquisition time for each pixel to avoid beam damage and sample drift effects. As a consequence, small fluctuations in the electron loss near-edge and extended fine structures, providing information about the electronic structure and bonding characteristics in the specimen, can be easily misinterpreted or even missed. To avoid that problem, advanced statistical technique of Multivariate Statistical Analysis is often used to analyze large SI datasets [61]. The application of this technique allows for the automatic extraction of statistically significant and dominant features without any bias and for the reconstruction of the noise-reduced data by subtracting random noise

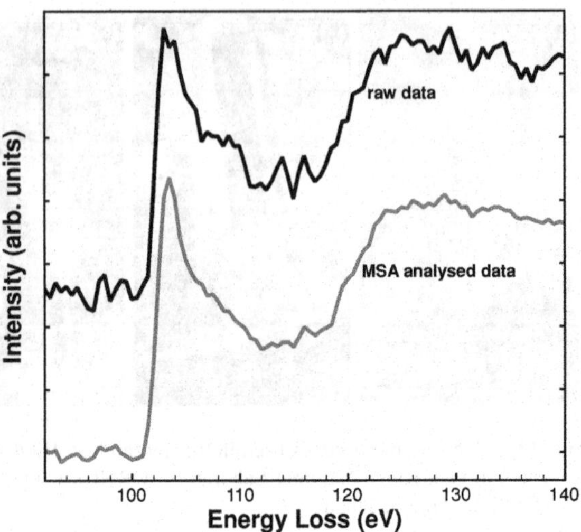

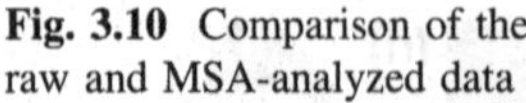
Fig. 3.10 Comparison of the raw and MSA-analyzed data

components. This is accomplished by identifying the most meaningful basis, assumed to be a linear combination of the original basis, to re-express a given dataset. The new basis minimizes redundancy in the dataset, measured by the magnitude of the covariance, and maximizes the signal, measured by the variance. Figure 3.10 shows the application of this technique to the SI datasets collected from bulk β-Si_3N_4. The upper spectrum corresponds to raw data, while the bottom one is MSA analyzed. Comparing the spectra, one can see a great reduction in the noise when MSA is applied. Also, the feature around 115 eV in the raw data, which could be easily misinterpreted for SiO_2 signature, is reduced with MSA. In this thesis, MSA will be applied to analyze data acquired from the Si_3N_4/SiO_2 interface.

References

1. Reimer, L.: Transmission Electron Microscopy. Springer, Berlin (1989)
2. Williams, D.B., Carter, C.B.: Transmission Electron Microscopy. Springer Science+Business Media, LLC, New York (1996)
3. Spence, J.C.H.: High-Resolution Electron Microscopy. Oxford University Press, New York (1986)
4. Egerton, R.F.: Physical Principles of Electron Microscopy. Springer, New York (2005)
5. Zhang, Z., Kaiser, U.: Structural imaging of β-Si_3N_4 by spherical aberration-corrected high-resolution transmission electron microscopy. Ultramicroscopy **109**, 1114–1120 (2009)
6. Haider, M., Rose, H., Uhlemann, S., Kabius, B., Urban, K.: Toward 0.1 nm resolution with the first spherically corrected transmission electron microscope. J. Electron. Microsc. **47**(5), 395–405 (1998)
7. Kabius, B., Haider, M., Uhlemann, S., Schwan, E., Urban, K., Rose, H.: First application of a spherical-aberration corrected transmission electron microscope in materials science. J. Electron. Microsc. **51**(5), S51–S58 (2002)

8. Kirkland, E.J.: Advanced Computing in Electron Microscopy. Plenum Press, New York (1998)
9. Frank, J.: The envelope of electron microscopic transfer functions for partially incoherent illumination. Optik **38**, 519–536 (1973)
10. Saxton, W.O.: Spatial coherence in axial high resolution conventional electron microscopy. Optik **49**, 81–92 (1977)
11. Fujimoto, F.: Periodicity of crystal structure images in electron microscopy with crystal thickness. Phys. Stat. Sol. A **45**, 99–106 (1978)
12. Kambe, K.: Visualization of bloch waves of high energy electrons in high resolution electron microscopy. Ultramicroscopy **10**, 223–227 (1982)
13. Yamazaki, T., Watanabe, K., Kuramochi, K., Hashimoto, I.: Extended dynamical HAADF STEM image simulation using the Bloch-wave method. Acta Cryst. A **62**, 233–236 (2006)
14. Findlay, S.D., Oxley, M.P., Allen, L.J.: Modeling atomic-resolution scanning transmission electron microscopy images. Microsc. Microanal. **14**(1), 48–59 (2007)
15. Pennycook, S.J., Jesson, D.E.: High-resolution incoherent imaging of crystals. Phys. Rev. Lett. **64**, 938–941 (1990)
16. Nellist, P., Pennycook, S.J.: Incoherent imaging using dynamically scattered coherent electrons. Ultramicroscopy **78**, 111–124 (1999)
17. Cowley, J.M., Moodie, A.F.: The scattering of electrons by atoms and crystals. Acta Cryst. **10**, 609 (1957)
18. Cowley, J.M., Moodie, A.F.: The scattering of electrons by atoms and crystals. III. Single-crystal diffraction patterns. Acta Cryst. **12**, 360–367 (1959)
19. Klie, R.F., Idrobo, J.C., Browning, N.D., Serquis, A., Zhu, Y.T., Liao, X.Z., Mueller, F.M.: Observation of coherent oxide precipitates in polycrystalline MgB_2. Appl. Phys. Lett. **12**, 3970 (2002)
20. Loane, R.F., Kirkland, E.J., Silcox, J.: Visibility of single heavy atoms on thin crystalline silicon in simulated annular dark-field STEM images. Acta Cryst. A **44**, 912 (1988)
21. Voyles, P.M., Grazul, J.L., Muller, D.A.: Imaging individual atoms inside crystals with ADF-STEM. Ultramicroscopy **96**, 251 (2003)
22. Jesson, D.E., Pennycook, S.J.: Incoherent imaging of crystals using thermally scattered electrons. Proc. R. Soc. Lond. A **449**, 273–293 (1995)
23. Möbus, G., Naufer, S.: Nanobeam propagation and imaging in a FEFTEM-STEM. Ultramicroscopy **96**, 285–298 (2003)
24. Pennycook, S.J.: Structure determination through Z-contrast microscopy. In: Merli, P.G., Vittor-Antisari, M. (eds.) Advances in Imaging and Electron Physics, vol. 123, p. 173. Academic Press, London (2002)
25. Ziegler, A., Idrobo, J.C., Cinibulk, M.K., Kisielowski, C., Browning, N.D., Ritchie, R.O.: Interface structure and atomic bonding characteristics in silicon nitride ceramics. Science **306**, 1768 (2004)
26. Muller, D.A.: Structure and bonding at the atomic scale by scanning transmission electron microscopy. Nat. Mater. **8**, 263–270 (2009)
27. Varela, M., Lupini, A.R., Van Benthem, K., Borisevich, A.Y., Chisholm, M.F., Shibata, N., Abe, E., Pennycook, S.J.: Materials characterization in the aberration-corrected scanning transmission electron microscope. Annu. Rev. Res. **35**, 539–569 (2005)
28. Browning, N., James, E.M., Kishida, K., Arslan, I., Buban, J.P., Zaborac, J.A., Pennycook, S.J., Xin, Y., Dusher, G.: Scanning transmission electron microscopy: an experimental tool for atomic scale interface science. Rev. Adv. Mater. Sci. **1**, 1–26 (2000)
29. Klie, R.F., Zhu, Y.: Atomic resolution STEM analysis of defects and interfaces in ceramic materials. Micron **36**(3), 219–231 (2005)
30. Pennycook, S.J.: Z-contrast imaging in the scanning transmission electron microscope. In: Zhang, Z.F., Zhang, Z. (eds) Progress in Transmission Electron Microscope 1: Concepts and Techniques, pp. 81–111. Springer, Tsinghyua (2001)
31. Bals, S., Kabius, B., Haider, M., Radmilovic, V., Kisielowski, C.: Annular dark-field imaging in a TEM. Solid State Commun. **130**, 675–680 (2004)

32. Cowley, J.M.: Electron diffraction phenomena observed with a high resolution STEM instrument. J. Electron. Microsc. Tech. **3**, 25–44 (1986)
33. Rodenburg, J.M., Macak, E.B.: Optimising the resolution of TEM/STEM with the Electron Ronchigram. Microsc. Anal. **90**, 5–7 (2002)
34. James, E.M., Browning, N.D.: Practical aspects of atomic resolution imaging and analysis in STEM. Ultramicroscopy **78**, 125–139 (1999)
35. Rose, H.H.: Historical aspects of aberration correction. J. Electron Microsc. **58**(3), 77–85 (2009)
36. Dellby, N., Krivanek, O.L., Nellist, P.D., Batson, P.E., Lupini, A.R.: Progress in aberration-corrected scanning transmission electron microscopy. J. Electron Microsc. **50**(3), 177–185 (2001)
37. Haider, M., Uhlemann, J.Z.: Upper limits for the residual aberrations of a high-resolution aberration-corrected STEM. Ultramicroscopy **81**, 163–175 (2000)
38. Krivanek, O.L.: Method for determining coefficient of spherical aberration from a single electron micrograph. Optik **45**, 97–101 (1976)
39. Krivanek, O.L., Dellby, N., Lupini, A.R.: Towards sub-Å electron beams. Ultramicroscopy **78**, 1–11 (1999)
40. Ramasse, Q.M., Bleloch, A.L.: Diagnosis of aberrations from crystalline samples in scanning transmission electron microscopy. Ultramicroscopy **106**, 37–56 (2005)
41. Egerton, R.F.: Applications of energy-loss spectroscopy. In Electron Energy-Loss Spectroscopy in the Electron Microscopy, 2nd edn., pp. 59–72. Plenum Press, New York (1996)
42. Idrobo, J.C.: A study of the electronic structure and the effects of oxygen on the superconducting properties of MgB_2 by electron energy loss spectroscopy. Ph.D. Thesis, University of California Davis, Davis (2004)
43. Petford-Long, A.K., Chiaramonti, A.N.: Transmission Electron Microscopy of multilayer thin films. Annu. Rev. Mater. Res. **38**, 559–584 (2008)
44. Browning, N.D., Chisholm, M.F., Pennycook, S.J.: Atomic-resolution chemical analysis using a scanning transmission electron microscope. Nature **366**, 143–146 (1993)
45. Rafferty, B., Pennycook, S.J.: Towards atomic column-by-column spectroscopy. Ultramicroscopy **78**, 141–151 (1999)
46. Cosgriff, E.C., Oxley, M.P., Allen, L.J., Pennycook, S.J.: The spatial resolution of imaging using core-loss spectroscopy in the scanning transmission electron microscope. Ultramicroscopy **102**, 317–326 (2005)
47. Klie, R.F.: Reaching a new resolution standard with electron microscopy. Physics **2**(85) (2009)
48. Scherzer, O.: Über Einige Fehler von Eletronenlinses. Z. Phys. **101**, 593–603 (1936)
49. Scherzer, O.: Spharische und Chromatische Korrektur von Elktronen-Linsen. Optik **2**, 114–132 (1949)
50. Scherzer, O.: The theoretical resolution limit of the electron microscope. J. Appl. Phys. **20**, 20–29 (1949)
51. Haider, M., Uhlemann, S., Schwan, E., Rose, H., Kabius, B., Urban, K.: Electron microscopy image enhanced. Nature **392**, 768 (1998)
52. Beck, V.D.: Hexapole spherical-aberration corrector. Optik **53**, 241 (1979)
53. Crewe, A.V., Kopf, D.: Sextupole system for the correction of spherical-aberration. Optik **55**, 1–10 (1980)
54. Hawkes, P.W., Kasper, E.: Principles of Electron Optics. Academic Press, London (1989)
55. Hardy, D.F.: Combined magnetic and electrostatic quadrupole electron lenses. Ph.D. Thesis, Cambridge University, Cambridge (1967)
56. Kisielowski, C., Freitag, B., Bischoff, M., van Lin, H., Lazar, S., Knippels, G., Tiemeijer, P., van der Stam, M., von Harrach, S., Stekelenburg, M., Haider, M., Uhlemann, S., Mueller, H., Hartel, P., Kabius, B., Miller, D., Petrov, I., Olson, E.A., Donchev, T., Kenik, E.A., Lupini, A.R., Bentley, J., Pennycook, S.J., Anderson, I.M., Minor, A.M., Schmid, A.K., Duden, T., Radmilovic, V., Ramasse, Q.M., Watanabe, M., Erni, R., Stach, E.A., Denes, P., Dahmen, U.: Detection of single atoms and buried defects in three dimensions by aberration-corrected

electron microscope with 0.5-angstrom information limit. Microsc. Microanal. **14**(5), 469–477 (2008)
57. Sawada, H., Hosokawa, F., Kaneyama, T., Ishizawa, T., Terao, M., Kawazoe, M., Sannomiya, T., Tomita, T., Kondo, Y., Tanaka, T., Oshima, Y., Tanishiro, Y., Yamamoto, N., Takayanagi, K.: Achieving 63 pm resolution in scanning transmission electron microscope with spherical aberration corrector. Jpn J. Appl. Phys. **46**(23), L568–L570 (2007)
58. Erni, R., Rossell, M.D., Kisielowski, C., Dahmen, U.: Atomic-resolution imaging with a sub-50-pm electron probe. Phys. Rev. Lett. **102**, 096101 (2009)
59. Varela, M., Findlay, S.D., Lupini, A.R., Christen, H.M., Borisevich, A.Y., Dellby, N., Krivanek, O.L., Nellist, P.D., Oxley, M.P., Allen, L.J., Pennycook, S.J.: Spectroscopic imaging of single atoms within a bulk solid. Phys. Rev. Lett. **92**, 095502 (2004)
60. Medlin, D.L., Howitt, D.G.: The role of sputtering and displacement damage in the electron scribe process. Phil. Mag. Lett. **64**, 133–141 (1999)
61. Watanabe, M., Okunishi, E.K.I.: Analysis of spectrum-imaging datasets in atomic-resolution electron microscopy. Micros. Analys. **23**(7), 5–7 (2009)

Chapter 4
Structural Energetics of β-$Si_3N_4(10\bar{1}0)$ Surfaces

This chapter summarizes results on and analyses of first-principles calculations for the structural energetics of two terminations of β-$Si_3N_4\,(10\bar{1}0)$ surface, the open-ring and the closed-ring, in the presence of O, Ce, and Lu. I start with a discussion of these terminations and their (1×2) reconstructions. Next, the results from an extensive set of calculations for the preferred bonding sites and configurations of O on both surface terminations are presented as a function of coverage and surface stoichiometry. This is followed by the discussion of Ce adsorption on bare surfaces of Si_3N_4. Lastly, I present the computational models for Ce and Lu adsorption sites on the oxygenated Si_3N_4 surfaces and compare them to the experimental results.

4.1 Introduction

As discussed in Chap. 1, metallic additives used in the sintering of Si_3N_4 reside largely within the amorphous, integranular films (IGF), which separate the crystalline grains from each other. Some of these species can segregate to grain surfaces and adsorb there. After densification, these grain-boundary phases are known to influence the resulting microstructure and properties of the ceramic, such as its $\alpha \rightarrow \beta$ phase transformation, grain growth, grain morphology, and ultimately its fracture toughness. Therefore, it is important to understand the mechanism and effects of dopant adsorption at Si_3N_4/IGF interfaces.

Various experimental and theoretical approaches have been used in the past to characterize and model the behavior of the sintering dopants at the Si_3N_4/IGF interface. The most relevant to this work is the systematic ab initio study of Si_3N_4 surfaces initiated by the Materials Modeling Group at UIC a few years ago focusing on the atomic and electronic structures of different *bare* β-Si_3N_4 surfaces [1]. In that work, Idrobo *et al.* showed that the stoichiometric open-ring termination of

W. Walkosz, *Atomic Scale Characterization and First-Principles Studies of Si_3N_4 Interfaces,* Springer Theses, DOI: 10.1007/978-1-4419-7817-2_4,

β-Si_3N_4 $(10\bar{1}0)$, which has been consistently observed in the recent STEM experiments at the Si_3N_4/REO interfaces, was, in fact, higher in energy than other stoichiometric as well as some non-stoichiometric terminations. Based on this somewhat puzzling result obtained for the bare surface, the authors argued that oxygen, rare-earth elements, and/or the other sintering additives must be changing the relative stability of the prismatic plane terminations in Si_3N_4.

In my very recent experimental study (to be described in detail in Chap. 5), I provided conclusive evidence for the presence of oxygen at the interface [2]. Furthermore, my experiments suggested that the characteristic open rings in the contact layer of the Si_3N_4 matrix grains were most likely oxygen-terminated, as the onset of the O signal was measured to occur $\sim$2 Å before the Ce signal in going from the grain to the IGF. Motivated by this plausible suggestion as well as by my recent experimental results, which indicate the presence of interfacial oxygen, here I continue with the ab initio studies of the interface by focusing first on the structural energetics of oxygen on β-Si_3N_4 $(10\bar{1}0)$ surfaces and later on the adsorption mechanisms of the RE elements. In the case of oxygen on β-Si_3N_4, I consider a large pool of initial structures corresponding to various stoichiometric and non-stoichiometric surface configurations, with and without surface reconstructions, and identify the lowest-energy structures at four different O coverages. Later, I use these surfaces as starting configurations to determine Ce and Lu preferred adsorption sites at the Si_3N_4 $(10\bar{1}0)$ open-ring surface.

4.2 Computational Methods and Parameters

All calculations for the bare and oxygenated surfaces of Si_3N_4 were performed within the framework of density functional theory using the projector augmented wave method as implemented in the Vienna Ab-initio Simulation Package (VASP) code [3, 4]. For exchange-correlation, the Perdew–Wang (PW) parametrization [5] of the generalized gradient approximation (GGA) was used. For the low-energy structures obtained, I repeated the computations using the Perdew–Burke–Ernzerhof parametrization (PBE) [6] of GGA. I observed that the total energy differences between different structures were quite insensitive to the choice of the exchange-correlation functional (PW versus PBE), remaining within less than 0.1 eV of each other. The hexagonal bulk unit cell of β-Si_3N_4 contains 14 atoms and six structural parameters: a, c, and four internal parameters. The surface calculations were performed using the theoretical structural parameters $(a, c, x_{Si}, y_{Si}, x_N, y_N) =$ (7.667 Å, 2.928 Å, 0.1752, 0.7692, 0.3299, 0.031), which are in good agreement with the experimental values of (7.608 Å, 2.911 Å, 0.1733, 0.7694, 0.3323, 0.0314). Previous convergence tests for surface energies [1] showed that the $(10\bar{1}0)$ surface of β-Si_3N_4 can be modeled accurately with a *symmetric* 5-layer slab (with equivalent top and bottom surfaces) and a vacuum region of 10 Å. Since I performed a large number of first principles structural optimization computations at various coverages for different initial adsorbate positions ($\sim$400 different starting configurations), I reduced the computational demand by working with *asymmetric* 3-layer slabs, in which the atoms of the bottom unit cell were fixed at their bulk positions and their dangling bonds were passivated by hydrogens. Such asymmetric 3-layer slabs model the $(10\bar{1}0)$ surface quite well, as I found the binding energies calculated with symmetric versus asymmetric slabs to be within less than 0.1 eV of each other. Still, in order to ensure accuracy of the final results, for low-energy Si_3N_4/O configurations found at each coverage (approximately ten configurations), I repeated the computations with 7-layer symmetric slabs and the vacuum size increased to 14 Å, which are reported here. The (1×1) unit cell of the $(10\bar{1}0)$ surface of β-Si_3N_4 has dimensions of $a \times c$. Dictated by the coverages I investigated and taking into account possible reconstructions, I also considered (1×2) surface unit cells (doubled along the $c-$direction). In all computations, I used an energy cutoff of 270 eV and a $1 \times 3 \times 8$ Monkhorst–Pack grid (for a 1×1 surface unit cell) for **k**–point sampling, which have been shown in the previous studies to provide well-converged structural energetics [7–9].

To model the surface of Si_3N_4 in the presence of Ce, I employed the GGA+U formalism of Dudarev *et al.* [10] to account for the strong on-site Coulomb repulsion amongst the localized Ce 4f electrons. An effective U_{eff} parameter of 4.5 eV was added for these states, as determined from my first-principles calculations for the lattice parameters of various cerium oxides and the values reported in the literature [11, 12]. In the case of Lu, the f electrons were treated as part of the core, thus no GGA+U treatment was necessary. The vacuum size in the calculations with RE elements was increased to 15 Å .

4.3 Silicon Nitride in the Presence of Oxygen

Before I examine the low-energy configurations of oxygen on the $Si_3N_4\,(10\bar{1}0)$ surface at various coverages, I briefly summarize the findings regarding the nature of the atomic relaxations and structural energetics of the bare (oxygen-free) stoichiometric $Si_3N_4\,(10\bar{1}0)$ surface. There are two different stoichiometric terminations of the $Si_3N_4\,(10\bar{1}0)$ surface; the open-ring and the "closed-ring" surfaces as shown in Fig. 4.1. On the unrelaxed open-ring (1×2) surface, there are one Si and one N atom, which have one dangling bond each. These are represented by Si_2 and N_1 in Fig. 4.1a. Upon relaxation (Fig. 4.1b), Si_2 undergoes a large relaxation ($\sim$0.8 Å) and forms a new Si–Si bond (at 2.58 Å) with Si_4. As a result, the dangling bond of Si_2 is saturated (at the expense of over-coordination of Si_4 with two resonant bonds), while N_1 still remains with a dangling bond. On the unrelaxed closed-ring (1×1) surface, there are also one Si (Si_4) and one N atom (N_5), both of which have one dangling bond. When relaxed (Fig. 4.1c), Si_4 moves inward considerably (by $\sim$0.64 Å) and achieves an sp^2-type bonding in a nearly planar configuration with three N atoms. Total energy calculations with PW or

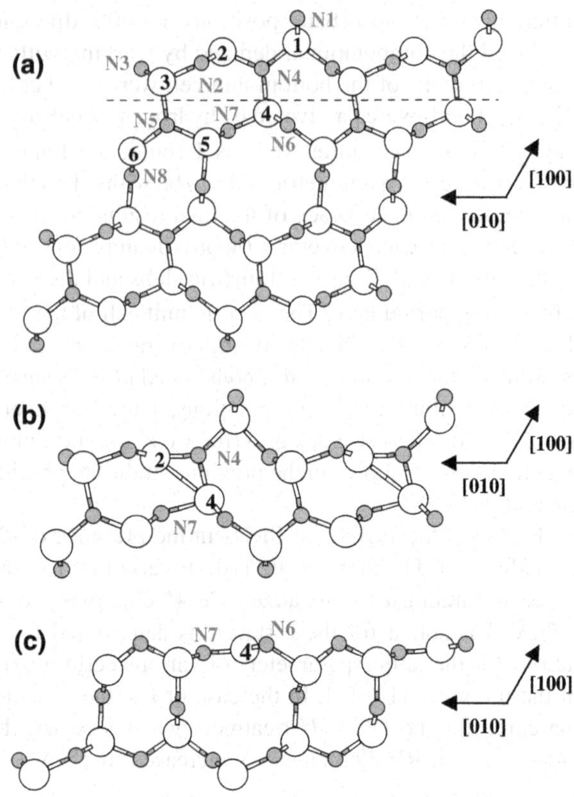

Fig. 4.1 **a** [001] projected, double-unit-cell view of a 2-layer unrelaxed $(10\bar{1}0)$ bare *open-ring* surface. If the atoms above the *dashed line* are removed, the unrelaxed *closed-ring* surface is obtained. **b, c** show the [001] projected, double-unit-cell views of the top-layers of the relaxed *open-ring* and *closed-ring* surfaces, respectively. The indices for the N atoms (*gray circles*) are given in *gray* next to them and the indices for the Si atoms (*white circles*) are shown inside the *white circles*. For example, the index 1 inside a *white circle* denotes the Si_1 atom mentioned in the text. Adapted with permission from [1]

PBE exchange-correlation functionals indicate that the relaxed closed-ring surface is 0.17 eV (per 1×1 surface unit cell) lower in energy than the relaxed open-ring surface, which has been consistently observed in all recent STEM experiments at the REO interfaces. This suggests that oxygen, the rare-earth elements, and/or other sintering additives most likely change the relative stability of the terminations of β-Si_3N_4 surfaces. When (1×2) surface unit cells are considered the energy difference between the open-ring and the closed-ring terminations increases to 0.23 eV as the latter reconstructs to minimize the number of dangling bonds. Figure 4.2 shows such a closed-ring surface at two different orientations. The top figure shows the structure in the [001] projection, while the bottom figure shows the structure slighty rotated. In both views, the distortion of the Si_4 and N_8 columns can be clearly seen as the alternating atoms in these column form bonds among themselves. Consequently, the number of dangling bonds on the surface is minimized, but at a cost of over-coordinating some of the subsurface N atoms (in the N_8 column) and departing from the nearly-planar configurations of Si_4 atoms with their neighboring atoms. This result indicates that unsaturated surface atoms play an important role in the stability of β-Si_3N_4, with possible consequences for promoting specific adsorption sites for dopants.

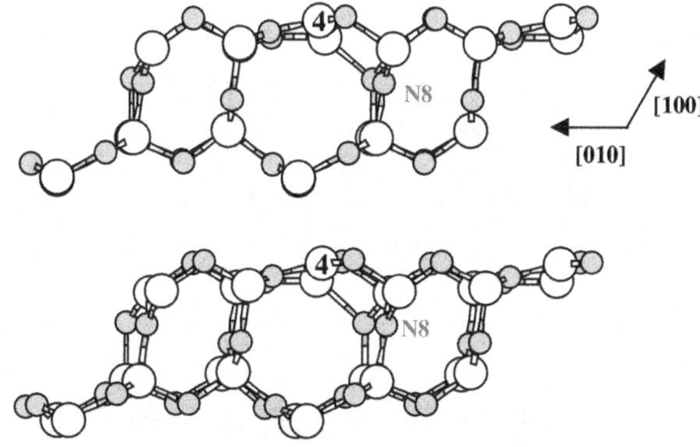

Fig. 4.2 *Top*: [001] projected, double-unit-cell view of a (1×2) reconstructed *closed-ring* surface. *Bottom*: *Oblique view* (slightly off the [001] projection) of this surface

As mentioned above, the unrelaxed $(10\bar{1}0)$ surfaces (open-ring or closed-ring) exhibit two dangling bonds per (1×1) surface unit cell. Based on this, I define a 1 monolayer (ML) oxygen coverage ($\Theta = 1$ ML) as a configuration in which two oxygen atoms per (1×1) unit cell are present on the $(10\bar{1}0)$ surface. In my study, I considered oxygen coverages of $\Theta = \frac{1}{4}, \frac{1}{2}, \frac{3}{4}$, and 1 ML and surface unit cells of (1×1) and (1×2). Using these cells, $\frac{1}{4}$ and $\frac{3}{4}$ ML coverages can only be achieved with (1×2) unit cells, while $\frac{1}{2}$ and 1 ML coverages can be obtained with reconstructed (1×2) cells or can be constrained to (1×1) surface unit cells. Below, I examine the low-energy configurations which resulted from my studies at each coverage.

4.3.1 $\Theta = \frac{1}{4}$ ML Coverage

The two most stable configurations corresponding to the $\Theta = \frac{1}{4}$ ML coverage are shown in Fig. 4.3. For the open-ring configuration (Fig. 4.3a), one O replaces every second N atom (N_1 in Fig. 4.1), which connects the Si atoms in a zigzag chain along the [001] direction. The replaced N, in turn, bridges two Si atoms (Si_2 in Fig. 4.1), thus saturating their dangling bonds. This N atom also forms a bond along the [010] direction (parallel to the surface along the a−axis) with the N atom positioned in-between the bridging O. Ideally, as in bulk silicon oxynitride (Si_2N_2O), Si, N, and O atoms would prefer to be fourfold, threefold, and twofold coordinated, respectively. Indeed, the relatively high binding energy (BE), which I find to be 4.83 eV per O atom (referenced to molecular oxygen) for the open-ring configuration at $\Theta = \frac{1}{4}$ ML, is due to the fact that all atoms in this structure are ideally coordinated without any dangling bonds or over-coordination.

A similar bridging oxygen (Si–O–Si) bonding pattern occurs for the most stable closed-ring configuration as shown in Fig. 4.3b. In this case, the O atom similarly replaces every second N atom with a dangling bond (N_5 in Fig. 4.1), and the replaced N forms two bonds with the Si atoms (Si_4 in Fig. 4.1), thus saturating

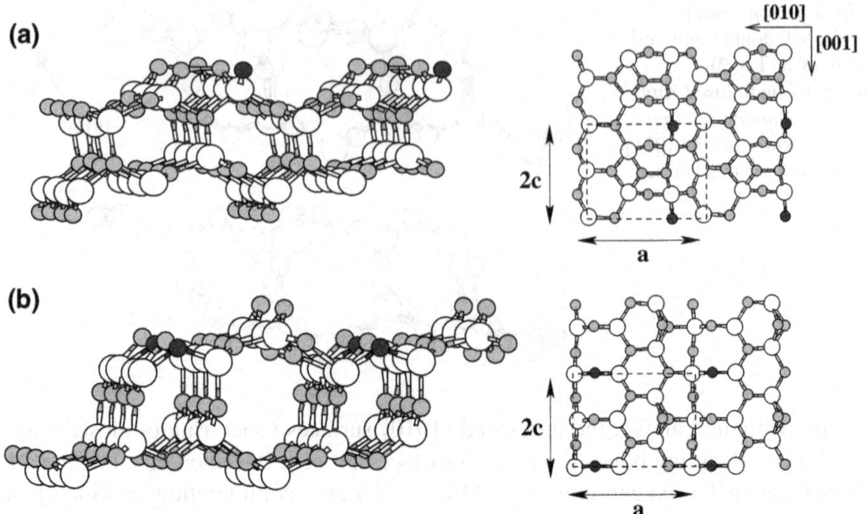

Fig. 4.3 **a** *Oblique* (slightly off the [001] projection) and *extended* (surface unit cell doubled along [010] and [001] directions) *view* of the lowest-energy *open-ring* configuration at $\Theta = \frac{1}{4}$ ML oxygen coverage. On the *right*, a *top view* is provided showing the surface unit cell with a *dashed rectangle*. **b** The same for the lowest-energy *closed-ring* configuration. The oxygen atoms are shown with *blue circles* (*dark*)

their dangling bonds. However, unlike the open-ring structure this closed-ring configuration does not result in an ideal coordination for all atoms, since two N atoms per (1×2) cell remain with a dangling bond each. As a result, the closed-ring surface at $\Theta = \frac{1}{4}$ ML oxygen coverage is energetically less stable than the open-ring surface, as the BE per O atom drops by 0.90–3.93 eV. These observations show that it is indeed possible for the relative stability of two different surface terminations to change upon oxygenation, which might provide the answer for why the open-ring termination has been consistently observed at the Si_3N_4/REO interfaces.

4.3.2 $\Theta = \frac{1}{2}$ ML Coverage

The characteristic Si–O–Si bridging oxygen configuration, resulting from the replacement of a surface N atom with O, is also observed for the most stable $\Theta = \frac{1}{2}$ ML stoichiometric configurations. If we constrain our search at this coverage to (1×1) surfaces (corresponding to one O atom in the unit cell), we find that the lowest energy open-ring configuration is the structure shown in Fig. 4.4a, which has one over-coordinated Si atom (Si_2) and one N atom (bonded with two Si_2 atoms) with a dangling bond. The calculated BE per O atom for this configuration

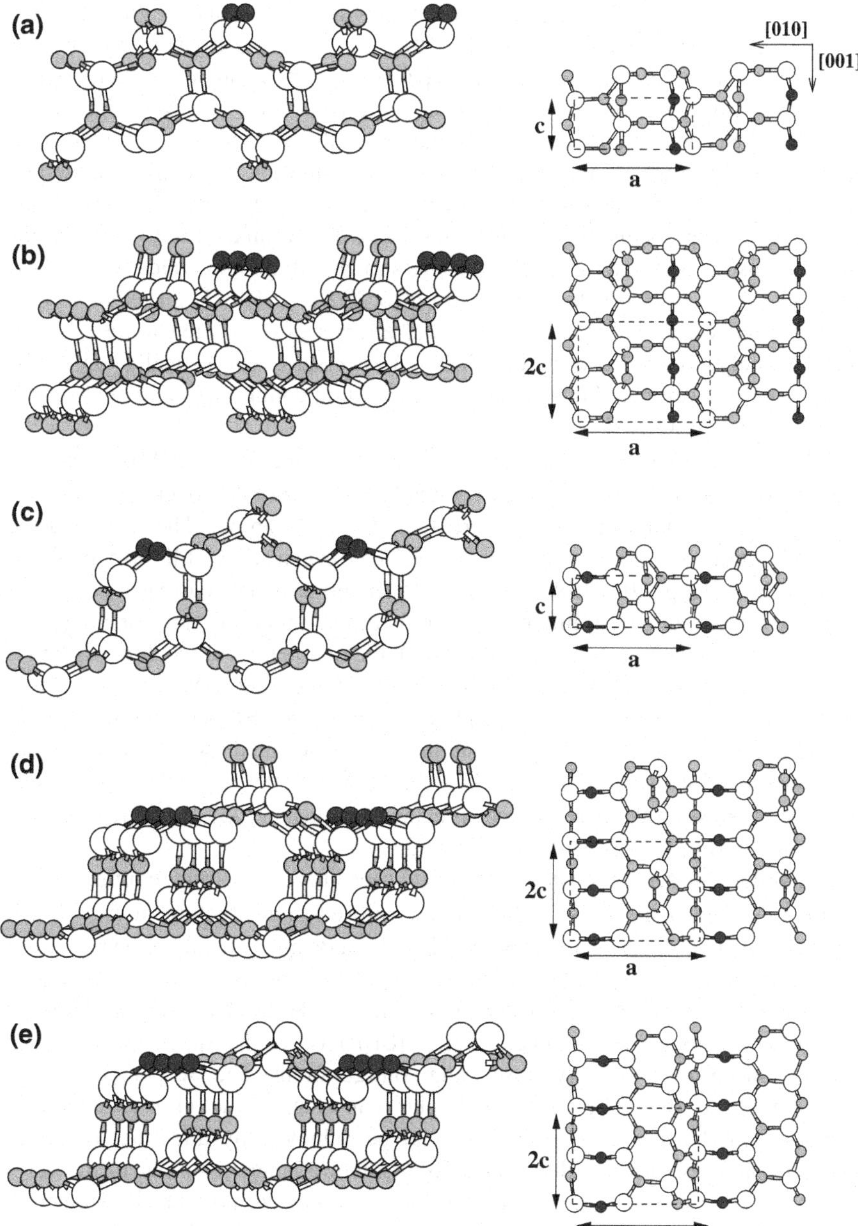

Fig. 4.4 **a** *Oblique* (slightly off the [001] projection) and *extended* (surface unit cell doubled along [010] and [001] directions) *view* of the lowest-energy unreconstructed *open-ring* configuration at $\Theta = \frac{1}{2}$ ML oxygen coverage. On the *right*, a *top view* is provided showing the surface unit cell with a *dashed rectangle*. **b–d** show the same for the (1×2) reconstructed *open-ring* surface, the unreconstructed *closed-ring* surface, and (1×2) reconstructed *closed-ring* surface, respectively. **e** A low-energy, non-stoichiometric [with 2 N atoms missing per (1×2) cell], and reconstructed *closed-ring* configuration. The oxygen atoms are shown with *blue* (*dark*) *circles*

is 3.06 eV. Here, I would like to point out that while both O and N atoms connect Si atoms in a zigzag fashion along [001], their particular positions in the unit cell are crucial. Namely, O connects Si_1 type atoms which are connected to two subsurface nitrogen atoms (N_3 and N_4 in Fig. 4.1), while the N bridges over-coordinated Si_2 type atoms which are connected to three subsurface nitrogen atoms (in Fig. 4.1, these are N_4 and two N_2 atoms in neighboring unit cells separated by c). If the positions of O and N were switched, which would result in Si_1–N–Si_1 and Si_2–O–Si_2 zigzag chains, Si_2 atoms would be over-coordinated with three N and two O atoms (as opposed to five N atoms, as shown in Fig. 4.4a), and the BE per O atom for the relaxed structure would drop to 2.28 eV. This is an example of a general trend we observed in the structural energetics of oxygenated Si_3N_4 surfaces: Namely, fivefold coordinated Si is not necessarily energetically too costly, especially if the resonant bonds of the corresponding configuration are with two N atoms rather than O.

If the possibility of a (1×2) surface reconstruction at $\Theta = \frac{1}{2}$ ML oxygen coverage is allowed, one obtains an energetically more favorable open-ring configuration (Fig. 4.4b) than the (1×1) structure discussed above. The reconstructed surface is similar to the unreconstructed one: The oxygen atoms still form Si_2-O-Si_2 zigzag chains along [001], however, the nitrogen atoms attached to Si_1 type surface atoms now form dimers. While both N atoms in the dimer still have one dangling bond each (as in the unreconstructed surface), the main effect of the dimerization is to prevent over-coordination of Si_1 type atoms, which are now ideally fourfold coordinated. As a result of this (1×2) reconstruction, the BE per O atom increases to 4.38 eV, which is 1.32 eV larger than the value calculated for the lowest-energy unreconstructed configuration (Fig. 4.4a) found at this coverage. This shows one example of the several oxygen-induced reconstructions on β-Si_3N_4 $(10\bar{1}0)$ surfaces I observed during the course of my studies.

A similar surface reconstruction occurs for the closed-ring configurations. Figure 4.4c shows the most stable (1×1) closed-ring structure at $\Theta = \frac{1}{2}$ ML coverage. O again replaces the N atom (N_5) with a dangling bond on the unrelaxed closed-ring surface, and each replaced N is attached to two Si atoms (Si_4 type with one initial dangling bond) and over-coordinates the latter. The bridging Si–O–Si configuration along the a−direction (i.e. [010]) is locally quite stable and is observed in all lowest-energy closed-ring configurations. The BE per O atom for this structure is 3.31 eV, which is larger than that of the (1×1) open-ring configuration. Further gain in the BE can be achieved by allowing a (1×2) reconstruction, as shown in Fig. 4.4d. This structure is again similar to the unreconstructed one in Fig. 4.4c, except that the Si_4 atoms are now ideally (fourfold) coordinated, and the N atoms attached to them dimerize along [001] to reduce the number of their dangling bonds. This simple reconstruction caused by the dimerization of the N atoms gives rise to a large energy gain of 1.35 eV resulting in a BE of 4.66 eV per O atom.

So far I have presented several examples which show that O has a rather strong tendency to replace N substitutionally, rather than attach to the surface epitaxially or interstitially. In most cases, the replaced N stays on the surface and typically helps saturate dangling Si bonds. In some cases, however, I observed that the N atoms would form dimers and diffuse away from the surface into the vacuum region, while the rest of the system was structurally optimized. Such behavior suggested that sub-stoichiometry on the N sublattice could potentially play an important role in the stability of oxygenated β-Si_3N_4 surfaces. Pursuant to these observations, I extended my search to non-stoichiometric surfaces by recalculating total energies with such N dimers removed, as well as trying new N-deficient initial structures (based on general bonding trends and minimization of dangling bonds) that could potentially yield low-energy structures. As will be shown later, when the total energies of such non-stoichiometric structures are compared with those of stoichiometric ones using the allowed ranges of the chemical potentials involved, we find that most of the lowest-energy structures that appear in the phase diagram of O/Si_3N_4 are indeed non-stoichiometric. An example of a low-energy non-stoichiometric surface at $\Theta = \frac{1}{2}$ is shown in Fig. 4.4e. The starting point for this optimized structure is the (1×2) closed-ring configuration discussed above (Fig. 4.4d) with the N dimers removed. Although one might think that the reconstruction induced by the N dimers would disappear when they are removed, symmetry-unrestricted reoptimization of the structure yields a new (1×2) reconstruction. The reconstruction is now due to alternating positions of the two neighboring Si_4 type atoms along the [001] direction. When the N dimer is removed, one of the Si_4 atoms remains with a dangling bond, as in the *unrelaxed* bare closed-ring surface shown in Fig. 4.1a, while the other moves toward the surface and exhibits an sp^2–type bonding in a nearly planar configuration with three N, as in the *relaxed* bare closed-ring surface shown in Fig. 4.1c. In addition, in spite of the significant distortion due to the Si_4 atom, both N_6 type nitrogen atoms (in neighboring unit cells along [001]) keep their preferred nearly planar coordination with three Si atoms. Since this structure is non-stoichiometric, it is not straightforward to compare the BE per O atom for this configuration with those for the stoichiometric ones, even at the same coverage. For the moment, I note here that (as will be shown later), this rather peculiar structure with one Si dangling bond and no N dangling bonds is indeed one of the lowest-energy structures in the N-poor region of the phase diagram. It is also one of the two low-energy structures (out of ~350 structures I considered), in which the dangling bond of a Si atom has *not* been saturated via sp^2 or over-coordinated resonant bond distortions or by achieving the ideal sp^3 configuration by N or O attachment.

4.3.3 $\Theta = \frac{3}{4}$ ML Coverage

From a large pool of stoichiometric and non-stoichiometric structures investigated at the $\Theta = \frac{3}{4}$ ML oxygen coverage (corresponding to 3 O atoms on 1×2 surfaces), the structures shown in Fig. 4.5a, b are by far the most stable open-ring and closed-ring configurations, respectively. Both structures are non-stoichiometric with two

N atoms per (1×2) cell missing. The two-thirds ratio between the number of missing N atoms (nominally in the -3 charge state) and the number of added O atoms (nominally in the -2 charge state) is crucial for the enhanced stability of these surfaces, which results in no dangling bonds or over-coordinated atoms. In fact, it is possible to predict, based on the general bonding trends observed so far, that the structures shown in Fig. 4.5a, b are likely to be quite stable. For example, for the open-ring surfaces, we have already seen several examples where it is energetically favorable for O to replace the N_1 type atoms and bridge the Si_1 atoms. This requires 2 O atoms, and results in two missing N atoms per (1×2) cell. These N atoms, for stoichiometric surfaces, would typically bond to Si_2 atoms, but this results in over-coordination of Si_2 (Fig. 4.4a) or dangling N bonds (Fig. 4.4b). Instead, if these two N atoms are removed from the surface, and the third O is used to saturate the dangling Si_2 bond (fully saturating both Si_2 *and* O bonds), the resulting structure, as shown in Fig. 4.5a after structural optimization, is expected to be quite stable. Similarly for the closed-ring configuration, we have seen that it is energetically favorable for O to replace the N_5 type atoms and bridge two Si atoms (Si_5 and Si_6 in Fig. 4.1), resulting in a Si–O–Si bridge along the a–direction. If this is done for two (1×1) unit cells (2 added O and 2 removed N atoms) in addition to saturating the Si_4 dangling bonds with a third O atom (thereby saturating both bonds of the added oxygen), the resulting structure, as shown in Fig. 4.5b after structural optimization, is expected to be quite stable.

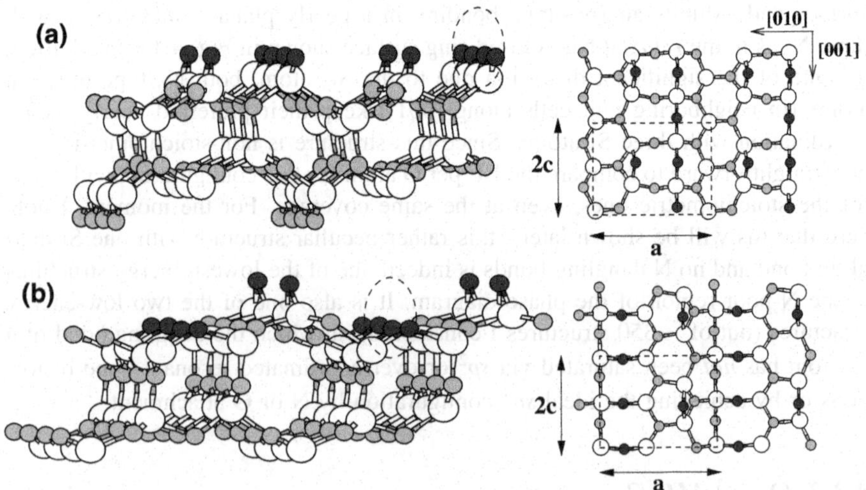

Fig. 4.5 **a** *Oblique* (slightly off the [001] projection) and *extended* (surface unit cell doubled along [010] and [001] directions) *view* of the lowest-energy *open-ring* configuration at $\Theta = \frac{3}{4}$ ML oxygen coverage. On the *right*, a *top view* is provided showing the surface unit cell with a *dashed rectangle*. **b** The same for the lowest-energy *closed-ring* configuration. The oxygen atoms are shown with *blue* (*dark*) *circles*. The *dashed ellipses* show the Si–O–Si bridges mentioned in the text

As will be shown later, this is indeed the case, as this structure stands out as the lowest-energy structure for a large portion of the calculated phase diagram.

The local environments for O in the two most stable structures at $\Theta = \frac{3}{4}$ ML coverage, shown in Fig. 4.5a, b, are similar. In spite of the similarities, I find that the closed-ring surface is lower in energy by 0.56 eV per (1×2) cell than the open-ring structure. To understand this, we should first note that the local environments around the oxygens saturating the Si_2 and Si_4 atoms in Fig. 4.5a, b, respectively, are almost identical. As such, it is natural to attribute the energetic favorability of the closed-ring configuration to the differences in the geometric constraints imposed on the Si–O–Si bridges, shown by dashed ellipses in Fig. 4.5a, b. In particular, the distance between Si_5 and Si_6 along the a−direction is $\sim$0.1 Å larger than the Si_1–Si_1 separation (i.e. the lattice parameter c) along the [001] direction. This extra room allows the Si atoms bonded to the bridging O in the closed-ring configuration to achieve nearly perfect tetrahedral coordination (with the bridging O and three N atoms), while still maintaining a large Si–O–Si bond angle of $\sim 129°$, close to the 142° bond angle observed in bulk SiO_2. In the open-ring configuration, on the other hand, the relaxed Si_1–O–Si_1 bond angle is only 121°, significantly smaller than the preferred bulk value. The reason that this angle cannot be made larger (with the O relaxing more inward, for example) is that such a configuration would cost more energy overall, since it would considerably distort the nearly tetrahedral coordination of the Si_1 atoms. As such, the energetic favorability of the closed-ring configuration results from the delicate balance it can sustain between the tetrahedral bonding angle of Si and the bridging angle of O between two Si atoms in this strongly covalent material.

4.3.4 $\Theta = 1$ *ML Coverage*

Since the $\Theta = 1$ ML coverage can be achieved in two ways (2 oxygens on (1×1) or 4 oxygens on reconstructed (1×2) surface cells) and there are more possibilities for the relative positioning of the oxygen atoms on the surface, I explored a large number of configurations at this coverage. As expected, most of the low-energy surfaces turned out to be N-deficient non-stoichiometric structures. Two of them are shown in Fig. 4.6. The open-ring configuration shown in Fig. 4.6a is fully-oxygenated, and has a (1×2) surface reconstruction induced by the dimerization of oxygens which saturate the dangling bonds of Si_2 type atoms. As in most low-energy open-ring structures, the N atoms connecting Si_1 chains along [001] have been replaced by O atoms, resulting in two missing N atoms per (1×2) surface unit cell. The closed-ring configuration, shown in Fig. 4.6b, is also non-stoichiometric with 2 N atoms missing per (1×1) cell, but it does not have a surface reconstruction. Reminiscent of the structure shown in Fig. 4.4e, Si_4 atom has a dangling bond as in the unrelaxed bare closed-ring surface and the O atoms connected to Si_4 are over-coordinated in a nearly planar configuration with three Si atoms. In spite of these peculiar features, this closed-ring configuration can be

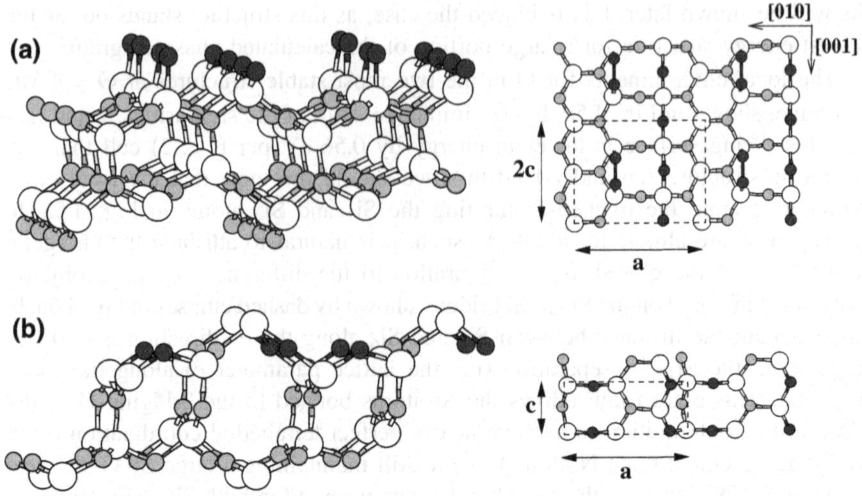

Fig. 4.6 **a** *Oblique* (slightly off the [001] projection) and *extended* (surface unit cell doubled along [010] and [001] directions) *view* of a low-energy (1 × 2) reconstructed and non-stoichiometric [with 2 N atoms missing per (1 × 2) surface cell] *open-ring* configuration at $\Theta = 1$ ML oxygen coverage. On the *right*, a *top view* is provided showing the surface unit cell with a *dashed rectangle*. **b** The same for a low-energy, unreconstructed, and non-stoichiometric [with 2N atoms missing per (1 × 1) surface cell] *closed-ring* configuration. The oxygen atoms are shown with *blue* (*dark*) *circles*

competitive in energy in the N-poor region of the phase diagram, as will be discussed next.

4.3.5 Phase Diagram of β-Si_3N_4 $(10\bar{1}0)$ Surface

Now I compare the total energies of all low-energy surfaces with different amounts of oxygen coverage and nitrogen stoichiometry. The total energy for a given surface configuration can be written as [13]

$$E_{surface} = E_{slab} - n_{Si}\mu_{Si} - n_N\mu_N - n_O\mu_O, \tag{4.1}$$

where E_{slab} is the total energy of the corresponding slab, n_i is the number of atoms of type i in the slab, and μ_i's are the corresponding chemical potentials, which satisfy $3\mu_{Si} + 4\mu_N = \mu_{Si_3N_4,bulk}$. From the upper limits on the N and Si chemical potentials ($\mu_{Si} \leq \mu_{Si,bulk}$ and $\mu_N \leq \frac{1}{2}\mu_{N_2}$) we find the allowed range of μ_N to be $-10.29\,eV \leq \mu_N \leq -7.67\,eV$. The upper limit for the oxygen chemical potential can be made more strict than the condition given by $\mu_O \leq \frac{1}{2}\mu_{O_2} = -4.56$ eV (PW) by preventing the precipitation of bulk Si_2N_2O, namely $2\mu_{Si} + 2\mu_N + \mu_O \leq \mu_{Si_2N_2O,bulk}$. Combining this criterion with the condition of equilibrium with

Table 4.1 The coverage (Θ), the figure label, and the total energy per (1×1) unit cell (referenced to the bare open-ring surface) of all the low-energy oxygen configurations on the β-Si_3N_4 $(10\bar{1}0)$ surfaces discussed in the text

Θ	Figure	Total energy (eV)	N_{db}	N_{oc}
$\frac{1}{4}$	Fig. 4.3a	$-4.69 - \frac{1}{2}\mu_O$	–	–
	Fig. 4.3b	$-4.24 - \frac{1}{2}\mu_O$	1	–
$\frac{1}{2}$	Fig. 4.4a	$-7.60 - \mu_O$	1	1
	Fig. 4.4b $-R$	$-8.92 - \mu_O$	1	–
	Fig. 4.4c	$-7.85 - \mu_O$	1	1
	Fig. 4.4d $-R$	$-9.20 - \mu_O$	1	–
	Fig. 4.4e $-R$	$-0.63 - \mu_O + \mu_N$	1	–
$\frac{3}{4}$	Fig. 4.5a	$-5.00 - \frac{3}{2}\mu_O + \mu_N$	–	–
	Fig. 4.5b	$-5.28 - \frac{3}{2}\mu_O + \mu_N$	–	–
1	Fig. 4.6a $-R$	$-7.84 - 2\mu_O + \mu_N$	–	–
	Fig. 4.6b	$0.75 - 2\mu_O + 2\mu_N$	1	1

To obtain the total energies referenced to the lower-energy bare closed-ring surface (which are used to produce the phase diagram in Fig. 4.7), 0.17 eV should be added to the energies reported below. The number of dangling bonds (N_{db}) and the number of over-coordinated atoms (N_{oc}) per (1×1) cell are also provided. The letter R next to some of the figure labels in the second column indicate that the surface has a (1×2) reconstruction

bulk Si_3N_4, we get the condition $\mu_O - \frac{2}{3}\mu_N \leq -2.48$ eV, which limits the maximum of the oxygen chemical potential to -9.34 eV (N-poor region) and -7.59 eV (N-rich region).

Table 4.1 shows the total energy expressions per (1×1) surface unit cell for all surface configurations at different coverages discussed so far as a function of the oxygen and nitrogen chemical potentials. Although it is not the lowest-energy bare surface, since the open-ring configuration is the one observed at the REO interfaces, I have referenced all surface total energies with respect to the bare open-ring surface (a negative surface total energy at a given μ_O and/or μ_N means that it is energetically more favorable for that surface to form under those conditions compared to the bare open-ring surface). The table also provides a summary of the number of dangling bonds and the number of over-coordinated atoms per (1×1) cell.

Based on the calculated energies, I show in Fig. 4.7 the zero-temperature phase diagram of the oxygenated β-Si_3N_4 $(10\bar{1}0)$ surface as a function of μ_N and μ_O. In the allowed ranges of the chemical potentials, there are three oxygenated lowest-energy structures, shown in Figs. 4.4d, e and 4.5b. The structure displayed in Fig. 4.6b also appears as the lowest-energy structure under O-rich and N-poor conditions, but this region should be excluded, as it would favor the formation of bulk Si_2N_2O. Non-stoichiometry (arising from N deficiency) plays an important role in the structural energetics, as mentioned before, with three of the four lowest-energy surfaces shown in Fig. 4.7 corresponding to non-stoichiometric surfaces. It is also interesting to note that all the lowest-energy surfaces are based on the

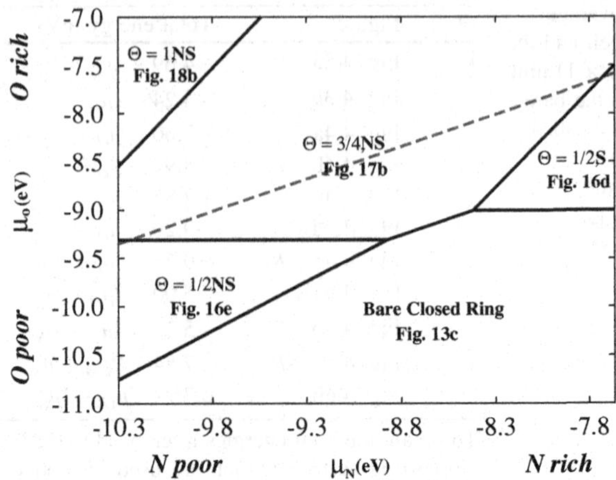

Fig. 4.7 The zero-temperature phase diagram of the oxygenated β-Si_3N_4 $(10\bar{1}0)$ surface as a function of the N and O chemical potentials showing the lowest-energy structures, their corresponding oxygen coverages and figure labels, and whether they are stoichiometric (S) or non-stoichiometric (NS). The *dashed line* shows the saturation condition with bulk Si_2N_2O, i.e. $\mu_O - \frac{2}{3}\mu_N = -2.48$ eV, as discussed in the text. The region of the phase space above this line would favor the formation of bulk Si_2N_2O

closed-ring configuration. A large portion of the phase diagram is occupied by the closed-ring configuration at $\Theta = \frac{3}{4}$ ML oxygen coverage, shown in Fig. 4.5b. In fact, when μ_O achieves its maximum value (just before the formation of bulk Si_2N_2O), this configuration is the lowest-energy structure among all structures for almost the entire range of the N chemical potential. As discussed above, the phase diagram shown in Fig. 4.7 shows the lowest-energy structure at given values of the O and N chemical potentials. In most cases, many low-energy structures compete with the lowest-energy structure. In order to show the richness in the structural energetics of the β-Si_3N_4 $(10\bar{1}0)$ surface, I also plot in Fig. 4.8 the energies of the various surfaces shown in Figs. 4.3, 4.4, 4.5, and 4.6 as a function of μ_O at three different values of μ_N corresponding to N-poor, N-rich, and intermediate N stoichiometry. The figure shows that the closed-ring $\Theta = \frac{3}{4}$ structure is quite low in energy, even when it is not the lowest-energy structure (Fig. 4.8a, b). I also observe that of all the open-ring configurations, the one that is the most competitive in energy with closed-ring configurations is the $\Theta = \frac{1}{4}$ ML structure (Fig. 4.3a), as shown in Fig. 4.8b. If we focus on the maximum solubility limit of oxygen, as shown by the dashed line in Fig. 4.7, the corresponding coverage is $\Theta = \frac{3}{4}$ ML, and the most stable surface has a closed-ring configuration which is now 0.28 eV (per (1×1) surface cell) lower in energy than the open-ring configuration at the same coverage, up from the value of 0.17 eV computed for the bare surfaces. This finding is puzzling in light of the fact that in all β-Si_3N_4/rare-earth oxide interfaces imaged in the recent STEM experiments, it is the open-ring

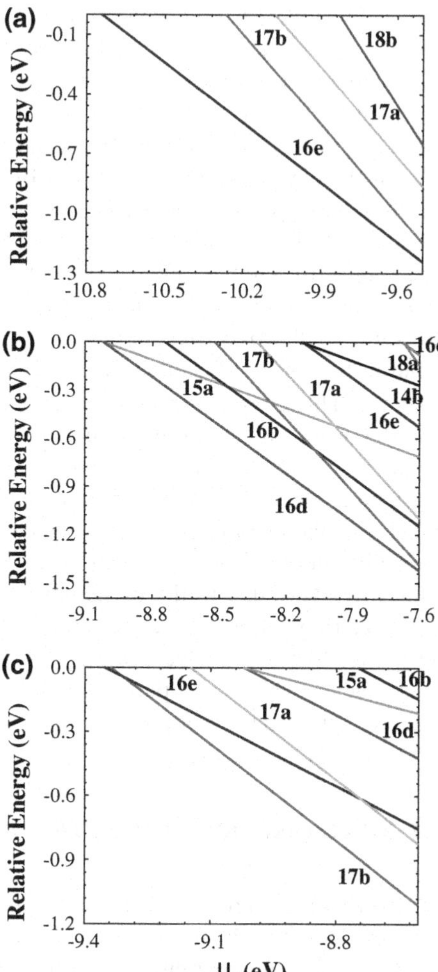

Fig. 4.8 The formation energies (relative to the bare *closed-ring* configuration) of various surfaces discussed in the text (as shown with their figure labels) as a function of the O chemical potential μ_O at three different N chemical potential values corresponding to **a** $\mu_N = -10.29$ eV (N-poor), **b** $\mu_N = -7.67$ eV (N-rich), and **c** $\mu_N = -8.90$ eV (intermediate N-stoichiometry) conditions. At each value of μ_N, the largest μ_O value is dictated by preventing precipitation of bulk Si_2N_2O, and the lowest μ_O value is chosen so that the surfaces have lower energies than the bare *closed-ring* configuration

termination of the $(10\bar{1}0)$ surface that is observed consistently. I, therefore, conclude that the particular presence of the rare-earth elements or other sintering cations (such as Mg or Al) likely influence the relative stability of the termination of β-Si_3N_4 matrix grains in their sharp interfaces with rare-earth oxides.

4.3.6 Summary

I presented results on and analyses of first principles calculations for the structural energetics of β-Si_3N_4 $(10\bar{1}0)$ surfaces at $\Theta = \frac{1}{4}, \frac{1}{2}, \frac{3}{4}$, and 1 ML oxygen coverages.

The study shows that (1) oxygen has a strong tendency to replace nitrogen in a bridging configuration between two Si atoms, rather than attach to the surface epitaxially or interstitially, (2) oxygenated β-Si_3N_4 $(10\bar{1}0)$ surface exhibits (1×2) reconstructions, and (3) most lowest-energy surfaces predicted to be stable in the phase diagram are N deficient non-stoichiometric surfaces. These findings, combined with the results from my recent EELS studies at the $Si_3N_4/CeO_{2-\delta}$ interface [2] strongly suggest that the Si_3N_4 open rings are oxygen-terminated. I find that the structural stability of most low-energy surfaces is driven by the tendency of surface atoms to saturate their dangling bonds, while preserving close-to-ideal coordination (tetrahedral, planar, bridging) or bond angles. The lowest-energy structure which covers the largest portion of the equilibrium phase diagram is found to be a non-stoichiometric closed-ring configuration at $\Theta = \frac{3}{4}$ ML oxygen coverage. This closed-ring structure is 0.28 eV lower in energy, per (1×1) cell, than the open-ring structure at the same coverage. In fact, I find that there is no region of the phase diagram where a surface with an open-ring termination is observed to be the lowest-energy structure. This finding indicates that the relative stability of the open-ring termination over the closed-ring termination, as experimentally inferred from consistent observations of the former in all recent STEM studies, is likely due to the particular presence of the rare-earth element and/or other sintering cations at the interface. A deeper understanding of the atomic and electronic structures of the ceramic will naturally require further ab initio modeling studies which incorporate oxygen, rare-earth elements, as well as other sintering cations at the interface.

4.4 Silicon Nitride in the Presence of Cerium

As shown in the previous section the presence of oxygen alone on the Si_3N_4 surfaces does not change the relative stability of the two surface terminations, i.e. the closed ring continues to be energetically more stable than the open ring. In order to check whether the rare-earth elements could reverse that trend, I modeled the two surface terminations of Si_3N_4 in the presence of Ce. In this work, I considered (1×2) surface cells with one Ce atom in the 4+ valence state [11]. Figure 4.9 shows the results obtained for both the (1) open-ring and the (2) closed-ring terminations. In both configurations, Ce atoms place themselves close to the initially unstaturated N atoms and form bonds with them. Their preference for adsoprtion close to N (as opposed to the unsaturated Si atom) is due to the larger electronegativity of N as compared to Si. Because of their large size, however, they are able to reach out to some of the Si atoms with dangling bonds, and to saturate them as well. What is interesting to see is that this tendency of Ce to form bonds with N is strong enough to afford over-coordination of other elements like Si atoms or even Ce itself (Fig. 4.9b). The results also suggests that size effects are secondary to the chemical bonding

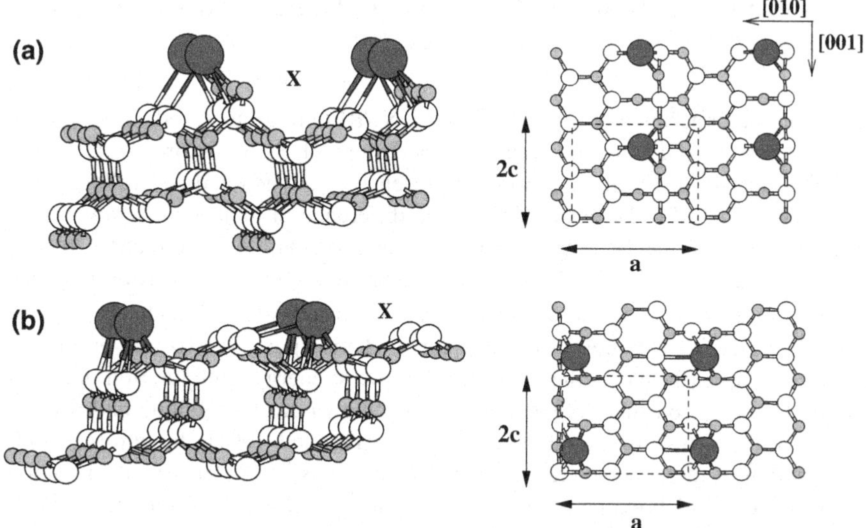

Fig. 4.9 **a** *Oblique* (slightly off the [001] projection) and *extended* (surface unit cell doubled along [010] and [001] directions) *view* of the lowest-energy *open-ring* configuration in the presence of three O and 2 Ce atoms. On the *right*, a *top view* is provided showing the surface unit cell with a *dashed rectangle*. **b** The same for the lowest-energy *closed-ring* configuration. The O and Ce atoms are shown with *blue* and *purple* circles respectively

factors as the former would place the adsorbates near the center of large open volumes, marked with **X** in Fig. 4.9. This is in agreement with previous studies [7]. Lastly, these calculations demonstrate that the presence of Ce itself does not change the relative stability of the two terminations of Si_3N_4 $(10\overline{1}0)$ surface, i.e., the closed-ring configuration is still energetically more stable than the open-ring. The energy differece between these structures is $\sim$0.5 eV per (1×2) unit cell. Just by considering the number of dangling bonds and coordination of the atoms, the obtained result seems counterintuitive, as the predicted low energy structure based on the closed-ring termination has more over-coordinated atoms than the calculated higher energy structure. To explain this discrepancy, a detailed understanding of the bonding mechanism at the surface is required. A possible approach would be to look at the charge density plots for these structures and carry out a detailed analysis of the charge distribution along the Si_3N_4 surface. To understand why only the open ring termination is observed experimentally at the Si_3N_4 interfaces, in turn, necessitates modeling of the two surface terminations in the presence of rare-earth elements, oxygen and other sintering additives in the IGF. Because of a multitude of possible starting configurations when all elements are considered, a molecular dynamics approach can be ideal for this purpose.

4.5 Silicon Nitride in the Presence of Oxygen and Rare-Earth Elements

Based on the experimental images of the Si_3N_4/REO interfaces, indicating that in the presence of IGF only the open-ring structure of Si_3N_4 is preferred, here I modeled this surface termination in the presence of oxygen and two different RE elements: Ce and Lu. Because of the large difference in their atomic radius, this study allowed further testing of atomic size considerations in regards to the observed differences in the RE elements adsorption sites. As a starting configuration I used the lowest energy open-ring structure corresponding to $\Theta = \frac{3}{4}$ ML coverage, described earlier in the chapter, and incorporated two rare-earth elements. Figure 4.10a shows the result for the lowest energy configuration obtained for the Ce atoms on the oxygenated surface. Shown are the oblique and the top views of the structure doubled along the [010] and [001] directions. Close examination of the structure reveals that the Ce atoms form a zig-zag chain along the [001] direction, bonding with Si, O, and among themselves. Their coordination is either threefold or twofold. As for the other surface atoms, all of them are in the ideal coordination state except for the surface Si atom, initially coordinated with O (see Fig. 4.1). In the presence of Ce the characteristic Si–O–Si bridging

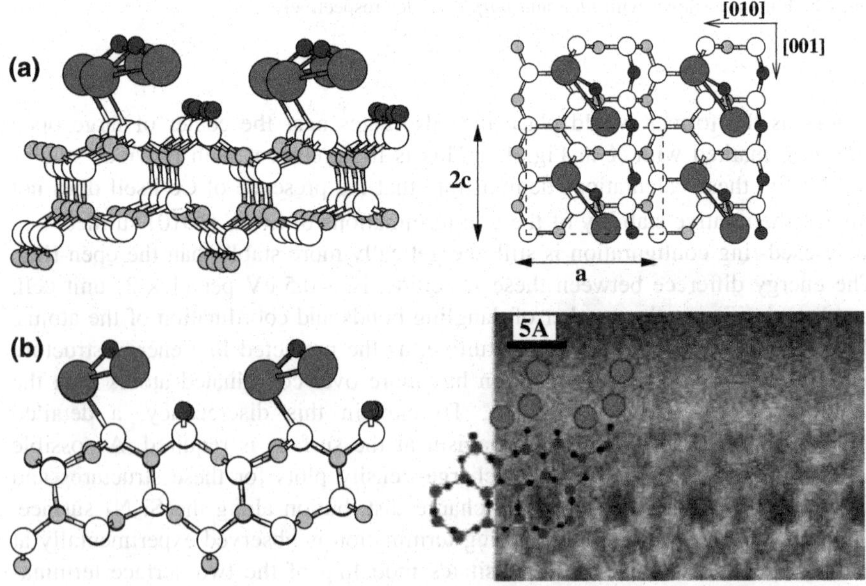

Fig. 4.10 **a** *Oblique* (slightly off the [001] projection) and *extended* (surface unit cell doubled along [010] and [001] directions) *view* of the lowest-energy *open-ring* configuration in the presence of three O and two Ce atoms. On the *right*, a *top view* is provided showing the surface unit cell with a *dashed rectangle*. **b** [001] projected double-unit-cell view of the same structure along with experimental Z-contrast image on the *right* [2]

configuration, however, gets disrupted as Ce wants to bond to Si without allowing for its overcoordination by O. The latter is driven out of the surface of Si_3N_4 to a position above the Ce atoms. The measured Ce–O bond-lengths ($\sim$2.1 Å) are found to be very close to those found in the Ce_2O_3 compound ($\sim$2.3 Å), suggesting that Ce tries to achieve a similar structure. The predicted configuration of Ce on Si_3N_4 in the presence of O is in remarkable agreement with the experimentally observed structure as shown in Fig. 4.10b and discussed in detail in Chap. 5. The small tilt of the Ce-pair observed in my experimental Z-contrast images is captured in these calculations. This is an important result because earlier studies at the $Si_3N_4/CeO_{2-\delta}$ interface had predicted a different arrangement for Ce than the one found in my experimental study (see Chap. 5 for a detailed discussion) [2].

Figure 4.11a shows the lowest energy structure of Lu on the oxygenated Si_3N_4 surface. Looking at these figures one can see that Lu atoms arrange themselves in a zigzag fashion along the [001] direction. When viewed in this projection, the axis of the Lu pairs is parallel to the [010] direction, in contrast to the Ce-pair's arrangement with its axis slightly tilted. The Lu atoms form bonds with the Si atoms (Si_2 in Fig. 4.1), saturating their dangling bonds, as well as with the O atoms in their vicinity. As a result every other O in the N1 column (see Fig. 4.1) is overcoordinated. Similarly to the Ce case, it is interesting to observe that the other O,

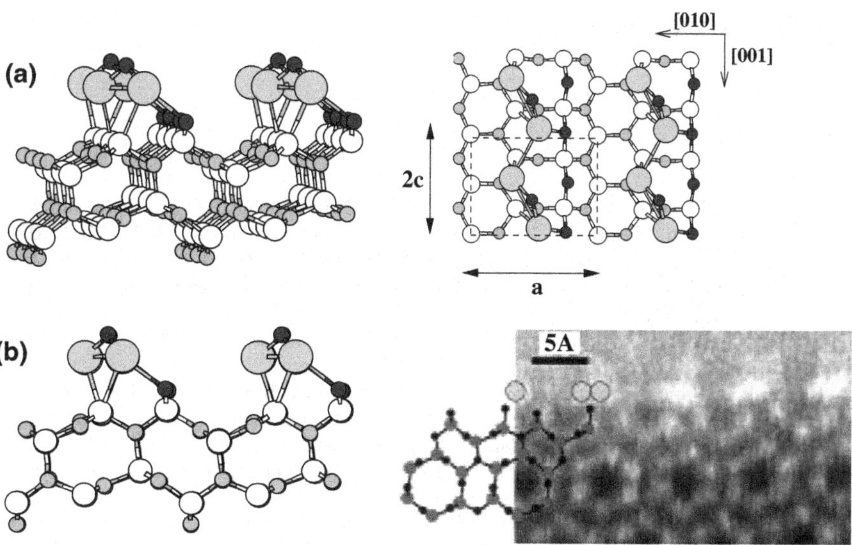

Fig. 4.11 **a** *Oblique* (slightly off the [001] projection) and *extended* (surface unit cell doubled along [010] and [001] directions) *view* of the lowest-energy *open-ring* configuration in the presence of three O and 2 Lu atoms. On the *right*, a *top view* is provided showing the surface unit cell with a *dashed rectangle*. **b** [001] Projected double-unit-cell view of the same structure along with experimental Z-contrast image on the *right* (adapted with permission from [14]

initially bridging two Si atoms (Si_2 in Fig. 4.1), in the presence of Lu is driven out of the Si_3N_4 layer to a position above the two Lu atoms. It retains its original preference for a bridging configuration, however, now pairing up with the Lu atoms to achieve its ideal twofold bonding configuration. As for the Lu atoms, they form bonds with four atoms: two O atoms, one Si and one Lu. The bond-lengths between Lu and O atoms ($\sim$2 Å) are close to the ideal Lu–O measurements in the Lu_2O_3 compound ($\sim$2.1 Å). The obtained structure is in a very good agreement with the experimental image as is evident in Fig. 4.11b showing the [001] projection of the predicted structure and the Z-contrast image with the schematic drawing of Si_3N_4 superimposed.

4.5.1 Summary

The results of first-principles calculations for the structural energetics of the β Si_3N_4 $[10\bar{1}0]$ surfaces in the presence of Ce only , Ce and O, and Lu with O were presented in detail. The first study shows that the absorption sites of Ce dopants are determined by the unsaturated surface N atoms providing active bonding sites for the electropositive Ce atoms. The subsequent studies show that in the presence of O, the rare-earth elements' arrangements on the Si_3N_4 surface is driven by their preference to achieve bonding configurations similar to those found in Ce_2O_3 and Lu_2O_3 compounds. These results are in a very good agreement with the experimentally observed structures.

References

1. Idrobo, J., Iddir, H., Öğüt, S., Ziegler, A., Browning, N.D., Ritchie, R.O.: Ab initio structural energetics of β-Si_3N_4 surfaces. Phys. Rev. B (Rapid Communications) **72**, 241301 (2005)
2. Walkosz, W., Klie, R.F., Öğüt, S., Borisevish, A., Becher, P.F., Pennycook, S.J., Idrobo, J.C.: Atomic resolution study of the interfacial bonding at $Si_3N_4/CeO_{2-\delta}$ grain boundaries. Appl. Phys. Lett. **93**, 053104 (2008)
3. Kresse, G., Hafner, J.: Ab initio molecular dynamics for liquid metals. Phys. Rev. B **47**(1), 558–561 (1993)
4. Blöchl, P.E.: Projector augmented-wave method. Phys. Rev. B **50**(24), 17953–17979 (1994)
5. Perdew, J.P., Wang, Y.: Accurate and simple analytic representation of the electron-gas correlation energy. Phys. Rev. B **45**(23), 13244–13249 (1992)
6. Perdew, J.P., Burke, K., Ernzerhof, M.: Generalized gradient approximation made simple. Phys. Rev. Lett. **77**(18), 3865–3868 (1996)
7. Painter, G.S., Averill, F.W., Becher, P.F., Shibata, N., Van Benthem, K., Pennycook, S.J.: First-principles study of rare earth adsorption at beta-Si_3N_4 interfaces. Phy. Rev. B **78**(21), 214206 (2008)
8. Iddir, H., Komanicky, V., Öğüt, S., You, H., Zapol, P.: Shape of platium nanoparticles supported on $SrTiO_3$:experiment and theory. J. Phys. Chem. C **111**, 14782–14789 (2007)
9. Tsetseris, L., Pantelides, S.T.: Large impurity in rubrene crystals: first-principles calculations. Phys. Rev. B **78**, 115205 (2008)

10. Dudarev, S.L., Botton, G.A., Savrasov, S.Y., Humphreys, C.J., Sutton, A.P.: Electron-energy loss spectra and the structural stability of nickel oxide: an LSDA+U study. Phys. Rev. B **57**, 1505–1509 (1998)
11. Skorodumova, N.V., Ahuja, R., Simak, S.I., Abrikosov, I.A., Johansson, B., Lundqvist, B.I.: Electronic, bonding, and optical properties of CeO_2 and Ce_2O_3 from first principles. Phys. Rev. B **64**(11), 115108 (2001)
12. Da Silva, J.L.F.: Stability of the Ce_2O_3 phases: a $DFT + U$ investigation. Phys. Rev. B **76**(19), 193108 (2007)
13. Qian, G.X., Martin, R.M., Chadi, D.J.: First-principles study of the atomic reconstructions and energies of Ga- and As-stabilized GaAs(100) surfaces. Phys. Rev. B **38**(11), 7649–7663 (1988)
14. Ziegler, A., Idrobo, J.C., Cinibulk, M.K., Kisielowski, C., Browning, N.D., Ritchie, R.O.: Interface structure and atomic bonding characteristics in silicon nitride ceramics. Science **306**, 1768–1770 (2004)

Chapter 5
Atomic-Resolution Study of the Interfacial Bonding at $Si_3N_4/CeO_{2-\delta}$ Grain Boundaries

In this chapter, I describe results of the atomic-resolution Z-contrast imaging and electron energy-loss spectroscopy (EELS) study of the interface between Si_3N_4 and $CeO_{2-\delta}$ intergranular film (IGF). First, I give an overview of the experimental setup, followed by a discussion of the obtained Z-contrast images of the interface, and compare them to the existing images in the literature. Next, I present the EELS results regarding the presence of oxygen in direct contact with the terminating Si_3N_4 open-ring structures, changes in the Ce valence state across the IGF, and concentration profiles of Si and N in the film. A brief summary of the results is given at the end of the chapter.

5.1 Experimental Setup

The polycrystalline sample used in this work was synthesized from the powder mixture of Ube E-10 Si_3N_4 (Ub Industries, Ltd. Tokyo) with an additive composition of 8 wt % $CeO_{2-\delta}$ and 2 wt% MgO (Alfa Aesar, Ward Hill, MA). In order to blend the powders, isopropanol was added to the mixture, which was then attritor-milled for 2 h using 2-mm silicon nitride balls. Next, the alcohol-powder slurry was poured through a 90 μm sieve to separate milling media with the powder-alcohol slurry captured in a stainless steel tray. The slurry was then air- dried in a hood for 24 h, after which it was gently ground and passed through a 425 μm sieve. A desired amount of the mixture was placed into a graphite die, and a uniaxial pressure of *sim* 20 MPa was applied via the graphite rams. The loaded die assembly was then placed into a hot press, which was evacuated and backfilled with nitrogen three times. Larger billets were densified at $1,800^\circ$ C for 1 h under an applied uniaxial pressure of 22 MPa with a continuous flow of nitrogen through the chamber. The experimental Z-contrast images and EELS spectra were acquired with the JEOL2010F microscope at the

W. Walkosz, *Atomic Scale Characterization and First-Principles Studies of Si_3N_4 Interfaces*, Springer Theses, DOI: 10.1007/978-1-4419-7817-2_5,

University of Illinois at Chicago. The lens conditions in the microscope were set up for a probe convergence angle of 15 mrad, an inner detector angle of 45 mrad, and a collection angle of 28 mrad (for EELS acquisition), resulting in a probe size of 1.4 Å for Z-contrast imaging and 2 Å for EELS with an energy resolution of 1.0 eV. These conditions provide sufficient electron-beam current to obtain statistically significant information and assure that the Z-contrast image is incoherent for direct imaging of the interfacial structure. The EELS spectra were acquired with 2 s exposure time to minimize electron-beam damage and sample drift. A supplementary set of Z-contrast images was taken in a dedicated 300 kV cold field-emission VG HB603U STEM with an aberration corrector (Nion Co.) located at Oak Ridge National Laboratory.

5.2 Z-Contrast Images of $Si_3N_4/CeO_{2-\delta}$ Interfaces

Figure 5.1 shows the atomic-resolution Z-contrast image of two Si_3N_4 grains separated by a ceria-IGF with an average width of 1.2 ± 0.3 nm. The Si_3N_4 crystal structure forms characteristic hexagonal rings in the [0001] orientation exhibiting open-ring structures along $(10\bar{1}0)$ in the terminating plane which can be identified in the image. The bright band above the Si_3N_4 c-axis oriented grain is the IGF with three distinctive Ce atomic columns periodically arranged in two layers. In the first layer, which is in direct contact with the [0001] oriented Si_3N_4 grain, Ce atoms segregate to the middle (B) and the end of the open rings (A), and in the second layer they sit in between the open rings (C). While the present Z-contrast images indicate a two-layer periodic arrangement of Ce atoms, similar to the structure observed by Ziegler et al. [1], there are significant differences between the Ce positions in the two studies (Fig. 5.1). In particular, the main difference between the structure observed by Ziegler et al. and the one reported here is the presence of Ce atoms at B. I find that the Ce atoms attach at the middle of the open rings (B), as previously reported for Sm [2], La [3], and Gd [4] based IGFs, whereas Ziegler et al. observed this site to be vacant. Since the surface termination of this grain is identical in the two studies, I speculate that this structure variation can be attributed to the different local cations and oxygen environments surrounding the Ce atoms. The La, Gd, Sm and our ceria based IGF were prepared with 2 wt % MgO, while the sample of Ziegler et al. incorporated 1 wt % Al_2O_3. I also find a slight difference in the position of the Ce atoms (C) in the second layer with respect to Ziegler et al. observation. Another reason for the observed differences between the two studies could be the thickness of the IGF, which in the experiment conducted by Ziegler et al. was 0.7–0.8 nm wide, thus much thinner than the one observed in my study. In such thin IGFs, the relative orientation of the second (misoriented) grain would impose greater geometric constraints for the atomic arrangement and could perhaps be responsible for the observed difference. These observations suggest that the oxide additive used in the sample preparation (typically MgO, Al_2O_3, or Si_2), and possibly the thickness of the IGF, play important roles in controlling the structure of the IGF and should be taken into account when modeling the Si_3N_4/REO interfacial structure.

Understanding the underlying mechanisms for the observed preferential segregation of the RE atoms at the Si_3N_4/IGF interface requires a comprehensive study of the atomic composition and chemical bonding across the IGF. As mentioned earlier, this can be achieved experimentally with atomic-resolution EELS studies in combination with Z-contrast imaging. In the only existing EELS study of the Si_3N_4/Sm_2O_3 interface, Ziegler et al. argued for the presence of oxygen bonded to Si at the interface by comparing the Si L-edge with reference spectra of SiO_2 and Si_2N_2O [2]. However, no direct O K-edge measurements were reported to substantiate this reasonable suggestion. On the other hand, in recent computational studies for modeling the Si_3N_4/REO interfaces, N-terminated Si_3N_4 surfaces have been assumed to be in direct contact with the RE element, without consideration for the presence of oxygen at the interface. The lack of detailed work on the interface has so far left these experimental suggestions and theoretical assumptions unchallenged, providing a timely motivation for atomic-resolution EELS studies to be performed.

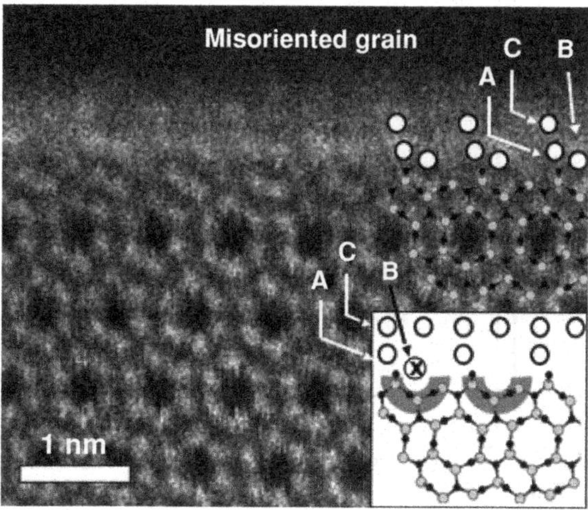

Fig. 5.1 High-resolution Z-contrast image of the Si_3N_4 $(10\bar{1}0)$ interface with the $CeO_{2-\delta}$ IGF. The Ce atoms (*circled*) in the nominally amorphous IGF are visible as bright contrast spots. The superimposed atomic structure shows Si and N atoms with light and dark circles, respectively. The actual arrangement of N atoms at the interface shown bonded with Si atoms in the schematic may differ when oxygen is present at the interface. The inset shows a schematic representation of the interfacial structure as found in the study by Ziegler et al. [1]

5.3 EELS Experiments

In my EELS analysis of an equivalent Si_3N_4/IGF/Si_3N_4 interface, I obtained spectra containing both the O and Ce core-loss signals acquired simultaneously. Each spectrum was recorded every 0.1 nm. Figure 5.2a shows the background-subtracted oxygen *K*-edges at 532 eV and the cerium *M*-edges at 880 eV as a function of the electron probe position across the interface. I find a clear signal of the O *K*- and Ce *M*-edges coming from the IGF, although the former exhibits a lower signal-to-noise ratio than the latter. Figure 5.2b shows the O and Ce signals integrated over 50 eV windows starting at their respective energies. A nonlinear fit using a model with two Gaussians for the integrated O and Ce signals reveals that the FWHM of the O signal is 0.60 nm wider than that of the Ce signal (see Fig. 5.2c). Moreover, the onset of the O signal is 0.18 nm before the Ce signal at the interface between the Si_3N_4 open-ring surface and the IGF, strongly suggesting that the Si_3N_4 open rings can also be oxygen-terminated (Removing the O background before normalizing the O signal does not change the conclusions. The oxygen FWHM is 0.46 nm broader than that of the cerium signal. The onset of the O signal occurs 0.1 nm before the Ce signal). This view is also supported by the earlier findings of Ziegler et al. [2] and my first-principles calculations, described in the previous chapter, for the bonding sites and structural energetics of oxygen at the Si_3N_4 surface [5]. As shown in these calculations, I find that for a wide range of surface coverage, oxygen always binds to Si for the lowest-energy configurations, typically bridging two Si atoms (Si-O-Si) much like the atomic

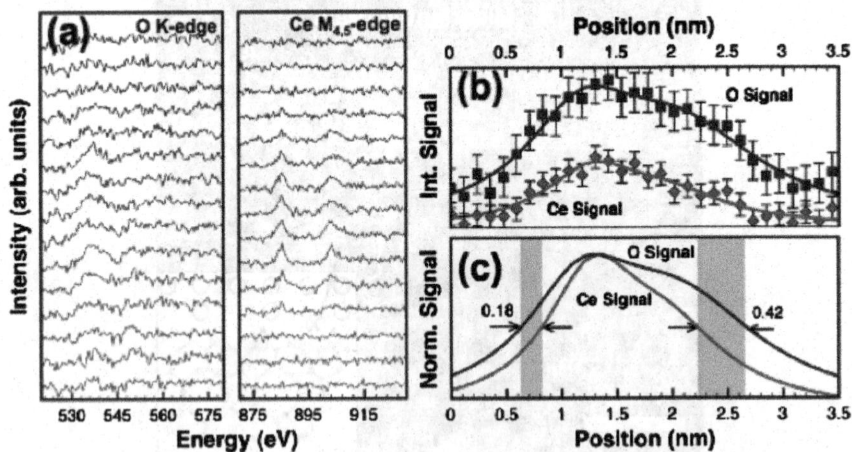

Fig. 5.2 **a** EELS line scan after background subtraction showing the O *K*-edge and the Ce *M*-edge as a function of position across the IGF. Only one out of every second spectrum is shown for visualization purpose. **b** O *K*-edge and Ce *M*-edge signals integrated over 50 eV windows and fitted to a model with two Gaussians (*solid lines*) as a function of position across the film. The error bars indicate the standard error of the concentration quantification. **c** Normalized O and Ce signal fits

configuration in SiO_2. In Fig. 5.2b, I also notice that the integrated signal across the IGF does not have a symmetric profile. For instance, the Ce signal has a larger intensity at the interface with the Si_3N_4 open-ring structure (left-hand side of Fig. 5.2b). This effect is most likely the result of a sharper projection of the [0001] oriented grain compared to the misoriented grain. Moreover, this effect is also observed in the O signal, as would be expected if they were in direct contact with the open-ring surface of the [0001] oriented grain.

In a subsequent experiment, I obtained the Ce *M*-edges to determine the Ce valence and its spatial variation. Figure 5.3a shows Ce $M_{4,5}$ edge intensities and edge onset for spectra taken at two different positions within the IGF; one from the center of the film and the other from the interface with the Si_3N_4 grain. It can be clearly seen that the M_4/M_5 intensity ratio and the onset energy of the Ce *M*-edge are lower for the spectra taken in the middle of the IGF compared to those taken at the interface. These two quantities have previously been shown to be directly related to the Ce valence state [6]. The inset in Fig. 5.3a shows reference spectra obtained for CeO_2 and $Ce_2(WO_4)_3$, corresponding to valences of Ce^{4+} and Ce^{3+}, respectively. Comparing the reference spectra with my experimental results, I find that the valence state of Ce at the interface is close to Ce^{3+}, while the valence state of Ce in the middle of the IGF is close to Ce^{4+}. This change in the Ce valence is most likely due to the change in the local oxygen environment around Ce, which is consistent with my finding above that oxygen diffuses into the Si_3N_4 grain and bonds to the Si at the interface.

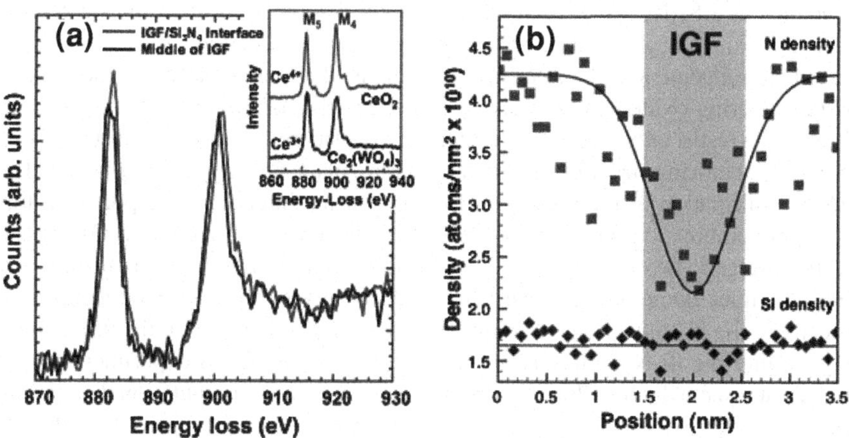

Fig. 5.3 **a** Near edge fine structure of the Ce *M*-edge taken from the middle and at the interface of the $Si_3N_4/CeO_{2-\delta}$. The inset shows reference spectra (from Ref. [6]) obtained for CeO_2 and $Ce_2(WO_4)_3$, corresponding to valences of Ce^{4+} and Ce^{3+}, respectively. **b** Si and N concentration profiles across the $Si_3N_4/CeO_{2-\delta}$ interface extracted from an EELS line scan. The Si *L*-edge and the N *K*-edge were background subtracted, integrated over a 40 eV window, and multiplied by the corresponding scattering cross-sections

Finally, I computed the atomic densities for Si and N from the measured EEL spectra. The atomic densities were extracted from an EELS line scan of $\sim$3.5 nm length. In each background-subtracted spectrum, the spectrum intensity under the Si *L*- and the N *K*-edges was integrated over a 40 eV window and multiplied by the corresponding scattering cross-sections. The resulting densities are shown in Fig. 5.3b as a function of position across two Si_3N_4 grains with the ceria IGF in between. It can be clearly seen that the N concentration decreases significantly when going across the Si_3N_4/CeO_2 interface, while the Si concentration remains nearly constant throughout. I find that the FWHM of the inverted Gaussian formed by the N signal corresponds to the IGF thickness measured directly from the Z-contrast image (shadow region shown in Fig. 5.3b). This is another signature that N is being replaced by oxygen not only within the IGF but also few tens of namometers into the Si_3N_4 grains as it has been previously discussed. Furthermore, the presence of a constant Si signal across the whole interface not only means that there is Si at the IGF, it also indicates that Si is not completely replaced by rare-earth elements at the termination of the Si_3N_4 open rings as has been modeled for so many years. As such, this has to be taken into consideration in future computational modeling of the Si_3N_4/IGF interface to accurately account for its atomic and electronic structures.

5.4 Summary

In summary, I have presented results on and analyses of the atomic and electronic structures at a $Si_3N_4/CeO_{2-\delta}$ interface using a combination of atomic-resolution Z-contrast imaging and EELS. I have observed that the Ce atoms form a two-layer periodic arrangement at the interface, the details of which are affected most likely by the sintering oxide and possibly by the orientation of the second (misoriented) Si_3N_4 grain, especially when thinner IGFs are considered. I have demonstrated that the Si_3N_4 open-ring structures in the terminating plane are oxygen-terminated. The direct confirmation of the presence of oxygen at the interface, as reported in this study, in combination with (1) previous Si *L*-edge comparisons with SiO_2 [2] and (2) the energetic favorability of Si-O-Si bridging configurations, as obtained in my first-principles computations (see Chap. 4), strongly suggests that oxygen bonds directly to the Si atoms in the terminating open rings. I believe that the diffusion of oxygen into the Si_3N_4 grain, as revealed in the present EELS experiments, is an important issue, which will need to be addressed in future computational modeling studies of the Si_3N_4/REO interfaces. Furthermore, I have shown that the Ce valence changes across the IGF (from +3 at the interface to almost +4 inside the film), suggesting that the semicrystalline part of the IGF is oxygen deficient, which might be either due to interfacial strain or the segregation of oxygen into the open-ring structures of the Si_3N_4 terminating planes. Lastly, I have shown that the Si concentration is constant across the whole width of the IGF, while the N concentration decreases in the film.

References

1. Ziegler, A., Idrobo, J.C., Cinibulk, M.K., Kisielowski, C., Browning, N.D., Ritchie, R.O.: Atomic-resolution observations of semicrystalline intergranular thin films in silicon nitride. Appl. Phys. Lett. **88**(4), 041919 (2006)
2. Ziegler, A., Idrobo, J.C., Cinibulk, M.K., Kisielowski, C., Browning, N.D., Ritchie, R.O.: Interface structure and atomic bonding characteristics in silicon nitride ceramics. Science **306**, 1768 (2004)
3. Winkelman, G.B., Dwyer, C., Hudson, T.S., Nguyen-Manh, D., Döblinger, M., Satet, R.L., Hoffmann, M.J., Cockayne, J.H.: Arrangement of rare-earth elements at prismatic grain boundaries in silicon nitride. Philos. Mag. Lett. **84**, 755–762 (2004)
4. Shibata, N., Painter, G.S., Satet, R.L., Hoffmann, M.J., Pennycook, S.J., Becher, P.F.: Rare-earth adsorption at intergranular interfaces in silicon nitride ceramics: subnanometer observations and theory. Phys. Rev. B **72**(14), 140101 (2005)
5. Walkosz, W., Idrobo, J.C., Klie, R.F., Öğüt, S.: Reconstructions and nonstoichiometry of oxygenated β -Si_3N_4 $[10\bar{1}0]$ surfaces. Phys. Rev. B **78**(16), 165322 (2008)
6. Wu, L., Wiesmann, H.J., Moodenbaugh, A.R., Klie, R.F., Zhu, Y., Welch, D.O., Suenaga, M.: Oxidation state and lattice expansion of CeO_2—x nanoparticles as a function of particle size. Phys. Rev. B **69**, 125415 (2004)

Chapter 6
Atomic-Resolution Study of β-Si_3N_4/SiO_2 Interfaces

In this chapter I describe results of the experimental study of Si_3N_4/SiO_2 interface. This work was carried out in an effort to examine the morphology of Si_3N_4 $(10\bar{1}0)$ surface in the absence of rare-earth elements to better understand their importance in stabilizing the ceramics open-ring termination discussed in Chap. 4. The chapter commences with an introduction to Si_3N_4/SiO_2 interfaces, followed by a short summary of the experimental methods used in this study. Next, the results of the investigation are presented in detail, focusing in particular on the composition and bonding characteristics across the interface.

6.1 Introduction

As shown in the previous chapter, in the presence of rare-earth oxides only the open-ring termination of Si_3N_4 $(10\bar{1}0)$ surface is observed experimentally. This suggests that the presence of RE elements, and/or other cations, could play an important role at the ceramic's surface, promoting the change of its termination from the closed-ring structure, predicted to be of lowest energy for bare and oxygenated surfaces, to the open-ring Fig. 4.1. To validate this claim, a set of experiments was designed at the Si_3N_4 interface with no RE elements present. Specifically, the interface between crystalline Si_3N_4 and amorphous SiO_2 IGF was examined in detail to determine the termination of Si_3N_4 $(10\bar{1}0)$ surface in the absence of RE elements. The characterization of this interface is not only relevant to this work, but it also carries a broader impact. Indeed, there has been considerable effort in recent years in characterizing amorphous/crystalline interfaces at the atomic scale because of their important role in many technological areas, ranging from semiconductor devices to advanced ceramics. While it has been known that an abrupt transition from an ordered crystal to an amorphous structure is not plausible, limited experimental data exist confirming the presence of the interfacial transitional structures at the amorphous/crystalline

W. Walkosz, *Atomic Scale Characterization and First-Principles Studies of Si_3N_4 Interfaces*, Springer Theses, DOI: 10.1007/978-1-4419-7817-2_6,

interfaces and identifying their bonding characteristics. As mentioned in Chap. 1, the latter is especially important in the case of sintered silicon nitride (Si_3N_4) ceramics, whose intrinsic brittleness can be controlled by the atomic and electronic structures at the grain boundaries. Moreover, with the promising new possibility of using the Si_3N_4/SiO_2 layers in high-speed memory devices [1, 2, 3, 4, 5] and in optical waveguide applications [6], information about the composition characteristics across this interface at the atomic scale is of both fundamental and technological interest.

6.2 Methods

The silicon nitride sample used in this work was synthesized by uniaxial hot pressing of a finely mixed and dispersed 95% α-Si_3N_4 (SSA of 11 m^2/g) powder with pure SiO_2 (SSA 4 m^2/g, 1000 total ppm metal impurities) powder in a 71/29 weight ratio. The consolidated sample had a density of 2.86 g/cm^3 and was substantially pore-free based on polished optical and SEM micrographs. The experimental Z-contrast images were acquired with a FEI Titan S operated at 300 kV and a field-emission VG HB603U located at Oak Ridge National Laboratory. To minimize the radiation dose at any given point near the interface, the spectra were collected by rastering the electron beam over a rectangular area of $54.8 \times 2.7 Å^2$ for 0.5 s. The spectra were collected using a 23 mrad convergence semiangle, ~35 mrad collection semiangle, and an electron beam current of ~50 pA. An additional set of images and EELS spectra was taken in the aberration-corrected JEOL-ARM200F microscope (Tokyo, Japan) equipped with a Schottky field-emission gun with variable accelerating voltage. This feature of the microscope was highly advantageous to this study because of the poor electron beam resistance of SiO_2 requiring lower voltages, especially for spectroscopy. Thus, the operating voltage for imaging and EELS line-scan acquisition was set to 120 kV, while the EELS images were taken at 80 kV. The lens conditions in the microscope were set up at a probe convergence angle of 30 mrad for imaging and 35 mrad for spectroscopy. The inner detector angle for the high-angle annular dark-field detector was set to 90 mrad, while the collection angle for EELS was 64 mrad. The resulting probe size for imaging and EELS was ~1.25 and 1.35 Å, respectively. The EELS energy resolution was ~0.8 eV and the spectra were acquired with 0.05-s exposure time. In addition to these standard methods, annular bright-field (ABF) imaging was employed in this study. This technique is very suitable for imaging light elements with low scattering amplitudes (within the 11–22 mrad angular range).

6.3 Imaging of the Si_3N_4/SiO_2 Interface

Figure 6.1 shows the atomic-resolution Z-contrast image (a), and the corresponding bright-field image (b), acquired simultaneously with the aberration-corrected FEI Titan S at 300 kV. The contrast of the bright-field image was

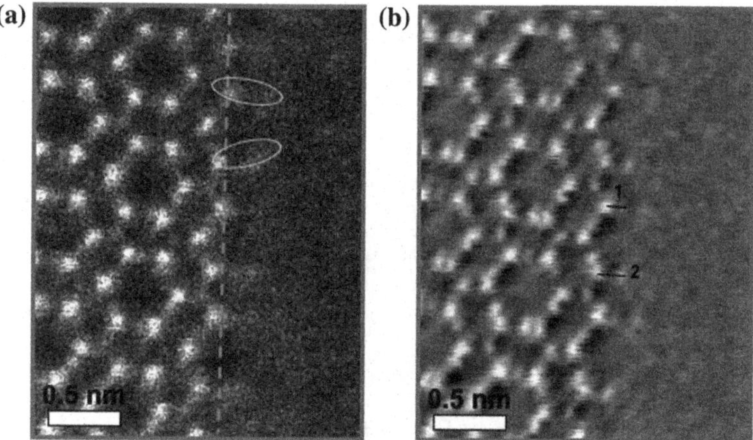

Fig. 6.1 Atomic-resolution: **a** Z-contrast image, **b** bright-field image of the Si_3N_4/SiO_2 interface taken with the aberration-corrected FEI Titan S at 300 kV. The *dotted line* marks the *open-ring* termination of Si_3N_4, while the ellipses highlight atomic structures extending out into the amorphous region

inverted for illustration purposes to enhance the crystal structure of Si_3N_4. In both images, the characteristic hexagonal rings of the ceramic in the [0001] orientation formed by the Si atoms (bright circles in Fig. 6.1a) and the N atoms (dimmer circles) are clearly visible. The dotted line in Fig. 6.1a marks the terminating surface of the ceramic as observed in the experiments at the Si_3N_4/REOs interfaces, i.e. the open-ring termination. With no RE elements present in the IGF, the occurrence of atomic structures (in ellipses) above the open rings can be clearly seen. The structures are more visible in the coherent bright-field image (Fig. 6.1b), giving a better contrast for lighter elements. Intensity profiles, shown in Fig. 6.2, acquired from the terminations of the open rings show that the projected two-dimensional distances from the interfacial Si atoms to their corresponding nearest neighbors in the IGF are approximately 0.09 and 0.17 nm. Interestingly, these distances are in accord with the Si–N distances in the theoretically predicted closed-ring termination of Si_3N_4 described in Chap. 4. This suggests that the observed short-range arrangement in the IGF could be part of the closed-ring structure, which in the presence of cations, especially heavier RE elements, changes to the open-ring. In order to identify the compositions of these atomic structures, atomic resolved EELS will be performed.

Due to the low electron beam tolerance of SiO_2, a supplementary set of images at the interface between Si_3N_4 and the SiO_2-IGF was acquired with the JEOL ARM-200F operated at 120 kV to check for the possibility of beam-induced changes at the interface. Figure 6.3 shows the obtained atomic-resolution Z-contrast image (Fig. 6.3a) and the annular bright-field image (Fig. 6.3b) of the Si_3N_4/SiO_2 interface acquired simultaneously. Similarly to the Z-contrast image taken with FEI Titan S, the Z-contrast image acquired at 120 kV shows the

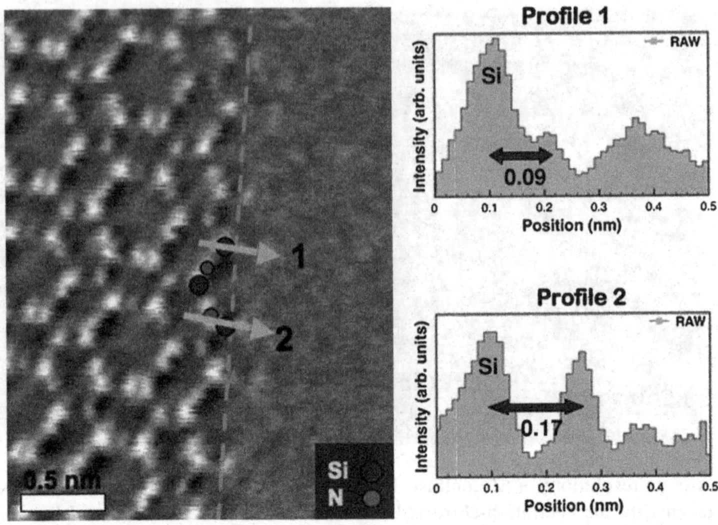

Fig. 6.2 *Left* The (inverted) bright-field image of the Si_3N_4/SiO_2 interfaces taken with the aberration-corrected FEI Titan S at 300 kV. The *dotted line* marks the open-ring termination of Si_3N_4, while the *arrows* show the direction of the intensity profiles shown to the *right*

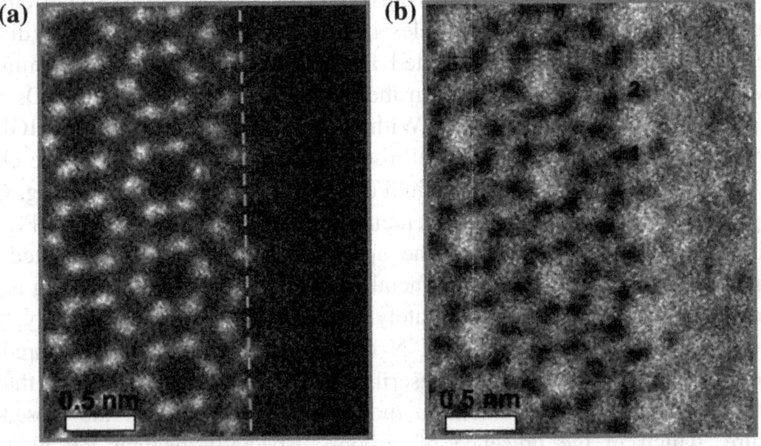

Fig. 6.3 Atomic-resolution: **a** Z-contrast image, **b** annular bright-field image of the Si_3N_4/SiO_2 interface taken with the aberration-corrected JEOL ARM-200F operated at 120 kV. The *dotted line* marks the *open-ring* termination of Si_3N_4. The atomic structures past the *open rings* are clearly visible

N columns in the Si_3N_4 structure. The result demonstrates a remarkable progress in aberration-corrected electron microscopy techniques and instrumentations able to overcome the challenge of imaging light atoms, while providing satisfactory resolution limit at low voltages. Moreover, the images in Fig. 6.3 confirm the presence of atomic ordering following the open rings demonstrated in the earlier figures. The structures are clearly visible in the annular bright-field image (Fig. 6.3b). The ordering extends at least 0.5 nm out into the amorphous region and takes on the shape of the closed-ring termination discussed in Chap. 4. The distortions in the closed-ring patterns, with the biggest ones being the removal of the top atom from some of the closed-ring structures, are, most likely, due to the contact of the Si_3N_4 grain with the IGF. The calculated distances from the Si atoms in the open rings to the nearest neighbors in the IGF are approximately (0.97 ± 0.016) and $(0.177 \pm 0.016$ nm) at the corresponding positions 1 and 2 in Fig. 6.3b. These distances fall in the range of the Si–N and Si–O bond length measurements for the bare and oxygenated Si_3N_4 surfaces, suggesting that the observed atomic arrangement at the Si_3N_4/SiO_2 IGF could involve either N and/or O atoms, i.e. the closed-ring structures could be partially oxygenated. Indeed, my theoretical results (discussed in detail in Chap. 4) have shown that O has a tendency to replace N on the surface of Si_3N_4. To identify the atoms forming the observed closed-ring structures, atomic resolved EELS is needed. The observations of the atomic ordering in the glassy film in this investigation are in agreement with the results of Winkelman et al. who studied Si_3N_4 in the presence of yttrium silicate IGF [7]. The averaged HRTEM images and power spectra reported by the authors show that the Si_3N_4 structure extends past the open rings, forming partially closed-ring patterns as the ones observed in this study. The similar atomic arrangements of atoms at the interface in the two studies, differing in the IGF composition, suggest that the observed arrangement following the open rings is, most likely, part of the predicted closed-ring termination of Si_3N_4, rather than the crystallization of the amorphous phase at the interface.

6.3.1 Spectroscopy

To identify the composition and bonding at the Si_3N_4/SiO_2 interface, the atomic-resolution spectrum imaging in EELS was performed. Figure 6.4a and b show the Z-contrast image of the interface with the spectrum image box enclosing the area used for the analysis, and the resulting ADF signal collected by the HAADF detector as the probe scanned the boxed area, respectively. The schematic drawing of Si_3N_4 crystal structure, with the N atoms in black and the Si atoms in gray, and the terminating open rings are superimposed in the ADF image. The dark band to the left from the open rings is the SiO_2 IGF of approximately 1.4 nm thickness. The ADF image is 50×50 pixels in area, with each pixel associated with an energy spectrum from 45 to 715 eV. The resulting 2,500 different spectra thus, provide very detailed information regarding the region of interest and can be used to generate elemental maps showing the concentration of a given element as a function of position in the ADF image. Figure 6.4d–f shows such elemental maps. The maps were acquired by integrating intensities under the background-subtracted N *K*- and O *K*-edges over a 40 eV window starting at their corresponding edge onsets

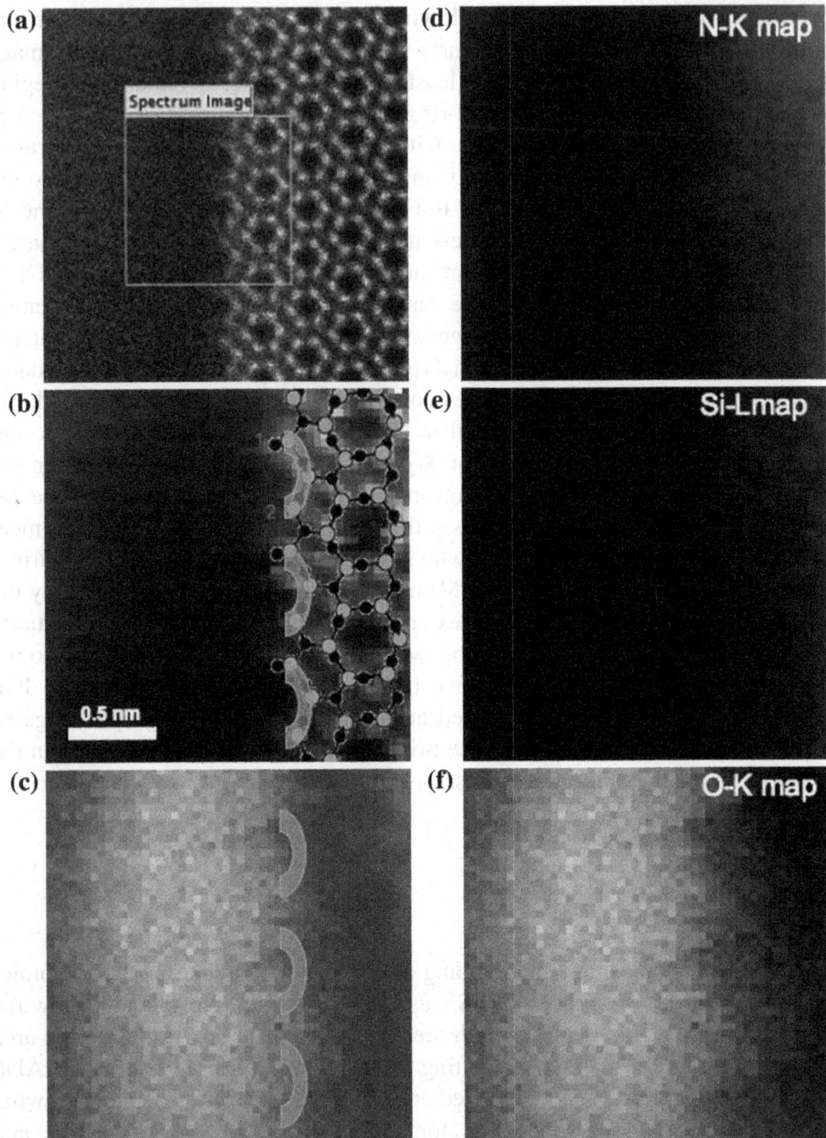

Fig. 6.4 Compositional imaging of the Si_3N_4/SiO_2 interface at the atomic scale: **a** Z-contrast image showing the area used in the analysis (spectrum image box), **b** the magnified High-Angle ADF signal from the boxed area with superimposed Si_3N_4 crystal structure with N and Si atoms (*dark* and *light spheres*, respectively) and the *open rings*, **c** combination of the elemental maps shown in **(d–f)**

(401 and 532 eV, respectively). In the case of the Si-L map, the signal integration over a 50 eV window was performed at 300 eV, rather than at its energy onset, to account for the signal delocalization, which increases from the sub-nanometer scale

to a few nanometers with decreasing energy loss, but it also depends on the atomic spacings in the specimen. The delocalization factors, d, can be calculated with a simple formula proposed by Egerton [8]:

$$(d)^2 \approx (0.5\lambda/\theta_E^{3/4})^2 + (0.6\lambda/\beta)^2 \tag{6.1}$$

where λ is the electron wavelength, θ_E is the characteristic scattering angle, and β is the collection angle. The calculated delocalization factors [8] of the Si–L_{23} and the N–K edges at 80 kV are 0.53 and 0.2 nm, respectively. On the other hand, the diagonal and the nearest interatomic distance of the Si columns are 0.56 and 0.31 nm, respectively, as shown in Fig. 6.5. Since the delocalization of the Si L_{23}-edge is larger than the interatomic distance between the neighboring Si atoms, its elemental mapping suffers considerable delocalization. Wang et al. showed that in this case one should place the energy window as far above ionization threshold as the background subtraction procedure allows [9].

Figure 6.6a–c shows the Si–L_{23} core-loss images obtained with different energy onsets for the intensity integrations. The atomic columns are resolved only in the high-energy loss domain (300–350 eV). This result is in agreement with the observations of Kimoto et al. who studied the localization of elastic and inelastic scatterings using Si_3N_4 samples, but with different image acquisition conditions [10]. Looking at the maps of Fig 6.4, in particular at the RGB map representing the sum of the individual elemental maps (Fig. 6.4c), one can see that O is present at the surface and even at the subsurface of Si_3N_4. At the interface, it is mostly concentrated in between the open rings, which is in agreement with the findings at the Si_3N_4/Sm_2O_3 interface [11]. With respect to the terminations of the open rings themselves, O is more concentrated at position 1 than at position 2 (see Fig. 6.4b), except for the last ring in the figure, which shows a different O signal distribution. The signal is found to go much deeper into the Si_3N_4 region, as far as $\sim$0.5 nm.

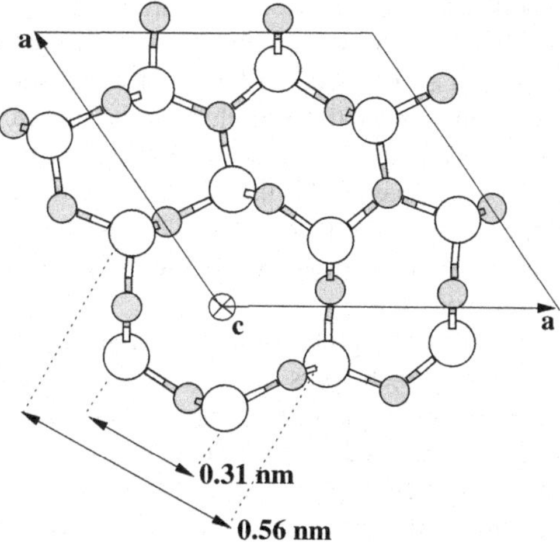

Fig. 6.5 β-Si_3N_4 crystal structure projected along [001]. The c-axis is marked in the image with a cross. The incident electron beam is parallel to this axis

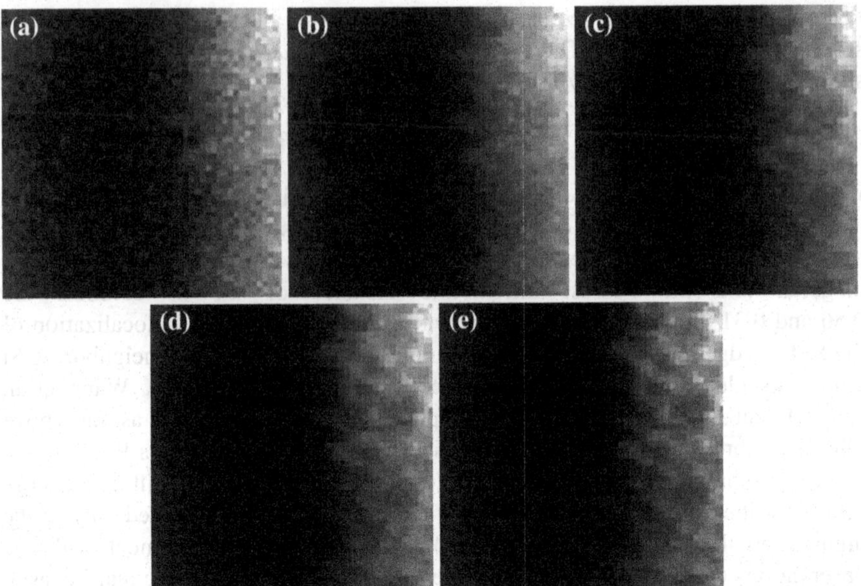

Fig. 6.6 Si-L_{23} core-loss images obtained within the following energy ranges: **a** 101–151 eV, **b** 151–201 eV, **c** 201–251 eV, **d** 251–301, and **e** 300–350 eV. Atomic-columns are resolved most clearly in the last energy domain

Looking at the corresponding regions in the N and Si maps, one can see that it is probably replacing N there as the latter shows a decrease in the intensity with the increase in the O signal intensity. Although energetically favorable, the O $\rightarrow$ N replacement in the vicinity of the last ring is, most likely, actuated by the electron beam since the oxygen penetration is much deeper there than in the case of the other rings. Previous studies on glassy films and N compounds have shown that N can be ejected out of the samples by the beam very easily. Van Bethem et al. reported that the electron beam irradiation of Gd-doped Si_3N_4 induces dislodging of nitrogen atoms from the glass, while oxygen atoms remain unaffected [12]. Mkhoyan et al. observed knock-on type damage with ejection of primarily nitrogen atoms from the sample in wurtzite InN [13]. The results of Levin et al. revealed radiation-induced N segregation to the Si/SiO_x and SiO_x/poly-Si interfaces accompanied by the diffusion of oxygen into nitride in the silicon-oxide–nitride-oxide multilayers [14]. The N–K map (Fig. 6.4d), on the other hand, shows that nitrogen, although mostly present in the bulk, can also be found throughout the whole SiO_2 IGF.

Detailed observations of the Si-L_{23} edge at each pixel in the spectrum image of Fig. 6.4b, which splits into two peaks (separated by $\sim$0.15 nm) in the vicinity of the interface, suggest the presence of distinct Si phases in that region. As opposed to the signal used to form elemental maps, this near edge fine structure signal is highly localized, providing a direct means of probing electronic structures, local coordinations and valence states via the use of "spectral fingerprints" [15, 16, 17].

Figure 6.7 shows such spectra, taken from different positions at the interface, i.e., (a) the closed-ring like structure in the IGF described earlier and (b) the terminating open ring. The insets show exact locations of these spectra, with each spectrum corresponding to the sum of four spectra (4 pixels), and thus providing average information about the $\sim 0.1 \times 0.1\,\mathrm{nm}^2$ area. Also, for comparison, a spectrum from the bulk region of Si_3N_4 in the image and that from pure SiO_2 compound are provided. Comparing the spectra with each other, one can identify the first peak in the spectra from the interface with a Si–N bond, whereas its second peak can be associated with a Si–O bond because their positions match those of the Si_3N_4 and SiO_2 peaks, respectively. This suggests that the atomic environment in the upper portion of the terminating open ring is abundant in both N and O since the heights of both peaks are similar. This conclusion is supported by the elemental maps in Figure 6.4. In contrast, the spectra obtained from the bottom of this ring (spectra 3, C in Fig. 6.7b) show a very prominent second peak, indicating Si–O type of bonds. The same type of bonding is also predominant in the film as reflected in the corresponding spectra (4–6, D-F in Fig. 6.7a). Using the near-edge fine structure of the Si-L_{23} edge one can thus identify specific atomic bonding configurations at the Si_3N_4/SiO_2 interface, as done in the past at the Si_3N_4/Sm_2O_3 interface [11].

To achieve this goal at any position in the whole spectrum image of Fig. 6.4b, we performed the multiple linear least-square (MLLS) fitting analysis in Gatan's Digital Micrograph Software Package [18]. This program generates a model function consisting of a linear combination of the specified spectra and then fits that model to the selected spectrum by adjusting the coefficient of each linear term to minimize the square deviation between the model and the selected spectrum. Here, I used the Si-L_{23} edge of SiO_2 and Si_3N_4 from Fig. 6.7 for the reference spectra (with their integrated signals kept the same) and I fitted them to the spectrum image over the 120 eV energy range.

Figure 6.8 shows the resulting fit-coefficient maps for the two reference phases measuring their concentrations in the whole spectrum. The coefficients were normalized with respect to the total concentration, i.e. the sum of the individual fit-coefficients at every pixel in the image. The intensity of each pixel in these images thus represents the fit-coefficients for the specific phase, with dark pixels standing for smaller coefficients than the bright one. It is interesting to note in the two figures that the right end of the open rings (position 1 in Fig. 6.4b) shows more of the Si–N type of bonding than its left side. This suggests that the nearest neighbor of the terminating Si atom, visible in all Z-contrast images, at position 1 is most likely N and the short-range arrangement is indeed the closed-ring termination predicted theoretically. The rest of the atoms in the observed closed-ring structures, and specifically, their bonding characteristics are mostly of the SiO_2 character. These observations indicate that the open rings are asymmetric with respect to their composition, or more specifically, with respect to the O $\rightarrow$ N replacements. As expected, the IGF shows a strong SiO_2 character as well, although it also contains N as noted first in the elemental maps. Table 6.1 quantifies the fit-coefficients of

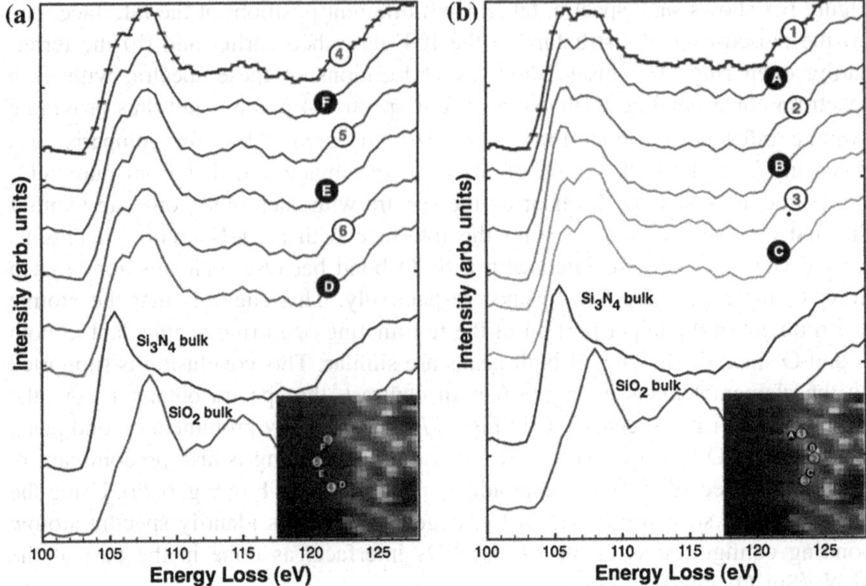

Fig. 6.7 EELS spectra taken from different positions **a** in the terminating *open ring*, **b** the *closing ring* at the termination of Si_3N_4, as marked in the inset, along with the Si_3N_4 and SiO_2 bulk spectra. Comparison of these spectra with the bulk measurements can be used to identify bonding characteristics at these locations. The first peaks in the spectra can be associated with a Si–N bond, whereas the second peaks characterize a Si–O bond because their positions match those of the Si_3N_4 and SiO_2 peaks, respectively. Although all spectra taken at the interface show two peaks in the 106–108 eV energy range, the relative heights of these peaks are different. This signifies different bonding patterns at each location. The top spectrum in each figure shows the error bars

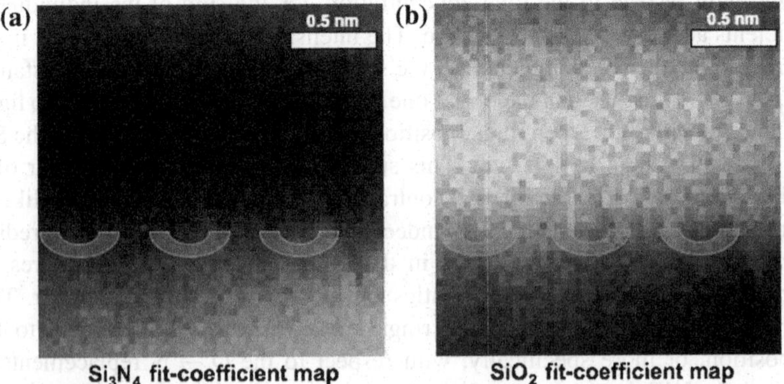

Fig. 6.8 Fit-coefficient maps for **a** Si_3N_4 and **b** SiO_2 showing the distribution of the two phases in the ADF image

Table 6.1 The (normalized) Si_3N_4 and SiO_2 fit-coefficients calculated for the spectra shown in Fig. 6.7. The numbers indicate the concentration of each phase in the given spectrum under assumption that only these two phases could be present

Spectrum	Si_3N_4 fit-coeff.	SiO_2 fit-coeff.
A	0.40	0.60
1	0.48	0.52
B	0.37	0.63
2	0.36	0.63
C	0.49	0.51
3	0.48	0.52
D	0.71	0.29
4	0.65	0.35
E	0.68	0.32
5	0.71	0.29
F	0.60	0.40
6	0.63	0.37

the spectra shown in Fig. 6.7. The listed coefficients were normalized under the assumption that only these two phases are present at the interface.

6.4 Thicker Intergranular Films

Atomic-column resolved EELS study was also performed using the VG HB603U electron microscope operated at 300 keV. As described earlier in the methods section, area scans were performed to minimize the radiation damage on the sample. Figure 6.9 shows the acquired O *K*- and N *K*-edges along with the Z-contrast image indicating the regions of the area scans. Each of the spectra shown below the image corresponds to signal enclosed by the corresponding white rectangle superimposed in the image and thus provides the average information of the enclosed area (measuring $54.8 \times 2.7\text{Å}^2$ in the actual experiment). The white arrow in the image indicates the sequence of the area scans (from 1 to 17). By looking at the O *K*-edge spectra one can see that O starts to peak in the 9th spectrum, which is at the subsurface of Si_3N_4. As for N, its signal is very high in the bulk, and it goes down when approaching the SiO_2 IGF. However, it does not vanish until the 13th spectrum corresponds to a region deep in the SiO_2 IGF. While this study and the results obtained are in agreement with the previously described EELS results, here the effect of the thickness of the IGF can be examined. The experiment involves a much thicker IGF (> 23 Å) providing less geometrical constraints on the distribution of atoms in the film, as evidenced in the case of the nitrogen signal. In the previous study N was found to be uniformly distributed throughout the whole IGF measuring approximately 1.4 nm whereas in this study it is found only close to interface. Deeper in the IGF (approximately 0.81 nm from the terminating open rings), its signal vanishes. This suggests that the thickness also plays an important role in the composition and morphology of the IGF which, in turn, can influence properties of the ceramic. Further studies are necessary to assess its exact role.

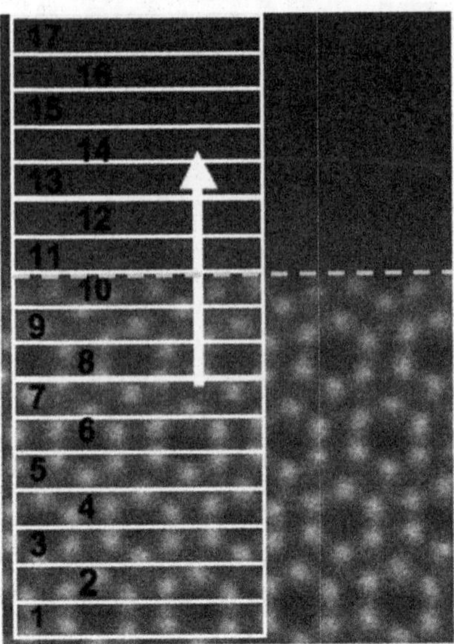

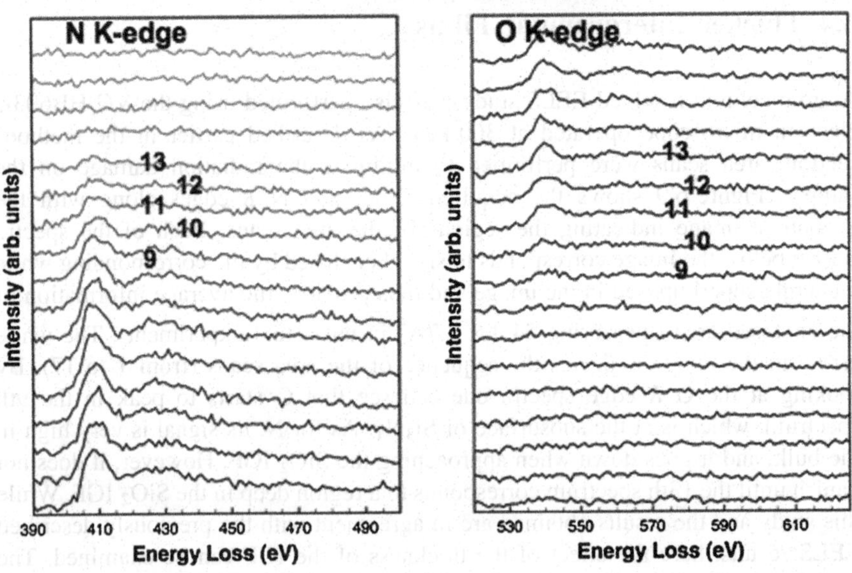

Fig. 6.9 Z-contrast image of the Si_3N_4/SiO_2 interface (*top*) and the N *K*- and O *K*-edges (*bottom*) extracted from the area scans. The rectangles (horizontal length is not shown at scale) drawn on the Z-contrast image show the individual scan-areas, with the *arrow* showing the sequence of the scans

Lastly, the atomic densities were computed for O and N from the measured EEL spectrum shown in Fig. 6.10. They were extracted from an EELS line-scan of $\sim$2.0 nm length. In each background-subtracted spectrum, the spectrum intensity

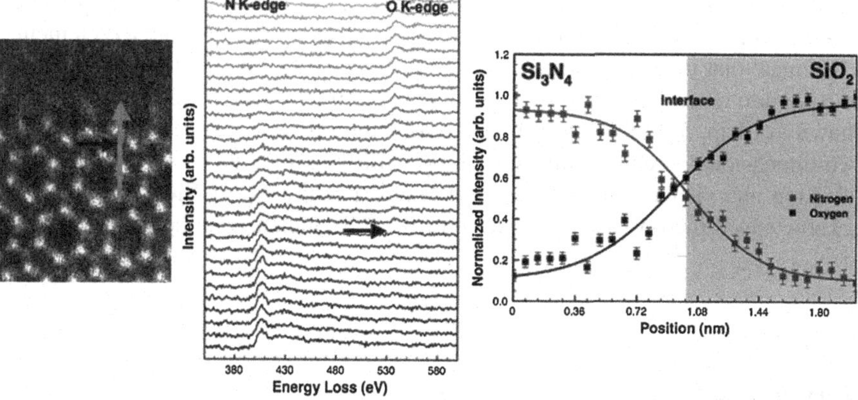

Fig. 6.10 *Left* Z-contrast image of the Si_3N_4/SiO_2 interface showing the region from which the EELS line-scan was taken. *Middle* The EELS line-scan after background subtraction showing the N K- and O K-edges. *Right* The computed concentration profiles over 50 eV windows and fitted to a Gaussian model (*solid lines*) as a function of position across the film. The error bars indicate the standard error of the concentration quantification. The horizontal *arrows* indicate the onset of the O signal

under the N K and O K-edges was integrated over a 40 eV window and multiplied by the corresponding scattering cross-sections. The resulting concentrations were normalized assuming that the maximum concentrations of N and O are in the bulk Si_3N_4 and SiO_2 regions, respectively, and fitted to a Gaussian model. It can be clearly seen from the figure that in the vicinity of the interface, both O and N signals are rather strong. The former starts to peak in the Si_3N_4 region, confirming earlier results that O replaces N at the surface of Si_3N_4.

6.5 Discussion of the Results and Conclusions

In summary, I have shown the presence of atomic structures extending past the open rings. Detailed atomic-resolution studies indicate that these structures are indeed remnants of the closed-ring termination predicted to be energetically favorable for bare surfaces. The presence of the SiO_2 IGF modifies these closed rings, replacing some of their constituent N atoms with O. This conclusion is supported by my theoretical calculations on the oxygenated Si_3N_4 surfaces, discussed in Chap. 4, showing that O tends to replace N on the surface of Si_3N_4 as well as my observations at the Si_3N_4/SiO_2 interface discussed in the previous section indicating the presence of O at the surface, and even at the subsurface of Si_3N_4. Existing calculations have shown that oxygen along grain boundaries in Si_3N_4 has a destabilizing effect on the bonding characteristics and serves as a trap for sintering atoms (in particular, the cations) to migrate toward the boundaries [19]. In the case of small cations, like Si atoms in the SiO_2 film, the terminating closed-ring structures are mostly unaffected as can be seen in the acquired STEM images at the Si_3N_4/SiO_2 interface. However, for heavy cations, like RE elements, I speculate that their presence has a more profound effect on the terminating layer causing it to reconstruct into the experimentally observed open-ring termination as shown in Chap. 5. In this study I have also demonstrated atomic-column resolved two-dimensional chemical imaging along with the simultaneous elemental identification and visualization of the local bonding at the Si_3N_4/SiO_2 interface and acknowledged the possible importance of the IGF thickness in the polycrystalline Si_3N_4.

References

1. Liu, L., Xu, J.P., Chen, L.L., Lai, P.: A study on the improved programming characteristics of flash memory with Si_3N_4/SiO_2 stacked tunneling dielectric. Microelectron. Reliab. **49**, (2009)
2. Saraf, M., Akhvlediani, R., Edrei, R., Shima, R., Roizin, Y., Hoffman, A.: Low thermal budget $SiO_2/Si_3N_4/SiO_2$ stacks for advanced SONOS memories. J. Appl. Phys. **102**(054512), (2007)
3. Berberich, S., Godignon, P.E.M., Fonseca, L. J. M., Hartnagel, H.L.: Electrical characterisation of Si_3N_4/SiO_2 double layers on p-type 6H-SiC. Microelectron. Reliab. **40**, 833–836 (2000)
4. Wang, Y.Q., Hwang, W.S., Zhang, G., Yeo, Y.C.: Electrical characteristics of memory devices with a High-*k* HfO_2 trapping layer and dual SiO_2/Si_3N_4 tunneling layer. IEEE Trans. Electron Dev. **54**(10), 2699–2705 (2007)
5. Santussi, S., Lozzi, L., Passacantando, M., Phani, A.R., Palumbo, E., Bracchitta, G., De Tommasis, R., Alfonsetti, R., Moccia, G.: Properties of stacked dielectric films composed of $SiO_2/Si_3N_4/SiO_2$ tunneling layer. J. Non-Cryst. Solids **245**, 224–231 (1999)
6. Kazmierczak, A., Dortu, F., Schrevens, O., Giannone, D., Vivien, L., Marris-Morini, D., Bouville, D., Cassan, E., Gylfason, K.B., Sohlstrom, H.S.B., Griol, A., Hill, D.: Light coupling and distribution for Si_3N_4/SiO_2 integrated mutichannel single-mode sensing system. Opt. Eng. **48**(1), (2009)
7. Winkelman, G.B., Dwyer, C., Marsh, C., Hudson, T.S., Nguyen-Manh, D., Döblinger, M., Cockayne, J.H.: The crystal/glass interface in doped Si_3N_4. Mater. Sci. Eng. A **422**, (2006)
8. Egerton, R.F.: Applications of energy-loss spectroscopy. In: Electron Energy-Loss Spectroscopy in the Electron Microscopy, 2nd edn. pp. 59–72, Plenum Press, New York (1996)
9. Wang, P., D'Alfonso, A.J., Findlay, S.D., Allen, L.J., Bleloch, A.L.: Contrast reversal in atomic-resolution chemical mapping. Phys. Rev. Lett. **101**(236102), (2008)
10. Kimoto, K., Ishizuka, K., Matsui, Y.: Decisive factors for realizing atomic-column resolution using STEM and EELS. Micron. **39**, (2008)
11. Ziegler, A., Idrobo, J.C., Cinibulk, M.K., Kisielowski, C., Browning, N.D., Ritchie, R.O.: Interface structure and atomic bonding characteristics in silicon nitride ceramics. Science **306**(1768), (2004)
12. Van Benthem, K., Painter, G.S., Averill, F.W., Pennycook, S., Becher, P.F.: Experimental probe of adsorbate binding energies at internal crystalline/amorphous interfaces in Gd-doped Si_3N_4. Appl. Phys. Lett. **92**, (2008)
13. Mkhoyan, K.A., Silcox, J.: Electron-beam-induced damage in wurtzite InN. Appl. Phys. Lett. **82**(6), (2002)

14. Levin, I., Leapman, R.D., Kovler M., Roizin, Y.: Radiation-induced nitrogen segregation during electron energy loss spectroscopy of silicon-oxide-nitride-oxide stacks. Appl. Phys. Lett. **83**(8), (2003)
15. Batson, P.E.: Simultaneous STEM imaging and electron energy-loss spectroscopy with atomic-column sensitivity. Nature **366**, 727–728 (1993)
16. Browning, N.D., Chisholm, M.F., Pennycook, S.J.: Atomic-resolution chemical analysis using a scanning transmission electron microscope. Nature **366**, 143–146 (1993)
17. Brydson, R.: Electron energy-Loss spectroscopy and energy dispersive X-ray analysis. In: Kirkland, A.I., Hutchison, J.L. (eds.) Nanocharacterisation, pp. 94–136. The Royal Society of Chemistry, Cambridge (2007)
18. Gatan: Gatan Digital Micrograph. Gatan. Inc. http://www.gatan.com/software (2001)
19. Benco, L.: Chemical bonding at grain boundaries: MgO on β-Si_3N_4. Surf. Sci. **327**, 274–284 (1995)

Chapter 7
Imaging Bulk α-Si_3N_4

In this chapter I describe results on bulk α-Si_3N_4. In particular, the problem regarding the stability of this polymorph is addressed in detail with aberration-corrected Scanning Transmission Electron Microscopy (STEM) and ab initio methods. I start this chapter with an introduction to α-Si_3N_4, followed by the discussion of the obtained Z-contrast images and the Bloch-wave Z-contrast image simulations. Next, I present my Density Functional Theory (DFT) calculations along with the Electron Energy-Loss Spectroscopy (EELS) results confirming the presence of interstitial O in bulk α-Si_3N_4.

7.1 Introduction

α-Silicon nitride (α-Si_3N_4) is a promising material for the next generation of structural ceramics because its strength is similar to that of β-Si_3N_4 and its hardness is surpassed only by boron carbide and diamond [1]. However, as mentioned in Chap. 1, α-Si_3N_4 tends to transform into β-Si_3N_4 at high temperatures, which limits its use as structural components. For over 40 years, impurities like oxygen or iron have been believed to control the α- to β- phase transformation of the ceramic, but no direct evidence for their presence, location, or even their identity has been provided thus far [2–5]. More specifically, although there exists evidence that α-Si_3N_4 grains are covered by a thin oxide layer (of composition close to SiO_2), no direct confirmation of the presence of oxygen in the bulk α-Si_3N_4 has been given. The reason for this is that α-Si_3N_4 is rather dense, containing 28 atoms per unit cell as opposed to 14 in the β phase, and hence much higher resolution is required to distinguish the individual atoms. Figure 7.1a shows the atomic resolution Z-contrast image of α-Si_3N_4 taken with the JEOL2010F microscope (with no aberration corrector) having $\sim$0.14 nm resolution. For comparison, a schematic structure of bulk α-Si_3N_4 in the [0001] crystalline orientation is also given (Fig. 7.1b). One can clearly see from here that no N nitrogen columns can be resolved in the Z-contrast image because of their closeness to the

W. Walkosz, *Atomic Scale Characterization and First-Principles Studies of Si_3N_4 Interfaces*, Springer Theses, DOI: 10.1007/978-1-4419-7817-2_7,

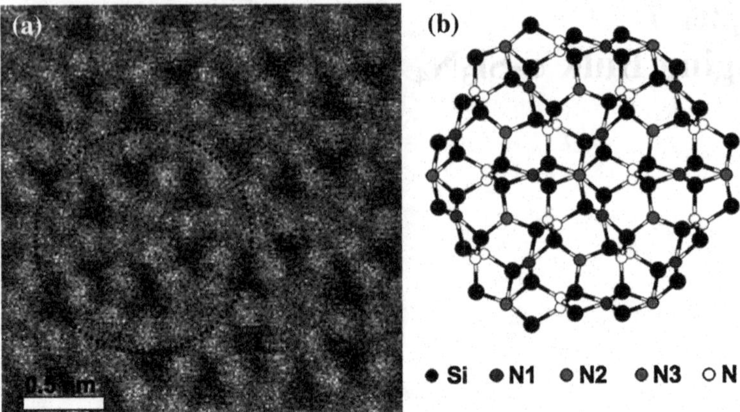

Fig. 7.1 **a** Annular dark field Z-contrast STEM image (raw data) of bulk α-Si_3N_4 taken with the JEOL2010F located at the University of Illinois at Chicago. *Dashed circle* highlights the crystal structure shown to the *right*. **b** Schematic structure of α-Si_3N_4 in the [0001] crystalline orientation

neighboring columns of Si atoms, which have higher atomic numbers. Recent advancements in the aberration-corrected STEM methods, which permit probing of various atomic structures in a sub-Ångstrøm range, have made it possible to finally study this polymorph. Here I report results from the work done in collaboration with the STEM Group at Oak Ridge National Laboratory on the presence of oxygen impurities in α-Si_3N_4 as investigated with the aberration-corrected FEI Titan S and the VG HB603U, as well as ab initio calculations [6].

7.2 Experimental Setup

The silicon nitride sample used in this study was prepared according to the procedure described in Chap. 6. Z-contrast STEM images were acquired with a FEI Titan S having 0.063 nm resolution at 300 kV. Electron energy-loss spectra (EELS) were acquired from the grain of the same sample with a thickness of 100 nm using a VG Microscopes HB603U operated at 300 kV. The spectra were collected using a 23 mrad convergence semiangle, 35 mrad collection semiangle, and an electron beam current of $\sim$50 pA. An acquisition time of 50s was used for collecting the spectra. The atomic-resolved EEL spectrum, shown in Sect. 7.2.3 , is the result of nine individual spectra with an acquisition time of 1 s.

7.2.1 Imaging α-Si_3N_4

As shown in the schematic structure in Fig. 7.1b, in the [0001] crystalline orientation α-Si_3N_4 has three different nitrogen atomic columns (N1, N2, and N3) that are spatially separated in the image plane with distances larger than 70 pm, which means they can be directly resolved by the FEI Titan S operated at 300 kV. Figure 7.2 shows the obtained Z-contrast image of a α-Si_3N_4 grain of ~78 nm thickness as calculated with the absolute log-ratio method [7]. The line through the image marks the region of a horizontal intensity line profile (formed by summing over a width of 0.04 nm) shown below the image. One can see from it that the N3 columns can be clearly identified, even though theoretically they have only 50% of the nominal occupancy of other N columns.

Additionally, Bloch-wave Z-contrast image simulations were carried out with thermal diffuse scattering using an Einstein model [8] and compared with the experimental Z-contrast images as shown in Fig. 7.3. The measured intensities of the N1 and N2 columns in the simulations (Fig. 7.4) were found to be in excellent agreement with those in the experimental images. However, the intensity of the N3 column was underestimated by ~25% in the simulations. The discrepancy was present for grains with different thickness. To ensure that this disagreement between the simulated and measured intensities in the N3 columns was not due to incoherence, various Gaussian and Lorentzian blurs were applied to the images and the simulations were performed at different defocus values. However, the intensity ratio between atomic columns would not change significantly, indicating that the result was insensitive to the uncertainties in the probe parameters.

7.2.2 DFT Calculations

The possibility that the deviant intensity could arise from interstitial atoms located along the N3 columns was explored with first-principles calculations based on

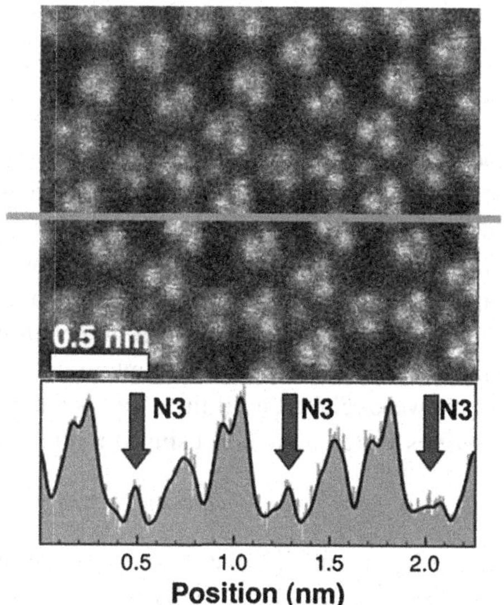

Fig. 7.2 Annular dark field Z-contrast STEM image (raw data) of bulk α-Si_3N_4 taken with the FEI Titan S located at the Oak Ridge National Laboratory. The line through the image marks the position of a horizontal intensity line profile formed by summing over a width of 0.04 nm. *Filled* (*solid line*) *region* in the intensity line profile shows the raw (low-pass filtered) data, respectively [6]

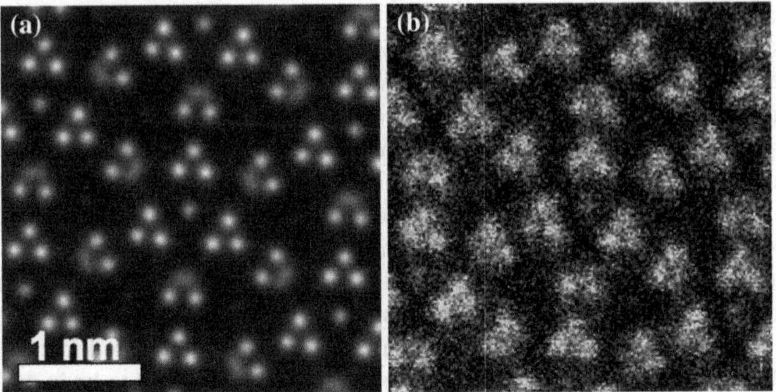

Fig. 7.3 **a** Bloch-wave Z-contrast simulation. **b** The experimental Z-contrast image (raw data) of bulk α-Si_3N_4 taken with the FEI Titan S located at the Oak Ridge National Laboratory [6]

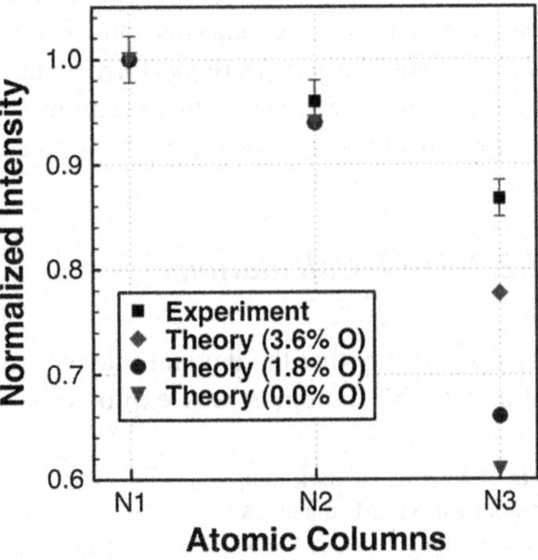

Fig. 7.4 Normalized intensities of the N1, N2, and N3 atomic columns for the experimental and Bloch-wave simulated images with and without O presence in the α-Si_3N_4 lattice [6]

Density Functional Theory. Formation energies (FE) of several likely interstitial atoms were calculated with the projected augmented wave method as implemented in VASP [9, 10]. The generalized gradient approximation (GGA) given by Perdew [11] was used in the calculations. They were performed using a plane wave energy cutoff of 270 eV and $(3 \times 3 \times 2)$ **k**-point grid. The Si, N, and O interstitial impurities were placed between two N atoms along the N3 columns in $(1 \times 1 \times 2)$ and $(1 \times 1 \times 4)$ supercells (containing 57 and 113 atoms, respectively). All the atoms were relaxed until their forces were smaller than 10 meV/ Å. The formation energies were calculated using the relation [12]

$$E_{form} = E_{tot} - n_j \mu_{\alpha Si_3N_4} - n_i \mu_i \tag{7.1}$$

where E_{tot} is the calculated total energy of α-Si_3N_4 with impurities, n_j and n_i are the total number of unit cells of bulk α-Si_3N_4 and impurities, respectively, while

$\mu_{\alpha Si_3N_4}$ and μ_i are the corresponding chemical potentials of bulk α-Si_3N_4 and the impurities. The chemical potentials for Si, N, and O were defined such that $\mu_{Si} = \mu_{Si-bulk}$, $\mu_N = \mu_{\frac{1}{2}N}$, and $\mu_O = \mu_{\frac{1}{2}O}$. These values correspond to the upper limit of the Si, N, and O chemical potentials and therefore the interstitial formation energies are the lowest. The calculations showed that interstitial N and Si atoms, one N or Si atom every four N3 atoms (equivalent to a total overall concentration of 1.8 %), have a large FE of 5.0 and 6.1 eV, respectively. For the same concentration, interstitial oxygen has a FE of only 1.0 eV. Doubling the oxygen concentration to 3.6 % (one O every two N3 atoms) increases the FE to 2.2 eV. When O in this concentration is incorporated in the simulations, the N3 column intensity increases to the level of reaching a similar overall agreement with experiment as shown in Fig. 7.4.

7.2.3 EELS Experiments

In order to test the hypothesis that O impurities are responsible for the excess intensity, atomic resolution EELS was performed. Figure 7.5 shows the acquired signal from a $1 \times 1\,nm^2$ area of the sample, as well as the spectrum from the N3 column. The area scan (in black) does not show any trace of O, indicating that there is no oxide layer covering the surface of the grain. In contrast, the signal from the N3 columns (in red) peaks around 532 eV, confirming the presence of O in that column.

7.3 Summary

In summary, this chapter described the results of the aberration-corrected STEM study and the Bloch-wave image simulations to detect the presence of excess

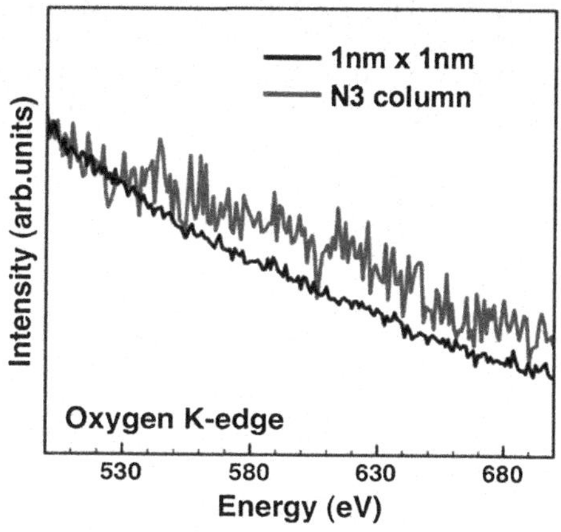

Fig. 7.5 EELS signal obtained from a $1 \times 1\,nm^2$ area and a single N3 column showing the presence of O in that column [6]

atoms in stoichiometric α-Si_3N_4. The initial discrepancy between the experimental and simulated images suggested the presence of light elements (N, Si, or O) in the bulk α-Si_3N_4. DFT calculations were then used to rule out N and Si on energetic grounds and accept O as a candidate. Subsequent EELS experiments confirmed the presence of the latter in the N3 columns. Lastly, we observed consistently a larger number of α-Si_3N_4 than β-Si_3N_4 grains in the studied sample. Since oxygen is always present in the sintering process and our studies show that oxygen indeed segregates into the bulk of α-Si_3N_4, we infer that oxygen is likely to play a role in the stabilization of α-Si_3N_4.

References

1. Chen, I.W., Rosenflanz, A.: A tough SiAlON ceramic based on alpha-Si_3N_4 with a whisker-like microstructure. Nature **789**, 701–704 (1997)
2. Campos-Loriz, D., Riley, F.L.: Factors affecting the formation of the α- and β-phases of silicon nitride. J. Mater. Sci. **13**, 1125–1127 (1978)
3. Peuckert, M., Greil, P.: Oxygen distribution in silicon nitride powders. J. Mater. Sci. **22**, 3717–3720 (1987)
4. Okada, K., Fukuyama, K., Kameshima, Y.: Characterization of surface-oxidized phase in silicon nitride and silicon oxynitride powders by x-ray photoelectron spectroscopy. J. Am. Ceram. Soc. **78**, 2021–2026 (2005)
5. Wang, C.M., Pan, X., Rühle, M., Riley, F.L., Mitomo, M.: Silicon nitride crystal structure and observations of lattice defects. J. Mater. Sci. **31**, 5281–5298 (1996)
6. Idrobo, J.C., Oxley, M.P., Walkosz, W., Klie, R.F., Öğüt, S., Mikijelj, B., Pennycook, S.J., Pantelides, S.T.: Identification and lattice location of oxygen impurities in alpha-Si_3N_4. Appl. Phys. Lett. **95**, 164101 (2009)
7. Egerton, R.F.: Applications of energy-loss spectroscopy. In: Electron energy-loss spectrosocy in the electron microscopy, 2nd edn, pp. 59–72. Plenum Press, New York (1996)
8. Allen, L.J., Findlay, S.D., Oxley, M.P., Rossouw, C.J.: Lattice-resolution contrast from a focused coherent electron. Probe Part I. Ultramicroscopy **96**(1), 65–81 (2003)
9. Kresse, G., Hafner, J.: Ab initio molecular dynamics for liquid metals. Phys. Rev. B **47**(1), 558–561 (1993)
10. Blöchl, P.E.: Projector augmented-wave method. Phys. Rev. B **50**(24), 17953–17979 (1994)
11. Perdew, J.P., Burke, K., Ernzerhof, M.: Generalized gradient approximation made simple. Phys. Rev. Lett. **77**(18), 3865–3868 (1996)
12. Qian, G.X., Martin, R.M., Chadi, D.J.: First-principles study of the atomic reconstructions and energies of Ga- and As-stabilized GaAs(100) surfaces. Phys. Rev. B **38**(11), 7649–7663 (1988)

Chapter 8
Conclusions and Future Work

The goals of this Ph.D. study were to integrate quantum modeling using first principles techniques and atomic scale characterization methods in the aberration-corrected STEM, to characterize Si_3N_4/rare-earth oxides (REO) interfaces, and to address the outstanding problems regarding the role of light elements at these interfaces. The experimental techniques such as Z-contrast imaging and electron energy loss spectroscopy in the STEM were used to provide information on the structure, composition, and bonding characteristics at the interface with a sub-Å ngstrøm resolution. The theoretical calculations based on Density Functional Theory, on the other hand, served to model the interface focusing on the question of how the RE elements and lighter atoms bond to the grains of Si_3N_4. Since the morphology of the ceramics grains as well as its overall properties depend upon the selection of the RE elements and their bonding characteristics at the Si_3N_4/REO interfaces, this study is of both fundamental and technological interest.

In order to understand the observed preferential attachment of the REO to the interface with the ceramic and to elucidate the role of lighter elements at the interface, I worked with a Si_3N_4 sample prepared with $CeO_{2-\delta}$. The choice of cerium as the rare-earth element was the key factor in this study because cerium has the ability to easily change its oxidation state (and so its radius) in different local oxygen environments. This allowed for testing both the role of the RE element as well as oxygen in controlling the Si_3N_4/REO interface.

My experimental results on $Si_3N_4/CeO_{2-\delta}$ interface can be summarized as follows: I have shown that Ce atoms segregate to the interface in a two-layer periodic arrangement, which is significantly different compared to the structure observed in a previous study [1]. This can be attributed to the different additives (in addition to $CeO_{2-\delta}$) used in the samples, or more specifically different oxygen concentrations, different sintering conditions, and possibly the differences in the thickness of the intergranular films in the two studies. Furthermore, I have shown for the first time that the observed Si_3N_4 open-ring structures in the terminating plane are in direct contact with oxygen, and not with the rare-earth elements, as

W. Walkosz, *Atomic Scale Characterization and First-Principles Studies of Si_3N_4 Interfaces*, Springer Theses, DOI: 10.1007/978-1-4419-7817-2_8,

previously assumed. This unambiguous finding of direct oxygen contact at the interface has significant implications for the theoretical modeling studies. I have also found that while the concentration of N decreases in the IGF, the Si concentration remains constant across the whole width of the film. This surprising finding has never been reported before and has potentially important implications for future theoretical models. Lastly, I have shown that the Ce valence changes across the IGF (from +3 at the interface to almost +4 inside the film), suggesting that the semi-crystalline part of the IGF is oxygen-deficient.

Having established that oxygen is in contact with the Si_3N_4 terminating layer, I have modeled the surface of Si_3N_4 in the presence of various oxygen concentrations to examine the effect of oxygen on Si_3N_4. In these calculations two stoichiometric Si_3N_4 $(10\bar{1}0)$ surface terminations, the open-ring as well as the closed-ring terminations, were considered. The work has resulted in several new findings on the structural energetics of oxygen-covered Si_3N_4 surfaces. In particular, based on my first-principles calculations, I have predicted (1×2) surface reconstructions of the oxygenated Si_3N_4 $(10\bar{1}0)$ surface and have shown that most energetically favorable surfaces are nitrogen-deficient, as O has a strong tendency to replace N and minimize the number of unsaturated dangling bonds. Furthermore, I have observed that the structural stability of most low-energy surfaces is driven primarily by the tendency of silicon to saturate its dangling bonds and the tendency of oxygen to bridge two Si atoms similar to the bonding of SiO_2, while preserving the tetrahedral bonding of Si.

My theoretical observations correlate well with previous experimental studies at the Si_3N_4/Sm_2O_3 interface [2] as well as with my STEM work at the Si_3N_4/SiO_2 interface [3]. The latter was carried out in an effort to examine the morphology of Si_3N_4 surfaces in the absence of RE elements to better understand the importance of RE elements in stabilizing various Si_3N_4 surface terminations. In this work, I have shown that O is replacing N at the interface between the two materials causing N to diffuse into the SiO_2 film in agreement with my theoretical calculations. Moreover, I have observed the presence of atomic columns completing the open rings, which have not been observed experimentally at Si_3N_4/REO interfaces, but have been predicted theoretically on bare Si_3N_4 surfaces. Prior calculations showed that oxygen along grain boundaries in Si_3N_4 serves as a trap for sintering atoms (in particular, the cations) to migrate toward the boundaries [4]. In the presence of small cations, like Si atoms in the SiO_2 film, the terminating closed-ring structures remain mostly unaffected as shown in my STEM images at the Si_3N_4/SiO_2 interfaces. However, for much heavier cations, like the RE elements, I speculate that their presence has a more profound effect on the oxygenated layer causing it to reconstruct into the experimentally observed open-ring.

Existing experimental studies have shown that the abundance and ordering of RE elements on the terminated prismatic surfaces of Si_3N_4 are element-specific and have a significant impact on the ceramics properties and performance. To assess the role of the rare-earth elements in the stability of the two surface terminations of the Si_3N_4, I have modeled them in the presence of Ce. In this study, I considered the N-terminated surfaces only. The study has shown that the chemical bonding factors play an important role in determining the adsorption sites for Ce on the N-terminated Si_3N_4 surfaces rather than its size, as the latter consideration would place Ce in regions with greater open volumes. Instead, Ce atoms were found to position themselves near the N atoms having dangling bonds. Similar conclusions were reached by Painter et al. who studied the adsorption of Lu, Gd, and La with a first-principles cluster and surface slab models [5]. Ziegler et al., however, argued that the size of the rare-earth elements plays the primary role in their attachment and bonding patterns at the interface [2]. Moreover, I have shown that Ce by itself does not change the stability of the two surface terminations discussed earlier; i.e., the closed-ring termination continues to be energetically more stable than the open ring in the presence of Ce alone. This again suggests that oxygen must play an important role at the surface of Si_3N_4, not only modifying its morphology and bonding arrangements as shown in Chap. 4, but perhaps also promoting the close-ring to open-ring surface reconstructions in the presence of RE elements. Other possibilities for the discrepancy between the experimental and theoretical results so far include (i) the inadequacy of the exchange-correlation functionals used in the present study to capture the subtle energy difference between competing phases (which might necessitate the use of hybrid functionals) or (ii) more complicated surface reconstructions, which could stabilize the open-ring terminations in the presence of Ce and/or O alone, requiring larger supercells than the one I considered.

Guided by the experimental results showing that in the presence of the REO IGF only the oxygenated open-ring termination is promoted, I have modeled such a surface in the presence of two different RE elements: Ce and Lu. Because of the large difference in their atomic radius, this study allowed further testing of atomic size considerations in regard to the observed differences in the RE elements adsorption sites. While the theoretical approach used in this work served only as a starting point for computational models of the IGF that should include all of its constituent atoms, the predicted adsorption sites for Ce and Lu compare remarkably well with the available experimental data. In particular, the positions of the Lu and Ce pairs in contact with Si_3N_4 surfaces were well reproduced in my calculations. The RE elements were found to form bonds with Si, O, and themselves, and arrange in one-layer pattern along the [001] direction. The small tilt of the Ce pair observed in my experimental Z-contrast images was also captured in the calculations. This is an important result because earlier studies at the $Si_3N_4/CeO_{2-\delta}$ interface had predicted different Ce arrangements than the one found in my experimental study (see Chap. 5 for a detailed discussion) [1]. The measured Ce–O and Lu–O bond lengths in the calculations were found to be similar to those in stable Ce_2O_3 and Lu_2O_3, respectively. Because the calculated adsorption sites of Ce and Lu on the Si_3N_4 surface were similar, I conclude that the size of the RE elements is not a predominant factor in their distribution at the interface. Although important, other considerations like bonding characteristics are also essential. Moreover, my studies show that the presence of secondary sintering additives does not affect the distribution of the RE elements at the interface in a significant way as the adsorption sites of the studied RE elements were accurately predicted even in the absence of other sintering cations.

Lastly, in collaboration with the Microscopy Group at Oak Ridge National Laboratory, I have studied the α phase of Si_3N_4. In particular, possible factors stabilizing this phase were examined with the aberration-corrected STEM, multislice image simulations, and the DFT calculations. The study has shown that oxygen is present in the bulk alpha-Si_3N_4 in one of the N columns. Since the studied sample contained more α-Si_3N_4 than β-Si_3N_4 grains and oxygen is always present in the sintering process, I conclude that oxygen is likely to play a role in the stabilization of α-Si_3N_4.

While this Ph.D. work has resulted in many important findings on β-Si_3N_4/REO interfaces, a deeper understanding of the atomic and electronic structures of the ceramic will naturally require further ab initio modeling studies, which incorporate oxygen, rare-earth elements, as well as other sintering cations at the interface. For this purpose, molecular dynamics simulations could be a perfect tool. In particular, I would like to study the effect of the thickness of the IGF, which provide a natural constraint to the atomic arrangement. Additionally, charge density plots at the interface could yield important information regarding the bonding character of the RE elements with the surface atoms and possibly explain the observed variations in the RE attachments. Lastly, I would like to model the closed-ring termination in the presence of the rare-earth elements and oxygen to establish energetic grounds for its reconstruction mechanisms into the observed open-ring termination. Experiments involving samples prepared with different techniques, on the other hand, could assess the role of sintering methods in the observed RE elements bonding patterns. As differences in microstructural evolution and properties induced by the rare earths (and likely other dopants) depend on their type, the characterization of their adsorption mechanisms at the interface with Si_3N_4 will continue to be an active research area.

References

1. Ziegler, A., Idrobo, J.C., Cinibulk, M.K., Kisielowski, C., Browning, N.D., Ritchie, R.O.: Atomic-resolution observations of semicrystalline intergranular thin films in silicon nitride. Appl. Phys. Lett. **88**(4), 041919 (2006)
2. Ziegler, A., Idrobo, J.C., Cinibulk, M.K., Kisielowski, C., Browning, N.D., Ritchie, R.O.: Interface structure and atomic bonding characteristics in silicon nitride ceramics. Science **306**, 1768–1770 (2004)
3. Walkosz, W., Klie, R.F., Öğüt, S., Borisevish, A., Becher, P.F., Pennycook, S.J., Idrobo, J.C.: Atomic resolution study of the interfacial bonding at $Si_3N_4/CeO_{2-\delta}$ grain boundaries. Appl. Phys. Lett. **93**, 053104 (2008)
4. Benco, L.: Chemical bonding at grain boundaries: MgO on β-Si_3N_4. Surf. Sci. **327**, 274–284 (1995)
5. Painter, G.S., Averill, F.W., Becher, P.F., Shibata, N., Van Benthem, K., Pennycook, S.J.: First-principles study of rare earth adsorption at beta-Si_3N_4 interfaces. Phy. Rev. B **78**(21), 214206 (2008)

Appendix A
Projector Augmented Wave Method

This derivation of the Projector Augmented Wave (PAW) method follows the work of Rostgaard [1]. The PAW method is based on a linear transformation T that maps the true valence wavefunctions $\psi_n(\mathbf{r})$ with their complete nodal structure onto smooth auxiliary wavefunctions $\tilde{\psi}_n(\mathbf{r})$ having a rapidly convergent plane wave expansion. The transformation is assumed to be in unity except for a sphere centered on the nucleus (so-called augmentation region), whose cut-off radius r_c^a should be chosen such that there is no overlap between neighboring augmentation spheres. The transformation can be written as

$$\hat{T} = 1 + \sum_a \hat{T}^a \tag{A.1}$$

where a is an atom index. Inside the augmentation spheres, the true wavefunction is expanded in partial waves ϕ_i^a, and for each of these partial waves a corresponding auxiliary smooth partial wave $\tilde{\phi}_i^a$ is defined such that

$$|\phi_i^a\rangle = (1 + \hat{T}^a)|\tilde{\phi}_i^a\rangle \Leftrightarrow \hat{T}^a|\tilde{\phi}_i^a\rangle = |\phi_i^a\rangle - |\tilde{\phi}_i^a\rangle. \tag{A.2}$$

for all i, a. This completely defines $\hat{T}^a$, given ϕ and $\tilde{\phi}$. Since $\hat{T}^a$ does not change anything outside the augmentation sphere, the partial wave and its smooth counterpart are identical outside the augmentation sphere, i.e.,

$$\phi_i^a(\mathbf{r}) = \tilde{\phi}_i^a(\mathbf{r}) \tag{A.3}$$

for $r > r_c^a$, where $\phi_i^a(\mathbf{r}) = \langle \mathbf{r}|\phi_i^a\rangle$, and likewise for ϕ_i^a. If the smooth partial waves form a complete set inside the augmentation spheres, the smooth all-electron wavefunctions can be expanded within each sphere as

$$|\tilde{\psi}_n\rangle = \sum_n c_{ni}^a |\tilde{\phi}_i^a\rangle \tag{A.4}$$

where c_{ni}^a are the expansion coefficients. Similarly, the expansion

$$|\psi_n\rangle = \hat{T}|\tilde{\psi}_n\rangle = \sum_n c_{ni}^a |\psi_i^a\rangle \tag{A.5}$$

has identical expansion coefficients, c_{ni}^a.

Since the transformation T is required to be linear, these coefficients must be linear functionals of $|\tilde{\psi}_n\rangle$, i.e.

$$c_{ni}^a = \langle \tilde{p}_i^a | \tilde{\psi}_n \rangle = \int d\mathbf{r} \tilde{p}_n^{a*}(\mathbf{r} - \mathbf{R}^a)\tilde{\psi}_n(\mathbf{r}) \tag{A.6}$$

where $\tilde{p}_i^a$ are some fixed functions (called smooth projector functions), which probe the local character of the auxiliary wavefunction in the atomic region.

As there is no overlap between the augmentation spheres, the one center expansion of the smooth all-electron wavefunction, $|\tilde{\psi}_n^a\rangle = \sum_i c_{ni}^a |\tilde{\phi}_i^a\rangle$ reduces to $|\tilde{\psi}_n\rangle$ itself inside the augmentation sphere defined by a. Thus, the smooth projector functions must satisfy

$$\sum_i |\tilde{\phi}_i^a\rangle\langle\tilde{p}_i^a| = 1 \tag{A.7}$$

inside each augmentation sphere and be orthonormal to the smooth partial waves inside the augmentation spheres, i.e., $\langle \tilde{p}_{i_1}^a | \tilde{\phi}_{i_2}^a \rangle = \delta_{i_1,i_2}$. There are no restrictions on $\tilde{p}_i^a$ outside the augmentation spheres, so for convenience they can be taken as local functions, i.e., $\tilde{p}_i^a = 0$ for $r > r_c^a$. Note that neither the projector functions nor the partial waves need to be orthogonal among themselves.

Using the completeness relation given in Eq. A.7 the transformation can be written as

$$\hat{T}^a = \sum_i \hat{T}^a |\tilde{\phi}_i^a\rangle\langle\tilde{p}_i^a| = \sum_i (|\phi_i^a\rangle - |\tilde{\phi}_i^a\rangle)\langle\tilde{p}_i^a| \tag{A.8}$$

where the first equality is true in all of the spaces, since Eq. A.7 holds inside the augmentation spheres and outside $\hat{T}^a$ is zero. The second equality is due to Eq. A.2. Thus,

$$\hat{T} = 1 + \sum_a \sum_i (|\phi_i^a\rangle - |\tilde{\phi}_i^a\rangle)\langle\tilde{p}_i^a| \tag{A.9}$$

The all-electron Kohn Sham wavefunction can thus be written as

$$\psi_n(\mathbf{r}) = \tilde{\psi}_n(\mathbf{r}) + \sum_a \sum_i (\phi_i^a(\mathbf{r}) - \tilde{\phi}_i^a(\mathbf{r}))\langle\tilde{p}_i^a|\tilde{\psi}_n\rangle \tag{A.10}$$

where the smooth (and thereby numerically convenient) auxiliary wavefunction $\tilde{\psi}_n(\mathbf{r})$ is obtained by solving the transformed KS equations

$$\hat{T}^{\dagger}\hat{H}\hat{T}|\tilde{\psi}_n\rangle = \epsilon_n \hat{T}^{\dagger}\hat{T}|\tilde{\psi}_n\rangle. \tag{A.11}$$

In summary, the decomposition given in Eq. A.11 separates the original wavefunctions into auxiliary wavefunctions, which are smooth everywhere, and a contribution, which has rapid oscillations, but only contributes in certain, small, areas of space. Having separated the different types of waves, these can be treated individually. The localized atom-centered parts (indicated by a superscript a) can efficiently be represented on atom-centered radial grids, while the delocalized parts (with no superscript a) are all smooth, and can thus be represented on coarse Fourier or real space grids.

Appendix B
Incoherent Imaging

The proof of incoherent imaging in STEM given in this section follows the derivation given by Nellist and Pennycook [2]. Consider a scattering event from the initial partial plane wave, $\mathbf{K}_i$, into a final plane wave, with wavevector transverse (perpendicular to the optic axis) component $\mathbf{K}_f$. The intensity I measured in the far field is

$$I(\mathbf{K}_f, \mathbf{R}_o) = \left| \int A(\mathbf{K}_i) \exp[-i2\pi \mathbf{K}_i \cdot \mathbf{R}_o] \Psi(\mathbf{K}_f, \mathbf{K}_i) \mathrm{d}\mathbf{K}_i \right|^2 \tag{B.1}$$

where $\Psi(\mathbf{K}_f, \mathbf{K}_i)$ is a complex multiplier due to the magnitude and phase change of the wave as it passed through the specimen, $A(\mathbf{K}_i)$ is the aperture function, and $\mathbf{R}_o$ denotes the position of the electron probe. Expanding Eq. B.1 gives the double integral

$$\begin{aligned} I(\mathbf{K}_f, \mathbf{R}_o) = \int \int & A(\mathbf{K}_i) A^*(\mathbf{K}_i') \exp[-i2\pi(\mathbf{K}_i - \mathbf{K}_i') \\ & \cdot \mathbf{R}_o] \Psi(\mathbf{K}_f, \mathbf{K}_i) \Psi^*(\mathbf{K}_f, \mathbf{K}_i) \mathrm{d}\mathbf{K}_i \mathrm{d}\mathbf{K}_i'. \end{aligned} \tag{B.2}$$

Taking the Fourier transform of Eq. B.2 results in

$$\tilde{I}(\mathbf{K}_f, \mathbf{Q}) = \int A(\mathbf{K}_i) A^*(\mathbf{K}_i + \mathbf{Q}) \Psi(\mathbf{K}_f, \mathbf{K}_i) \Psi^*(\mathbf{K}_f, \mathbf{K}_i + \mathbf{Q}) \mathrm{d}\mathbf{K}_i \tag{B.3}$$

which is the Fourier transform of the image intensity that would be recorded for a point detector at position $\mathbf{K}_f$ in the far field. It is clear from Eq. B.3 that the contributions to an image spatial frequency, $\mathbf{Q}$, come from pairs of incident partial plane waves separated by the reciprocal space vector $\mathbf{Q}$. These two partial waves are scattered by the specimen into the same final wavevector, $\mathbf{K}_f$, where they interfere.

The image that would be recorded on an annular dark-field detector can now be formed simply by integrating Eq. B.3 over some detector function, $D(\mathbf{K}_f)$, so that

$$
\begin{aligned}
\tilde{I}(\mathbf{Q}) &= \int D(\mathbf{K}_f) \int A(\mathbf{K}_i)A^*(\mathbf{K}_i+\mathbf{Q})\Psi(\mathbf{K}_f-\mathbf{K}_i)\Psi^*(\mathbf{K}_f-\mathbf{K}_i-\mathbf{Q})\mathrm{d}\mathbf{K}_i\mathrm{d}\mathbf{K}_f \\
&= \int A(\mathbf{K}_i-\mathbf{Q}/2)A^*(\mathbf{K}_i+\mathbf{Q}/2) \\
&\quad \times \int D(\mathbf{K}_f)\Psi(\mathbf{K}_f-\mathbf{K}_i+\mathbf{Q}/2)\Psi^*(\mathbf{K}_f-\mathbf{K}_i-\mathbf{Q}/2)\mathrm{d}\mathbf{K}_i\mathrm{d}\mathbf{K}_f
\end{aligned}
\tag{B.4}
$$

with the shift of $\mathbf{K}_i$ by $-\mathbf{Q}/2$ allowed because the integral has an infinite limit. If $D(\mathbf{K}_f)$ has a geometry that is much larger than the objective aperture, the dependance of the $\mathbf{K}_f$ integral on $\mathbf{K}_i$ becomes very small allowing the integrals to be separated. The intensity measured by the ADF detector can now be written as

$$
\begin{aligned}
\tilde{I}_{ADF}(\mathbf{Q}) &= \int A(\mathbf{K}_i-\mathbf{Q}/2)A^*(\mathbf{K}_i+\mathbf{Q}/2)\mathrm{d}\mathbf{K}_i \\
&\quad \times \int D(\mathbf{K}_f)\Psi(\mathbf{K}_f-\mathbf{K}_i+\mathbf{Q}/2)\Psi^*(\mathbf{K}_f-\mathbf{K}_i-\mathbf{Q}/2)\mathrm{d}\mathbf{K}_f \\
&= T(\mathbf{Q})\tilde{O}(\mathbf{Q})
\end{aligned}
\tag{B.5}
$$

where $\mathrm{T}(\mathbf{Q})$ is the transfer function for incoherent imaging, containing information about the objective lens defocus, aberrations, and the aperture. $\tilde{O}(\mathbf{Q})$ is the Fourier transform of the object function, providing information about the specimen scattering properties and the detector. The Fourier transformation of $\tilde{I}_{ADF}$, representing the image intensity in real space, can be written as

$$
I_{ADF}(\mathbf{R}_o) = |P(\mathbf{R}_o)|^2 \otimes O(\mathbf{R}_o) \tag{B.6}
$$

which is essentially the definition of incoherent imaging. Therefore, the image intensity can be interpreted as being the convolution between an object function $O(\mathbf{R}_o)$ and a positive-real point-spread function $P(\mathbf{R}_o)$ that is simply the intensity of the illuminating probe. The convolution integral is summing over the intensity of the probe, because interference effects between spatially separated parts of the probe are no longer observed, just as if the specimen were self-luminous.

References

1. Rostgaard, C.: http://arxiv4.library.cornell.edu/abs/0910.1921
2. Nellist, P.D., Pennycook, S.J.: The Principles and interpretation of annular dark-field Z-contrast imaging. Adv. Imag. Elect. Phys. **113** (2000)

Index